D1234338

Cellular and Molecular Bases of Biological Clocks

Leland N. Edmunds, Jr.

Cellular and Molecular Bases of Biological Clocks

Models and Mechanisms for Circadian Timekeeping

With 156 Illustrations

Springer-Verlag
New York Berlin Heidelberg
London Paris Tokyo

Leland N. Edmunds, Jr.
Department of Anatomical Sciences
School of Medicine, Health Science Center
State University of New York
at Stony Brook
Stony Brook, New York 11794, USA

Library of Congress Cataloging-in-Publication Data
Edmunds, Leland N.
Cellular and molecular bases of biological clocks.
Bibliography: p.
Includes indexes.
1. Circadian rhythms. I. Title.
QP84.6.E36 1988 574.1'882 87-16632

Media conversion by David E. Seham Associates, Inc., Metuchen, New Jersey.
Printed and bound by R.R. Donnelley & Sons, Harrisonburg, Virginia.
Printed in the United States of America.

9 8 7 6 5 4 3 2 1

ISBN 0-387-96559-9 Springer-Verlag New York Berlin Heidelberg
ISBN 3-540-96559-9 Springer-Verlag Berlin Heidelberg New York

To the women in my life
Danielle,
Sunja, Kira, Alissa,
Hélène, Isabelle,
Galipette

Preface

As Bünning (1973) noted in his prefatory remarks, to suggest the existence of an endogenous diurnal rhythm was generally regarded even as late as 1955 as subscribing to a mystical or metaphysical notion. One might argue justifiably that the field of biological rhythms emerged as a discipline (often designated as chronobiology) with the Symposium on Biological Clocks held in 1960 at Cold Spring Harbor, New York. The tens of thousands of papers that have been published since this important meeting attest to the fact that the periodic behavior of living systems, rather than being in some sense pathological—or at the very least abnormal—and confined to a very small number of organisms and tissues, is the normal state of affairs, the rule rather than the exception.

A particularly intriguing class of biological periodicity is reflected in the large number of well-documented persisting circadian rhythms, having periods of approximately 24 hours, that occur at every level of eukaryotic organization. Despite their importance, many reviews of rhythmic phenomena stop short of a detailed consideration of these characteristically longer-period rhythms, perhaps because a convincing explanation of this very same property, together with that of temperature compensation of the free-running period, has not as yet yielded to experimental analysis. Small wonder then that a lead editorial in *Nature* some years ago (231:97–98, 1971) somewhat plaintively queried why so little was known about the biological clock and suggested that it was time to wind it up. Present-day grant review panels, some 15 years later, seem to echo this theme even more stridently!

Nevertheless, progress has been made. The Dahlem Conference on The Molecular Basis of Circadian Rhythms, held in Berlin over a decade ago (Hastings and Schweiger, 1976), admirably summarized in its unique format the most exciting current trends at the time (membranes were hot) and is still quite useful as a reference work. This monograph attempts to update this treatment and consider some of the more provocative developments in the field (as, for example, those obtained with the powerful approach of molecular genetics). Its title, nevertheless, is a bit presumptuous. We still cannot clearly delineate the molecular basis of circadian rhythmicity, and we will have to expand our treatment somewhat to include progress made at the cellular and biochemical levels as well.

The layout of the book emulates that of a kind of sextet; the outer two movements, consisting of a brief introduction to circadian rhythms and a final section treating general theoretical and applied aspects, flank an inner quartet of chapters comprising the core of the book. The first of these chapters (chapter 2) surveys circadian organization at the cellular level and describes the most important eukaryotic microorganisms that have served as experimental material for the biochemical and molecular analysis to follow. Chapter 3 discusses in some detail the interaction of cell division cycles and circadian oscillators. Chapter 4 treats the results obtained by several major experimental lines of attack on circadian clock mechanisms. Finally, the various biochemical and molecular models for circadian oscillators that have been constructed on the basis of the data presented earlier are outlined in chapter 5, along with their formal predictions and the degree to which they have been validated or disconfirmed.

Where appropriate, ultradian and other noncircadian periodicities are discussed (such as the glycolytic cycle—perhaps the best understood biochemical oscillator, and for that reason, instructive as a possible mechanism whereby longer periods might be generated), and parallels are drawn to the recent explosion of work on neuronal oscillators (Berridge et al., 1979; Carpenter, 1981, 1982). Although the chapters are designed to be read sequentially, particularly by the potential initiate to the field, it is unlikely that they can compete successfully with a good concert, movie, or home videotape. Consequently, those readers at least somewhat versed in this arcane area should also find it profitable to skip among the various sections. To this end, there is extensive cross-referencing and a rather large bibliography is appended (literature search ended June 1987) with name and date citations' being given in the text.

In a work of this scope, sins of omission and commission unfortunately are inevitable. I will appreciate learning of such errors, distortions and misinterpretations. I thank all those authors who have given me permission to use their published illustrations; Dr. Colin S. Pittendrigh, who introduced me to the field; the National Science Foundation, which has supported my own research throughout the years; Prof. Régis Calvayrac, Directeur de la Laboratoire des Membranes Biologiques, Université Paris VII, who provided a home away from home for the finishing touches on the manuscript; Dr. Charles F. Ehret, who read the entire work; and Dr. Danielle L. Lavel-Martin, who aided in the production of the manuscript. I am grateful to all my students in my course on Biological Clocks over the past 20 years who helped me justify my existence, particularly those special classes at the University of Tel Aviv and the Université Paris VII who endured shortened versions of this monograph during my tenure as visiting professor there.

Stony Brook, New York — Leland N. Edmunds, Jr.

Contents

Abbreviations

LL	continuous illumination
DD	continuous darkness
LD	light-dark cycle
LD: x,y	light-dark cycle comprising x hours of light and y hours of dark
WC: x,y	temperature cycle comprising x hours of warmer and y hours of colder temperatures (°C)
T	period of an LD cycle or other periodic *Zeitgeber* (environmental cue)
τ	average period of a free-running rhythm in constant conditions
ϕ	phase of a rhythm
ϕ_r	phase reference point, or phase marker
$+\Delta\phi, -\Delta\phi$	change (advance, delay) in phase (phase shift)
ZT	environmental (*Zeitgeber*) time (where ZT 0 corresponds to the onset of light)
CT	circadian time (CT 0 indicates the phase point of a free-running rhythm that has been normalized to 24 hours and corresponds to that occurring at the onset of light in a LD: 12,12 reference cycle)
PRC	phase-response curve [plot of the phase shift of a free-running circadian rhythm engendered by a perturbing light (or other) signal as a function of the circadian time at which it was applied]
g	average generation (doubling) time of a population of cells
ss	average step-size, or factorial increase in cell concentration (plateau to plateau) after a phased, or synchronized division step
CAM	crassulacean acid metabolism
CAP	compound action potential
CDC	cell division cycle

G_q	fundamental quantal cell cycle
PA	photosynthetic activity
PC	photosynthetic capacity
PSI(II)	photosystem I (II)
SCN	suprachiasmatic nuclei

AC	adenyl cyclase
ANISO	anisomycin
cAMP	cyclic AMP
CAP	chloramphenicol
CCCP	carbonyl cyanide *m*-chlorophenyl hydrazone
CPZ	chlorpromazine
CHX	cycloheximide
DCCD	*N*, *N*′-dicyclohexylcarbodiimide
DCMU	diuron, 3-(3,4-)dichlorophenyl 1)-1, 1-dimethyl urea
DES	diethylstilbestrol
DNP	dinitrophenol
FCCP	*p*-trifluoromethoxyphenylhydrazone
FUdR	5-fluoro-2′-deoxyuridine
MDH	malate dehydrogenase
NAD(H)	nicotinamide adenine dinucleotide
NADP(H)	nicotinamide adenine dinucleotide phosphate
PDE	phosphodiesterase
PEPC	phosphoenolpyruvate carboxylase
PFK	phosphofructokinase
PUR	puromycin
TFP	trifluoperazine
VIP	vasoactive intestinal polypeptide
VP	vasopressin

I trifle with my papers from time to time;
It is one of the lesser frailities.

Horace

Were I to await perfection,
My book would never be finished.

Tai T'ung, 13th Century Chinese scholar;
The Six Scripts: Principles of Chinese Writing

A clock is a clock is a clock.

Science News, 1983, 124(22) : 346

1 Introduction

1.1 Temporal Organization

As Pittendrigh noted in 1961 in an engaging essay, biologists are confronted with a continuously reproducing and evolving set of highly organized living systems. An organism that has thrived by differential reproductive success is said to be "adapted," and its adaptation is reflected in its total organization. This organization is strongly history-dependent, having arisen through the twin processes of natural selection and adaptation. Biological problems, therefore, pivot on the complexities of biological organization.

It is almost self-evident that the spatial organization and the functioning of living forms are inextricably intertwined. Of equal importance, however, is the temporal dimension: at the physiological level, for example, not only must the *right* amount of the *right* substance be at the *right* place, but also this must occur at the *right* time (Halberg, 1960). This is true also for the organism itself, which often must be positioned in time in favorable biotic or physical conditions. Since the environment is highly periodic with respect to many of its variables, it would not be surprising (indeed, it would be essential) for the organism to adapt to these cyclicities.

Organisms can and do measure astronomical time in some manner—as opposed to the purely private timekeeping reflected in such variable-period physiological rhythms (see Section 1.3) as heartbeat and alpha-brain waves. This is demonstrated explicitly by four categories comprising diverse phenomena occurring throughout the animal and plant kingdoms: (1) persistent rhythms, having daily (circadian), tidal, lunar (monthly), and yearly (circannual) periods, (2) the *Zeitgedächtnis,* or time sense, of bees and humans, (3) seasonal photoperiodism, wherein many organisms perform a certain function at a specific time of the year by what may be essentially a daily measurement of the length of the day (or night), and (4) celestial orientation and navigation, in which the sun, moon, or stars are used as direction givers, implying a timing system to compensate for their continuously, but predictably, shifting positions (Bünning, 1973; Palmer et al., 1976; Brady, 1982). All four types of timekeeping, or functional bio-

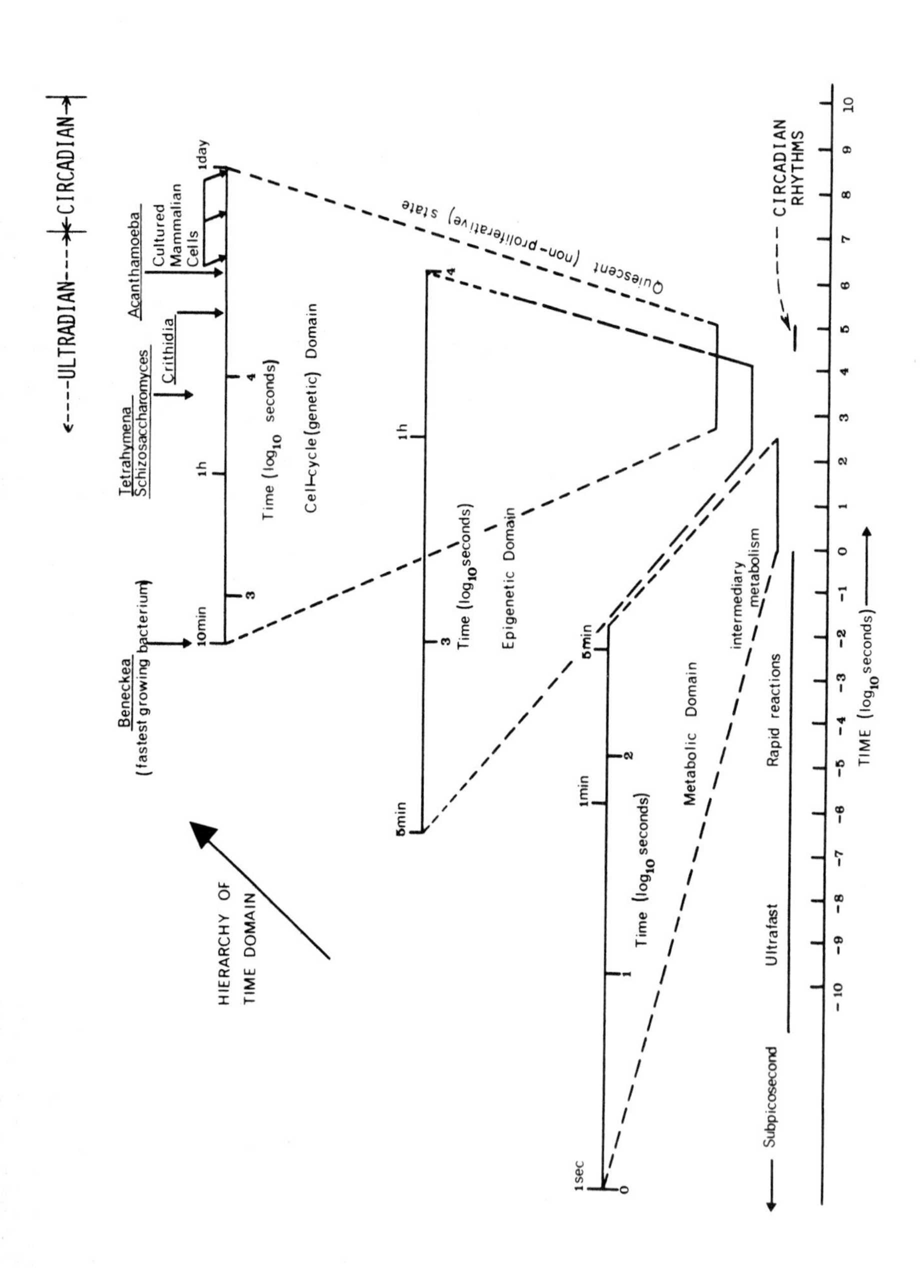
ULTRADIAN
CIRCADIAN
Beneckea
(fastest growing bacterium)
Tetrahymena
Schizosaccharomyces
Crithidia
Acanthamoeba
Cultured
Mammalian
Cells
10min
3
1h
4
1day
Time ($\log_{10}$ seconds)
Cell-cycle (genetic) Domain
Quiescent (non-proliferative) state
5min
3
1h
4
Time ($\log_{10}$ seconds)
Epigenetic Domain
1sec
0
1
1min
2
5min
Time ($\log_{10}$ seconds)
Metabolic Domain
intermediary
metabolism
Subpicosecond
Ultrafast
Rapid reactions
CIRCADIAN
RHYTHMS
-10 -9 -8 -7 -6 -5 -4 -3 -2 -1 0 1 2 3 4 5 6 7 8 9 10
TIME ($\log_{10}$ seconds)
HIERARCHY OF
TIME DOMAIN

chronometry, have external correlates (generated by the movements of the earth, moon, and sun) to which the organism has adapted. Although the last three kinds commonly are found only in higher organisms and probably are relatively recent, more sophisticated variations on a more ancient evolutionary theme, the first category of persistent rhythms, are displayed commonly in most, if not all, eukaryotic (but not prokaryotic) unicells (however, see Section 6.1.1). An understanding of the physiological and biochemical bases of these simpler clocks, therefore, may be crucial to the elucidation of the higher-level phenomena.

The underlying biological clocks that generate the foregoing types of rhythmicity all possess considerably longer periods than those that give rise to ultrafast and rapid chemical reactions, to biochemical rhythmicities of intermediary metabolism (e.g., glycolytic oscillations), and to those rhythms (such as cellular respiration) commonly observed in the epigenetic and genetic time domains (Lloyd et al., 1982b; see Section 1.3). The time domain, therefore, of circadian rhythms—the primary subject matter of this review—lies at the interface (Fig. 1.1) between the upper border of the genetic domain comprising cell division cycles and those even longer periodicities in the temporal hierarchy of living systems (*cf.* Fig. 17, p. 72 in Ehret, 1974).

1.2 General Properties of Circadian Rhythms

Circadian rhythmicities, having a period of about 1 day, have been documented throughout the plant and animal kingdoms at every level of eukaryotic organization. Their general characteristics are summarized in Figure 1.2. Typically, they can be synchronized (entrained) by imposed diurnal light or temperature cycles to precise 24-h periods and can be predictably phase-shifted by single light and temperature signals. Yet they are able to free-run for long timespans as persisting rhythms under conditions held constant with respect to most environmental time cues *(Zeitgeber),* with a natural period close to but seldom exactly 24 h. (Unless otherwise noted, we will always use the term "circadian" in this restricted sense.) Furthermore, the free-running period (τ) is remarkably well compensated for changes in the ambient temperature within the physiological range, as might be expected of an accurately functioning oscillator or clock.

FIGURE 1.1. Time domains of living systems. Note that the domain of circadian rhythms, which display periodicities of about a day, lies at the interface between ultradian rhythms (having periods less than 24 h) and longer infradian rhythms (with periods greater than 24 h), and encroaches upon the (genetic) domain of the cell division cycle. (Adapted from Lloyd et al., 1982b, with permission of Academic Press)

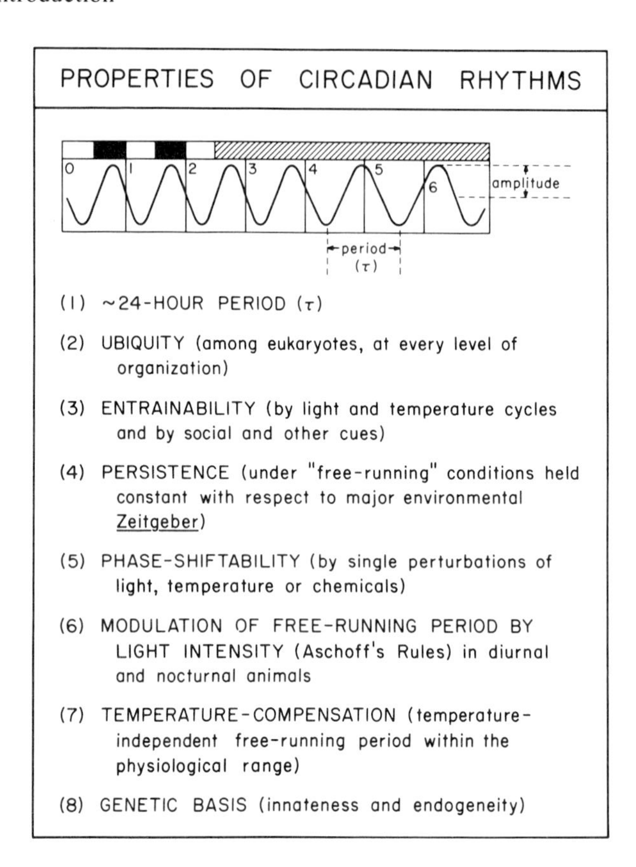

FIGURE 1.2. Properties of circadian (sensu stricto) rhythms. A generalized sinusoidal oscillation under entraining (LD) and free-running (dim LL) conditions is shown at the top over a timespan of 7 days. *Key:* white areas, light intervals; black bars, darkness; hatched areas, continuous dim illumination. Note that the period (τ) of the oscillation is longer in dim LL than it is in LD, where it exactly matches that (T) of the synchronizing LD cycle. (Reprinted from Edmunds, 1984b, with permission of CRC Press, Inc.)

Our central thesis is that there is a selective premium for temporal adaptation, especially to solar periodicities, and that these adaptive features have been attained by organisms through some sort of timing mechanism: in particular, an endogenous, autonomously oscillatory biological clock that is responsive to and can be reset and otherwise modulated by those environmental periodicities that the organism has encountered throughout its evolutionary history.

A number of books and symposia provide comprehensive coverage of circadian rhythms. The following texts afford more general treatment: Sweeney (1969a), Bierhuizen (1972), Bünning (1973), Palmer et al. (1976),

Saunders (1976), Brady (1982), Moore-Ede et al. (1982). Major symposia and edited works with similar breadth include Frisch (1960—*Cold Spring Harbor Symposia on Quantitative Biology*, vol. 25); Aschoff (1965, 1981), Menaker (1971), DeCoursey (1976), Hastings and Schweiger (1976), Naylor and Hartnoll (1979), Suda et al. (1979), Wever (1979), von Mayersbach et al. (1981), Follett and Follett (1981), Aschoff et al. (1982), Edmunds (1984a), Hiroshige and Honma (1985), Rensing and Jaeger (1985), and the *Proceedings of the International Society for Chronobiology* (which has met biennially since 1971). In addition, the journals *Chronobiologia, Chronobiology International, Journal of Biological Rhythms, International Journal of Chronobiology*, and *Journal of Interdisciplinary Cycle Research* are useful reference sources in the field of chronobiology. Finally, the important monographs by Pavlidis (1973), Goodwin (1963, 1976), and Winfree (1980, 1987) provide particularly relevant mathematical and theoretical background for some of the considerations raised in subsequent chapters.

1.3 Analogies with Shorter Periodicities

Despite the emphasis on circadian rhythms (and other astronomical periodicities) in this book, it is important not to overlook the great number of biological and biochemical oscillations with ultradian periods (0.5 h$<\tau<$20 h). In fact, we consider in Chapter 5 the possibility that high-frequency ticks might generate lower-frequency, circadian tocks by the use of some sort of counter, by coupling among oscillators, or by other mechanisms (Edmunds, 1978). (It is beyond the scope of this book to examine the possibility that circadian oscillations, in turn, might contribute to the generation of even longer periodicities, such as infradian estrous cycles and circannual rhythms.) At the very least, it is instructive to compare the formal properties of the oscillators underlying the various types of rhythmic output.

1.3.1 An Atlas of Cellular Ultradian Periodicities

In support of the proposition that periodic behavior is not confined to a limited number of cell types but rather is a common property of most biological systems, Rapp (1979) has catalogued systematically an atlas of cellular oscillations with periods of a few hours or less. Oscillations in purely chemical systems, such as the Belousov-Zhabotinskii reaction (Winfree, 1974; Tyson, 1976) were not included. A few selected examples from the 450 citations to experimental papers are given in Table 1.1 to illustrate the variety of different classes of ultradian oscillator. Let us now briefly examine the formal parallels between two ultradian systems taken from this compendium and circadian oscillators.

TABLE 1.1. Some examples of ultradian cellular periodicities with periods of a few hours or less.

Class	Organism or preparation
Oscillations in enzyme-catalyzed reactions	
Ammonium efflux (1–3 min)	*Scenedesmus*
ATPase activity (1–2.5 min)	Kidney and brain microsomes
Creatine kinase (3–10 min)	Heart muscle extract
Glycolytic intermediates (2 s–3 h)	Intact cells or cell-free extracts of *Saccharomyces*
Lactate dehydrogenase activity (1–3 min)	*Acetabularia*
Oxidation, catalyzed (1 min)	Horseradish peroxidase, purified
Ion movements in mitochondria, organellar volume, and respiration rate (1–30 min)	Mitochondria from rat liver and pigeon heart
Photosynthesis (dark reactions and oxygen evolution)	*Chlorella* and other algal cells
Respiration (1 h)	*Acanthamoeba castellanii*
Oscillations in protein synthesis	
Aspartate and ornithine transcarbamylases, dehydroquinase, histidase (1 h)	*Bacillus subtilis*
β-Galactosidase, pyruvate-synthesizing enzymes (50–60 min)	*Escherichia coli*
α-Glucosidase, glutamate dehydrogenase (1.5–6.5 h)	*Saccharomyces cerevisiae*
Lactate and glucose-6-P-dehydrogenases, aldolase (3–4 h)	Chinese hamster cells
Protein synthesis (0.5–1 h)	Sea urchin embryo
Protein, total cellular (1 h)	*Acanthamoeba castellanii*
Oscillations in cell membrane potential	
Membrane potential (6 s–2 min)	Mouse fibroblast L cells; *Hydrodictyon; Neurospora; Nitella*
Hyperpolarization, spontaneous (1 s)	Mammalian macrophage
Oscillations in secretory cells	
Membrane potential (1 s–5 min)	Rat anterior pituitary, adenohypophysis; rabbit adrenal glands; mouse pancreatic islet; blowfly salivary gland
Periodic discharge (0.5 s)	Fleshfly neurosecretary cells
Neural oscillators	
Neurotransmitter content and release (2 s–5 min)	Electric organ of *Torpedo;* neuromuscular junction of frog
Membrane potential in single neurones (0.1–20, s)	Blowfly; *Aplysia* abdominal ganglion; *Procambrus* stretch receptor; jellyfish; hermit crab; lobster; shrimp; leech
Central nervous system (electrocorticogram, EEG, temperature, thalmic discharge, evoked potentials) (20 ms–48 s)	Mammalian cerebral cortex; cat olfactory bulb and prepyriform cortex; rabbit hypocampus; cat and monkey
Muscle oscillations	
Skeletal muscle: contractions (0.1–70 s)	Skinned muscle fibers; cultured chick muscle; frog and rabbit fiber bundles; giant waterbug; locust

TABLE 1.1. Continued.

Class	Organism or preparation
Cardiac muscle: mechanical and electrical activity (0.1–2 s)	Cultured chick heart; Purkinje fibers of sheep, calf, dog; atrial muscle of frog, carp, guinea pig, rabbit; ventricular muscle of cat, dog, frog
Smooth muscle: mechanical and electrical activity (2 s–2.5 h)	Digestive tract of human, monkey, guinea pig, cat, dog, rabbit; urogenital tract of human, rabbit, guinea pig
Oscillations in cell growth and development	
Periodic movement during aggregation; cyclic AMP and GMP synthesis (5–10 min)	*Dictyostelium; Polyspondylium*
Current pulses in developing eggs (0.2–1 h)	*Pelvetia* (seaweed)
Mitosis (8–12 h)	*Physarum*
Spore release and growth (6–16 h)	Ascomycetes (*Nectria, Penicillium*)
Neural fold closure (5 min)	*Trituris, Ambystoma*
Action potential and contractions during regeneration (4–25 min)	*Acetabularia; Tubularia*
Miscellaneous	
Transpiration and water uptake (30 min)	*Avena*
Periodic flashing (0.5–1 s)	Fireflies

For a detailed, systematic catalog, see the atlas compiled by Rapp (1979).

1.3.2 The Glycolytic Oscillator

The glycolytic oscillations observed in the yeast *Saccharomyces* represent one of the best understood biochemical oscillators and is considered in more detail in Section 5.2.1. If yeast cells or yeast extracts are given a continuous addition of glucose as substrate, sustained oscillations occur in NADH fluorescence (which can be continuously monitored) and in other elements of the glycolytic system (Fig. 1.3). The enzyme phosphofructokinase (PFK) has been found periodically to generate its products, ADP and fructose-1,6-bisphosphate. The cyclic changes in PFK activity, in turn, are propagated along the entire enzymatic reaction sequence of glycolysis through the adenylate system. The allosteric properties of PFK and the positive feedback exerted on it by a reaction product appear to be responsible for the resulting temporal dissipative structure (Hess and Boiteux, 1971; Goldbeter and Caplan, 1976; Goldbeter and Nicolis, 1976).

Experimental studies on the yeast system (Boiteux et al, 1975; Hess, 1976) reveal that it has a number of properties that are formally similar to those characterizing circadian systems (Fig. 1.2). The self-sustained oscillations in NADH absorbance typically have a free-running period of 2 to 70 min (both τ and amplitude are flux-rate dependent, and oscillations are observed only at critical flux range). The rhythm can be entrained by

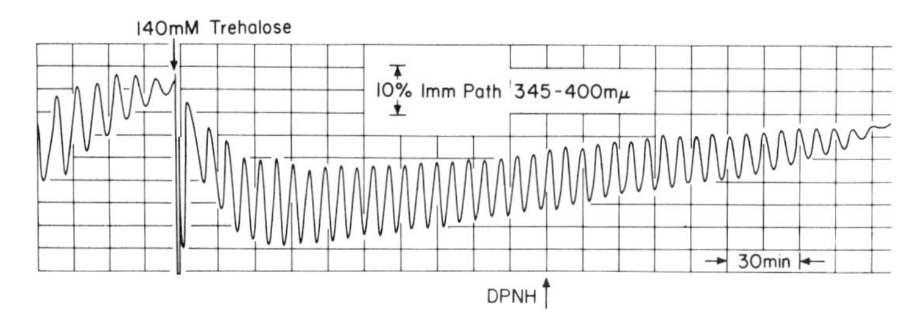

FIGURE 1.3. The generation of sustained glycolytic oscillations in NADPH (DPNH) in a cell-free extract of yeast by an addition of trehalose. The level of NADPH was continuously monitored with a dual-beam spectrophotometer, and percentage transmission was recorded directly. Time proceeds from left to right (1 unit on the abscissa corresponds to 15 min). The train of 42 oscillations lasted over 4.5 h and had a stable frequency of 0.15 min^{-1} and a period of about 7 min. (Reprinted from Pye, 1969, with permission of the National Research Council, Canada.)

a periodic source of substrate that is sinusoidally applied. Indeed, entrainment by the fundamental frequency of substrate input occurs only within limits corresponding to the domain $0.75 < T/\tau < 1.25$. This range of entrainment for glycolysis is comparable to that observed for light cycles in circadian systems, in which T of the driving *Zeitgeber* typically must fall between 18 and 30 h (see Bruce, 1960; Pittendrigh, 1965). Finally, the glycolytic oscillations can be phase-shifted: the addition of 0.7 mM ADP to a yeast extract generates a delay ($\Delta\phi = -1.5$ min). On the other hand, the temperature sensitivity of the glycolytic oscillator is high, unlike the temperature compensation of τ observed for circadian rhythms. We consider this system in more detail in Section 5.2.1.

1.3.3 Rhythmic Flashing of Fireflies

The rhythmic flashing behavior of fireflies is analogous in many ways to circadian systems (Fig. 1.2), yet differs in at least two respects: fireflies can synchronize their emissions with each other (requiring that both interval and phase be matched with the driving signals), and the period is much shorter, usually about 1 sec (although τ ranges from 400 msec to 5 sec in species known to synchronize), suggesting control by a cellular pacemaker or oscillator (Buck et al., 1981a,b; Hanson, 1978, 1982). These similarities and differences warrant a brief discussion here because they conveniently introduce neural events and the concept of populations of interacting oscillators.

Male fireflies (*Pteroptyx* spp., *Luciola* spp.) from New Guinea and Malaysia often congregate in groups of thousands in bushes and trees, flashing in unison (Buck and Buck, 1968). This synchronous flashing behavior seems to function as a social attraction mechanism to facilitate mating (Buck and Buck, 1978). These impressive, stable displays, documented by videocinemaphotography, resemble the blinking lights of a Christmas tree (*Pteroptyx malaccae* emits a double flash about once per second). Although one might initially hypothesize that the synchrony arises from a slightly more rapidly flashing male serving as a leader for the other fireflies comprising the population, this appears not to be the case. A minimum latency of only 60 to 100 msec between stimulus and flash has been measured electrophysiologically—much shorter than the 930 msec-interval observed between naturally occurring flashing events (Hanson, 1978, 1982).

Alternatively, a type of anticipatory synchrony could result from mutual entrainment among individuals (Buck and Buck, 1968), in which the average mass flash would serve as the driving oscillator, occurring once each cycle. The theoretical and mathematical aspects of such a phenomenon have been treated by Pavlidis (1969, 1973) and are considered in more detail in Sections 5.2.3 and 6.2.2. Experimentally, fireflies will entrain to appropriate flashes from a flashlight or light-emitting diode, either leading or lagging the artificial signal. This technique has facilitated a series of experiments designed to characterize the pacemaker responsible for entrainment (Hanson, 1978, 1982).

The ultradian rhythm of flashing is clearly endogenous and self-sustaining in these species. Photometer recordings in the laboratory of isolated, spontaneously flashing *Pteroptyx cribellata* have shown that flashes are emitted about once per second. In some cycles, a flash is skipped, but without affecting the timing of the next emission (Hanson, 1978). The free-running rhythm can be phase-shifted by a single, short, light signal. Advances or delays are obtained (with attendant shortenings or lengthenings of τ over one or more cycles, commonly referred to as "transients") depending on the time during the interflash interval at which the pulse is administered. Such phase responsiveness of a putative oscillator is a formal requirement for entrainment to occur (Fig. 1.2; see Pittendrigh, 1965).

Indeed, Hanson (1978) has derived phase-response curves (PRCs) for the effects of light signals that were applied systematically at different circadian times (CT) in three species of firefly (Fig. 1.4). The PRC for *P. cribellata* was of the strong type 0 (Winfree, 1980), with high-amplitude phase shifts ($-\Delta\phi = -600$ to -800 msec; $+\Delta\phi = +200$ msec), whereas the PRCs for *P. malaccae* and *Luciola pupilla* were weak, small-amplitude (type 1) ones. As might be predicted (Pittendrigh, 1965) from their respective PRCs, *P. cribellata* has a greater range of entrainment (following pacemaker T of 800 to 1600 msec, where $\tau \cong 1000$ msec) and a shorter period of transients (only one cycle required for entrainment) than does *P. malaccae* (which entrains to $T = 770$ to 1030 msec for $\tau \cong 900$ msec, and only after many cycles) or *L. pupilla*. Artificial pacemaker T outside these limits (i.e., for $800 > \tau > 1600$ msec in *P. cribellata*) results either

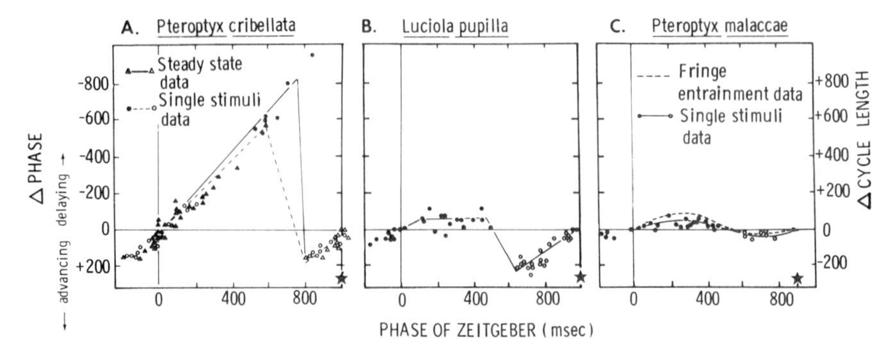

FIGURE 1.4. Phase-response curves (PRCs) for three species of synchronously flashing fireflies. A. *Pteroptyx cribelleta*. B. *Luciola pupilla*. C. *Pteroptyx malaccae*. *Ordinates:* Changes in cycle length (right axis) resulting in a change in phase (left axis) of succeeding flashes relative to what they would have been had a signal not been imposed. Solid symbols denote intervals culminated by flash (designated by a star) at the end of the depicted period; open symbols denote intervals of the next succeeding cycle. *Abscissa:* Time (ϕ) during the interflash interval at which the stimulus was given (intervals normalized to the species-specific τ of ~1000 msec). (Reprinted from Hanson, 1982, with permission of John Wiley & Sons, Inc.)

in inhibition of flashing or in a reversion to $\tau \cong 1000$ msec. Finally, *P. cribellata* and *L. pupilla* both maintain a constant phase angle difference in leading or lagging a pacer having a different T, whereas *P. malaccae* does not (and therefore τ itself must vary).

These formal differences among pacemakers are reflected in the manner by which synchrony is attained in groups of the three species of firefly (Hanson, 1978, 1982), whose displays presumably are effective only when the insects flash coincidently. *Pteroptyx cribellata,* which has a constant τ, synchronizes with other fireflies having the same intrinsic period (at the same temperature) by adjusting each cycle length so that $\tau - T = \Delta\phi$. True synchrony would obtain with pacing at $T \cong \tau$, where the phase angle difference is zero; thus, this species matches period but not phase. In a population, the mass flash would thus constitute a stable oscillator with period T driving individual fireflies having instrinsic periods τ. In contrast, *P. malaccae* matches both τ and ϕ and consequently can maintain synchrony with other individuals having τ considerably different from its own. If the pacer changes, this species requires several cycles to reattain synchrony.

Thus, a group of fireflies may be thought of as a population of interacting oscillators (see Section 5.2.3 and Section 6.2.2) in which the period of the group flash should depend on the number of interacting units and the type and strength of coupling among them (Pavlidis, 1969, 1973). If the interaction is excitatory, τ should decrease with the addition of oscillators (i.e., fireflies), whereas the converse holds true if the interaction is inhibitory. Indeed, populations of *P. malaccae* have higher flashing rates than have isolated individuals (Hanson, 1982). Either interacting insects may modify continuously each other's cycle length by flashing during the sensitive phase, or their τ's may shorten.

This comparative study of the formal entrainment process in three species of synchronously flashing fireflies has revealed that the underlying ultradian pacemaker system is in many ways analogous to circadian oscillators, but with the additional capability of synchronizing (τ and ϕ) its flash emissions with the output of the other oscillators comprising the population. In fact, the firefly system has circadian components as well. Buck (1937) and Dreisig (1978) have demonstrated a free-running rhythm of bioluminescence in, respectively, the common North American firefly, *Photinus pyralis,* and the European glowworm, *Lampyris noctiluca,* maintained under dim continuous illumination (LL). It is quite possible that cellular circadian systems similarly may reflect interacting populations of oscillators (Section 6.2.2). Finally, the fact that the ultradian firefly rhythms display a τ of $\sim$1 sec suggests the plausibility of control by a single-celled pacemaker.

2
Circadian Organization in Eukaryotic Microorganisms

Circadian organization is not restricted to higher plants and animals or even to multicellular organisms. Over the past 30 years, overt persisting circadian rhythms have been documented in a number of eukaryotic microorganisms (Table 2.1) for a wide spectrum of behavorial, physiological, and biochemical variables. These periodicities include mating type reversal, phototaxis, pattern formation, stimulated bioluminescence, photosynthetic capacity, cell division, chloroplast shape and ultrastructure, susceptibility to noxious agents and treatments, enzymatic activities, concentrations of metabolic intermediates, energy charge carriers, biogenic amines, and many others.

Clearly, virtually all levels of cellular organization display a circadian time structure. Furthermore, in many of these various microorganisms, several different circadian rhythms have been observed concurrently—in some cases even in isolated, individual cells—with the attendant implication that many or all of the overt rhythms may represent the hands of an underlying master pacemaker (see Sections 4.1, 6.2). These organisms, therefore, constitute attractive systems for the experimental investigation of the finer details of circadian rhythms (see Section 4) and the mechanism(s) whereby they are generated (see Section 5)—the core of this monograph.

In this general review, it is both undesirable and impossible to consider anywhere close to a majority of the reports of circadian periodicities (now probably numbering in the thousands) in microorganisms. Rather, only representative rhythms in several of the more intensively studied protozoans and unicellular and lower algae and fungi have been selected to illustrate the dominant themes and current problems in the field and to set the stage for the experimental studies of the biochemistry and molecular genetics of biological clocks to be treated in detail in Section 4. Fortunately, there are a number of earler detailed reviews and comprehensive treatments available that provide discussions of this area: Bruce and Pittendrigh (1957), Hastings (1959), Bruce (1965), Ehret and Wille (1970), Vanden Driessche (1970a), Sweeney (1972), Ehret (1974), Edmunds (1975),

Schweiger and Schweiger (1977), Wille (1979), Edmunds and Halberg (1981), Feldman (1982), Edmunds (1982, 1983), Feldman and Dunlap (1983), Sweeney (1983), Edmunds (1984c), Jerebzoff (1986) and Schweiger et al. (1986).

2.1 Circadian Rhythms in Protozoa

2.1.1 *Tetrahymena* spp

Rhythms of Cell Division

One of the most intensively investigated unicellular circadian systems is that of the ciliate *Tetrahymena pyriformis* (reviewed by Ehret and Wille, 1970; Wille, 1979) (Table 2.2; see Table 3.1). Earlier studies (Wille and Ehret, 1968a) focused on photoentrainment of batch cultures of the W strain. Population increase could be characterized by a short (1 to 3 h) lag period, followed by an exponential growth phase, the ultradian mode (with typical doubling times of 3 h at 27°C), and then by a period of extended slow growth, the infradian growth mode (Szyszko et al., 1968). This ultradian growth exhibited a high degree of temperature dependence. Growth curves for cultures in the infradian mode reflected a monotonically decreasing function, in contrast to the so-called stationary phase commonly found in batch cultures of bacteria. Wille and Ehret (1968a) discovered that a sudden increase or decrease in irradiance, such as that afforded by either a continuous darkness/continuous illumination (DD/LL) or an LL/DD transition, could initiate a free-running, circadian (τ = 19 to 23 h) rhythm of cell division in axenic populations of *T. pyriformis* provided that the switch-up or switch-down occurred at a critical time in the late ultradian growth mode as the cells began to enter the infradian mode. Similar observations were made in continuous cultures achieved by growing cells in an electronically controlled nephelostat, which operated like a chemostat, maintaining a constant cell titer by a programmed dilution with nutrient medium (Wille and Ehret, 1968a). Photoentrainment by a light-dark cycle (LD): 10,14 regimen was consistently observed when the washout rate was infradian (e.g., when it matched a generation time, *g*, of 40 h) but was not possible for ultradian growth rates.

These studies were extended by Edmunds (1974) in an investigation of the phasing effects of light on cell division in exponentially increasing cultures of *T. pyriformis* (strains W and GL) grown at relatively low temperatures (see Section 3.2.1). Populations maintained on proteose peptone-liver extract in DD or LL (850 lx) at 10 ± 0.05°C exhibited values of *g* of about 20 or 30 h, respectively. Imposed diurnal LD cycles (e.g., LD: 12,12 or LD: 6,18) phased cell division so that doublings of cell number occurred once every 24 h and were confined primarily to the dark intervals. Finally, long trains of 24-h oscillations in apparent cell number could be

TABLE 2.1. Some eukaryotic microorganisms in which persisting circadian rhythms have been demonstrated.

Microorganism	Reference
Protozoans	
Paramecium aurelia	Karakashian (1968) [2.1.2]
Paramecium bursaria	Ehret and Wille (1970) [2.1.2]
Paramecium multimicronucleatum	Barnett (1966) [2.1.2]
Tetrahymena pyriformis	Wille (1979); Edmunds (1983, 1984c) [2.1.1]
Algae	
Acetabularia crenulata	Terborgh and McLeod (1967) [2.2.1]
Acetabularia major	Sweeney and Haxo (1961)
Acetabularia mediterranea	Schweiger and Schweiger (1977); Vanden Driessche (1980); Edmunds (1983, 1984c) [2.2.1; 4.3]
Amphidinium carteri	Brand (1982)
Asterionella glacialis	Brand (1982)
Biddulphia mobiliensis	Brand (1982)
Cachonina niei	Eppley et al. (1968)
Ceratium furca	Weiler and Eppley (1979); Adams et al. (1984)
Chaetoceros sp.	Brand (1982)
Chlamydomonas sp.	Brand (1982)
Chlamydomonas eugametos	Demets et al. (1987); Tomson et al. (1987)
Chlamydomonas reinhardi	Bruce (1970); Bruce and Bruce (1981); Edmunds (1984c) [2.2.2; 3.1; 4.4.1]
Chlorella fusca var. *vacuolata*	Wu et al. (1986)
Chlorella pyrenoidosa	Pirson and Lorenzen (1958); Hesse (1972); Chen and Lorenzen (1986a, b)
Coscinodiscus rex	Harding et al. (1981)
Cylindrotheca fusiformis	Brand (1982)
Cylindrotheca signata	Round and Palmer (1966) [2.2.3]
Dunaliella tertiolecta	Brand (1982)
Emliania huxleyi	Chisholm and Brand (1981); Brand (1982)
Euglena gracilis	Edmunds (1982, 1983, 1984c); Edmunds and Halberg (1981) [2.2.3; 3.2; 4.3]
Euglena mutabilis	Round and Palmer (1966) [2.2.3]
Euglena obtusa	Round and Palmer (1966) [2.2.3]
Glenodinium sp.	Prézelin et al. (1977)
Gonyaulax polyedra	Hastings and Krasnow (1981); Sweeney (1981, 1983); Wille (1979); Edmunds (1983, 1984c) [2.2.4; 4.3]
Gonyaulax tamarensis	Brand (1982) [2.2.4]
Gymnodinium splendens	Hastings and Sweeney (1964)
Gyrodinium dorsum	Foward and Davenport (1970)
Hantzschia virgata	Palmer and Round (1967)
Hymenomonas carterae	Chisholm and Brand (1981)
Navicula salinarum	Round and Palmer (1966)
Nitzschia tryblionella	Round and Palmer (1966)
Olisthodiscus sp.	Chisholm and Brand (1981)
Pavlova lutheri	Brand (1982)
Phaeodactylum tricornutum	Palmer et al. (1964); Brand (1982)
Prorocentrum micans	Brand (1982)
Pyraminonas sp.	Chisholm and Brand (1981)
Prymnesium parvum	Brand (1982)
Pyrocystis cf. *acuta*	Hardeland (1982) [4.1.2]

TABLE 2.1. Continued.

Microorganism	Reference
Pyrocystis elegans	Hardeland (1982) [4.1.2]
Pyrocystis fusiforms	Sweeney (1981, 1982) [3.2.1]
Pyrocystis lunula	Swift and Taylor (1967); Töpperwien and Hardeland (1980) [4.1.2]
Pyrocystis noctiluca	Hardeland and Nord (1984) [4.1.2]
Scrippsiella trochoidea	Brand (1982)
Skeletonema costatum	Brand (1982); Østgaard and Jensen (1982); Østgaard et al. (1982)
Skeletonema menzelii	Brand (1982)
Stephanopyxis turris	Harding et al. (1981)
Surirella gemma	Hopkins (1965)
Symbiodinium microadriaticum	Fitt et al. (1981)
Thalassiosira pseudonana	Brand (1982)
Fungi	
Candida utilis	Wille (1974)
Neurospora crassa	Feldman (1982); Feldman and Dunlap (1983); Edmunds (1983, 1984c) [2.3.1; 4.4.1; 4.4.3]
Saccharomyces cerevisiae	Edmunds et al. (1979); Edmunds (1984c) [2.3.2; 4.2.1]

Updated from Edmunds, 1984c, with permission of Academic Press. Only selected references are cited (to review articles if possible); citations in brackets indicate the sections in which more detail is given.

TABLE 2.2. Some persisting circadian rhythms documented for the ciliate *Tetrahymena pyriformis*.

Circadian rhythm	Reference
Physiological	
Autotaxis (pattern formation)	Wille and Ehret (1968b)
Cell death	Meinert et al. (1975)
Cell division	Wille and Ehret (1968a); Ehret and Wille (1970); Edmunds (1974); Meinert et al. (1975); Readey (1987)
Oxygen induction of ultradian growth mode (Pasteur effect)	Ehret et al. (1974, 1977)
Respiration	Ehret et al. (1977)
Biochemical	
Cyclic AMP (cAMP)	Dobra and Ehret (1977)
ATP	Dobra and Ehret (1977)
DNA, total cellular	Ehret et al. (1974)
Glycogen, total cellular	Dobra and Ehret (1977); Ehret et al. (1977)
RNA, total cellular	Ehret et al. (1974)
RNA, transcriptotypes	Barnett et al. (1971a, b); Wille et al. (1972)
Tyrosine aminotransferase	Dobra and Ehret (1977); Ehret et al. (1977); see Kämmerer and Hardeland (1982)

Updated from Edmunds (1984c), with permission of Academic Press.

obtained in semicontinuous culture in LD cycles, and the entrained rhythm persisted for at least 6 days with a circadian period, τ = 23.8 to 24.4 h. Thus, once again, cell populations in the infradian (but not ultradian) growth mode were capable of showing circadian periodicities. In this case, however, the infradian state was achieved by lowering the temperature rather than by nutrient depletion.

These findings led Wille and Ehret (1968a) to formulate the circadian–infradian rule (Ehret and Wille, 1970), which simply states that when an exponentially growing culture switches from the ultradian ($g < 24$ h) to the infradian ($g > 24$ h)growth mode, the cells comprising it are invariably capable of light-synchronizable, circadian outputs (not only of cell division). This hypothesis is discussed again in Section 3.2.2.

Physiological Rhythms Observed During the Infradian Growth Mode

A number of other persisting circadian rhythms (Table 2.2, Fig. 2.1) have been documented for cultures of *Tetrahymena* spp. that have reached the slow-growing, infradian growth mode (Ehret and Wille, 1970; Wille, 1979). Thus, dense cultures of this ciliate maintained in a shallow dish exhibited very regular, hexagonally packed cell aggregates. An endogenous circadian rhythm (τ = 21 h) in the rate of pattern formation could be initiated by a single DD/LL switch-up that had a specific phase relationship to the similarly initiated cell division rhythm (Wille and Ehret, 1968b).

Likewise, a circadian rhythm in respiration (measured continuously as CO_2 produced) has been found in cells cultured on the surface of solid, enriched proteose-peptone agar plates. Maximum values were attained

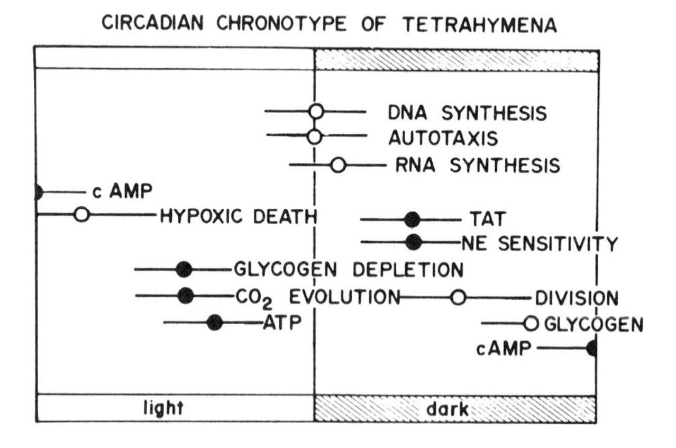

FIGURE 2.1. Temporally characteristic molecular and physiological properties of *Tetrahymena*, taken from experiments on LD: 12,12 and in liquid culture (open circles) and from experiments on LD: 8,16 and on solid agar (closed circles). The horizontal bars represent the range of values around the peaks (circles). (Reprinted from Ehret et al., 1977, with permission of Williams & Wilkins Co.)

during the latter part of the (subjective) light phase (Dobra and Ehret, 1977). Presumably, an inversely correlated rhythm might be anticipated for the level of oxygen present in liquid batch cultures. Indeed, a circadian rhythm in the capacity of sudden aeration (by bubbling) to induce cells in the infradian growth phase to return to ultradian growth has been demonstrated (Ehret et al., 1974). Thus, oxygen induction of increased cell division rates is under circadian phase control and appears to represent the Pasteur effect in *Tetrahymena*. It could be blocked with CO_2, amytal, and rotenone, all of which are classic inhibitors of aerobic respiration.

Finally, Meinert et al. (1975) have reported what might be construed to be the ultimate in physiological circadian rhythms: chronotypic death. In this ciliate (as well as in many other microorganisms), the attainment of a given cell titer is dependent on the level of dissolved oxygen in the medium. Asynchronous cell populations do not usually exhibit hypoxic death because the titer never exceeds the oxygen support level. In contrast, Meinert et al. (1975) found that in thermally (WC) entrained cultures [warmer (W) 7h, colder (C) 17h (W = 31 or 29°C, C = 26.5°C)] maintained in constant dim LL, cell death appeared to occur at each switch-up of temperature. This was manifested indirectly by a decrease in cell titer and was observed microscopically as a rupturing of a small percentage of cells in samples taken from the master culture at these phases alone of the WC cycle. These authors speculate that in synchronized cultures, overproduction of cells for a given level of oxygen occurs at the time of transition from the infradian to the ultradian growth mode, and cell death then ensues so as to lower the titer to an appropriate level. If this decrease were to be temporarily prevented, as, for example, by imposition of the WC cycle, sensitive cells would be induced to die synchronously. In other microorganisms, however, alternative mechanisms may generate these oscillations in cell number. Thus, a circadian rhythm in settling of cells out of the liquid phase of the culture and attachment to the vessel walls has been discovered in *Euglena* (Terry and Edmunds, 1970b; see Section 2.2.3).

Circadian Rhythmicity in Metabolism and Biochemistry

In addition to the many physiological variables discussed in the preceding section, a number of metabolic and biochemical chronotypes also exhibit circadian time structure in *Tetrahymena* (Fig. 2.1, Table 2.2). Thus, in a photoentrained continuous culture, synchronously dividing during the infradian growth mode (*g* was 3 days or more), both total deoxyribonucleic acid (DNA) and ribonucleic acid (RNA) synthesis (as measured by [^{3}H]thymidine or [^{3}H]uridine incorporation over a 36-h time span) displayed circadian rhythmicity (Ehret et al., 1974).

Subsequently, the question arose whether different RNA species were synthesized at different times of the circadian day, as predicted on the basis of the linear sequential transcription of the chronon postulated by

Ehret and Trucco (1967) in their model for circadian timekeeping (see Section 5.3.1). To test this hypothesis, *Tetrahymena* (as well as *Paramecium;* see Section 2.1.2) was synchronized by an LD: 12,12 cycle in the infradian mode and then pulse-labeled at various time of the day with [^{3}H]thymidine or ^{32}P to label RNA. Molecular hybridization of DNA-RNA was used to determine the binding capacity of the purified labeled RNA species with single-stranded homologous DNA immobilized on nitrocellulose membrane filters. The experiments were performed with annealing reactions in order to compare the kinetics, saturation, and competition behavior of the various pulse-labeled RNA stocks (Barnett et al., 1971a,b; Wille et al., 1972). The data from these pioneering studies showed clear evidence of the presence of chronotypically characteristic RNA species, or transcriptotypes, and thus are consistent with the prediction of the chronon model that temporally differentiated RNA molecules are synthesized during the circadian cell division cycles that are present during the infradian growth mode, during which many other overt circadian rhythms also are observed (but see Section 5.3.2).

More recently, an entirely different approach has been taken to circadian regulation in *T. pyriformis,* with primary emphasis being placed on metabolism and energy use (Ehret et al., 1977). In highly oxygenated (21% O_2) plate cultures of this ciliate, concentrations of glycogen (a storage product, or energy depot) reached maximum values during the ultradian growth phase and then underwent periodic depletion during the ensuing infradian growth. These stepwise decreases were circadian, occurring both during photoentrainment by LD cycles and during a subsequent free-run. Further, the decrease in glycogen content was in phase with the increase in the production of CO_2 (see previous subsection).

Since other workers had demonstrated that epinephrine and serotonin, both neurotransmitter substances, occur in *T. pyriformis* (Janakidevi et al., 1966a,b) and that glycogen levels could be altered by reserpine, dichloroisoproterenol, and aminophylline (Blum, 1967), Dobra and Ehret (1977) investigated the possibility that the regulation of the circadian rhythm of glycogen metabolism in *Tetrahymena* species might resemble that found in rodent liver, in which glycogen (Haus and Halberg, 1966) and certain enzymes associated with its metabolism, such as tyrosine aminotransferase (TAT) (Ehret and Potter, 1974), oscillate with a circadian period also. Further, Dobra and Ehret (1977) studied the control of associated pathways for glycogen storage and release, which in the hepatic system involves changes in the intracellular levels of cyclic AMP (adenosine 3′,5′-monophosphate) through norepinephrine-activated, membrane-bound adenylate cyclase. Indeed, circadian rhythms were found (Dobra and Ehret, 1977) in all these components; TAT, AMP, and adenosine triphosphate (ATP) showed highly significant oscillations (Table 2.2). The rhythm of AMP peaked just before the times of greatest glycogen depletion and CO_2 production, the concentration of ATP underwent maximal increases toward the end of the day-phase, during times of highest glycogen utilization, and TAT activity was greatest during the night, preceding the increases in AMP concentrations and glycogen metabolism. The phase relation of the TAT rhythm to that of the adenylate system suggested a regulatory role of biogenic amines in glycogen utilization in *Tetrahymena* (as for the rat), since this enzyme is known to influence their rate of synthesis when the substrate is limiting. This hypothesis was further supported by the finding that norepinephrine, topically administered to entrained cultures on agar plates, maximally suppressed TAT activity at the time when TAT is known to be synthesized at the highest rate (Ehret et al., 1977).

These findings logically led Ehret and coworkers to propose an energy reserve escapement mechanism for circadian clocks (see Section 5.2.2), which intimately associates glycogen metabolism with the circadian–infradian rule (Ehret and Dobra, 1977; Ehret et al., 1977; see Section 3.2.2). In this unifying formulation, a gene-action circadian oscillator, producing phase-specific molecular transcripts and chronotypic enzymes, would ultimately drive overt circadian rhythms, coupling to them through causally interconnected oscillations in glycogenolysis and in biogenic amine metabolism. Glycogen would provide the reliable energy source needed for the clock as well as for infradian intermediary metabolism (Fig. 2.2). This energy would be meted out discontinuously, on a daily basis, by an escapement component comprising either the indoleamine pathway (from tryptophan to serotonin) or the catecholamine pathway (tyrosine to epinephrine).

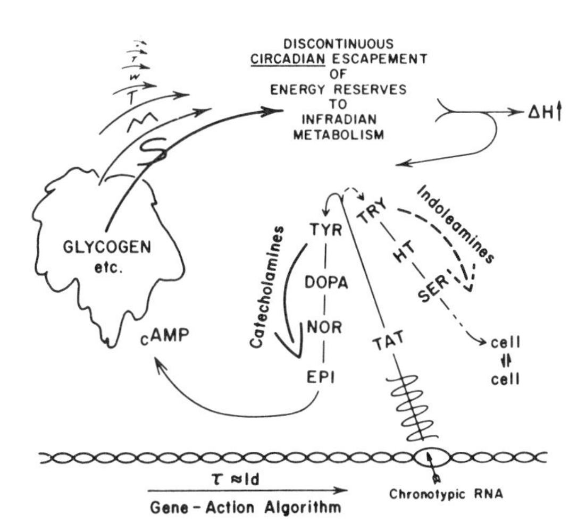

FIGURE 2.2. The circadian clock is pictured as an energy reserve escapement with alternate path options coupled to a gene-action circadian oscillator that controls the path of choice. This scheme results in long-term conservation of energy reserves through circadian parceling during infradian growth. (Reprinted from Ehret and Dobra, 1977, with permission of Publishing House Il Ponte.)

Ehret et al. (1977) have speculated provocatively that this exclusively eukaryotic mechanism allows cells to adjust to the feast or famine conditions that they encounter. During times of plentiful nutrients, the gene-action clock mechanism programs the ultradian cell division cycle (CDC) to run as rapidly as possible (temperature compensation might well be sacrificed for maximal cell proliferation). Under much more commonly encountered restrictive conditions, the eucell clock would permit the CDC to be uncoupled and the cell to enter the circadian–infradian growth mode while continuing to program the circadian cycle of the daily rationing of energy reserve stores and ultimately to generate the multitude of other overt periodicities.

In this regard, Kämmerer and Hardeland (1982) addressed the question of whether there is any relationship or common element between ultradian and circadian oscillator mechanisms and their separate or joint control of the CDC and other rhythmicities. Cultures of *T. pyriformis* were synchronized in the ultradian growth mode by a 5-h temperature cycle [WC: 4,1 (W = 20°C; C = 0°C)]. On transfer to a constant temperature of 20°C, the cells showed a free-running rhythm of TAT activity (as well as of cell division) having a τ of ~5 h. This rhythmicity persisted even when cell growth was inhibited with cycloheximide. The authors concluded that an independent ultradian oscillator was operating, which, although not identical with the CDC per se, might control it under normally permissive conditions. The data seem to suggest that a separate circadian oscillator controls both the CDC and TAT activity in the circadian–infradian growth mode, although it is not excluded that the ultradian oscillator might play some role in circadian timekeeping (see Section 3.2.2). This hypothesis has been further corroborated by the findings that the ultradian rhythm of TAT activity persisted in *Tetrahymena thermophila* (= *T. pyriformis* syngen 1) in the presence of 1 m*M* emetine, which inhibited both protein synthesis and cell division, and in the presence of 250 m*M* hydroxyurea, which suppressed cell growth to a high degree (Thiel et al., 1985; see Fig. 5.7), and that the free-running period of the enzyme rhythm was temperature-compensated (the value of the Q_{10} for τ was much closer to 1.0 than 2.0) over a range of constant temperatures of 10 to 30°C (Michel and Hardeland, 1985; see Fig. 5.6).

Similarly, Readey (1987) recently demonstrated that ultradian LD cycles can entrain cultures of *T. pyriformis* (strain GLC) growing in the ultradian mode at low cell titers in a nephelostat as long as the period of the synchronizer does not exceed the nearest modal value of *g* observed in free-running single cells. Thus, LD cycles (in fact, bright L alternating with dim L) having *T*-values of 3, 5, or 6 h (i.e., LD: 1,2; LD: 1.5,3.5; LD: 2,4) induced synchronous division (as well as changes in mean cell volume) in populations having mean cell generation times of 5.63 to 5.98—values corresponding closely to the value (5.4 h) found for individual cells in the same growth medium at the same temperature. Under subsequent dim LL, the cultures free-ran for several cycles with τ-values of ~3 or ~6 h.

2.1.2 *Paramecium* spp.

One of the first nongreen cells shown to possess a circadian clock was the ciliated protozoan *Paramecium bursaria,* which had been deprived of its symbiotic chlorellae (Ehret, 1951). Mating reactivity between cells of different mating types (syngens) occurred at midday in LD: 12,12 in populations of nongrowing cells. This rhythm persisted for as long as a week in DD and could be phase-shifted by brief light pulses. In the visible region, an action spectrum was obtained (Ehret, 1960) for phase-shifting (advances), with peaks in the red (650 nm), blue (440 nm), and near ultraviolet (350 nm). Furthermore, Ehret (1959) found that far ultraviolet also was highly efficient in inducing phase shifts, which could be reversed (photoreactivated) by exposure of the cells to white light. [These and other observations led to the suggestion that nucleic acids were involved in the clock mechanism and to the chronon model for circadian timekeeping (Ehret and Trucco, 1967; see Section 5.3.2)]. The rhythm of mating reactivity now has been demonstrated within single cells of *P. bursaria,* even in individuals taken from arrhythmic populations in which circadian rhythmicity had damped out after 2 weeks in dim LL of 1000 lx (Miwa et al., 1987). Thus, population arrhythmicity appears to be the result of desynchronization among the constituent oscillators. The fact that a 9-h dark pulse was able to restore rhythmicity in a population could be explained by its differential phase-shifting action (a dark-pulse PRC was derived) on the individual cells, the phases of which were randomly distributed at the time of the signal.

A similar circadian rhythm of mating type reactivity has been demonstrated in *Paramecium aurelia* (syngen 3) by Karakashian (1965, 1968). In DD at 17°C, τ was about 22 h and was relatively temperature-independent. Manifestation of the rhythm seemed to occur only in slowly growing cells (e.g., g = 56 h). The system, therefore, adheres to the circadian–infradian rule.

One of the most unique cellular circadian rhythms, and one also amenable to genetic analysis, is the rhythm of mating-type reversals, reported first in *Paramecium multimicronucleatum* (Sonnenborn and Sonnenborn, 1958). Although the mating type of a cell is usually inherited by its asexual progeny, with every cell line's normally having a specifically assignable mating type, certain stocks of syngen 2 (mating types III and IV) exhibit a fascinating exception to this generalization: the same individual expresses one mating type for part of each day and the complementary type during the remainder. Sonnenborn and Sonnenborn (1958) and Barnett (1959) found that this rhythm could be entrained by LD cycles (e.g., LD: 6,18) and that it persisted for at least 3 days in DD.

Later, in an intensive study of the mating-type reversal rhythm in three separate stocks of *P. multimicronucleatum,* Barnett (1961, 1965, 1966) found that each displayed a characteristic phase in its switch-over time (III to IV, and vice versa) in an entraining LD: 8,16 cycle. The rhythm persisted for as long as 6 days in an ensuing DD free-run. The phase of the switch-over time characterizing a given stock was identical in all sublines as long as the progenitor of that clone derived its macronucleus from one of the segregating macronuclear anlage produced at conjugation. Barnett (1961,1966) discovered that the capability of the stock to cycle depended on the presence of a dominant allele, *C* (cycler). Cells homozygous for the recessive allele, *c* (acyclic), did not reverse mating type but were either type III or IV as a consequence of macronuclear differentiation. Thus, the nucleus controlled not only the phase of the circadian rhythm (as shown in the alga *Acetabularia*; see Section 2.2.1), but also its ability to be expressed.

Photoentrainable, persisting circadian rhythms of cell division have been reported in *P. bursaria* (Volm, 1964) and in *P. multimicronucleatum* (Barnett, 1969; see Table 3.1). A circadian rhythm of locomotor behavior has been discovered both in populations (Hasegawa et al., 1984) and in single cells (Hasegawa and Tanakadate, 1984) of the latter species by means of a microcomputerized closeup video/photoamplifier system (Tanakadate and Ishikawa, 1985). This rhythm, in which individuals swam fast and unidirectionally during the day but more slowly and with frequent turning and avoidance responses during the night, was light-entrainable and could be phase-shifted by LD cycles, free-ran in either DD or LL, and was temperature-compensated over a 15 to 25°C temperature range. Recently, Miwa et al. (1986) have observed a circadian rhythm of photoaccumulation in *P. bursaria,* which is expressed even in sexually immature cells that do not exhibit a rhythm of mating activity.

Finally, the duration of the CDC in individual cells of *Paramecium tetraurelia* in the ultradian growth mode appears to be a discrete multiple of the ultradian cycle of motility, whose τ of about 1 h is temperature-compensated over the range 18 to 33°C (Kippert, 1985b). The number of motility subcycles varied in stepwise fashion, from 3 (for short *g*s at 30°C) to 12 (for long *g*s at 18°C). In addition, divisions occurred at the same phase of the motility rhythm, a fact that suggests that ultradian rhythms function in gating the CDC in a manner similar to that of their circadian counterparts (see Sections 3.2.1, 3.2.2).

2.2 Circadian Rhythms in Unicellular Algae

2.2.1 *Acetabularia* spp.

Although there may be those who balk at the notion of viewing the green alga, *Acetabularia mediterranea* (or *Acetabularia crenulata*), even as a *maxi*-microorganism, it is a unicellular system (at least, until its months-long cell developmental cycle eventually culminates in the formation of a cap), whose large size is one of its strengths, facilitating the monitoring of various functions directly in an individual cell. Inasmuch as *Acetabularia* has proved useful to research on circadian clocks [having recently given rise to the coupled translation–membrane model (Schweiger and Schweiger, 1977; see Section 5.4.2)], the salient features of its physiological circadian time structure and recent advances with this organism are briefly reviewed. Biochemical and molecular aspects of these rhythmicities are discussed in detail in Section 4.3.

A number of persisting circadian rhythms have been documented for *Acetabularia mediterranea* (Table 2.3). Most of these are associated with the rhythms of photosynthesis, which have been investigated intensively (Mergenhagen, 1980). Thus, the number and shape of the chloroplasts, their ultrastructure, their carbohydrate content and RNA synthetic abil-

TABLE 2.3. Some persisting circadian rhythms exhibited by the unicellular green alga *Acetabularia* spp.

Circadian rhythm	Reference
Cyclic AMP	Vanden Driessche et al. (1979)
ATP content (chloroplastic)	Vanden Driessche (1970b)
Chloroplast migration	Koop et al. (1978); Broda et al. (1979); Broda and Schweiger (1981b); Schmid and Koop (1983); Borghi et al. (1986); Schmid (1986)
Chloroplast number	Vanden Driessche (1973)
Chloroplast shape	Vanden Driessche (1966a)
Chloroplast ultrastructure	Vanden Driessche and Hars (1972a,b)
Electric potential	Novák and Sironval (1976); Broda and Schweiger (1981a,b); Borghi et al. (1986)
Hill reaction, activity	Vanden Driessche (1974)
Photosynthesis, activity	Schweiger et al. (1964a,b); Terborgh and McLeod (1967); von Klitzing and Schweiger (1969); Mergenhagen and Schweiger (1973, 1975a,b); Karakashian and Schweiger (1976a,b,c); Schweiger and Schweiger (1977); Vanden Driessche (1979); Broda and Schweiger (1981b)
Photosynthesis, capacity	Sweeney and Haxo (1961); Sweeney et al. (1967); Hellebust et al. (1967); Terborgh and McLeod (1967); Vanden Driessche (1966a, b; 1967); Vanden Driessche et al. (1970); Sweeney (1972, 1974b)
Polysaccharide content, chloroplastic	Vanden Driessche et al. (1970)
Protein synthesis (p230)	Hartwig et al. (1985, 1986)
RNA synthesis, chloroplastic	Vanden Driessche (1966b); Vanden Driessche and Bonotto (1969)

Updated from Edmunds (1984c), with permission of Academic Press.

ities, their Hill reaction capacity, and even their migration rate have all been shown to be subject to regulation by a circadian clock (see reviews by Vanden Driessche, 1973, 1975, 1980; Broda et al., 1979; Puiseux-Dao, 1984).

Photosynthetic functioning itself has been measured usually in one of two ways: (1) as photosynthetic capacity (PC), reflecting the ability of aliquots of cells to evolve O_2 or take up CO_2 under saturating light conditions at different times, or (2) as actual photosynthetic activity (PA), wherein photosynthetically produced O_2, which is released into the medium, is measured either continuously or over an extended time span under the given culture conditions [which typically are not saturating (e.g., LD or low-intensity LL)]. One of the first reports (Sweeney and Haxo, 1961) of a rhythm of PC was in populations (10 cells) of *Acetabularia major*, as well as in an individual cell, measured by the cartesian diver technique. The rhythm was entrained by LD: 12,12 (higher activity in the day) in both nucleate and anucleate cells (obtained by simply cutting off the nucleus-containing rhizoid) and persisted in LL for 2 days in the anucleate cells. Similarly, both nucleate and anucleate cells of *Acetabularia crenulata* could be entrained by 24-h LD cycles and their rhythms phase-shifted by the prolongation of the light interval by 12 h. These earlier results, therefore, suggested that the nucleus (and by implication, continuous transcription of the nuclear genome) was not necessary for entrainment or the maintenance of persisting circadian rhythms.

This conclusion was corroborated in an extension of these studies to *Acetabularia mediterranea,* in which a rhythm of PA was demonstrated in both nucleate and anucleate cells by measuring photosynthetically produced O_2 by an electrochemical method in LL (2500 lx) at 20°C (Schweiger et al., 1964a). The role of the nucleus in the generation of circadian oscillations was further elucidated by experiments in which rhizoids and nuclei were exchanged (Schweiger et al.,1964b). Two groups of 180° out-of-phase cells maintained in reversed LD: 12,12 cycles were enucleated by removal of their rhizoids. The rhizoids containing the nuclei were then grafted to the anucleate stalks exhibiting rhythms with opposite phase angles from the donor plants. The recipient cells were observed to gain gradually the phase of the rhizoid donor cells. Similar results were obtained by implanting isolated nuclei into anucleate cells instead of transplanting the entire rhizoid (Fig. 2.3A). Once again, it appeared that the nucleus was capable of determining the phase of the rhythm but was not necessary for its maintenance. Results consistent with this hypothesis were obtained by Vanden Driessche (1967), who was able to induce periodicity in anucleate cells of a strain of *A. mediterranea* that lacked rhythmicity by grafting on rhizoids from cells that displayed rhythmicity.

The development of flow-through techniques for continuous recording of photosynthetic oxygen production and the use of a highly sensitive platinum electrode (von Klitzing and Schweiger, 1969; Mergenhagen and

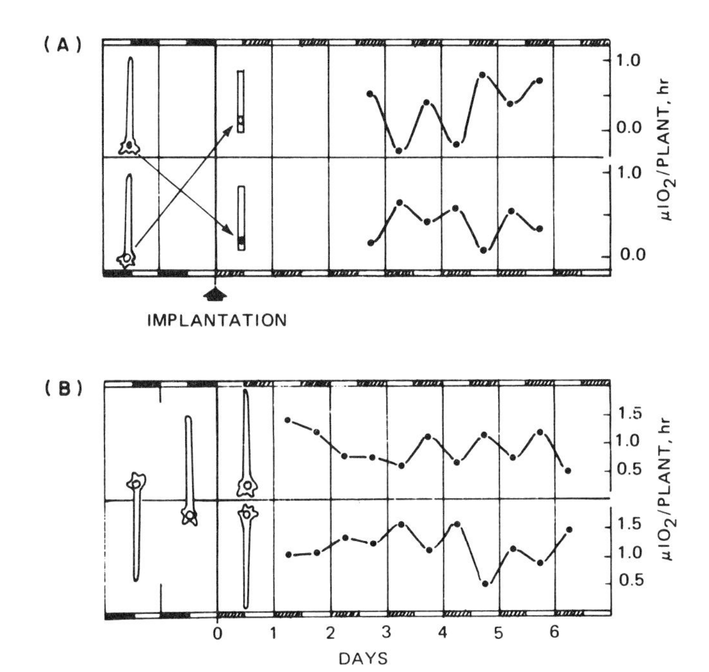

FIGURE 2.3. The role of the nucleus in the photosynthetic-capacity rhythm of *Acetabularia mediterranea*. (A) Test plants were maintained in normal and reversed light regimens. The nuclei were then exchanged between plants, and the photosynthetic rhythms were studied in constant conditions. The phases of the two rhythms were that of the nuclei they contained. (B) Portions of intact plants were subjected to opposite illumination cycles for 2 weeks, and the photosynthetic rates then were measured in constant conditions. The phase displayed was that of the nucleus-containing rhizoid rather than the cytoplasmic end of the plant. (Adapted from Schweiger et al., 1964b, with permission of American Association for the Advancement of Science.)

Schweiger, 1973) greatly facilitated the measurement and analysis of the PA rhythm in single cells. With this technique the free-running period in nucleated *Acetabularia* cells was estimated to be 22.6 ± 3.2 h and was essentially identical to that found for anucleate cells (Mergenhagen and Schweiger, 1973). A Q_{10} of 0.8 was calculated (Karakashian and Schweiger, 1976a), indicating an overcompensation similar to that found for *Gonyaulax polyedra* (Sweeney and Hastings, 1960; see Section 2.2.4). It was also possible to phase-shift the rhythm of PA using dark pulses to perturb the rhythm free-running in LL (Karakashian and Schweiger, 1976a; see Fig. 4.9).

Finally, by means of the flow-through technique, rhythms in PA were found in a variety of cell fragments. Isolated caps, basal nucleate and anucleate fragments, apical fragments with and without stalks, and small segments taken from the middle of the stalk all displayed circadian periodicity (Mergenhagen and Schweiger, 1975a). Taken at face value, these results would seem to provide dramatic evidence that there is no particular compartment of the cell that is responsible for the expression of the rhythm and no specific site of the clock.

Lest one think that the only circadian rhythms observed in *Acetabularia* spp. are those associated with photosynthesis and chloroplasts, one has merely to note the recent demonstration of a rhythm in electric potential along the longitudinal axis in individual cells (Broda and Schweiger, 1981a). The study of this rhythm, initially inspired by the discovery (Novák and Sironval, 1976; see Borghi et al., 1986) of a circadian periodicity of the transcellular current in regenerating enucleated posterior stalk segments, was undertaken to further substantiate the role of membranes in clock function, as called for by the coupled translation-membrane model (Schweiger and Schweiger, 1977; see Section 5.4.2). Gradual changes in τ were observed during long (100 days or more) free-runs in LL.

Assuming that this rhythm is shown to exhibit all the other attributes of a full-fledged circadian rhythm (Fig. 1.2), it would be interesting simultaneously to monitor it and one or two other periodicities (such as PA and chloroplast migration) over extended time spans to see if mutual dissociation occurs (Broda and Schweiger, 1981b; see Section 6.2). This should now be facilitated by the development of a fully automated, multitasking, time-sharing computer system that allows simultaneous data acquisition and analysis (Broda et al., 1983). Indeed, Schweiger et al. (1986) recently have observed that if one simultaneously monitors the rhythms of chloroplast migration and electric potential during their free-run and then imposes an 8-h dark pulse, occasionally one of the rhythms is phase-shifted while the other is not. These observations are consistent with the notion that at least two circadian clocks, usually tightly coupled, exist in a single cell.

2.2.2 *Chlamydomonas* spp.

The unicellular green alga, *Chlamydomonas reinhardtii*, like the algal flagellate *Euglena* (Section 2.2.3), has been well characterized physiologically and biochemically. Furthermore, because it can reproduce sexually, this microorganism has proved particularly attractive for genetic analysis of cellular behavior and its circadian regulation (Bruce, 1976; Mergenhagen, 1980; Feldman, 1982).

It has been shown that *C. reinhardtii* has a circadian clock that controls the phototactic response of slowly dividing (infradian) or nondividing cells (Bruce, 1970). This rhythm of phototaxis (photoaccumulation) can be initiated by a sudden, single shift from LL to DD, entrained by LD cycles, and phase-shifted by exposure to short dark perturbations. The free-running period at 22°C is about 24 h and is temperature-compensated over the range 18 to 28°C. Hastings et al. (1987), using the Okazaki large spectrograph (resolution: 1 nm = 1 cm), determined the action spectrum for light-induced phase resetting of the phototaxis rhythm. The effects of 6-h pulses given in dim white LL during the subjective night (when bright white light yielded large $-\Delta\phi$s of 6 to 12 h) were measured at 15- to 20-nm intervals from 300 to 800 nm. Blue (470 nm) and red (660 nm) light pulses generated $\Delta\phi$s, but yellow (550 nm) light pulses did not. Under these conditions, however, blue and red signals had qualitatively different effects: blue generated advances, red produced delays. Similar differences between the effects of the two wavelengths were recorded as a function of fluence and in the measurement of PRCs (they were displaced by 24 h). One interpretation of these findings is that two photoreceptors exist, each having a different effect on the clock. On the other hand, the action spectrum was similar to that for photosynthesis, and diuron (DCMU; 10 μM), an inhibitor of photosynthesis, was found to block $\Delta\phi$s by both blue and red light. Perhaps the response is mediated by photosynthesis. The clock is manifested also by a rhythm of the tendency of cells to stick to a glass surface (Straley and Bruce, 1979), reminiscent of the circadian rhythm in cell settling and adhesion discovered by Terry and Edmunds (1970b) in *Euglena gracilis* (Section 2.2.3). In addition, a circadian rhythm in cell division has been reported (Bruce, 1970) (but see Spudich and Sager, 1980; Section 3.1), with the possibility existing that the sexual cycle as well as vegetative growth may also be under clock control. However, recent findings indicate that the mating reactivity of the gametes is not so regulated (Bruce and Bruce, 1981). Nevertheless, the mating competence of gametes of *C. eugametos* is under circadian clock control, the rhythm being caused by endogenous changes in flagellar agglutinability (Demets et al., 1987). This latter rhythm is, in turn, attributable to circadian variations in the concentration of membrane-bound surface receptors (agglutinins), which appear to undergo cyclical synthesis and degradation (Tomson et al., 1987). This biological rhythm, therefore, lends itself to further exploration at the molecular level.

Subsequent to the discovery of the rhythm of phototaxis, clock mutants with both shorter and longer periods than the wild type have been isolated and characterized genetically (Bruce, 1972; Mergenhagen, 1984). These, together with a number of mutants with metabolic deficiencies and altered rhythm properties (Mergenhagen and Hastings, 1977), are treated in a discussion of the genetic dissection of the clock (Section 4.4).

Recently, the photoaccumulation rhythm has been examined in a *Zeitgeber*-free environment (Mergenhagen, 1986; Mergenhagen and Mergenhagen, 1987). Under 0 − gravity conditions in space, both the wild type and a short-period mutant (s^-) exhibited well-expressed circadian rhythms having average τs of 29.6 and 21.6 h, respectively. These values did not differ significantly from those of the ground-based controls (29.4 and 20.8 h, respectively), although the amplitudes were more pronounced and the rhythm damped much more slowly in space (probably because swimming in a low-gravity environment is less energy consuming). Thus, the circadian clock is truly endogenous and not dependent on terrestrial *Zeitgeber* [cf. results for the conidiation rhythm of *Neurospora* in space (Sulzman et al., 1984)].

2.2.3 *Euglena* spp.

The temporal organization of *Euglena gracilis* Klebs (strain Z) has been extensively studied in a number of laboratories over the past 25 years (see reviews by Edmunds, 1975, 1982, 1984a,c; Edmunds and Halberg, 1981; Wille, 1979), and this microorganism and *Gonyaulax* and *Tetrahymena* form a trio that has provided evidence for the so-called *G-E-T* effect, which is more formally embodied in the circadian–infradian rule formulated by Ehret and Wille (1970) (see Sections 2.1.1 and 3.2.2). This algal flagellate can be grown on a variety of completely defined media, either photoautotrophically in the presence of CO_2 and vitamins B_1 and B_{12} or organotrophically in the light or dark on carbon sources ranging from acetate and ethanol to lactic, glycolic, glutamic, and malic acids over a wide pH range. This versatility in growth mode and the fact that cell division can be synchronized easily by appropriate 24-h (and other) lighting schedules (Cook and James, 1960; Edmunds, 1965a) and temperature cycles (Terry and Edmunds, 1970a) have made *E. gracilis* an important experimental organism for a variety of physiological and biochemical investigations (Buetow, 1968a,b, 1982).

A large number of persisting circadian rhythms have been reported, (Table 2.4). These studies have been aided by the fact that *E. gracilis* can be maintained in the stationary phase of growth (or infradian growth mode) for days or even weeks with little or no net change in cell concentration. Circadian output can thus be monitored while divorced from the driving force of the CDC. Likewise, the fact that a number of photosynthetic mutants (or even completely bleached strains devoid of their chloroplast genomes) have been isolated that still exhibit light-entrainable circadian rhythms effectively has eliminated the problem of the dual use of imposed light spans and signals—as an energy source for growth, on the one hand, and as a timing cue for the underlying clock, on the other (Kirschstein, 1969; Jarrett and Edmunds, 1970; Mitchell, 1971; Edmunds et al., 1976).

Clearly, then, the *E. gracilis* system provides an excellent case for temporal differentiation: a large number of diverse behavorial, physiological, and biochemical activities are partitioned along the 24-h time axis, thus providing dimensions for both environmental adaptation and functional integration in time (Edmunds, 1982). This circadian time structure is perhaps best illustrated by an acrophase chart (Fig. 2.4), which provides a convenient way to indicate time relationships of periodicities (analyzed mathematically by cosinor or other mathematical techniques) both to a

TABLE 2.4. Circadian rhythms in *Euglena gracilis* Klebs.

Rhythm[a, b]	Strain	Phase marker	ϕ^d	Reference[c]
Physiological				
Cell division	Z	Onset	CT 11–13 (180–195°)	Edmunds (1964, 1965a, 1975, 1978, 1982; Edmunds and Funch (1969a,b); Edmunds and Laval-Martin (1984)
	P_4ZUL	Onset	CT 10–12 (150–180°)	Edmunds (1971, 1978); Edmunds et al. (1971, 1974, 1976); Jarrett and Edmunds (1970)
	P_7ZN_gL	Onset	CT 10–12 (150–180°)	Mitchell (1971)
	W_6ZHL	Onset	CT 10–12 (150–180°)	Edmunds (1978); Edmunds et al. (1971)
	W_nZUL	Onset	CT 10–12 (150–180°)	Mitchell (1971)
	Y_9ZN_a1L	Onset	CT 10–11 (150–165°)	Edmunds et al. (1976)
Flagellated cells (%)	Z (1224-5/9, Göttingen)	Maximum	CT 03 (45°)	Brinkmann (1966)
Motility, random (dark) *Dunkel-beweglichkeit*	Z (1224-5/9, Göttingen)	Minimum	CT 18–21 (270–315°)	Schnabel (1968); Brinkmann (1966, 1971); Kreuels and Brinkmann (1979); Kreuels et al. (1984) Martin et al. (1985)
	W_nZHL (1224–5/25, Göttingen)	Minimum	CT 12 (180°)	Kirchstein (1969)
pH (external medium)	Z			Brinkmann (personal communication, 1980)
Photokinesis (photo-activation of random motility)	Z	Maximum	CT 18–21 (270–315°)	Brinkmann (1976b)
Photosynthetic capacity				
$^{14}CO_2$ uptake	Z, ZR	Maximum	CT 06–12 (90–180°)	Walther and Edmunds (1973); Laval-Martin et al. (1979); Edmunds and Halberg (1981); Edmunds and Laval-Martin (1981)
O_2 evolution	Z	Maximum	CT 04–06 (60–90°)	Walther and Edmunds (1973); Lonergan and Sargent (1978, 1979); Lonergan (1983, 1986a,b)

TABLE 2.4. Continued.

Rhythm[a, b]	Strain[c]	Phase marker	φ[d]	Reference[e]
Phototaxis (capacity)	Z	Maximum	CT 04–08 (60–120°)	Pohl (1948); Bruce and Pittendrigh (1956, 1958); Brinkmann (1966); Feldman (1967); Feldman and Bruce (1972)
Settling	Z	Maximum	CT 21–09[f] (315–135°)	Terry and Edmunds (1970b)
			CT 15 (225°)	Kiefner et al. (1974)
Shape	Z	Maximum elongation	CT 06 (90°)	Lonergan (1983, 1984, 1985, 1986b); Lachney and Lonergan (1985)
	Z (1224-5/9, Göttingen)		CT 03–09 (45–135°)	Brinkmann (1976b)
Susceptibility				
Ethanol (pulses)	Z (1224-5/9, Göttingen)	Maximum	CT 03–06 (45–90°)	Brinkmann (1976a)
Trichloroacetic acid	Z (1224-5/9, Göttingen)	Maximum	CT 03–06 (45–90°)	Brinkmann (1976b)
Volume	Z	Maximum	CT 18–21 (270–315°)	K. Brinkmann and U. Kipry (unpublished)
Biochemical				
Amino acid incorporation into proteins	Z	Maximum	CT 10–12 (150–180°)	Feldman (1968)
			CT 08 (120°)	Quentin and Hardeland (1986a,b)
	T		CT 04 (60°)	Quentin and Hardeland (1986a,b)
	WTHL		CT 12	Quentin and Hardeland (1986a,b)
Gross metabolic variables[g]				
Carotenoids	Z	Onset	ZT 0 (0°)	Cook (1961b); Edmunds (1965b)
Chlorophyll *a*	Z	Onset	ZT 0 (0°)	Cook (1961b); Edmunds (1965b)
Dry weight	Z	Onset	ZT 0 (0°)	Cook (1961b); Edmunds (1965b)
Protein, total	Z	Onset	ZT 0 (0°)	Cook (1961b); Edmunds (1965b)
DNA, total	Z	Onset	ZT 08–09 (120–135°)	Cook (1961b); Edmunds (1964, 1965b)
RNA, total	Z	Onset	ZT 0 (0°)	Cook (1961b); Edmunds (1965b)
NAD^+, intracellular	Z	Peak	CT 18 (270°)	Goto et al. (1985)
Enzymatic activity[h]				
Acid phosphatase	Z	Peak	ZT 06–08 (90–120°)	Sulzman and Edmunds (1972); Edmunds (1975)

TABLE 2.4. Continued.

Rhythm[a, b]	Strain[c]	Phase marker	ϕ^d	Reference[e]
Alanine dehydrogenase	Z	Peak	CT 06–08 (90–120°)	Sulzman and Edmunds (1972, 1973); Edmunds et al. (1974)
Carbonic anhydrase	Z	Peak	ZT 06–08 (90–120°)	Lonergan and Sargent (1978)
Glucose-6-phosphate dehydrogenase	Z	Peak	ZT 06 (90°)	Sulzman and Edmunds (1972); Edmunds (1975)
Glutamic dehydrogenase	Z	Peak	ZT 06–08 (90–120°)	Sulzman and Edmunds (1972); Edmunds (1975)
Glyceraldehyde-3-phosphate dehydrogenase (NADP and NADPH-dependent)	Z	Peak	ZT 05–08 (75–120°)	Walther and Edmunds (1973); Lonergan and Sargent (1978)
Lactic dehydrogenase	Z	Peak	ZT 06 (90°)	Sulzman and Edmunds (1972); Edmunds (1975)
NAD^+ kinase	Z	Peak	CT 0 (0°)	Goto et al. (1985)
L-Serine deaminase	Z	Peak	ZT 06 (90°)	Sulzman and Edmunds (1972); Edmunds (1975)
L-Threonine deaminase	Z	Peak	ZT 08 (120°)	Sulzman and Edmunds (1972); Edmunds (1975)

Modified from Edmunds and Halberg (1981), with permission of Williams & Wilkins.

[a]Unless otherwise noted, all rhythms listed have been shown to persist with a circadian period in DD (or LL) and constant temperature after synchronization.

[b]All cultures were essentially nondividing (stationary, or long infradian), except those in which the cell division rhythm itself was monitored. Consult reference for precise culture conditions.

[c]The wild-type strain is designated as Z. Unless otherwise noted, it was originally obtained from the American Type Culture Collection (No. 12716) and maintained at the State University of New York at Stony Brook since 1965 or at Princeton University. Also available from the Algal Collection at Indiana University (No. 753) and the Algensammlung Pringsheim at Göttingen (No. 1224-5/9). Other strains are photosynthetic mutants or completely bleached strains incapable of photosynthesis.

[d]Phase given in circadian time (CT, hours or degrees after subjective dawn, when the onset of light would have occurred had the synchronizing light cycle been continued). Entraining light cycles were either LD: 12,12 or LD: 10,14 [except in the case of the settling rhythm in which the cultures were synchronized by a 12:12 temperature cycle (18/25°C, LL) before release into LL, and CT 0 denoted the onset of lower temperature]. In those instances were a free-run was not monitored, phase is given in *Zeitgeber* (synchronizer) time (ZT 0, onset of light).

[e]Only key references are given; no attempt is made to give all citations.

[f]Rhythm synchronized by a 12:12 temperature cycle (18/25°C) in either dividing or nondividing cultures maintained in LL before release into constant conditions (25°C, LL). CT 0, onset of lower temperature.

[g]These variables were monitored only in light-synchronized (LD: 14,10) dividing cultures. The phase marker here refers to the point when values start to increase.

[h]Although all of the enzymes indicated undergo oscillations in activity in nondividing cultures in LD: 10,14, only alanine dehydrogenase and NAD^+ kinase have been investigated in sufficient detail to demonstrate conclusively that they will persist for long time spans in DD and constant temperature (or other free-running conditions).

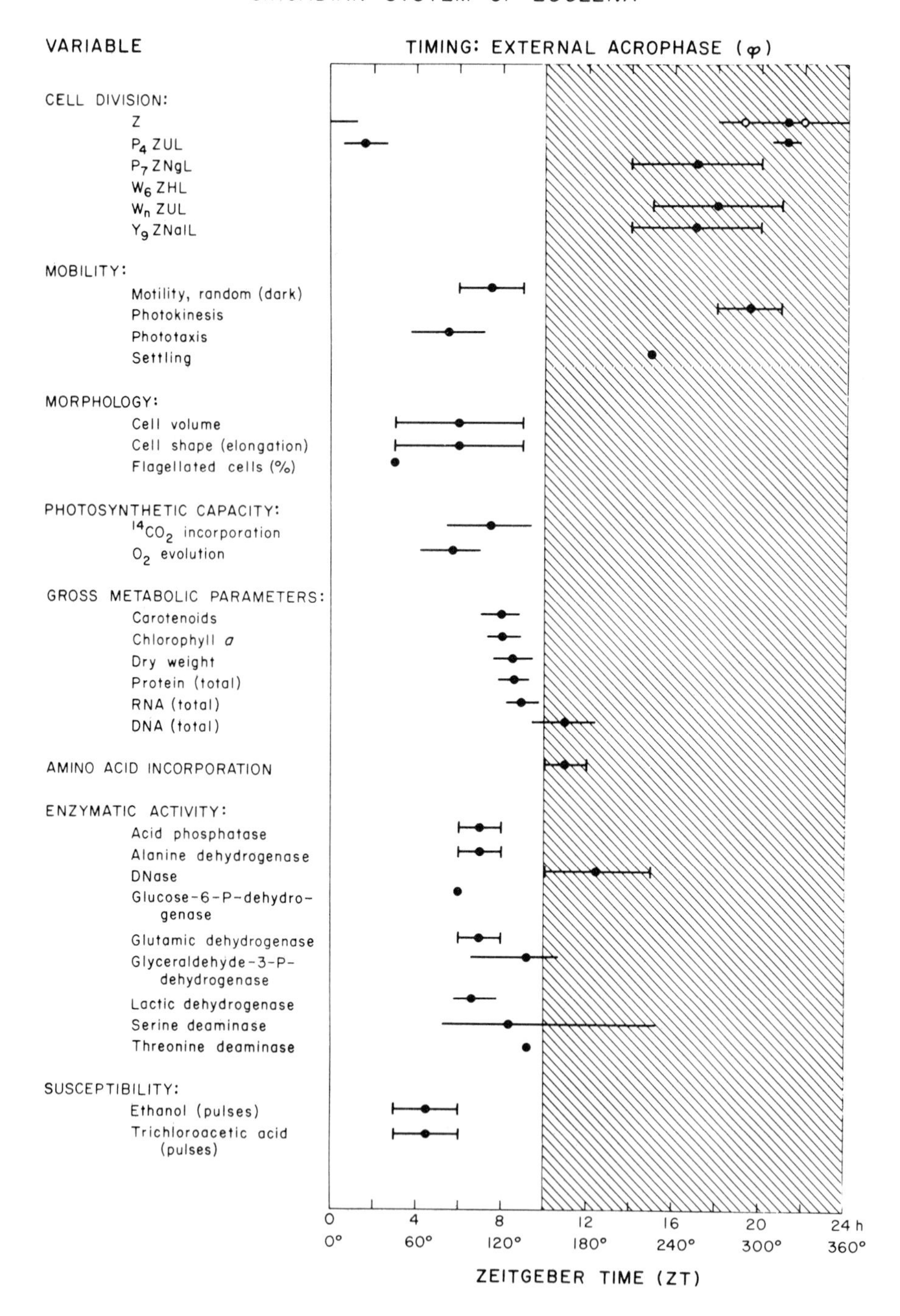
CIRCADIAN SYSTEM OF EUGLENA
VARIABLE
TIMING: EXTERNAL ACROPHASE (φ)
CELL DIVISION:
Z
P4 ZUL
P7 ZNgL
W6 ZHL
Wn ZUL
Y9 ZNaIL
MOBILITY:
Motility, random (dark)
Photokinesis
Phototaxis
Settling
MORPHOLOGY:
Cell volume
Cell shape (elongation)
Flagellated cells (%)
PHOTOSYNTHETIC CAPACITY:
14CO2 incorporation
O2 evolution
GROSS METABOLIC PARAMETERS:
Carotenoids
Chlorophyll a
Dry weight
Protein (total)
RNA (total)
DNA (total)
AMINO ACID INCORPORATION
ENZYMATIC ACTIVITY:
Acid phosphatase
Alanine dehydrogenase
DNase
Glucose-6-P-dehydrogenase
Glutamic dehydrogenase
Glyceraldehyde-3-P-dehydrogenase
Lactic dehydrogenase
Serine deaminase
Threonine deaminase
SUSCEPTIBILITY:
Ethanol (pulses)
Trichloroacetic acid (pulses)
0
4
8
12
16
20
24 h
0°
60°
120°
180°
240°
300°
360°
ZEITGEBER TIME (ZT)

synchronizing LD cycle (or other *Zeitgeber*) and to each other during either entrainment or free-running conditions (Edmunds and Halberg, 1981).

In the following subsections, we survey in greater detail selected classes of circadian rhythms itemized in Table 2.4 and examine some of the more recent developments in these areas.

Circadian Rhythms of Cell Division

The relationship between the CDC and circadian oscillators perhaps has been most intensively investigated in *E. gracilis* Klebs (Edmunds, 1978; Edmunds and Adams, 1981). Because of the importance of cell cycle regulation, Chapter 3 is devoted to this subject and the major lines of evidence (Edmunds and Laval-Martin, 1984) that implicate an endogenous, self-sustaining, circadian oscillator (see Fig. 1.2) in the control of the CDC in this and other microorganisms (see Table 3.1) are reviewed.

Circadian Rhythms of Cell Motility

Pohl (1948) first demonstrated a daily rhythm of phototactic response in nondividing (stationary or infradian phase) populations of autotrophically grown *E. gracilis* Klebs. The rhythm showed a maximal response during the day and a minimal response at night. Since then, this and associated motility rhythms in *Euglena* have been intensively studied.

Rhythm of Phototaxis

Bruce and Pittendrigh (1956, 1958) introduced a high degree of automation into the assay system for this rhythm. This allowed simultaneous recording of the kinetics and degree of aggregation of cells into a brief (30 min), narrow, vertical test light beam from a heat-compensated microscope lamp in a number of cultures maintained in transparent Falcon flasks otherwise

◁———

FIGURE 2.4. Acrophase (ϕ) chart for *Euglena gracilis* showing the time relations among the circadian rhythms listed in Table 2.4. The external acrophase relates the peak of the fitted curve to the onset of light in a synchronizing LD cycle and to subjective dawn in DD or LL (when the light would have come on had the LD cycle been continued). The bars extending to the right and the left of the acrophase points give the 95% confidence interval, as calculated by cosinor analysis. Single points or points with dashed lines and brackets indicate subjective estimates of the acrophase from published data without benefit of statistical analysis. The data for gross metabolic variables, although originally obtained in LD: 14,10, have been adjusted to correspond to a synchronizing LD: 10,14 cycle used for most of the other rhythms. (Reprinted from Edmunds and Halberg, 1981, with permission of Williams & Wilkins Co.)

kept in LD, DD (or other regimen). The pattern thus obtained was termed the "phototactic response," which was actually a composite of both general motile and light-oriented behaviors.

The response to the test beam during photoautotropic growth in synchronously dividing cultures in LD: 14,10 (Fig. 2.5) was minimal during the dark phase. The response began to decrease well before the actual L/D transition, as if the cells anticipated its coming (Edmunds and Halberg, 1981). Inasmuch as the cells all divide (and replicate their flagellae) during darkness (Edmunds, 1965a), one could hypothesize that the reduced response was due to the inability of the cells to swim toward the test beam. This explanation was effectively ruled out by the observations of Bruce and Pittendrigh (1956, 1958) and Feldman (1967), which indicated that (1) the rhythm continued to be entrained by LD: 12,12 even in the stationary phase, where little, if any, net increase in cell number occurred, and (2) the rhythm subsequently free-ran with a circadian periodicity (ranging from 23.6 to 24.3 h) for as long as 14 days in DD (except for the intermittent test light, each 2 h). The free-running period appeared independent of temperature or nearly so. Over a 15°C range (16.7 to 33.0°C), τ varied only between 26.2 and 23.2 h, yielding a Q_{10} of 1.01 to 1.10 (Bruce and Pittendrigh, 1956). This lack of dependence of a unicellular clock on temperature was considered to be achieved by virtue of a compensating mechanism within the organism, although component parts of the oscillator might well be temperature-dependent, as suggested by the observation (Bruce, 1960) that a diurnal sinusoidal temperature cycle (18 to 31°C) can modify the phase relation between the rhythm and a simultaneously im-

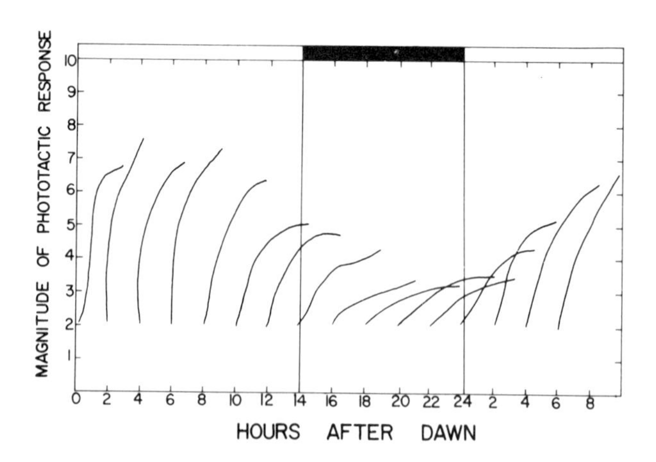

FIGURE 2.5. Rhythm of phototactic response of *E. gracilis* during synchronized photoautotrophic growth in LD: 10,14. The rhythm persists with a circadian period in DD. (Reprinted from Edmunds and Halberg, 1981, with permission of Williams & Wilkins Co.)

posed LD: 12,12 cycle in a manner dependent on the phase angle difference between the two *Zeitgeber*.

Bruce (1960) further investigated the entrainment of the phototaxis rhythm by 24-h LD cycles having various photofractions (i.e., ratio of light duration to entire cycle length). Entrainment to a precise 24-h period occurred for photofractions between 1/12 and 5/6, and the phase of the rhythm depended chiefly on the timing of the L/D transition; for photofractions greater than 5/6 (i.e., cycles having more than 20 h of light), the rhythm damped out. The system also appears to have relatively wide limits of entrainment, synchronizing to LD: 3,3, LD: 8,8, and even LD: 24,24 (where T = 6, 16 and 48 h, respectively). Bruce (1960) noted, however, that what might appear to be entrainment to a short period (for example, 6 h), might, in fact, be frequency demultiplication by separate subpopulations of cells to 24-h periods but with differing phase relationships. (Indeed, the rhythm exhibited frequency demultiplication to a 24-h period in LD: 2,10 and LD: 12,36.) This did not seem likely, however, in view of the fact that the rhythm free-ran in DD with a circadian period following direct entrainment by the high-frequency cycle.

Bruce and Pittendrigh (1956, 1957) and Feldman (1967) have demonstrated that the free-running phototaxis rhythm in DD could be reset by light perturbations ranging from 2 to 12 h duration. As has been found for the majority of circadian rhythms, the sign and magnitude of the steady-state phase shift ($\pm\Delta\phi$) thus engendered depended on the phase of the internal oscillation at which the pulse was given. In a PRC obtained for the action of 4-h light pulses (Fig. 2.6), the maximum delay phase shift was obtained at about CT 16 during the subjective night (Feldman, 1967).

Early attempts to affect the period of the phototaxis rhythm yielded negative results. The respiratory inhibitor potassium cyanide, the mitotic inhibitor phenylurethane, the adenine growth factor analog 2,6-diaminopurine sulfate, the pyrimidine and nucleic acid analog 2-amino-4-methylpyrimidine, the nucleic acid bases, and the growth factors gibberellic acid and kinetin were all without consistent effect on period and phase (Bruce and Pittendrigh, 1960). Deuterium oxide (^2H_2O), or heavy water, which has been shown to reversibly lengthen τ in a variety of organisms (Enright, 1971), similarly alters both the period and phase of the phototaxis rhythm in *E. gracilis*. Thus, a culture adapted to 95% ^2H_2O increased its period to about 27 h but returned to a τ of about 24 h when it was readapted to H_2O again (Bruce and Pittendrigh, 1960). The problem with such studies, of course, is that the action of ^2H_2O is general and nonspecific, and the period-lengthening effects are equally compatible with kinetic, transcriptional, and membrane models for circadian oscillators (see Chapter 5). Indeed, Kreuels and Brinkmann (1979) found in a comparative study of the effects of ^2H_2O on the cell-bound circadian oscillator underlying the motility rhythm in *E. gracilis* (see next subsection), on the cell-free glycolytic oscillator of yeast, and on the Belousov-Zhabontinsky chemical

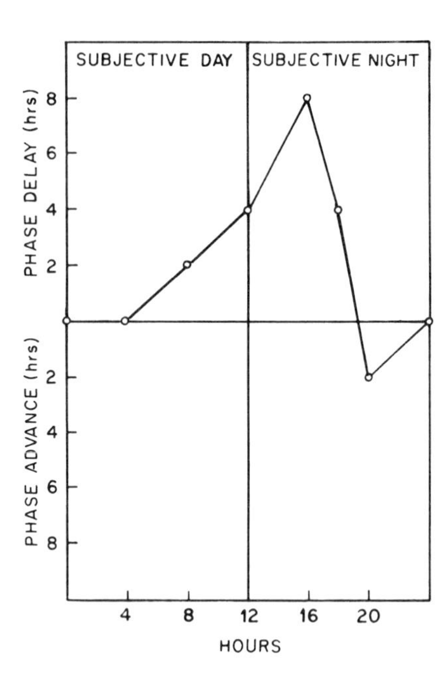

FIGURE 2.6. Phase-response curve for the rhythm of phototaxis in *E. gracilis* for 4-h light pulses given on the second day of DD. Points indicate the time of the beginning of the light pulse. Each point is based on the results of at least four independent experiments (except hour 18, for which there is only one experiment). Abscissa, circadian time; circadian time 0, subjective dawn. (Reprinted from Feldman, 1967, with permission of the author.)

reaction having a known network structure (Kreuels et al., 1979), that although the circadian and glycolytic oscillations were slowed down to an extent depending on the ^2H_2O concentration, the period of the chemical reaction was either lengthened or shortened, respectively, at high or low catalyst concentrations. They concluded, therefore, that the generalized period-lengthening effect of ^2H_2O can be explained only by a more complex network approach.

More specific (but still not well understood) perturbations have been obtained by altering the nutritional conditions. Feldman and Bruce (1972) reported that the addition of acetate (100 m*M*) to autotrophic cultures of *E. gracilis* lengthened the phototaxis rhythm to 27 h and that 10 m*M* pulses of acetate administered at different phases of the free-running rhythm induced phase shifts. Pulses of other carbon sources (succinate, lactate, pyruvate), however, caused a temporary cessation of the rhythm and then a resumption with variable phase shifts. They argue that the changes are probably caused by a general metabolic switch (i.e., from autotrophy to mixotrophy) rather than by any specific effect.

Finally, and perhaps most specifically, the period of the phototaxis rhythm has been shown to increase (reversibly) by the addition of cycloheximide (CHX), an inhibitor of protein synthesis on 80S ribosomes of the cytosol (Fig. 2.7; Feldman, 1967). The effects of this drug appeared to be on the clock itself rather than on some variable controlled by the clock (Fig. 2.8), as confirmed by assaying the position of the light-sensitive oscillation by 4-h resetting light signals after CHX addition in DD (as predicted from the PRC for light pulses already derived for this unicell). Yet even these results implicating protein synthesis in circadian clock function must be interpreted cautiously. CHX not only may have other primary effects besides the inhibition of protein synthesis but also may produce secondary effects that would modify only indirectly the circadian oscillator (Sargent et al., 1976).

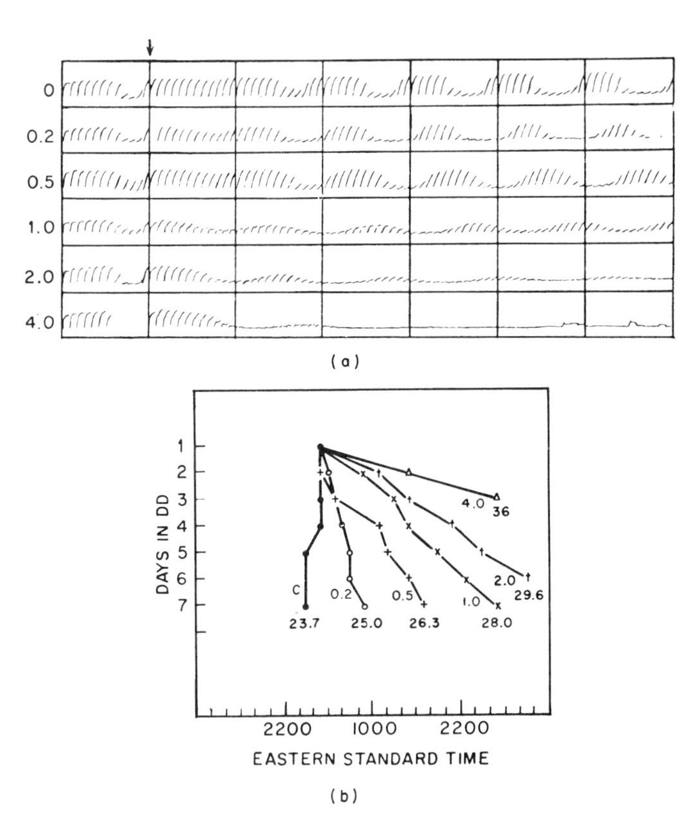

FIGURE 2.7. Period lengthening of the free-running rhythm of phototaxis in *E. gracilis* by cycloheximide. (a) Photograph of original records of phototatic response of a culture to which cycloheximide was added at subjective dawn of the second day of DD (indicated by the arrow). The numbers at the left are the concentrations (μg/ml) of drug added to each culture; control (C), no drug added. Vertical lines, 24 h apart, indicate 1000 EST. (b) The clock hours at which successive minima occurred are plotted for successive days for each of the cultures shown in (a). The numbers next to the data lines indicate the concentrations (μg/ml) of drug added; C, control. The numbers below the data lines indicate the period length (hours). There is some slight variability (up to about 1 h) in the exact amount of lengthening of the period induced by the highest concentrations of the drug (2 and 4 μg/ml). (Reprinted from Feldman, 1967, with permission of the author.)

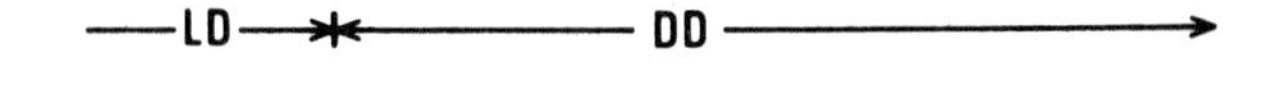

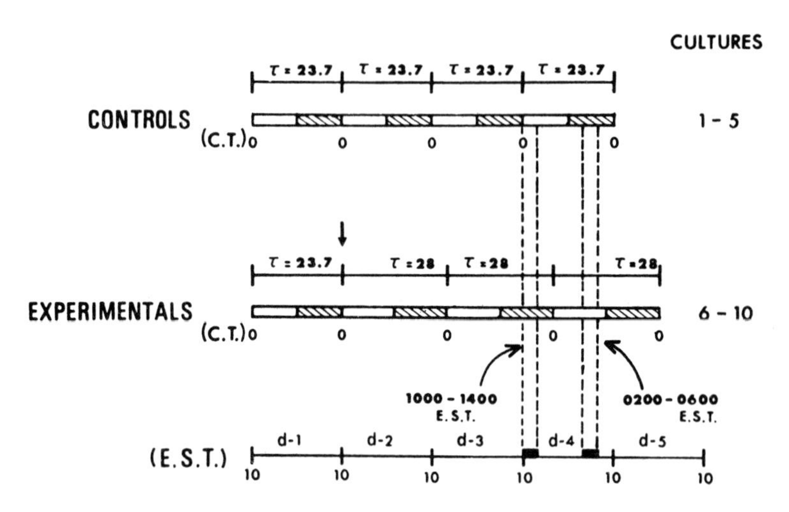

FIGURE 2.8. Testing for the lengthening of the free-running period (τ) of the rhythm of phototatic response in *E. gracilis* after the addition of cycloheximide in constant darkness. The phase of the underlying clock was assayed by measuring the phase-resetting effects of 4-h light pulses in reference to those predicted by the phase-response curve determined previously for this rhythm (Fig. 2.6). The bottom line indicates Eastern Standard Time (i.e., absolute elapsed time). In the prior light cycle, lights were on from 1000 to 2000 EST; d-1, d-2, and so on indicate successive days in constant darkness. The bar labeled "controls" indicates circadian times (CT) of subjective day (open bars) and subjective night (hatched bars) of the control cultures. Notations are similar for the experimentals, in which cycloheximide was added at subjective dawn of the second day of DD (arrow). Solid blocks on the EST scale at day 4 of DD give times at which each of the two 4-h light pulses were administered to separate cultures. Note that the first pulse strikes the control cultures in their early subjective day, whereas the second pulse hits during the subjective night, but the same pulses strike the experimentals with the reverse-phase relationship, indicating that the phase had been changed due to the lengthening of the period of each cycle by the inhibitor. (Reprinted from Feldman, 1967, with permission.)

Dark Motility Rhythm (Dunkelbeweglichkeit).

In stationary cultures of *E. gracilis* (strain 1224-5/9), a diffusion gradient in the vertical distribution of cells as indicated by differences in their sedimentation equilibrium has been observed (Fig. 2.9). It is apparently a function of random cell motility that itself fluctuates with a circadian periodicity in DD for as long as 3 months (Brinkmann, 1966, 1971; Schnabel, 1968). This dark mobility rhythm *(Dunkelbeweglichkeit)* damps out in LL but may be reinitiated by a single L/D transition. The assaying test-light

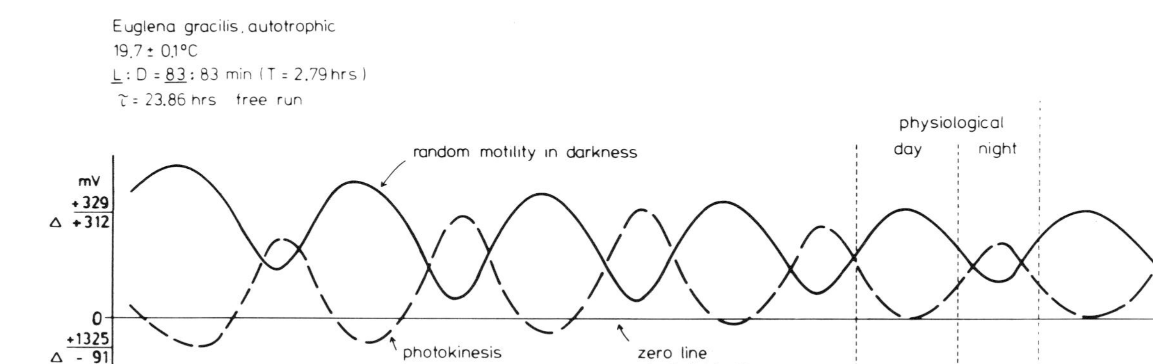

FIGURE 2.9. An original trace of the random motility rhythm in DD (τ = 23.9 h) and the photokinetic response in *E. gracilis* cultured autotrophically at 19.7°C and assayed with an LD: 83′,83′ cycle (T = 2.8 h). The photokinetic response smoothly passes the border of the positive and negative portions of the random motility curve several times, with an increasing and then decreasing level of reactivity. This may suggest the operation of a threshold system where the threshold itself fluctuates with a circadian cycle, as opposed to different reaction mechanisms for positive and negative photokinesis. (Reprinted from Edmunds, 1982, with permission of Academic Press, data courtesy of K. Brinkmann.)

cycle (20 min of light every 2 h) did not entrain the circadian rhythm either directly or by frequency demultiplication (Schnabel, 1968). In contrast to the composite rhythm of phototaxis response previously described, the random motility rhythm is independent of phototaxis and photokinesis and, consequently, allows one to distinguish more easily between the influence of light on the circadian oscillator and on the response itself.

Schnabel (1968) examined the effects of a wide variety of LD cycles on the motility rhythm in both autotrophic and mixotrophic cultures. Regimens with periods ranging from 16 to 48 h (LD ratios of 1:1) entrained the rhythm, which subsequently free-ran with a circadian period upon release into DD. Under shorter LD cycles, she found that a circadian component was exhibited, as if the rhythm was free-running, ignoring the imposed regimen. A colorless, obligatorily heterotrophic mutant (strain 1224-5/25) also displayed a light-entrainable, circadian rhythm of random motility in DD just as the green cultures did (Kirschstein, 1969). Finally, Schnabel (1968) determined the PRC for this rhythm using 6-h light pulses (1000 lx) during the DD free-runs (Fig. 2.10). The PRCs of both green and colorless strains were nearly identical.

The range of entrainment of the motility rhythm by sinusoidal temperature cycles, having driving periods from 4.8 to 55.7 h, has been explored as well (T. Kreuels and K. Brinkmann, personal communication; Edmunds, 1982). All cycles used synchronized the overt rhythm, which then reverted to its free-running τ upon removal of the temperature regimen only when the period of the latter had been in the range of approximately 19 to 36 h. Although passive enforcement did occur with the other temperature cycles (having T = 4.8 to 19 h, or 36 to 55.7 h), the rhythm damped out in DD and constant temperature. The phase angle difference between the temperature cycle and the synchronized rhythm changed within the range of entrainment (19 to 36 h) but was relatively constant in cycles having $T > 36$ h. Finally, a systems analysis of the masking effects and mutual interactions of the light response (photokinesis) and the temperature response (thermokinesis) in both the wild type and in a white, photosynthetic mutant of *E. gracilis* has been undertaken (Kreuels et al., 1984; Martin et al., 1985).

The nutritional mode also plays a role in the effects of constant temperatures on the rhythm of random motility. Brinkmann (1966, 1971) has reported that under autotrophic conditions (minimal medium), τ in DD was more or less independent (23.5° ± 0.3 h) of temperature in the range between 15 and 35°C, whereas a sudden temperature increase of 5°C or more caused a transitory increase in τ (Fig. 2.11). In contrast, τ increased with increasing constant temperature in mixotrophic cultures (complex nutrient medium), whereas the phase was not affected by a sudden step-up in temperature. A PRC for single 10°C steps-up given at different CT times to a free-running autotrophic culture was derived (Brinkmann, 1966, 1971) and was similar to that for 6-h light pulses (Schnabel, 1968; see Fig.

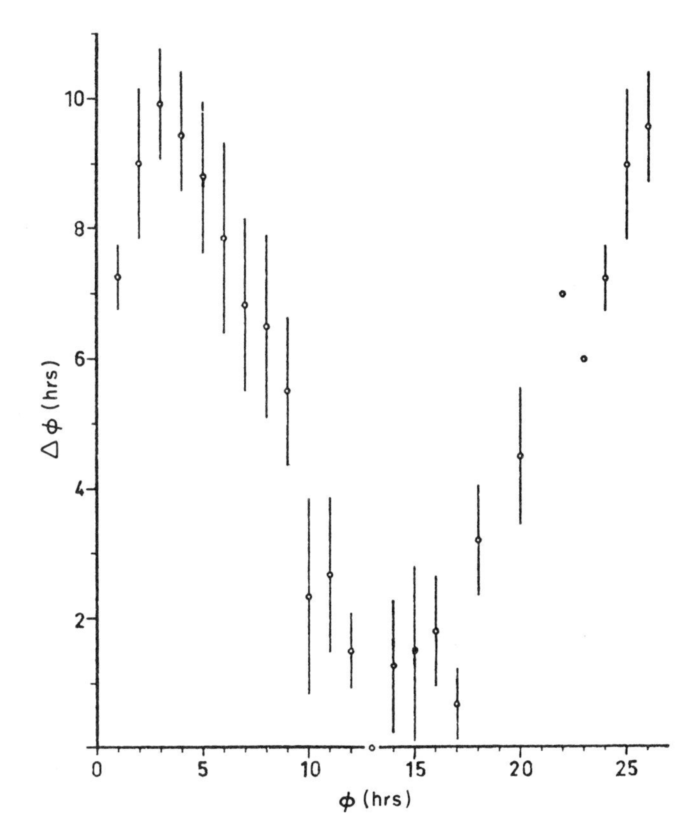

FIGURE 2.10. Phase-response curve for the effect of single 6-h light signals (1000 lx) on the circadian rhythm of random motility in autotrophic cultures of *E. gracilis* in the infradian growth phase. *Abscissa:* Hours after the minimum point of the free-running rhythm at which the light pulse was given. *Ordinate:* Phase shift, in hours, after reestablishment of the previous frequency. The final phase difference between the shifted rhythm and the control is plotted, neglecting the sign of the phase shift. (Reprinted from Schnabel, 1968, with permission of Springer-Verlag.)

2.10). The type of response could be switched from one to the other by the addition of either lactate or ethanol to the medium (Fig. 2.12). Brinkmann (1966, 1971) has attributed the different behavior of the two types of culture in response to different steady-state temperatures to the participation in the autotrophic strains of two types of reaction—those with a Q_{10} of 1.0 (photochemical) and those with a Q_{10} of 2.0 (dark reactions)—whereas in mixotrophic cells the dark reactions would predominate. It now appears (K. Brinkmann, personal communication; Edmunds, 1982) that the degree of dependence of energy conservation on temperature in turn determines the type of response of the circadian rhythm to different steady-state temperatures.

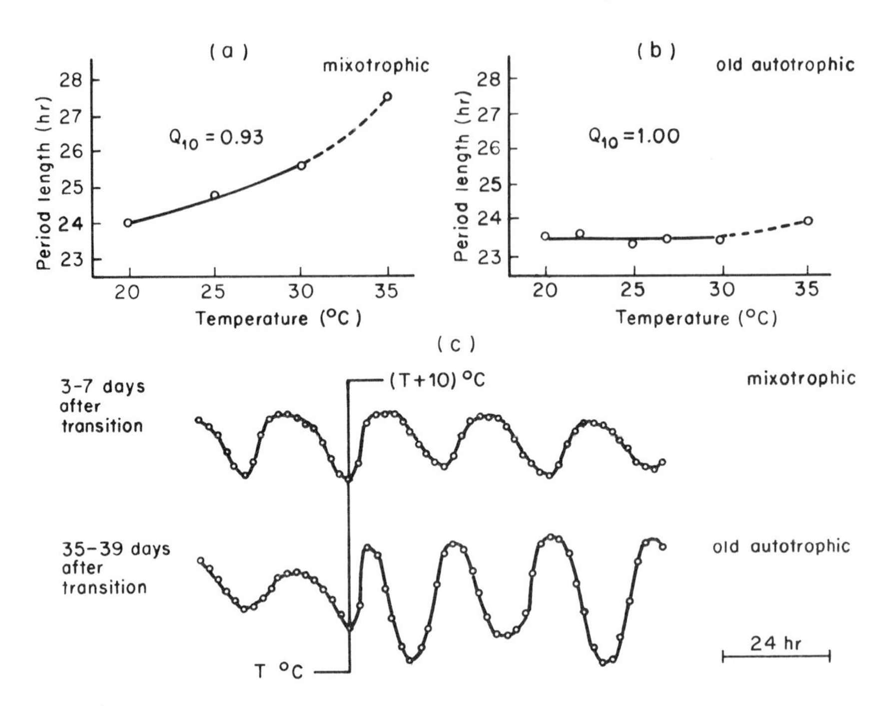

FIGURE 2.11. Types of temperature compensation in the circadian rhythm of random motility of *E. gracilis*. In the two upper graphs (a) and (b), the free-running frequencies in young mixotrophic and old autotrophic cultures are compared as a function of different constant temperatures. The lower graphs (c) illustrate the phase response of the two cultures after a single temperature jump of + 10°C given at a minimum of the cycle. The mixotrophic culture represents the frequency-sensitive type of temperature compensation, and the old autotrophic culture represents the phase-sensitive type of temperature compensation. Transition means transition from LD: 12,12 to the test-light program of LD: 20′,100′. (Reprinted from Brinkmann, 1971, with permission of the National Academy of Sciences.)

Finally, several other circadian rhythms that have some relevance to the rhythm of random motility have been investigated recently. Brinkmann and coworkers have measured the energy charge (EC) and the concentration of glucose-6-phosphate (G6P) under various sinusoidal temperature cycles within and beyond the limits of entrainment for the motility rhythm. In contrast to G6P, EC never synchronized (Fig. 2.13a), and neither showed rhythmicity in autotrophic cultures during a free-run at a constant 27.5°C (Fig. 2.13c), indicating that these variables cannot be responsible for generating the motility rhythm. Similarly, although the redox state of nicotinamide adenine dinucleotide phosphate ($NADP^+$), but not nicotinamide adenine dinucleotide (NAD^+), was synchronized by a temperature cycle (Fig. 2.14a), there was no significant oscillation in the redox state

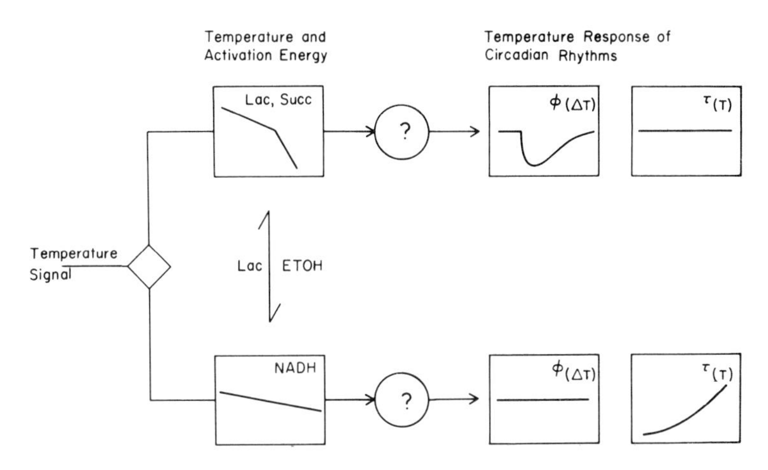

FIGURE 2.12. Metabolic switching between the two types of response to temperature step-ups [$\phi_{(\Delta T)}$] and changes in steady-state constant temperature [$\tau_{(T)}$] as correlated with the temperature dependence of energy conservation in cultures of *E. gracilis* by the addition of lactate (Lac) or ethanol (ETOH). Arrhenius plots for the oxidation of lactate and succinate (Succ) or NADH (see left insets) were obtained with isolated mitochondria; activation energies were 15 to 25 kcal/mole and 10 kcal/mole, respectively. (Reprinted from Edmunds, 1982, with permission of Academic Press; data courtesy of K. Brinkmann.)

of pyridine nucleotides during a free-run. The temperature-enforced NADPH cycle showed a peak coinciding with low G6P, indicating flux regulation of the pentose-phosphate cycle in response to temperature (K. Brinkmann, personal communication; see Edmunds, 1982). Recently, Kreuels and Brinkmann (personal communication; Edmunds, 1982) have found that the period of motility rhythm can be lengthened from 23.8 h to about 25.3 h by the addition of urea to the culture. Indeed, saturation was not obtained (above concentrations of 200 mm, no oscillations were observed, although *E. gracilis* could still grow). Urea penetrated the cells but was not metabolized, a fact suggesting that the period-lengthening effect perhaps was due to structural changes (possibly affecting hydrogen bonds of macromolecules or water structures, or by weakening hydrophobic interactions). Finally, a low-amplitude circadian oscillation of external pH in weakly buffered cell suspensions has been reported (Fig. 2.15). Preliminary evidence suggests that this rhythm is due to the activity of a plasmalemma proton pump and is not generated by the inverse-phase, circadian rhythm of photosynthesis (K. Brinkmann, personal communication; see Edmunds, 1982). A similar conclusion was reached by Hoffmans and Brinkmann (1979) for *C. reinhardtii*, which exhibited a pH rhythm even though photosynthetic activity was inhibited by 3-(3,4-dichlorophenyl)-1,1-dimethyl urea (DCMU).

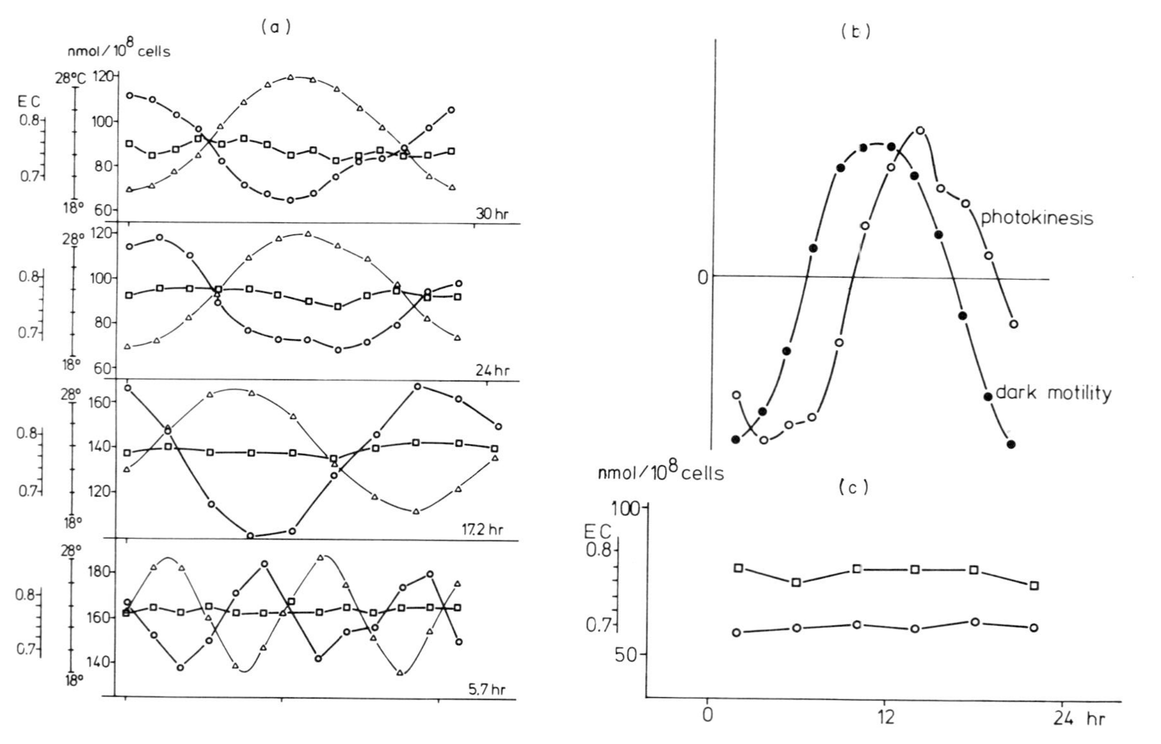

FIGURE 2.13. Energy charge (EC) (□) and glucose-6-phosphate (○) concentration during (a) various sinusoidal temperature cycles (△) and during a free-run at 27.5°C (c) when circadian rhythms ($\tau \cong 23$ h) of photokinesis (○) and dark motility (●) are exhibited (b) in autotrophic cultures of *E. gracilis*. (Reprinted from Edmunds, 1982, with permission of Academic Press; data courtesy of K. Brinkmann.)

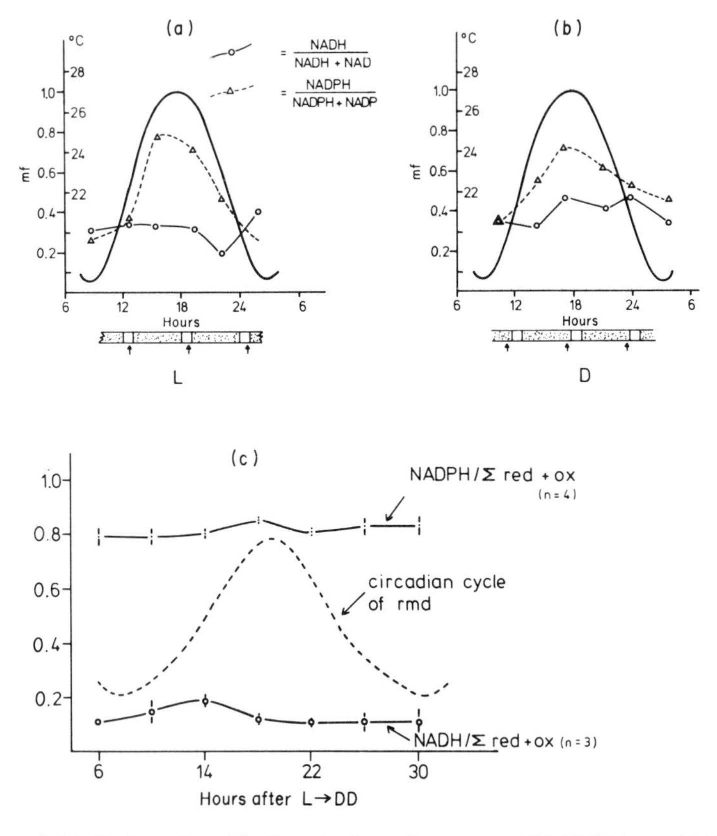

FIGURE 2.14. Enforced oscillations in the redox state of $NADP^+$ (a and b) but not of NAD^+, by an 18-h sinusoidal temperature cycle in late-stationary, mixotrophic (Glu-malate) cultures of *E. gracilis* grown at 21°C in LD: 20′,100′. The amplitude of the curve is greater when the cultures were assayed at the end of the light pulse (a) than at the end of the dark intervals (b). Solid curves: circadian cycle of random dark motility. (c) No significant free-running rhythm in the redox state of pyridine nucleotides at constant temperature, despite the presence of a circadian cycle of random dark motility (dashed curve). (Reprinted from Edmunds, 1982, with permission of Academic Press; data courtesy of K. Brinkmann.)

Rhythm of Cell Settling

In addition to the light-oriented rhythm of phototaxis and the dark random-motility rhythm, a circadian rhythm of cell settling has been discovered in *E. gracilis*, which may occur concurrently with, or in the absence of cell division (Terry and Edmunds, 1970a,b). This periodicity was reflected during the growth phase by an apparent change in cell concentration in autotrophically batch-cultured populations synchronized in LL by 24-h temperature cycles (18°C/25°C or 28°C/35°C, although the synchronous

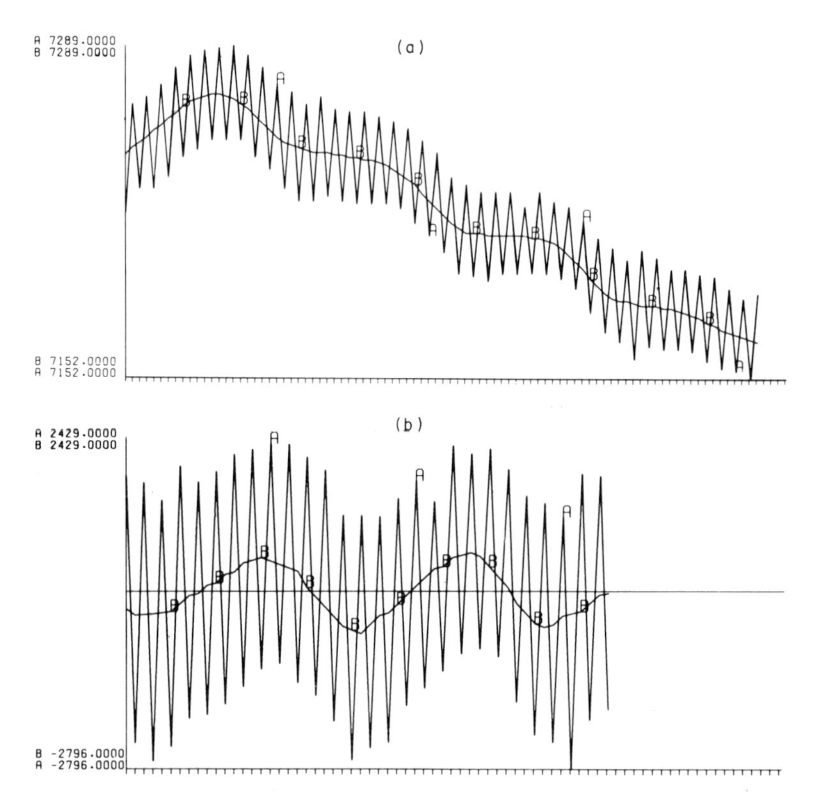

FIGURE 2.15. Circadian rhythm of external pH in weakly buffered (30 μ*M* phosphate) suspensions of *E. gracilis* autotrophically cultured at 20°C in LD: 1,1 following an LD: 12,12 cycle (the traces begin at the end of the last main photoperiod). (a) The original computer record showing direct responses (A) to the short light intervals (pH span 7.152 to 7.289) was put through a low-pass filter (solid line B) to remove the direct light response. (b) High-pass filtered version to remove trend remaining after low-pass filtering, to show the circadian component of the low-amplitude pH changes. (Reprinted from Edmunds, 1982, with permission of Academic Press; data courtesy of K. Brinkmann.)

division bursts tended to mask the response. Small dips were seen in the plateau portion of the growth curves. The settling rhythm maintained the same phase in relation to either entraining temperature cycle (i.e., maxima occurred during the warmer interval) regardless of whether this was the time of maximum cell division (which had a 180° phase difference between the two temperature regimens).

In temperature-cycled, stationary cultures, the maxima of the settling rhythm also occurred during the warmer phase of the imposed temperature-cycle. The rhythm persisted for as long as 9 days in infradian cultures released from the temperature cycle and held at a constant 25°C in LL

(Terry and Edmunds, 1970b). Neither experimental artifacts nor cell death (necessarily rhythmic) were responsible for this rhythm. Rather, the cells actually tended to settle out of the liquid phase, adhere to the vessel walls, and then subsequently detach themselves and reenter the medium. This was substantiated by automatically monitoring cell number at two different levels in the magnetically stirred, homogeneous cultures with a dual sampling system. Attachment could be prevented and the rhythmic dips in cell number abolished only by vigorous agitation of a rotary shaker (Terry and Edmunds, 1970b).

Rhythms in Photosynthetic Capacity

Several studies have reported a rhythm in photosynthetic capacity during the cell division cycle of *E. gracilis,* although the results were often conflicting (Lövlie and Farfaglio, 1965; Cook, 1966; Codd and Merrett, 1971). Walther and Edmunds (1973) demonstrated a clear diurnal rhythm in the capacity of this unicell to fix CO_2 (as measured by its ability to incorporate $NaH^{14}CO_3$ at saturating light intensities in aliquots taken from the master culture at different times) in dividing, autotrophic populations synchronized by an LD: 10,14 (L = 12,000 lx) at 25°C. Photosynthetic capacity (PC) was found to reach a peak value an hour or so before the onset of darkness, at which time cell division ensued. A similar pattern in O_2 evolution also was observed. This daily rhythm occurred in nondividing cultures and could thus be divorced from the CDC. At the time, the rhythm was found to be only weakly persistent under continuous dim LL (750 lx), under which conditions cell division was completely suppressed, the rate of CO_2 fixation was greatly reduced (making the assay of the rhythm difficult for longer time spans), and, by implication, the overall physiology of the cell was disturbed. (Higher intensities of LL caused the rhythm to damp out, and DD could hardly be employed in an autotrophic system.)

More recently, these problems have been overcome by monitoring the rhythm of PC and of chlorophyll (Chl) content in three substrains of *E. gracilis* synchronized by an LD: 12,12 cycle (L= 7000 lx) during both the exponential growth and stationary phases and then released into an LD: 20′,20′ (L= 7000 lx) regimen (Laval-Martin et al., 1979). This particular high frequency, 40-min cycle was selected so as to afford an amount of lighting during a 24-h timespan identical to that received in the full-photoperiod LD: 12,12 cycle. The cell division rhythm is known to persist with a circadian period under these conditions (Edmunds and Funch, 1969a,b; see Section 3.2.1). For all intents and purposes, the high-frequency cycle is perceived by the cells as continuous illumination, at least with regard to rhythmic output. Under these conditions, a non-damped, high-amplitude, persisting rhythm of PC was observed for some 5 to 6 days after the entraining 24-h LD cycle had been discontinued, whose τ was estimated to be about 27.5 h, corresponding closely to that of the cell division rhythm. Similar observations (Edmunds, 1980b; Edmunds and Laval-Martin, 1981) were made in an LD: 3,3 cycle imposed from the moment of inoculation of the culture (Fig. 2.16), which also elicited a circadian rhythm of cell division whose τ, for some 156 monitored oscillations, ranged between 27 and 34 h and averaged 30.2 ± 1.8 h (Edmunds et al., 1982).

Reproducible, high-amplitude circadian rhythms were also observed in total Chl in both dividing and nondividing cultures maintained in these high-frequency LD cycles, although there did not appear to be a close correspondence between the Chl rhythm and that of PC (Edmunds and

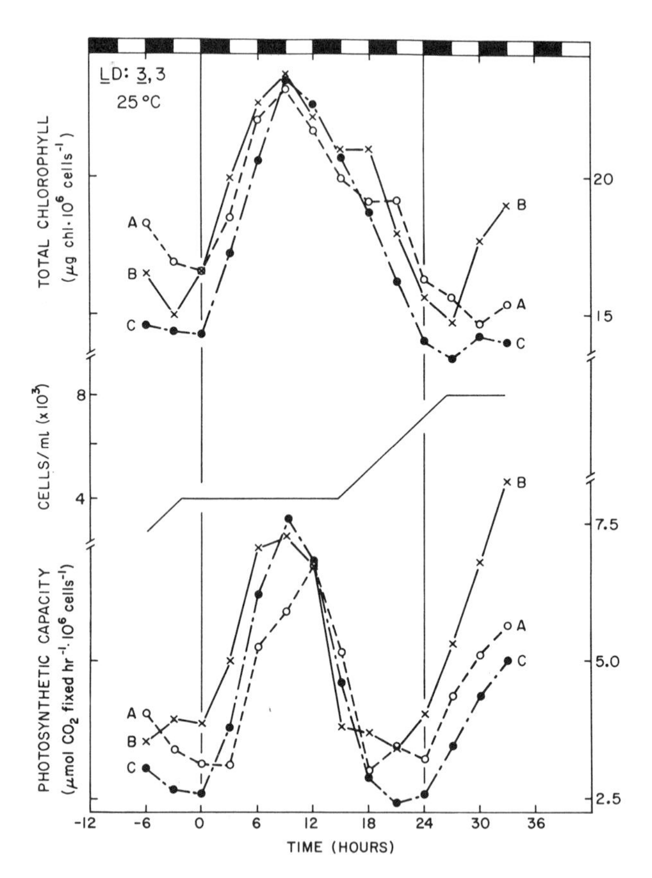

FIGURE 2.16. Free-running circadian photosynthetic rhythms in *E. gracilis* (Z) maintained in a short-photoperiod LD: 3,3 cycle during the exponential phase of growth. The data for each of the three replicate cultures (A,B,C) have been shifted in real time so that the peaks of the curves for total chlorophyll (Chl) are aligned in order to facilitate comparison. A generalized growth curve is also given in the center of the diagram. (Reprinted from Edmunds, 1982, with permission of Academic Press; see also Edmunds and Laval-Martin, 1981.)

Laval-Martin, 1981). Indeed, the Chl rhythm sometimes appeared as a bimodal circadian rhythm or even as an ultradian one having a period of about 13 h. Likewise, the amplitude of the Chl rhythm was usually considerably less than that for the rhythm in PC. The fact that the periods, and therefore the phase relations, of the Chl and PC rhythms varied not only appears to rule out any simple causality between the two but also suggests the possibility of desynchronization, or dysphasia, among different rhythms driven by the same clock. It even may suggest the existence of a multicellular oscillator system in a unicellular organism (Section 6.2.1) with, perhaps, cell division's acting as an entraining signal for a hierarchical clockshop in dividing cultures.

Lonergan and Sargent (1978) have also reported a circadian rhythm of O_2 evolution in *E. gracilis* (Z), that persisted for at least 5 days in dim LL and constant temperature but damped out in bright LL (Fig. 2.17A). The rhythm could be phase-shifted by light pulses, and the free-running period length was unchanged over a 10°C span of growth temperature. Although the O_2 rhythm was found in both exponentially dividing cultures and stationary phase cultures, CO_2 uptake was clearly rhythmic only in the latter. This may have been due to the sensitivity of the infrared CO_2 analyzer used for the assay. Recently, Lonergan (1986b) has isolated a stable, naturally occurring population of *E. gracilis* in which the photosynthesis reactions are uncoupled from the clock. The rate of oxygen evolution was influenced predominantly by the environmental growth parameters [Note the analogous loss of rhythmicity of cell division and its restoration by the addition of sulfur-containing amino acids to the medium (Edmunds et al., 1976)].

There are at least three nontrivial ways by which the rhythm in PC could be generated (Edmunds, 1980b, 1982): (1) rhythmicity in total Chl content; (2) rhythmicity in the activity or amount of enzymes mediating the dark reactions, or both, or (3) rhythmicity in the coupling between photochemical events within the two photosynthetic systems and electron flow between them. Although there is an endogenous circadian rhythm in Chl in nondividing cultures of *E. gracilis*, perhaps reflecting changes in the functional role of the pigment, it is not sufficient to explain the rhythm in PC because of the lack of quantitative relationship between the amplitudes of the two rhythms and the lack of correspondence in their periods and phases (Edmunds and Laval-Martin, 1981).

With regard to the second hypothesis (enzymatic variations), although a number of photosynthetic dark reactions comprising the Calvin scheme have been examined in *E. gracilis* and several other algal systems (Edmunds, 1980b), both during the CDC and in nondividing cultures, no enzyme has achieved a consensus as a clear candidate responsible for generating the rhythm in PC. For example, ribulose-1,5-biphosphate carboxylase, although sometimes showing fluctuations in activity, does not show sufficient correspondence to satisfy the rates of CO_2 fixation at

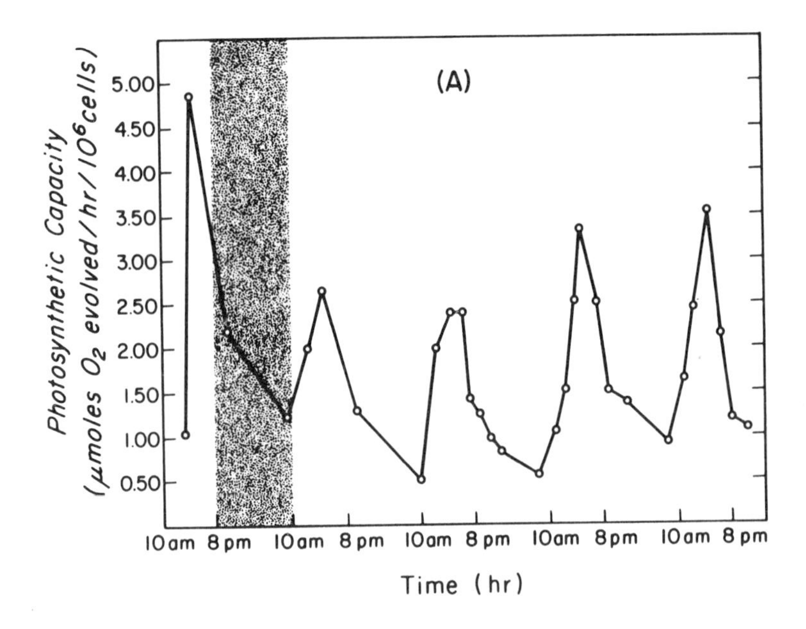

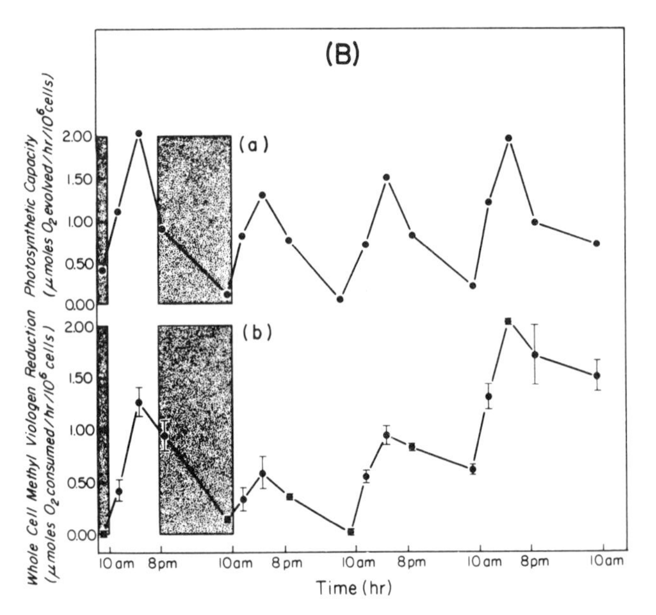

FIGURE 2.17. (A) Persistence of photosynthetic capacity rhythm (O_2 evolution) in *E. gracilis* in constant conditions. The division-synchronized culture was grown in LD: 10,14 (shade for darkness) and then exposed to dim LL (2×10^3 ergs cm^{-2} sec^{-1}) at 25°C after the last dark period. (Reprinted from Lonergan and Sargent, 1978; with permission of American Society of Plant Physiologists.) (B) Photosynthetic capacity and whole-chain photoelectron flow in intact cells: (a) photosynthetic capacity; (b) rate of whole chain electron flow as measured by a H_2O-to-methyl viologen assay in the presence of gramicidin D. Bars represent standard deviation for replicates. New cells were used for each replicate. (Reprinted from Lonergan and Sargent, 1979, with permission of American Society of Plant Physiologists.)

all stages investigated (Codd and Merrett, 1971; Walther and Edmunds, 1973). Similarly, although an earlier observation (Walther and Edmunds, 1973) of changes in glyceraldehyde-3-phosphate dehydrogenase suggested a possible mechanism for generating the rhythm in PC, this does not appear to be the case (Lonergan and Sargent, 1978). Finally, although carbonic anhydrase was found to be rhythmic in photoentrained cultures of *E. gracilis* in LD: 10,14, with peak activity occurring at the time of the highest rate of O_2 evolution, the rhythm in enzyme activity disappeared under constant conditions under which the rhythm in PC persisted (Lonergan and Sargent, 1978).

Concerning the third hypothesis, although the individual activities of photosystem I (PS I) and photosystem II (PS II) in *E. gracilis* do not appear to change significantly with time of day (Walther and Edmunds, 1973; Lonergan and Sargent, 1979), the rate of light-induced electron flow through the entire electron chain (water to methyl viologen) was rhythmic both in whole cells and in isolated chloroplasts (Fig. 2.17B), the highest rate of flow coinciding with the highest rate of O_2 evolution (Lonergan and Sargent, 1979). Evidence consistent with the notion that the coordination of the two photosystems may be the site of circadian control of PC rhythms in *E. gracilis* was obtained from studies of low-temperature fluorescence emission from PS I and PS II after preillumination, respectively, with light wavelengths of 710 nm or 650 nm, whereas there was no indication that changes in total Chl, the ratio of Chl *a* to *b*, or the size of the photosynthetic units were responsible (Lonergan and Sargent, 1979). Still further evidence supporting this type of mechanism comes from work with the dinoflagellate *G. polyedra* (Prézelin and Sweeney, 1977; Sweeney et al., 1979) and is discussed in Section 2.2.4.

Rhythm in Cell Shape

Euglena gracilis changes its shape two times per day when it is cultured in LD: 10,14 (Lonergan, 1983; see Brinkmann, 1976b). At the beginning of the light interval, when photosynthetic capacity is relatively low, the population of cells is largely spherical in shape (Fig. 2.18). Mean cell length increases to a maximum by the middle of the light period, when photosynthetic capacity is greatest, and then decreases for the remainder of the LD cycle. These changes persist under continuous dim LL up to 72 h; under these conditions, the circadian rhythm of cell division is either slowed considerably or completely arrested.

The involvement of respiratory and photosynthetic pathways in these changes in cell shape has been investigated with inhibitors of energy pathways (Lonergan, 1983). Antimycin A and sodium azide both inhibited the round-to-long and long-to-round transitions, implicating respiration. Atrazine and diuron (DCMU) inhibited the round-to-long shape change but not the converse, indicating that light-induced electron flow is necessary only for the former. The influence of the changes in cell shape on photosynthesis also was examined by altering cell shape with the cytoskeletal

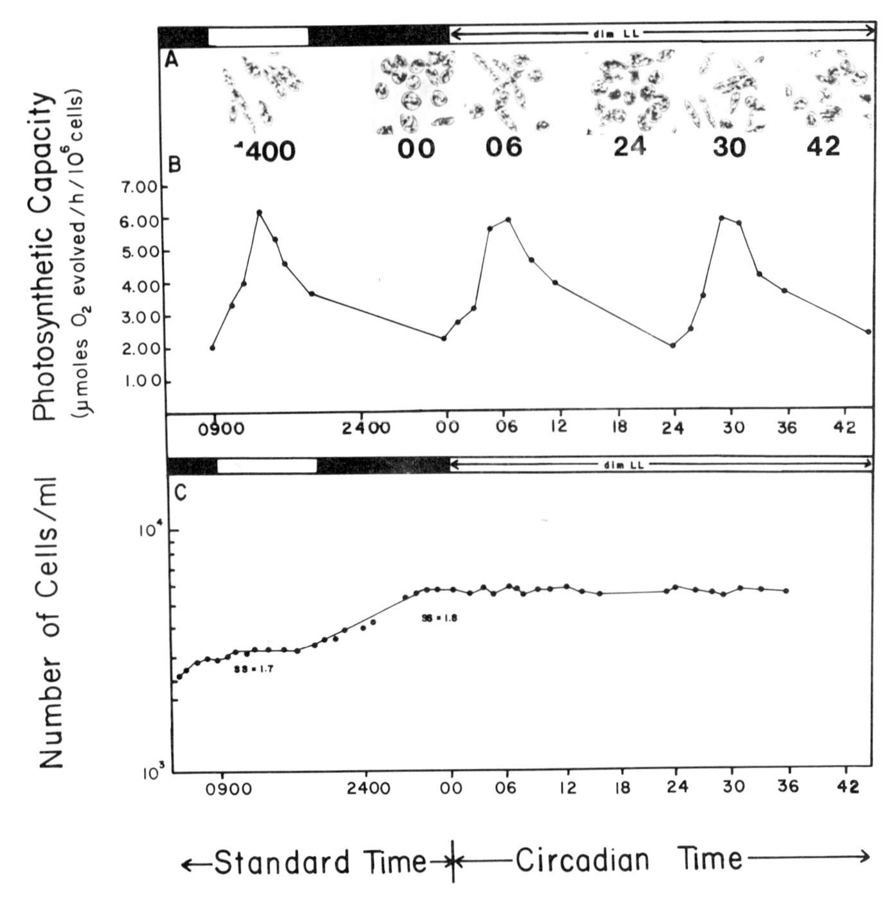

FIGURE 2.18. Cell shape changes, photosynthetic capacity, and cell number as a function of standard and circadian time in photoautotrophic cultures of *E. gracilis*. (A) Photomicrographs of cells at the minimum (ST 0900, CT 00, 24) and at the maximum (ST 1400, CT 06, 32) of the photosynthetic rhythm. (B) Photosynthetic capacity rhythm of cells grown in LD: 10,14 cycles and then placed in dim LL at CT 00, which represents the normal dawn. Lights came on at 0900 standard time (9 AM) and were turned off at 1900 (7 PM). Photosynthetic capacity is plotted as the rate of O_2 evolution. (C) Cell number, same growth conditions as (B). Cell number is nearly constant during the light portion of the cycle and approximately doubles by the end of the dark period. SS, step size (number of cells after a period of cell division/number of cells before the division; a value of 2.0 would reflect the division of every cell. (Reprinted from Lonergan, 1983, with permission of American Society of Plant Physiologists.)

inhibitors cytochalasin and colchicine. Both inhibitors blocked either type of shape transition. In contrast, cytochalasin B had minimal cytotoxic effects on the photosynthetic reactions, although colchicine did significantly inhibit light-induced electron flow and the in vivo expression of photosynthesis (Lonergan, 1983).

More recently, Lonergan (1984) has investigated the role of extracellular and intracellular concentrations of Ca^{2+} and calmodulin in the regulation of cell shape. When cultures of *E. gracilis,* usually grown in medium containing 180 μM Ca^{2+}, were resuspended in Ca^{2+}-free medium, cells assumed round shapes within 10 min; this effect was reversible by the restoration of Ca^{2+} to the medium. Furthermore, cultures grown in low concentrations of Ca^{2+} (10 μM) did not display the typical circadian rhythm in cell shape even though those of photosynthesis and cell division were unaffected. Elevating intracellular levels of Ca^{2+} by the addition of the ionophore A23187 prevented the cells from undergoing changes in shape, although the shape found at the time of the addition of the ionophore was maintained. In contrast, trifluoperazine (TFP) and chlorpromazine (CPZ), both calmodulin antagonists, always caused the cells to round, thereby also blocking the rhythm of daily shape change. Provocatively, pulses of TFP apparently do not elicit $\Delta\phi$s of the cell shape rhythm (Lonergan, personal communication), nor do 2-h pulses of TFP, CPZ, or the calmodulin antagonist W7 shift the phase of the rhythm of O_2 evolution (Lonergan, 1986a)—instead, photosynthesis is uncoupled from the clock—although Goto et al. (1985) have shown that pulses of W7 do phase-shift the rhythm of cell division in this unicell (see Section 5.2.5). Perhaps the same regulatory protein has more than one role in coordinating different rhythmicities in *Euglena*; alternatively, perhaps there is more than one master oscillator (see Section 6.2).

These observations on the ability of changes in Ca^{2+} concentration to induce changes in the shape in *Euglena* are particularly interesting from the point of view of the regulation of the contractile process by a biological clock. Indeed, Lachney and Lonergan (1985) have implicated stable, pellicle-associated microtubules in the modulation of cell shape, and Lonergan (1985) has localized by immunofluorescence techniques actin, myosin, calmodulin, and tubulin beneath the pellicle of *E. gracilis*. The fluorescence patterns remained intact during the daily shape changes, implying that the rhythm does not result from cycles of polymerization and depolymerization of the microtubules and microfilaments. The data are consistent, however, with an actomyosin contractile system controlled by calmodulin. Puiseux-Dao (1984) also has suggested the role of the cytoskeleton in the generation of circadian rhythmicity.

Oscillatory Enzymatic Activities

In addition to the numerous gross cellular constituents that have been temporally mapped across the cell divison cycle in *E. gracilis* (Cook,

1961a,b; Edmunds, 1965b) and other biochemical rhythmicities, such as the incorporation of amino acids into proteins (Feldman, 1968; Quentin and Hardeland, 1986a), the activities of a number of enzymes have been monitored both in synchronously dividing populations and in cultures in the stationary phase of growth (see Table 2.4 and Fig. 2.4). Certain of these enzymes might contribute to some of the overt physiological rhythms concomitantly observed, and, indeed, circadian enzymatic oscillations themselves can be considered validly as indices of an underlying biological pacemaker(s).

Although for technical reasons not all of the enzymes listed in Table 2.4 have been assayed rigorously in both dividing and stationary-phase populations under both entraining and free-running regimens, it would seem likely that most would continue to oscillate under LL (or DD) and constant temperature during the exponential growth phase if one can draw a parallel from the persistence of the circadian rhythm of cell division itself. To attack this problem and to divorce autogenous enzyme oscillations from those directly generated by the driving force of the CDC (whereby replication of successive genes would lead to an ordered, temporal expression of enzyme activities), light-synchronized, photoorganotrophically batch-cultured *E. gracilis* that had reached the stationary growth phase have been used successfully (Edmunds, 1975, 1982). Relatively large-amplitude oscillations were found in the activities of alanine, lactic, and G6P dehydrogenases, and L-serine and L-threonine deaminases, with maxima usually occurring during the light intervals (Sulzman and Edmunds, unpublished work; Edmunds, 1982). These rhythmic changes in enzyme activity, therefore, were effectively uncoupled from the CDC.

Even more interesting, however, was the finding (Sulzman and Edmunds, 1972, 1973; Edmunds et al., 1974) that the activity of alanine dehydrogenase continued to oscillate in these nondividing (infradian) cultures for at least 14 days in DD (but not in LL) and thus constituted an overt circadian rhythm in itself. The possibility that these oscillations in alanine dehydrogenase activity could be generated by fluctuations in pools of substrates or products that could change the stability of the enzyme during its extraction and trivially produce the rhythm was ruled out (Sulzman and Edmunds, 1973). Results from mixing experiments likewise did not suggest the presence of fluctuating pools of effector molecules that could generate the rhythm by altering the activity of the enzyme, nor were there differences in pH optimum, K_m value, or electrophoretic motility on polyacrylamide gel of enzyme extracted at different phases of the oscillation. On the other hand, activity determinations of alanine dehydrogenase extracted from the maximum and minimum points of the free-running rhythm and partially purified by ammonium sulfate fractionation and polyacrylamide gel electrophoresis suggested that periodic de novo synthesis and degradation may generate the observed variations in its activity (Sulzman and Edmunds, 1973).

2.2.4 *Gonyaulax* spp.

Along with *Tetrahymena* and *Euglena*, the marine dinoflagellate *Gonyaulax polyedra* is one of the most intensively investigated unicellular circadian systems (Wille, 1979; Mergenhagen, 1980; Sweeney, 1981). Of the many documented periodicities (Table 2.5), two categories of rhythm—bioluminescence and photosynthesis—are discussed in this section, with particular emphasis being placed on recent advances in these areas. The

Table 2.5. Some persisting circadian rhythms exhibited by the marine dinoflagellate *Gonyaulax polyedra*.

Circadian rhythm	Reference
Bioluminescence, glow	Sweeney and Hastings (1958); Hastings and Sweeney (1959); Sweeney et al. (1959); Hastings (1960); Hastings and Bode (1962); Karakashian and Hastings (1962, 1963); McDaniel et al. (1974); Taylor and Hastings (1979, 1982); Taylor et al. (1979, 1982a,b); Dunlap et al. (1980); Hastings and Krasnow (1981); Njus et al. (1981); Sulzman et al. (1982); Hobohm et al. (1984); Broda et al. (1986a,b); Cornélissen et al. (1986)
Bioluminescence, stimulated	Sweeney and Hastings (1957); Hastings and Sweeney (1958, 1959, 1960); Sweeney et al. (1959); Sweeney (1969b, 1974a, 1976b, 1979, 1981); Christianson and Sweeney (1972); Sweeney and Herz (1977); Walz and Sweeney (1979); Hastings and Krasnow (1981); Sweeney and Folli (1984)
Cell division	Sweeney and Hastings (1958); Hastings and Sweeney (1964); Homma (1987)
K^+ (intracellular)	Sweeney (1974a)
Luciferase activity	Hastings and Bode (1962); McMurry and Hastings (1972b); Dunlap and Hastings (1981); Dunlap et al. (1981); Johnson et al. (1984); Milos et al. (1987)
Luciferin activity	Bode et al. (1963); Dunlap et al. (1981)
Luciferin-binding protein	Sulzman et al. (1978); Morse et al. (1987a)
Membrane properties	Adamich et al. (1976); Scholübbers et al. (1984)
Photosynthetic capacity	Hastings et al. (1961); Sweeney (1960, 1965, 1969b); Prézelin and Sweeney (1977); Prézelin et al. (1977); Govindjee et al. (1979); Sweeney et al. (1979); Samuelsson et al. (1983); Sweeney and Folli (1984); Sweeney (1986)
Protein synthesis	Volknandt and Hardeland (1984a,b); Donner et al. (1985); Cornelius et al. (1985); Schröder-Lorenz and Rensing (1986, 1987)
RNA synthesis, ribosomal	Walz et al. (1983); Schröder-Lorenz and Rensing (1986)
Ultrastructure, chloroplast	Herman and Sweeney (1975)
Ultrastructure, thecal membranes	Sweeney (1976a)

Updated from Edmunds (1984c), with permission of Academic Press.

biochemical dissection of the clocks(s) underlying these rhythmicities are treated in detail in Sections 4.3 and 4.5.

Circadian Rhythms in Bioluminescence: Physiological Characteristics

There are at least two types of rhythmic bioluminescence in *Gonyaulax*, the induced flashing rhythm and the spontaneous glow rhythm (Hastings and Sweeney, 1958; Sweeney and Hastings, 1957; see Table 2.5). The assay in the former case is performed typically by removing an aliquot of culture, mechanically stimulating it by bubbling it with air (or by addition of acid), and measuring the light output (maximum emission at about 475 nm) with a photomultiplier tube. Bioluminescence capacity was greatest in the middle of the dark period in entraining LD cycles and persisted for many days (τ = 24.4 h) in dim LL (1200 lx). In more intense LL (3800 lx), τ was 22.8 h, and at 3800 lx, τ was reduced to 22.8 h, and the rhythm damped out after 4 days. Similarly, in DD, the amplitude of the free-running rhythm (τ = 23.0 to 24.4 h) progressively decreased over several days until it damped also (Sweeney and Hastings, 1957; Hastings and Sweeney, 1958).

The other rhythm of steady, dim, spontaneous glow (Sweeney and Hastings, 1957, 1958; Hastings and Sweeney, 1959) reached maximal intensity just at the end of the dark period in LD and, like the flashing rhythm, persisted in dim LL or DD. Here, the glow rhythm was cleverly monitored by placing small samples of culture in each of the many vials on the turntable of a liquid scintillation counter maintained in a programmed overhead LL or LD regimen (to provide energy for photosynthesis). At periodic intervals, the vials were then lowered by a platform elevator into a dark chamber for detection of light emission by the photomultiplier photometer. Not only did this system provide a graphic record of the glow rhythm, but also it made feasible the systematic (and statistical) treatment of the rhythm with a wide variety of chemicals (Hastings, 1960; Hastings and Bode, 1962; see Section 4.3). More recently, the assay system has reached new heights of sophistication, allowing the simultaneous monitoring of both rhythms (by either a phototube or a fiberoptic light pipe), with mechanical functions and data acquisition both being under software control provided by an Apple II microcomputer (Taylor et al., 1979, 1982a,b).

The rhythm of stimulated bioluminescence (luminescence capacity) was found to have wide limits of entrainment: LD cycles ranging from LD: 6,6 to LD: 16,16 (i.e., T = 12 to 32 h) all synchronized the rhythm, which then free-ran with a circadian period upon subsequent release into LL (2000 lx) (Hastings and Sweeney, 1958, 1959). The effects of single, 3-h light (14000 lx) perturbations given at different times during a free-run yielded a standard PRC. Signals in the late subjective day induced phase delays of as long as 5 h or more, whereas perturbations given in the latter part of the subjective night (after peak luminescence response) generated phase advances up to 8 to 10 h in magnitude (Hastings and Sweeney, 1958). Maxima in effectiveness for phase-shifting were found at 475 and 650 nm (Sweeney et al., 1959; Hastings and Sweeney, 1960). Short exposures (2 to 4 min) to ultraviolet light also reset the rhythm, yielding only phase advances, whose magnitude depended on the duration of the exposure and the CT time at which they were applied. These effects did not appear to be photoreversible (Sweeney, 1963). The free-running period of the rhythms of both stimulated and glow luminescence (as well as cell division; see Table 2.5) was temperature-(over)compensated, having a Q_{10} of about 0.85 (Hastings and Sweeney, 1958, 1959).

The fact that free-running rhythms of bioluminescence persist for long timespans in populations of *G. polyedra* implies that either the individual biological clocks that underlie the overt rhythmicity are highly accurate or that the clocks are coupled, in the sense that some sort of chemical (or even photobiological, light-flash) communication, or crosstalk, exists (Edmunds, 1971; see Section 6.2.2). In an attempt to test this hypothesis of cellular interaction, Hastings and Sweeney (1958) mixed equal volumes of two cultures of *G. polyedra* whose rhythms of luminescence capacity were 5 h (75°) out of phase and found no evidence for crosstalk after one cycle under constant conditions. The rhythm continued and the maximum of luminescence in the mixed cultures was quite similar to that obtained when the measured luminescence of the separate cultures was summated. Had interaction occurred, one might have expected that a curve would have been obtained representing the resultant of the summated peaks for two separate in-phase cultures or some other effect. These experiments were extended in more detail and for several cycles under constant conditions to the rhythm of glow bioluminescence—a system possessing sharp and reproducible waveforms so that even small changes could be detected (Sulzman et al., 1982). Once again, no cellular interaction was detected. The bioluminescent glow of the mixed cultures matched the algebraic sum of the independent control cultures (Fig. 2.19). Thus, intercellular exchange of temporal information did not play a significant role in the maintaining of synchrony in circadian rhythms of *G. polyedra*, implying that long-term persistence must arise from the accuracy of the individual cellular clocks. [It is possible, however, that some sort of interaction can be observed over longer timespans (Broda et al., 1986a; Hastings et al., 1985); see Section 6.2.2 and Fig. 6.5]. That sufficient accuracy indeed does exist has been demonstrated recently for the glow rhythm. It is accurate to within 2 min per day, and the variance in τ among individual cells is about 18 min (Njus et al., 1981). The very slow decay of rhythmicity necessarily entailed, in fact, was empirically observed. Finally, the cellular autonomy of the *G. polyedra* clock(s) has been directly demonstrated by the measurement of the rhythms of both stimulated and glow luminescence in individually isolated cells (Hastings and Krasnow, 1981)(see analogous results for the rhythm in photosynthetic capacity in the next subsection).

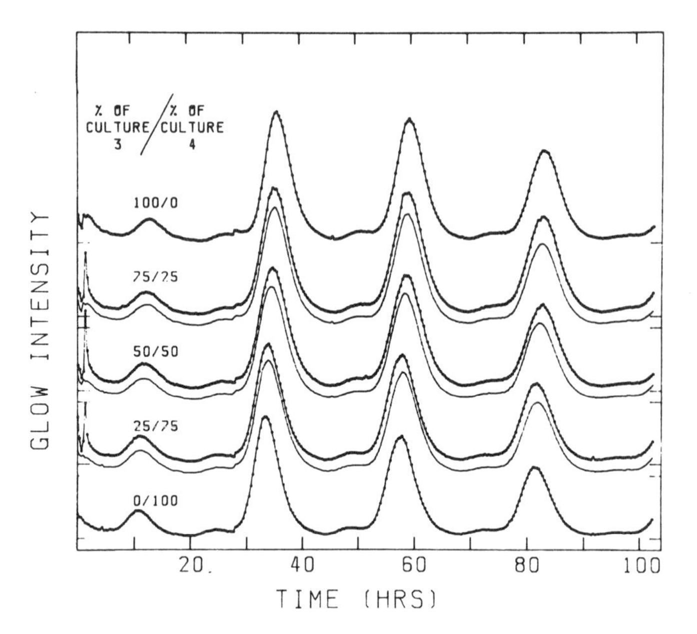

FIGURE 2.19. Bioluminescent glow patterns of various mixtures of two cultures that were about 2 h out of phase. The dotted lines (each dot represents a data point) are the averaged sums of replicate vials; the solid lines are the predicted curves determined by making appropriate linear averaged summations of the Culture 3 curve (100) and Culture 4 curve (100). Cells were transferred from the LD cycle at the end of the light period to dim LL. Time 0 refers to the time when Culture 4 was transferred to dim LL, immediately after mixing was done. The two curves for the observed and predicted patterns for the mixed cultures were virtually superimposable. The two curves have been displaced on the abscissa in order to show them distinctly. (Reprinted from Sulzman et al., 1982, with permission of Humana Press, Inc.)

Like cell division, they were phased, or gated; some cells emitted, others did not, but those that did emit did so at the allowed time.

It is a generalization (Bünning, 1973) that when rhythmic organisms, including *G. polyedra* (Sweeney and Hastings, 1957), are transferred from bright LL (in which circadian periodicity is lost in most cases) to DD, the rhythms recommence at the phase characteristic of the onset of the night, at CT 12 (Bünning, 1973). The inference is that the clock has been stopped at CT 12, the rhythms starting up again as this phase point when given DD. As Sweeney (1979) has noted, however, this is not the only interpretation; the underlying oscillator (as opposed to the overt rhythm) could be continuing to run undetected in bright LL and then could be shifted to CT 12 from whatever phase it was in by the LL/DD transition. Indeed, she observed this to be the case for the rhythm of acid-stimulated bioluminescence (as well as of cell division and photosynthetic capacity).

Bright light did not cause the circadian clock(s) to immediately stop; some 4 weeks were required under these conditions before all rhythms damped out, and when periodicity was restored by a transfer of the cells to DD after different lengths of exposure to LL, the rhythms were always reset to about CT 12 to 14.

Analogously, if cultures of *G. polyedra* were placed in either 11 or 4°C, the circadian rhythm of glow luminescence was lost, even though the cells were still capable of emitting light at these low temperatures. Rhythmicity was restored, however, when the cultures were returned to a higher temperature of 20°C, with the phase of the reinitiated rhythm apparently being determined by the time of transfer to the permissive temperature (Njus et al, 1977). The reinitiated circadian oscillation started up at CT 12, suggesting to these workers that critical temperatures stopped the clock and held it at this unique phase point (similar to the earlier hypothesis for LL). Further, rhyhmicity was found to be lost under combined, noncritical low temperatures (i.e., $T > 12.5$°C) and intensities of LL that otherwise would have been ineffective if individually imposed. Njus et al.(1977) term this lack of persistence at a critical temperature "conditional arrhythmicity," as opposed to historical arrhythmicity, which does not involve the inhibition of an existing rhythm (such as is observed when an organism is raised from the seed under constant conditions and where rhythmicity can be elicited by a single light or temperature perturbation).

Circadian Rhythms in Photosynthesis

In the study of transducing mechanisms from the putative circadian oscillator to observed rhythmicities, photosynthesis is particularly attractive because it comprises a number of relatively well understood processes that can be measured individually, and it was one of the first to be examined in detail in *G. polyedra*. Hastings et al. (1961) found that cells exhibited a rhythm of photosynthesis (as measured by incorporation of $^{14}CO_2$) at the eighth hour of the light interval during a synchronizing LD: 12,12 cycle. Cultures transferred to continuous dim LL (1100 lx) continued to exhibit only a circadian (τ = 26 h) rhythm of photosynthetic capacity (incorporation measured under 9600-lx illumination in aliquots taken from the master culture). Cultures transferred to bright LL (9600 lx), however, did not show any periodicity. Thus, conditions could be chosen where oscillations in capacity would persist although the actual rate of photosynthesis was constant.

Using the cartesian reference diver technique, Sweeney (1960) was able to measure (under saturating conditions) the evolution of O_2 over a 24-h timespan in single, isolated *Gonyaulax* cells maintained either in LD or in bright LL (8000 to 10,000 lx). The results were similar to those found for cell populations, thus suggesting that bright LL inhibits rhythmicity in the individual cell rather than merely generating asynchrony among rhythmic cells of the culture. [A similar situation obtains for the loss of circadian rhythmicity of cell division in isolated cells maintained in intense LL (Hastings and Sweeney, 1964)]. Further, these observations clearly demonstrate that a bona fide circadian rhythm may occur in a single eukaryotic microorganism without the necessity of intercellular coupling for its generation or maintenance (Section 6.2.2).

Attention then turned to the transducing mechanisms involved in the expression of the rhythm of PC. Earlier efforts concentrated on the dark reactions comprising the Calvin scheme (Sweeney, 1969b, 1972; Bush and Sweeney, 1972), but results were not promising. For example, although ribulose-1,5-biphosphate carboxylase sometimes showed fluctuations in activity, there was not sufficient correspondence to satisfy the rates of CO_2 fixation at all stages investigated, and, indeed, in *Acetabularia,* none of the Calvin cycle enzymes were found to have rhythmic activities (Hellebust et al, 1967; see Section 2.2.1).

Inasmuch as the enzymes responsible for carbon fixation did not vary over the circadian cycle, experimentation focused on the light reactions of photosynthesis, and results have made it clear that, indeed, they are regulated by the clock (Sweeney, 1981). Thus, from analysis of the rates of photosynthesis as a function of irradiance, a temporal change in relative quantum yield in dim LL was found, although total chlorophyll, half-saturation constants, and the size and number of particles on the thylakoid freeze-fracture faces were constant (Prézelin and Sweeney, 1977; Prézelin et al., 1977). More recently, Samuelsson et al. (1983) found that changes in PS I and PS II account for the circadian rhythm in PS in *G. polyedra.* Electron flow (O_2 uptake in cell-free extracts from late log cultures in LL for 3 days) through both PS II and PS I, and through PS II alone, was rhythmic, but flow through PS I alone (either including or excluding the plastoquinone pool) was not.

Finally, Govindjee et al. (1979) examined fluorescence transients at different times in *G. polyedra* cells maintained in either LD or LL, and no rhythmic changes were found. On the other hand, the intensity of Chl *a* fluorescence (both initial and peak values) was about twice as high during the day phase of the circadian cycle in LL than during subjective night, and these changes were positively correlated with the rhythm of O_2 evolution (Sweeney et al., 1979). This periodicity in fluorescence persisted in the presence of 10 μM DCMU, which blocks electron flow from PS II during photosynthesis, indicating that the cause was not due to a change in net electron transport between PS II and PS I. Although the rhythm of fluorescence could arise from differences in the efficiency of spillover energy from strongly fluorescent PS I, such spillover should occur unimpaired at 77°K, but this was not the case: the rhythmicity in PC was abolished at this temperature (Sweeney et al., 1979). The results could indicate that the nonradiative decay of Chl excitation is less during the day than at night, although the reason for such a change remains obscure. Perhaps circadian ion fluxes across the thylakoid membrane generate reversible conformational changes that would couple and uncouple entire photosynthetic units and their light-harvesting pigment-protein complexes and thus induce a circadian rhythmicity in PC (Prézelin and Sweeney, 1977; Sweeney, 1981).

2.3 Circadian Rhythms in Fungi

2.3.1 *NEUROSPORA* SPP.

As we shall see in our discussion of the molecular genetics of circadian clocks (Section 4.4), genetic analysis has great potential for identifying the mechanisms underlying circadian rhythmicity, just as it has proved to be a powerful approach to the elucidation of biochemical pathways, gene organization, and gene regulation, and, more recently, for the dissection of more complex systems such as bacteriophage assembly, cell division cycles, and behavorial responses. In addition to *Drosophila* spp. and *Chlamydomonas* spp., considerable amount of data on the genetics of rhythms has been accumulated in the filamentous bread mold, *Neurospora crassa*, which not only has a long history of research in genetics and biochemistry but also has proved to be an excellent system for studying circadian rhythms (reviewed by Feldman et al., 1979; Feldman, 1982; Feldman and Dunlap, 1983). In this section, we briefly examine the major morphological and biochemical circadian rhythmicities documented for this fungus (Table 2.6) in order to set the stage for the genetic dissection of the clock (Section 4.4).

The Rhythm of Conidiation

The most intensively examined circadian rhythm in *N. crassa* is that of conidiation, or asexual spore formation (Sargent et al., 1966). This periodicity is expressed in nearly all strains (Sargent and Woodward, 1969), although several mutants (such as *patch* and *band*) exhibit the rhythm much more clearly and under a wider variety of conditions than do the wild-type strains. The rhythm can be assayed quite easily on agar medium in either race tubes (usually 12 to 16 mm in diameter and 20 to 60 cm in length and bent upward at either end) or Petri dishes, in which the culture, inoculated at one end of the tube or at the edge of the dish, produces alternating areas of denser conidia (or bands) and sparser surface mycelia (or interbands)(Fig. 2.20). Once the mycelium develops on the agar, it does not fill in; thus because the culture leaves a permanent record of the position of the conidial band in time, no automated equipment is needed for continuously monitoring the cultures, and a great many experiments can be performed in a single incubator. The banding pattern itself can be measured after an experiment has been completed and recorded with a digitizer interfaced with a microcomputer system (Gardner and Feldman,

TABLE 2.6. Representative persisting circadian rhythms in the bread mold *Neurospora crassa*.

Circadian rhythm	Reference
Physiological	
CO_2 production	Woodward and Sargent (1973)
Conidiation	Brandt (1953); Pittendrigh et al. (1959); Stadler (1959); Sargent et al. (1966); Sargent and Briggs (1967); Sargent and Woodward (1969); Sargent and Kaltenborn (1972); Bitz and Sargent (1974); Brody and Martins (1979); West (1976); Dharmananda and Feldman (1979); Mattern and Brody (1979); Nakashima and Feldman (1980); Nakashima et al. (1981a,b); Mattern et al. (1982); Nakashima (1982a,b; 1983, 1984a,b, 1985); Nakashima and Fujimura (1982); Sulzman et al. (1984); Mattern (1985a,b)
Hyphal branching	Sussman et al. (1964)
Biochemical	
Adenylate and pyridine nucleotides	Brody and Harris (1973); Delmer and Brody (1975); (1975); Dieckmann (1980); Schulz et al. (1985)
Enzymatic activity	
Citrate synthase	Hochberg and Sargent (1974)
Glucose-6-phosphate dehydrogenase	Hochberg and Sargent (1974)
Glyceraldehyde-phosphate dehydrogenase	Hochberg and Sargent (1974)
Isocitrate lyase	Hochberg and Sargent (1974)
Nicotine adenine dinucleotide nucleosidase	Hochberg and Sargent (1974)
Phosphogluconate dehydrogenase	Hochberg and Sargent (1974)
Fatty acids	Roeder et al. (1982)
Ion content (K^+, Na^+)	Sato et al. (1985)
Nucleic acid metabolism	Martens and Sargent (1974)
Protein, heat-shock	Cornelius and Rensing (1986)
Protein, soluble	Hochberg and Sargent (1974)

See Section 4.3 (Tables 4.1 and 4.2) for detailed reference to the biochemical genetics of biochemical and clock mutants.

1980). Finally, a system for assaying the clock in liquid cultures has been devised (Perlman et al., 1981), involving the transfer of mycelial pieces from liquid medium to race tubes, where the phase of the rhythm of conidial banding reflects the phase in liquid cultures. This important development facilitated the carrying out of many biochemical experiments that had not been feasible (see Section 4.4.3).

Expression in Various Strains

Brandt (1953) first demonstrated the circadian rhythm of conidiation in *N. crassa* in a strain (a proline auxotroph) that he named *patch*. Subsequent experiments by Pittendrigh et al. (1959) showed that the conidiation

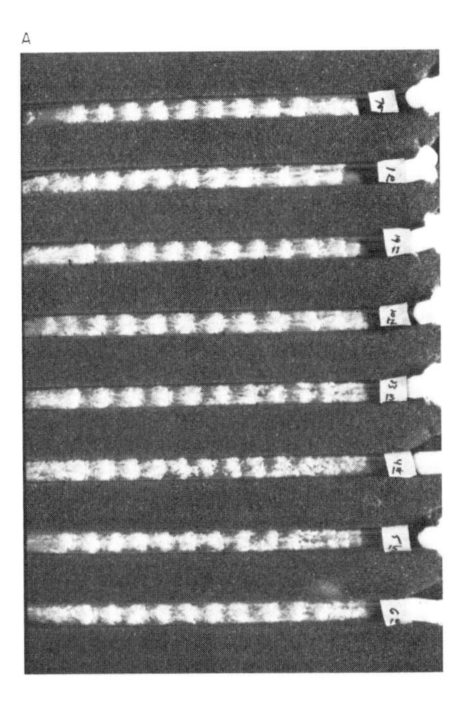

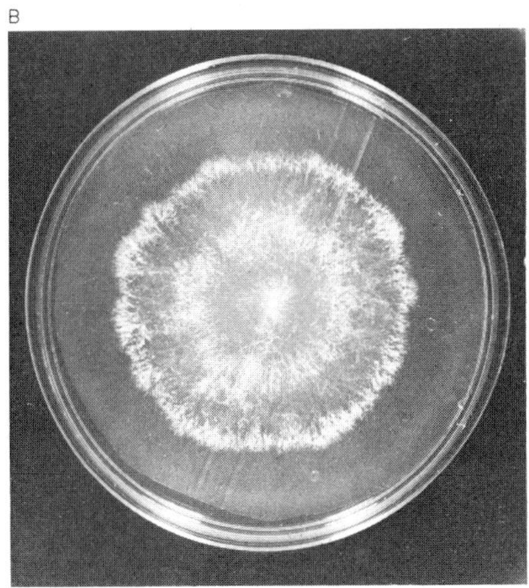

FIGURE 2.20. Circadian rhythm of conidiation of the band strain of *Neurospora crassa*. (A) Growth in race tubes. Each tube was inoculated on the left end, and the fungus grows toward the opposite end. The tubes were maintained for 2 days in LL at 27°C and then transferred to DD for 9 days. They were then placed in constant light for an additional day before photography. Ten conidiation bands are present in each tube. (Reprinted from Sulzman et al., 1984, with permission of American Association for the Advancement of Science.) (B) Growth in concentric rings on Petri plates. (Courtesy of S. Brody.)

rhythm exhibited three key characteristics of a bona fide circadian rhythm (see Fig. 1.2): a circadian period under free-running conditions, temperature-compensation, and susceptibility to phase-shifting by light. Stadler (1959) discovered that this particular strain contained a second mutation that was responsible for the expression of rhythmicity, and this mutant gene was then designated *patch*. Nevertheless, precise measurements of period and phase were difficult in this strain. Although another mutant (Sussman et al., 1964), termed *clock,* exhibited a hyphal branching rhythm whose τ under some conditions approximated 24 h, the period was highly dependent on temperature and on medium composition and could be neither entrained by LD or phase-shifted by light pulses. Indeed, even wild-type strains could be induced to exhibit the hyphal branching rhythm on appropriate concentrations of sorbose in the medium, but the properties of the rhythm in this phenocopy were found to be similar to those of the *clock* mutant (Feldman and Hoyle, 1974).

A major development occurred with the discovery (Sargent et al., 1966) of another strain of *N. crassa* that exhibited a circadian rhythm of conidiation. This strain, designated *timex,* itself comprised two mutations—*inv* (invertaseless) and *bd* (band)—although only *bd* was essential for expression of the rhythm (Sargent and Woodward, 1969). Furthermore, *bd* conferred two additional advantages as compared to *patch*: rhythmicity could be expressed on a defined medium containing glucose, salts, and arginine, and the conidial bands were much sharper, allowing precise determination of τ and ϕ. This latter characteristic may be attributable to the insensitivity of *band* to accumulation of CO_2 in the race tube, to which conidiation in wild-type strains is highly sensitive (Sargent and Kaltenborn, 1972). For these reasons, therefore, almost all recent work on circadian rhythms in *N. crassa* has used the *bd* mutation. Similarly, Brody and coworkers, often working with cultures grown in Petri dishes (Mattern et al., 1982), have inserted the csp^- mutation into *band* strains. This mutation facilitates manipulation of the cultures for various experiments by preventing the separation of conidia from the mycelium. [In fact, Cramer-Herold et al. (1986) recently have demonstrated that a circadian rhythm of conidiation can be induced in *csp-2* by the addition of the surfactant, sodium dodecylsulfate, to the medium.]

Light Effects

Sargent et al. (1966) demonstrated that the conidiation rhythm in the *band (timex)* strain of *N. crassa* could be entrained by LD: 9,15 to a 24-h period. Entrainment of the rhythm to LD cycles whose $T \neq 24$ h is also possible, although the limits of entrainment have not been ascertained. Single, 5-min pulses of light once per cycle (a one-point skeleton photoperiod; Pittendrigh, 1965) will also entrain to $T = 24$ h (G. F. Gardner, unpublished data).

Sargent and Briggs (1967), using single, white-light signals 45 min in duration (1430 foot-candles) imposed at different times (CT) during the free-running rhythm at 28°C, found that steady-state phase shifts were engendered whose sign and magnitude were dependent on the CT that the pulse was administered. The PRC thus derived was similar to that of many other organisms (Pittendrigh, 1965; Winfree, 1980), with $-\Delta\phi$ occurring during the late subjective day and early subjective night, and $+\Delta\phi$ occurring during the late subjective night and early subjective day, and has been corroborated by many other workers (Fig. 2.21, lower panel). The action spectrum for this phase-shifting effect had a strong peak in the blue (λ = 465 nm) region of the visible spectrum (Sargent and Briggs, 1967), typical of blue-light photoreceptors (see Section 4.2.1).

More recently, Nakashima and Feldman (1980) have shown that this light-induced phase-shifting is temperature sensitive. The amplitude of the PRC was drastically reduced as the temperature was increased from 25° to 34° C (Fig. 2.21) in two new mutants of *N. crassa*, designated *htb-1* and *htb-2* (*htb* = high-temperature banding). Phase advances were decreased more than phase delays. At 34°C there was little or no phase shifting by light, despite the fact that there was a clear expression of the circadian conidiation rhythm in these mutants even at 36°C. Finally, West (1976) has found that 5-min pulses of short-wavelength ultraviolet light (λ < 254 nm) could induce $+\Delta\phi$ as large as 11 to 12 h when given during the late subjective night.

Nakashima (1982a,b) has identified several inhibitors of light-induced

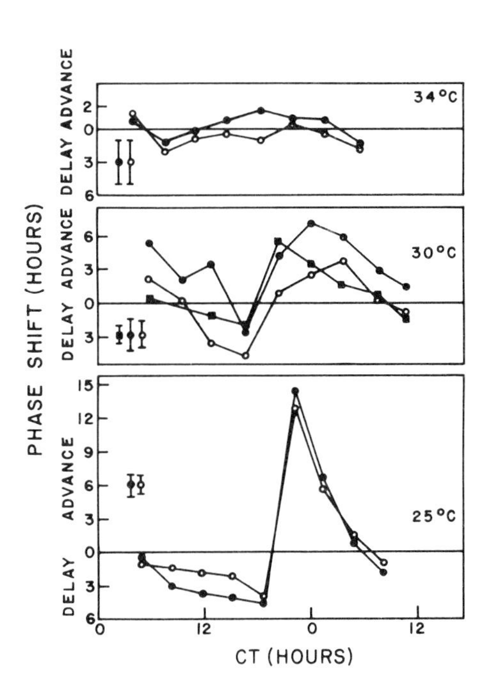

FIGURE 2.21. Phase-response curve for light perturbations at different temperatures in *N. crassa*. Three strains were inoculated into race tubes and cultured for 1 day at 25°C and then transferred to DD at different temperatures. All experiments were performed using medium with yeast extract. Circadian time 12 was defined as the time of transfer to DD. Light pulses, 5 min of fluorescent white light (500 lx). Vertical bars indicate the mean of the standard deviations for each time point. (■) *bd;* (●) *bd htb-1;* (○) *bd htb-2*. (Reprinted from Nakashima and Feldman, 1980, with permission of Pergamon Press.)

phase shifting in cultures of *N. crassa* grown in liquid culture. In particular, diethylstilbestrol and N,N-dicyclohexylcarbodimide, both inhibitors of the H^+-translocating plasma membrane ATPase, inhibited the effects of light pulses in cultures grown at pH 6.7 but not at pH 5.7. That the pH of the culture medium is important was also indicated by the observation that the overall light sensitivity (light duration and intensity necessary to evoke a saturating response) of the mycelium was about 8 times greater in cultures grown at pH 5.7 than in those cultured at pH 6.7. Likewise, Nakashima and Fujimura (1982) discovered that the transfer of mycelium from culture medium (of different pHs) to 10 m*M* Pipes buffer (also having various pHs) inhibited light-induced phase shifts for cultures grown at pH 6.7 but not at pH 5.7. However, the addition of ammonium salts (but not of K^+, Na^+, Cl^-, or NO_3^- ions) to the buffer restored light-induced phase advances. In contrast, inhibitors of mitochondrial ATPase, such as venturicidin and oligomycin, had no significant effect on light-induced phase shifting at either pH. (Although azide, which inhibits both types of ATPase, did inhibit the phase-shifting response evoked by light signals, the effect was attributed to an inhibition of plasma membrane enzyme, given the fact that the inhibition of respiration, and thus of mitochondrial ATPase, was dependent on the pH of the treatment medium.) These results collectively suggest an important role of ionic and electrochemical gradients across the plasma membrane in light-induced phase shifting of the *Neurospora* clock (see Section 4.3.3).

Temperature Effects

Although the growth rate of the *band* strain of *N. crassa* is quite dependent on temperature (Q_{10} = 2.35 over a temperature range of 16 to 35°C, with an optimum at about 35°C), the length of the free-running period of the conidiation rhythm was relatively constant between 18 and 35°C (Sargent et al., 1966). This temperature compensation of τ is reflected in the observed values for its Q_{10}: 0.95 over the 18° to 25°C interval and 1.21 over the 25° to 35°C interval. At temperatures below 18°C the production of conidia was so intense that the growth appeared to be almost continuous, whereas above 35°C, no conidia were produced, thereby making the assay of τ impossible. Similarly, τ of the conidiation rhythm was temperature-compensated below 30°C ($Q_{10} \cong 1$) in the *htb* mutants isolated by Nakashima and Feldman (1980) but not well compensated above 30°C (Q_{10} = 1.3 to 1.7). Furthermore, the value of Q_{10} was lowered by the addition of yeast extract in the higher temperature range (30 to 34°C) but not in the lower range (25 to 30°C). Such an alteration of the temperature dependence of τ by nutritional conditions has been found also in *E. gracilis* (Brinkmann, 1966; see Section 2.2.3).

The effects of temperature perturbations on the conidiation rhythm in *N. crassa* have been examined also (Francis and Sargent, 1979) and were generally similar to the responses that have been reported in other organisms: temperature steps-up induced $+\Delta\phi$s, steps-down generated $-\Delta\phi$s, and temperature pulses produced either $+\Delta\phi$s or $-\Delta\phi$s that were roughly equal to the algebraic sum of the phase shifts produced by the individual temperature steps which the pulse formally comprised (Pavlidis et al., 1968; Zimmerman et al., 1968).

Finally, we have already noted (see Fig. 2.21) that higher temperatures caused a dramatic change in the PRC for light pulses in the *htb* mutants of *N. crassa*, suppressing particularly light-induced phase advances (Nakashima and Feldman, 1980). It may be that some component or process between the photoreceptor and the clock is heat labile (temperature sensitive), and as a result, the light pulse would not cause much phase-shifting at high temperatures. Alternatively, the amount of photoreceptor material synthesized at higher temperatures may be decreased so that light exerts a smaller effect on the clock (Nakashima and Feldman, 1980).

Other Factors

The composition of the medium significantly affects the expression of the conidiation rhythm in the *band* strain, although the basic circadian properties (light-entrainability, period length, temperature compensation) themselves are not usually modified (Feldman and Dunlap, 1983). Similarly, the size and geometry of the culture vessel in which the conidiation rhythm is assayed are important. Although there have been reports of the effects of subtle geophysical variables (such as magnetic or electrical fields) on circadian rhythms in a number of organisms (Palmer et al., 1976), Bitz and Sargent (1974) failed to detect an influence of magnetic fields on the growth rate or conidiation rhythm of *N. crassa* cultured in race tubes. Despite these negative results, recent and provocative results from the STS 9 flight of the Space Shuttle Columbia revealed that although the period of the conidiation rhythm of this microorganism in DD was the same as that of earth-based controls, there was a marked reduction in amplitude of the rhythmicity, and even arrhythmicity, in some culture race tubes. The data were insufficient, however, to determine if the circadian timing mechanism itself was affected or only its expression (Sulzman et al., 1984). Finally, unpublished results from experiments by D.O. Woodward (as cited in Feldman and Dunlap, 1983, pp. 343–344), in which heterokaryons of *N. crassa* were created in DD in Y-shaped race tubes between two strains previously entrained by LD to different phases, unexpectedly suggested that a phasing substance may exist in the mycelium and that a significant part of the clock mechanism may be compartmentalized (see Section 6.2.2).

Rhythm of Carbon Dioxide Production

One of the first physiological circadian rhythms documented in *N. crassa* was that of the production of CO_2 (Woodward and Sargent, 1973). In race tube cultures, an amplitude of 20 ppm may be attained, with a peak occurring approximately midway through the formation of a conidial band, which probably reflects increased activity of the Krebs cycle associated with the increased rate of accumulation of mycelial mass.

Rhythms in Adenylate and Pyridine Nucleotides

On the hypothesis that the rhythm in CO_2 production might reflect changes in respiratory metabolism and mitochondrial function, Delmer and Brody (1975) sampled the edges of the growing front of a mycelial mat at different ages in the *band* mutant of *N. crassa* cultured in Petri plates and enzymatically assayed the levels of adenine nucleotides. Although the concentrations of ADP and ATP showed no obvious oscillation, that of AMP oscillated between 0.5 and 6.0 μmol/g (residual dry weight), with a peak at the midpoint of the conidiating phase (see Fig. 5.3). This oscillation had many of the properties of a true circadian rhythm: τ was 22 h, it could be phase-shifted by light, and it damped out in LL. In turn, the rhythm in the level of AMP gave rise to an oscillation (between 0.65 and 0.93; see Fig. 5.4) in overall cellular energy charge [calculated as (ATP + 0.5 ADP)/(ATP + ADP + AMP)], if one assumes the absence of compartmentalization of AMP (Atkinson, 1968). Delmer and Brody suggested that the underlying cause of the oscillation in AMP level could be a rhythmic, partial uncoupling of mitochondrial oxidative phosphorylation, although it might also be considered to be merely a reflection of the morphological conidiation rhythm and the accompanying cycle in the synthesis and processing of adenine nucleotides (Feldman and Dunlap, 1983; see Section 5.2.2). Recently, however, Schulz et al. (1985) have excluded conidiation as a causal factor: the concentrations of ATP and ADP (and thus total AEC) in mycelia kept in liquid culture (where rhythmic conidiation does not occur) underwent circadian changes in DD, exhibiting a maximum at CT 0 to CT 6—the time of maximal activity of Krebs cycle enzymes and CO_2 production.

Brody and Harris (1973) also have observed spatial differences in the levels of oxidized and reduced pyridine nucleotide levels (NAD^+, $NADP^+$, NADH, and NADPH) in the two morphologically distinct areas of the growing front of the *band* strain of *N. crassa* formed as a result of the rhythmic conidiation process. Although the total pyridine nucleotide content of these two areas was the same, the level of NAD^+ was higher in the conidiating area, and the levels of the other three nucleotides were lower. More recent data (Dieckmann, 1980) indicate that there is a significant overall decrease in NAD^+ in aging cultures without regard to circadian time and that changes in nucleotide levels at the growing front are small. Also, the redox ratio appeared to be constant across the circadian cycle, indicating that the spatial differences observed in pyridine nucleotide levels probably are not due to a clock-controlled rhythmicity in their levels.

Rhythms in Nucleic Acid Metabolism

The wild-type and *band* strains of *N. crassa* exhibit circadian rhythms of RNA and DNA content (normalized to mycelial mass) in the growth-front hyphae of cultures grown in large (19 by 30 cm) baking dishes and sampled over time (Martens and Sargent, 1974). The rhythm was in phase with that of conidiation. There was also a circadian rhythm of [^{3}H]uridine incorporation into the nucleic acids of the *band* strain. Maximum incorporation preceded by about 6 h the peaks in nucleic acid content that occurred during conidiation.

In order to determine whether these oscillations could be divorced from the large morphological changes embodied in the conidiation process itself, nucleic acid content also was monitored in the mutant *fluffy (fl)*, an aconidial strain that fails to complete septation even though it does produce aerial hyphae (Martens and Sargent, 1974). Rhythms in both DNA and RNA content were found, although their amplitudes were considerably less than those of *band*. Furthermore, the peaks for *fluffy* observed during the DD free-run were out-of-phase with those found for the *band* and wild-type strains. Indeed, the peaks of nucleic acids content in the latter two strains were often observed to have different phase relationships to their conidiation rhythm. All of these findings, therefore, indicate that the circadian rhythm of conidiation does not drive the circadian rhythms of nucleic acid metabolism, which appear to be coupled more directly to the underlying clock.

Rhythms in Enzymatic Activities

Using the same line of attack that was used with nucleic acid metabolism (see preceding section), Hochberg and Sargent (1974) examined the activities of a number of enzymes (Table 2.6) in the mycelial growth front of the *band* strain of *N. crassa* in an effort to determine if any observed fluctuations were part of the clock mechanism underlying rhythmic conidiation or merely manifestations of it. Of the 13 enzymes assayed, 6 evidenced circadian rhythmicity during DD free-runs (Fig. 2.22), as did soluble-protein content. The rhythmic enzymes associated with the Krebs and glyoxylate cycles were more active during conidiogenesis, whereas the activities of those associated with glycolysis and the hexose monophosphate shunt were reduced during this phase of the conidiation rhythm. For example, isocitrate lyase (EC 4.1.3.1), citrate synthase (EC 4.1.3.7), and nicotine adenine dinucleotide nucleosidase (NADase; EC 3.2.2.5) oscillated in phase with conidiation, whereas G6P dehydrogenase (G6PDH; EC 1.1.1.49) and phosphogluconate dehydrogenase (PGDH; EC 1.1.1.44) phase-led conidiogenesis by approximately 6 h, and glyceraldehyde-phosphate dehydrogenase (GAPDH; EC 1.2.1.12) oscillated a full 180° out of phase with the morphological rhythm.

Not all enzymes in these pathways oscillated; thus, glutamate dehy-

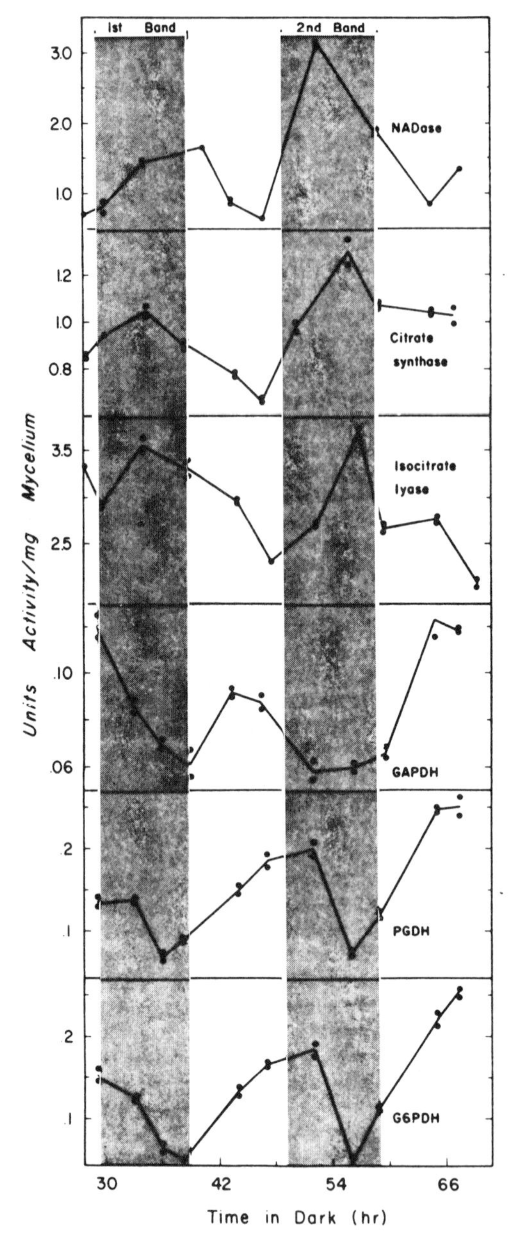

FIGURE 2.22. Enzyme activities in the growth front of the band strain in Petri plate cultures of *N. crassa*. Bars at top indicate times of conidiation. Each point represents one assay. (Reprinted from Hochberg and Sargent, 1974; with permission of the American Society for Microbiology.)

drogenase (EC 1.4.1.4), malate dehydrogenase (EC 1.1.1.37), constitutive (EC 3.1.3.1) and repressible alkaline phosphatases, constitutive acid phosphatase (EC 3.1.3.2), malate synthase (EC 4.1.3.2), and NAD-linked glutamate dehydrogenase exhibited no evidence of circadian rhythmicity. This lack of rhythmicity has been extended (Schmit and Brody,

1976) to the mycelial contents of glucosamine and galactosamine of *N. crassa*.

Hochberg and Sargent (1974) found that the enzymatic oscillations present in the *band* strain were absent in the nonconidiating mutant strain *fluffy*. Similarly, if the concentration of CO_2 were increased sufficiently to damp the conidiation rhythm, the activities of NADase, G6PDH, and PGDH also disappeared, although they were unaffected by lower concentrations of CO_2 that still permitted conidiation to occur. Oscillations of enzyme activity were restricted to the growth front and were not present in those portions of the surface of the agar plate previously traversed by rhythmic growth. In aggregate, these observations suggest that enzymatic rhythmicity (at least of those enzymes assayed) is a consequence of clock-generated periodicities in conidiogenesis itself.

Rhythm in Synthesis of Heat-Shock Proteins

Cornelius and Rensing (1986) have found that *N. crassa* shows a maximal capacity to synthesize three major heat-shock proteins (HSP)(99, 81, and 69 kDa) at about CT 12 (the phase at which the temperature decreases under normal environmental conditions). Mycelia in the stationary phase of growth were subjected to 3-h heat shocks (42°C) at various CTs. Samples were then incubated with [^{35}S]methionine and fractionated, and the labeled proteins were separated and identified by polyacrylamide gel electrophoresis and fluorography. HSP have been associated with acquisition of thermotolerance in many organisms (review: Nover, 1984), and many participate more generally in the control of cellular states, which the circadian oscillator may control endogenously by anticipating periodic environmental changes. Finally, it is interesting to note that many of the treatments that induce HSP synthesis also phase-shift CRs, such as that of conidiation (see Section 4.3.2).

Rhythms in Fatty Acids

Roeder et al. (1982) have demonstrated circadian oscillations in certain unsaturated fatty acids in the growing mycelial front of both the wild-type and *band* strains of *N. crassa*. The mole percentages of linoleic (18:2) and linolenic (18:3) acids [but not of palmitic (16:0), stearic (18:0), or oleic (18:1) acids, or of total lipid content] in the total lipids and in the phospholipids oscillated 180° out of phase with each other. It is not yet clear whether these oscillations in fatty acids have a role in the mechanism of the clock (as, for example, via mitochondrial uncoupling or alterations in membrane fluidity), although the possibility that they were merely part of the development of the rhythmicity itself was eliminated by the demonstration (Roeder et al., 1982) of similar oscillations in the *csp-1* strain, which did not express the conidiation rhythm under the growth conditions used (see Sections 4.3.3 and 4.4.3).

2.3.2 *Saccharomyces* spp.

The interplay among cell cycle controls, biological clocks, and membrane transport in microorganisms can be quite complex (Edmunds and Cirillo, 1974). Membrane processes may serve as pacemakers in endogenous biological rhythms (Njus et al., 1974; Sweeney, 1974b), and, conversely, self-sustaining biological clocks might well underlie oscillations in transport capacity and perhaps modulate the CDC itself (Edmunds, 1978). Inasmuch as the regulation of transport in the common yeast, *Saccharomyces cerevisiae,* has been intensively investigated, and the control of the CDC has yielded to a genetic approach using temperature-sensitive CDC mutants (Hartwell et al., 1974; Pringle, 1981), Edmunds et al. (1979) attempted to ascertain whether a circadian clock might generate overt rhythmicities in growth and cell division in this microorganism as it does in a variety of other unicellular algae and protozoans (Edmunds, 1978, 1984a; see Table 3.1 and Section 3.2.1) and, if so, whether such long-period oscillations could also be found in the uptake of amino acids in cultures that had attained the infradian (stationary) phase of population increase.

In addition to the direct inhibitory effects of visible light (cool-white fluorescent, <3500 lx) on growth rate and amino acid transport that had been previously reported for yeast (strain Y185 *rho*$^{+}$) grown at a relatively low temperature of 12°C (so that $g > 24$ h) in medium containing glucose, yeast carbon base, KH_2PO_4, ammonium ion, and proline (Woodward et al., 1978; Edmunds, 1980a), evidence was found for light-entrainable, autonomous, circadian, and ultradian oscillators underlying both cell division and transport capacity (Edmunds et al., 1979). Diurnal LD cycles (L $\cong$ 3000 lx), imposed on yeast cultures previously grown in the dark, phased or synchronized cell number increase to a 24-h period, with bud release's being confined primarily to the dark intervals (although not necessarily every cell divided during any given division burst). The observed division or budding rhythm free-ran with a circadian period ($\tau \cong 26$ h) for a number of days in constant darkness (DD) following prior entrainment by LD (see Section 3.2.1 and Fig. 3.30).

Further, a similar light-entrainable circadian rhythm in the uptake of [^{14}C]histidine or [^{14}C]lysine occurred in nondividing (or very slowly dividing) cultures synchronized by a 24-h LD cycle and then released into DD for as long as 10 days (Fig. 2.23). Such oscillations in transport capacity would be anticipated in dividing cultures, if for no other reason than that the CDC constitutes a driving force itself. Indeed, Carter and Halvorson (1973) have reported periodic changes in the initial uptake rates of a variety of amino acids at different stages of the cell developmental cycle in synchronous cultures of *S. cerevisiae* fractionated according to cell size and age by zonal centrifugation (although the value of g for these cultures was usually 1 to 3 h).

In some experiments (Edmunds et al., 1979), a bimodal (ultradian) pe-

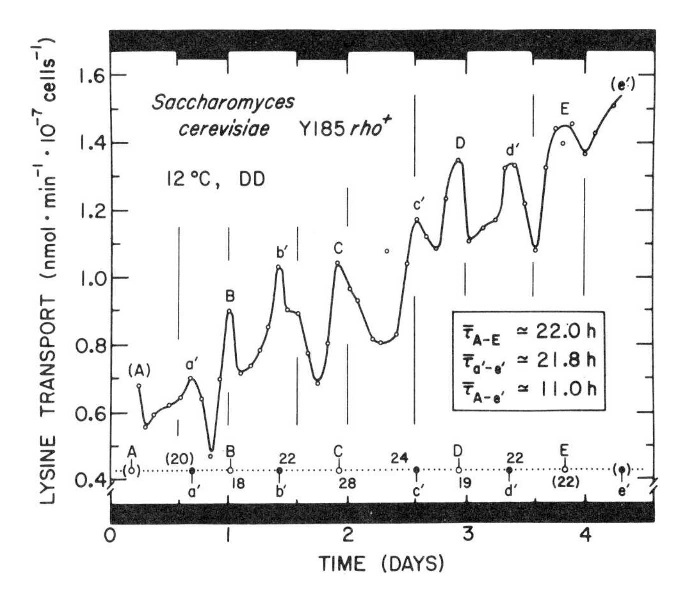

FIGURE 2.23. Persistence of a (bimodal?) circadian rhythm of [^{14}C]-lysine transport for 5 days during early-to-midstationary phase (g > 7 to 8 days) in cultures of *Saccharomyces cerevisiae* previously entrained by LD: 14,10 and released into DD. Subjective dark intervals are indicated to facilitate comparison with LD. Each point on the curve represents a rate determination from a time course assay of [^{14}C]-lysine transport. Major peaks are labeled (A, B, C, D, E); secondary peaks or shoulders are indicated by primed small letters (a', b', c', d' e'). Intervals between the major peaks are given on the dotted horizontal line below the curve; in LD, the period (τ_{LD}) approximates that of the imposed 24-h LD cycle. (Reprinted from Edmunds et al., 1976, with permission of Pergamon Press.)

riodicity in transport capacity was found in both LD and DD (Fig. 2.23), with secondary peaks, or shoulders, occurring at intervals of about 12 h, corresponding approximately to subjective dawn and dusk (although the beginning of the increase in capacity always anticipated the D/L or L/D transitions in the LD cycle). The occurrence of these peaks at intervals of about 11 to 12 h is open to several interpretations: (1) a true bimodal circadian rhythm existed, (2) there were two (or more) subpopulations of cells, each of which had a circadian rhythm in amino acid uptake but which were 180° out-of-phase, (3) the primary and secondary peaks were governed by two different circadian oscillators, keyed to dawn and dusk, or (4) the rhythm was truly ultradian, with a period of about 12 h. It is difficult to distinguish experimentally among these alternatives.

Some preliminary experiments were undertaken to determine the nature of the photoreceptor for these light effects. Previous results indicated that blue light (λ ≅ 410 nm) was the most effective wavelength of visible radiation for the inhibition of growth and transport in yeast and that a wide variety of petite mutants either partially or totally lacking cytochromes b or a/a_3, or both, evidence reduced photosensitivity, thus suggesting that these blue-absorbing chromophores are the primary photoreceptors for the inhibitory effects of visible light (Ułaszewski et al., 1979; Edmunds, 1980a). It is provocative that transport in cultures of the Y185 *rho*$^-$ petite mutant, lacking cytochromes a/a_3, b, and c_l, could not be synchronized by LD cycles (Edmunds et al., 1979), a finding that is consistent with, but by no means demands, the hypothesis that cytochromes may constitute photoreceptors for not only the direct inhibitory action of visible light but also the entrainment of biological oscillators by light in this microorganism.

A more rigorous test of the photosensitizing role of the heme chromophore in light inhibition might be to use yeast mutants incapable of synthesizing δ-aminolevulinic acid to regulate cytochrome levels and, by hypothesis, the degree of photoinhibition of growth and transport. Similarly, a more satisfying test of the photoreceptor for the putative biological clock would be the derivation of a high-resolution action spectrum for the phase-shifting effects of single light pulses given at different CTs of the rhythm free-running in DD in both the *rho*$^+$ and *rho*$^-$ strains. This would require, however, that the mutants be affected in only one (or very few) genes and that they exhibit some assayable clock function. If, indeed, yeast cytochromes are the primary photoreceptors for the inhibitory and entraining effects of light (especially by blue wavelengths) observed, they (together with the flavins) would fall into the larger class of blue-light receptors that are being reported with increasing frequency for a large number of biological phenomena (Senger, 1980).

3 Cell Cycle Clocks

The cell cycle, or perhaps more precisely, the cell division cycle (CDC), typically comprises a series of recurrent, relatively discrete, morphological and biochemical events (Fig. 3.1), although the specific elements may vary among different systems (Fig. 3.2). Ordinarily the CDC is loosely considered to begin with the completion of one cell division and end with the completion of the next; the time taken for one such cycle is termed generation time *(g)*. Nesting within its time domain are the ultrafast, the metabolic, and the epigenetic domains (Lloyd et al., 1982b; see Fig. 1.1). Progress through the CDC is usually monitored by observing the overt processes of DNA replication (synthetic, or S, period) and cell division (D; or mitotic, M, period). Thus, the CDC has been divided for more than 30 years (Howard and Pelc, 1953) into four consecutive intervals—G_1, S, G_2, and M—where G_1 and G_2, respectively, designate the time gaps between the completion of division and the onset of DNA replication and between the end of replication and the onset of mitosis. We now know that both G_1 and G_2 (as well as S and M) are characterized by many specific, additional events, such as bud emergence (in budding yeast), chloroplast replication (in plant cells), nuclear division, or the expression of a particular enzyme, and that the CDC perhaps can be still further subdivided. As a consequence, a veritable alphabet soup of cell cycle states has emerged, replete with A, B, G_o, and G_q in addition to G_1, G_2, M, and S [see books by Mitchison (1971), Prescott (1976), John (1981) and Lloyd et al. (1982b) for general treatments of the subject].

These sequences of periodic events (CDCs) in a population of dividing cells constitute clocks or timers of sorts (Mitchison, 1974) in the sense that they measure time—albeit imperfectly—under a given set of environmental conditions, ranging from approximately 10 to 30 min in certain prokaryotes (e.g., *Benekea natriegens, Escherichia coli*) and 2 to 3 h in the yeast *Saccharomyces cerevisiae* or the ciliate *Tetrahymena pyriformis,* to 15 to 20 h in cultured mammalian cells under ideal conditions or even several days in some dinoflagellates (e.g., *Pyrocystis fusiformis*). Of

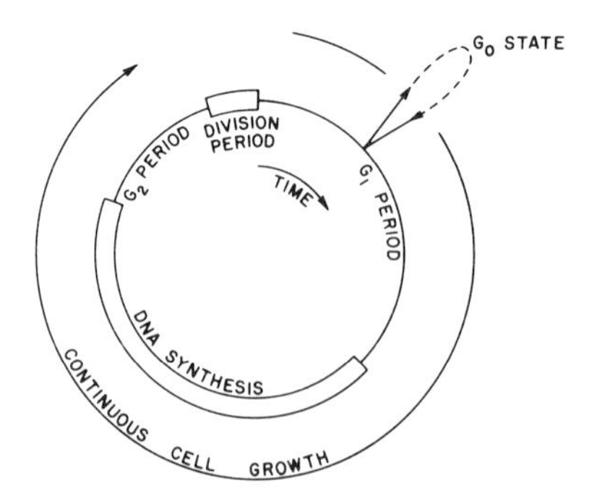

FIGURE 3.1. The major features of the cell life cycle. The relative proportions of the cycle may vary considerably from one kind of cell to another, but the reproduction of every cell consists of growth coupled with DNA replication followed by cell division. A mammalian cell growing in culture with a generation time of 16 h, for example, might have a G_1 of 5 h, S of 7 h, G_2 of 3 h, and D(M) of 1 h. G_0 is the state into which cells are postulated to move when the cell cycle is arrested in G_1 by various kinds of environmental conditions. (Reprinted from Prescott, 1976, with permission of Academic Press.)

course, there is a considerable degree of heterogeneity of *g* values within a cell population, arising from variations in the external microenvironment (e.g., temperature, nutrition) as well as from karyotypic variability, and for this reason individual CDCs do not seem to reflect the high degree of accuracy and precision of time measurement inherent in circadian oscillatory systems (Edmunds and Adams, 1981).

In addition, there is considerable evidence that the CDCs of unicellular algal and protistan populations, as well as of mammalian cells (both in vivo and in vitro), may themselves be governed by underlying oscillatory mechanisms or timing devices, which at one level are conceptually and operationally distinct from the CDC but, nevertheless, ultimately must be associated with the clocked division cycle in the sense that the timers are replicated with each round of division (Edmunds, 1978; Edmunds and Adams, 1981). This most amusing paradox presents a major challenge in the construction of formal and molecular models for cell cycle regulation (Edmunds, 1984a) and, because of the fundamental importance of the CDC in biological systems, is the focus of this chapter. After briefly reviewing some of the major notions of cell cycle regulation—a detailed treatment is beyond the scope of this monograph—we examine the evidence for treating the CDC as an oscillatory system (Section 3.1.2) and its interaction with autonomous (particularly circadian) cellular oscillators (Section 3.2.1).

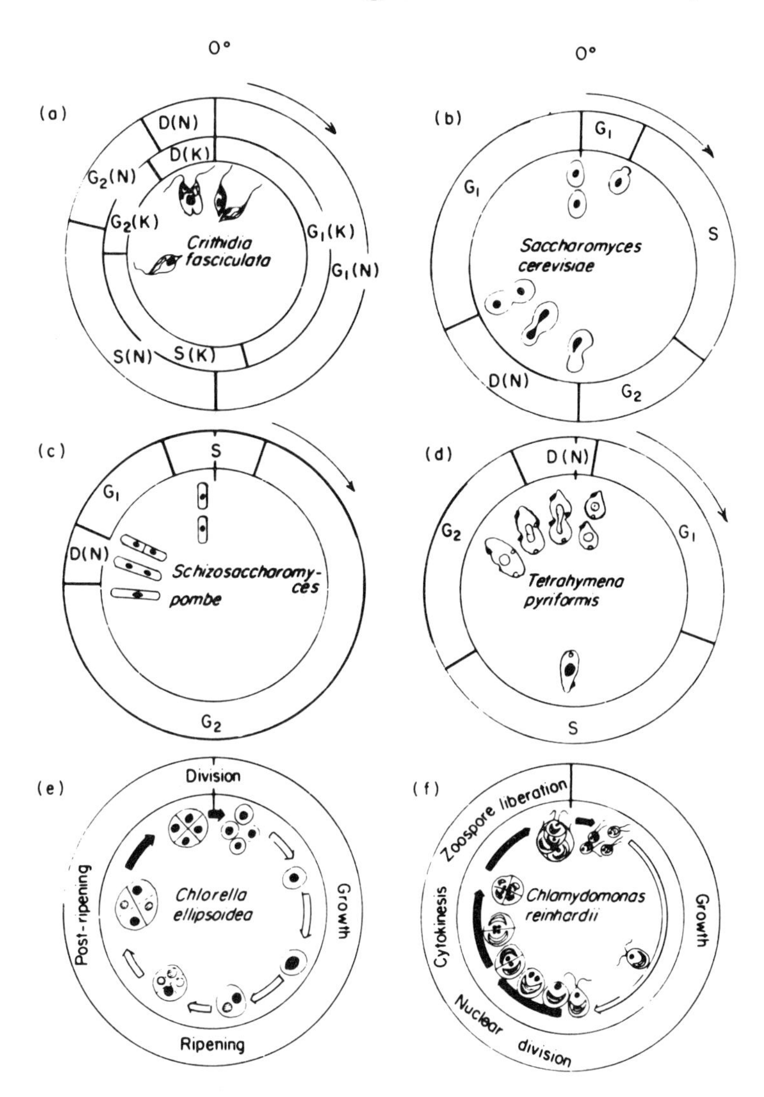

FIGURE 3.2. The CDCs of some lower eukaryotes. Cell division occurs at zero degrees and traverse is clockwise in all cases: (a) *Crithidia fasciculata*, showing events occurring in nucleus (N) and kinetoplast (K). (b) *Saccharomyces cerevisiae;* (c) *Schizosaccharomyces pombe;* (d) *Tetrahymena pyriformis,* showing nuclear events, division of contractile vacuole and stomatogenesis, (e) *Chlorella ellipsoidea;* (f) *Chlamydomonas reinhardii*. White arrows in (e) and (f) indicate processes in the light period, and black ones those in dark period. D (N), nuclear division; D (K), kinetoplast division. (Reproduced from Lloyd et al., 1982b, with permission of Academic Press.)

3.1 Regulation of the Cell Division Cycle

Advances in the field of cell cycles have not been confined to a mere filling in of the gaps in the CDC and subdividing them into smaller and smaller steps. Rather, recent experimental and theoretical work has emphasized the mechanisms controlling the cell cycle and, indeed, has indicated that a single cell cycle may be somethiong of a misnomer. [Indeed, perhaps there is no such thing as the CDC; the final event would be merely the end of a sequence of events and the beginning of nothing (Cooper, 1979).] Attempts to model the CDC and to account for the timing and coordination of cellular growth and division (for recent comprehensive treatments, see Lloyd et al., 1982b; Edmunds, 1984a) have been (1) indeterminate, or probabilistic, and (2) deterministic.

3.1.1 Probabilistic Models

Large variances (as great as 20% of the mean) in generation time are observed commonly in many systems, seemingly rendering timekeeping relatively imprecise and leading to the rapid decay of synchrony in phased cultures (Engelberg, 1968). This variability in traverse time of the complete CDC has been documented in bacteria (Koch and Schaechter, 1962; Kubitschek, 1971), yeasts (Nurse and Thuriaux, 1977; Fantes and Nurse, 1977), algae (Cook and Cook, 1962), and cultured animal cells (Kubitschek, 1962; Petersen and Anderson, 1964; Sisken and Morasca, 1965; Burns and Tannock, 1970; Fox and Pardee, 1970) and has been extended to other CDC stages, such as the entry of mammalian G_1 cells into S (Yen and Pardee, 1979). Even the durations of S and G_2 in many experimental mammalian systems exhibit important variabilities (Guiguet et al., 1984).

In an effort to explain this variability, some characterizations of the

FIGURE 3.3. (A) Distribution of generation times of various cell types in culture, where P denotes the transition probability of a cell's entering the determinate B phase (of duration T_B) leading to division, and α represents the proportion of the initial population (N) remaining in interphase. (a) Rat sarcoma: $P(h^{-1}) = 0.45$; $T_B(h) = 9.5 \pm 1.0$. (b) HeLa S3: $P = 0.32$; $T_B = 14 \pm 0.8$. (c) Mouse fibroblasts: $P = 0.30$; $T_B = 15 \pm 1.2$. (d) L5 cells: $P = 0.18$; $T_B = 22.5 \pm 1.4$. (e) HeLa: $P = 0.14$; $T_B = 23 \pm 0.8$. (Reprinted from Smith and Martin, 1973, with permission of the authors.) (B) Graphs of α against age at mitosis for cells of *Euglena gracilis* and of β against age difference at mitosis for sibling cells. β is the proportion of sibling pairs with age differences greater than that specified on the abscissa (see text for explanation). (Calculated from data published by Cook and Cook, 1962). The slope of the β line is the same as that of the linear part of the α curve. (Reprinted from Smith and Martin, 1974, with permission of Academic Press.)

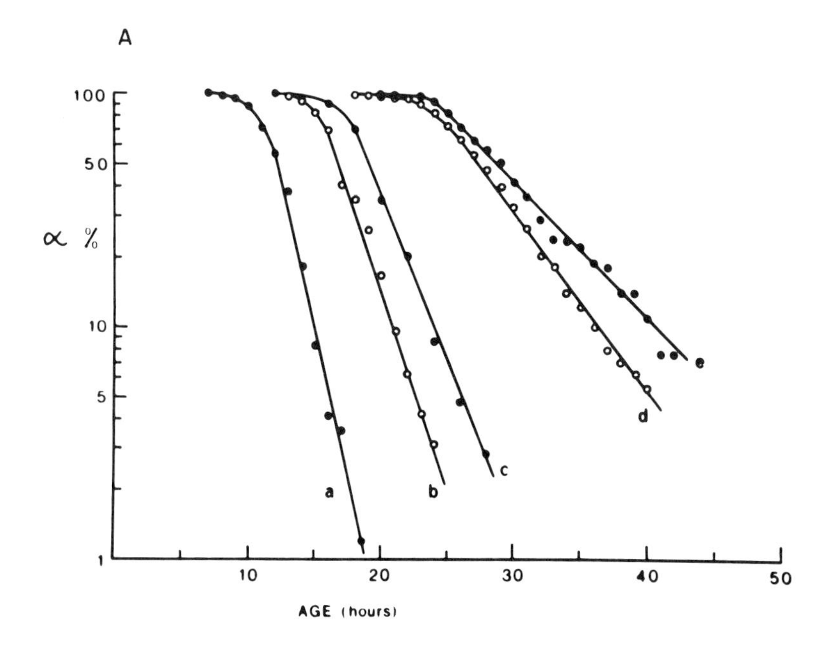
A
100
50
∝ %
10
5
1
a
b
c
d
e
10
20
30
40
50
AGE (hours)

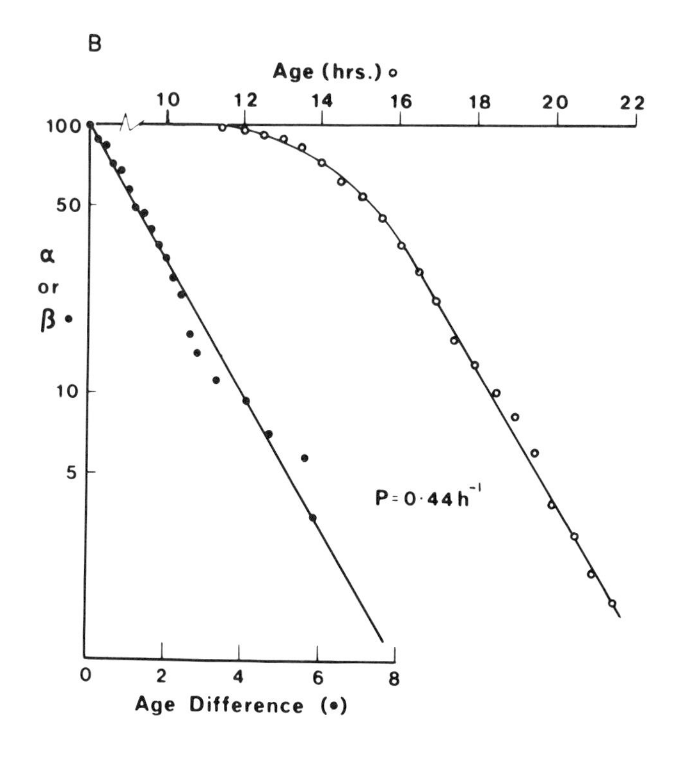
B
Age (hrs.) ○
10
12
14
16
18
20
22
100
50
α
or
β •
10
5
P = 0·44 h⁻¹
0
2
4
6
8
Age Difference (•)

cell cycle traverse have considered a portion of the CDC to be indeterminate (Temin, 1971) or have turned to probabilistic descriptions (Kubitschek, 1962; Burns and Tannock, 1970; Smith and Martin, 1973, 1974; Gilbert, 1974a, 1974b, 1978a, 1978b). Thus, Smith and Martin (1973), observing that the proportion of cells remaining in interphase decreases exponentially as time progresses (the so-called α-plot; see Fig. 3.3), have suggested that the S and G_2 portions of the CDC (which they term the B phase), are deterministic and invariant, whereas other parts are probabilistic (the A state, or waiting phase, of G_1). According to this transition probability model (Fig. 3.4) newly divided cells enter the A phase, in which their activity is no longer actively directed to proliferation. A cell may remain in this A state for any length of time, throughout which its probability of leaving A state remains constant. Upon exiting the A state, the cells reenter the B phase, wherein DNA synthesis and mitosis occur once more and the cells' activities are deterministic and directed toward replication. The initiation of cell replication processes would thus be random; cell population growth rates would be determined by the probability with which cells leave the A state, the duration of the B phase, and the rate of cell death. The probabilistic event might depend on a critical concentration of a key compound regulated by a number of feedback control loops and corresponds (Shilo et al., 1976) to the start point of the CDC of *S. cerevisae* and to the specific time in the G_1 phase of the mammalian CDC (the restriction point) at which cells escape the quiescent phase, brought on by suboptimal nutritional conditions or by other diverse blocks, to reenter the alternative proliferative state (Pardee, 1974). Gilbert has taken a similar theoretical approach (1974a), giving particular attention to various perturbations that may trigger the transition of the cell from the quiescent state into the more highly dynamic one of replication (1978b) and the possible relation to differentiation and cancer (1974a, 1978a).

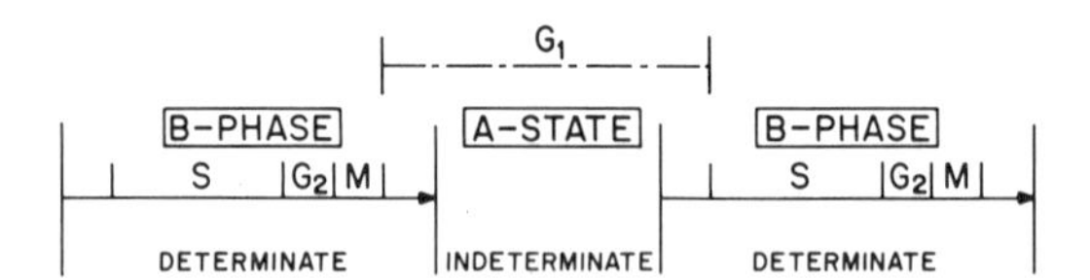

FIGURE 3.4. Transitional probability model for the regulation of cell proliferation. The intermitotic period is composed of an A state and a B phase. The B phase includes the conventional S, G_2, and M phases, and, perhaps, part of the G_1 phase. Some time after mitosis the cell enters the A state, in which it is not progressing toward division. It may remain in this state for any length of time, throughout which its probability of entering the B phase is constant. This transition probability *(P)* may be supposed to be a characteristic of the cell type but capable of modification by environmental factors. (Reprinted from Smith and Martin, 1973, with permission.)

The fact that the generation times of sister cells are also distributed exponentially (the so-called β-plot; see Fig. 3.3B) provides further evidence for a random transition in the cell cycle (Shields, 1977, 1978; Shields et al., 1978). More recent modifications of Smith and Martin's original model (1973) have added a second random transition and a cell size component (Brooks, 1981; Brooks et al., 1980) to take into account observations that subsequent cell generations do not seem to be influenced by their ancestry. Likewise, Svetina (1977) extends the transition probability model to incorporate the observation that the rate constant for the transition from the A state of the CDC is dependent on the length of time that a particular cell has spent in that state. On the other hand, Castor (1980) claims that a G_1 rate model can account for the cell cycle kinetics attributed to transitional probability. Similarly, Gilbert (1982a,b) has found that an oscillator model of the CDC (see Section 3.1.2) yields transition-probability α- and β-type curves (Fig. 3.3) and requires no first or second chance event. Klevecz (1976) has proposed a quantal subcycle, G_q, for mammalian cells, whose traverse time is about 3 to 4 h (see Section 3.1.2). This cycle would be appended to the deterministic $S + G_2 + M$ pathway at point *i*. The exit of a cell from G_q would be probabilistic in the sense that there could be an indefinite number of G_q cycles, depending on environmental conditions (e.g., cell density, nutritional variables, and mitosis-stimulating factors). In a similar fashion, Homma (1987) hypothesized that the CDC of *Gonyaulax polyedra* is divided into a cyclic part corresponding to G_1 and a noncyclic branch that constitutes the $S + G_2 + M$ sequence, which has a fixed duration of approximately 11 h. The former constitutes a circadian clock and, therefore, it takes one circadian period to traverse the subcycle. According to this scheme, cells that satisfy a minimal volume requirement (between CT 12 and CT 18) exit probabilistically to the replication-segregation sequence culminating in division, and reenter the cyclic portion a fixed time interval (11 h) afterwards.

3.1.2 Deterministic Models

A completely indeterminate view of cell cycle regulation does not sit well with many workers and goes against the grain of those searching for a more causal explanation at the biochemical and molecular levels. The field has obliged with a plethora of models (see books edited by Edmunds, 1984a; Nurse and Streiblová, 1984). Thus, as Mitchison (1974, 1984) has noted, there are two possible types of mechanism for ordering a fixed sequence of cell cycle events relative to each other (Fig. 3.5). There may be a direct causal connection between one event and the next so that it would be necessary for the earlier event in the CDC to be completed before the following could occur. Hartwell and co-workers (1974) have referred to this notion as the "dependent pathway" model (Fig. 3.6a) and have analyzed the circuitry of the CDC of the budding yeast, *S. cerevisiae*,

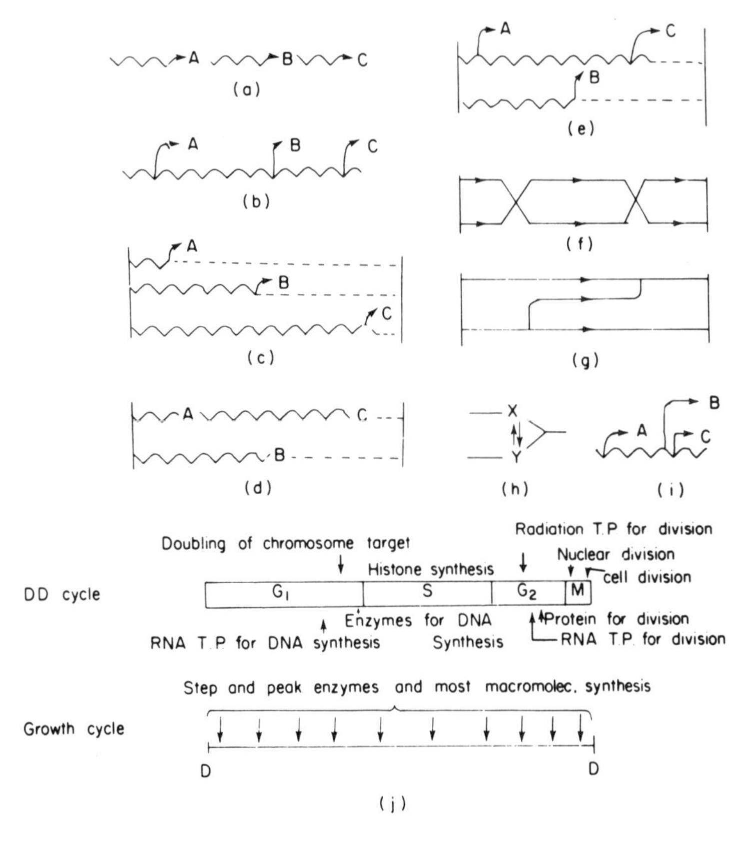

FIGURE 3.5. Models for sequential events during the cell cycle. The wavy lines indicate the progress of timers. (a) Dependent sequence. (b) Independent, single timer (IST) sequence. (c) Independent, multiple timer (IMT) sequence. (d) Two parallel dependent sequences. (e) Two parallel IST sequences. (f) Two sequences with two checkpoints. The events of each sequences before a checkpoint have to be completed if there is to be progress beyond the checkpoint. (g) Two sequences in which an early event in the lower sequence has to have been completed before a late event in the upper sequence can occur. (h) Interdependency of two events. (i) Delay in the full expression of an event B in an IST sequence may alter its order. (j) The DNA-division cycle and the growth cycle. T.P., transition point; D, division. (Reprinted from Mitchison, 1974, with permission of Academic Press.)

using temperature-sensitive mutants whose CDC was blocked at various stages (see next subsection). In contrast to this sequential type of approach, there is the possibility that no direct causal connection exists between any two events but that they are ordered by some master timing mechanism (Fig. 3.5b,c, Fig. 3.6b) that operates on one or more key events (control points) of the CDC, such as the initiation of DNA synthesis or mitosis. In this independent pathway model (Hartwell et al, 1974), the accumulation

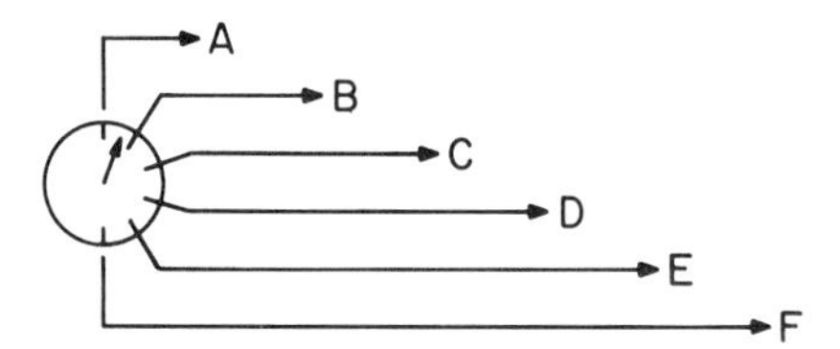

FIGURE 3.6. Two models to account for the ordering of cell cycle events. (Reprinted from Hartwell et al., 1974, with permission of American Association for the Advancement of Science.)

of a mitogen or other substance, the completion of a division protein structure, or the attainment of a critical size or ratio of DNA to mass or nuclear volume may initiate a new CDC state. Obviously, it is possible, even likely, that the CDC is controlled by a combination of these types of mechanism. A brief outline follows, in which yeasts [recently reviewed by Hanes et al. (1986) and Hayles and Nurse (1986)] and slime molds [Tyson and Sachsenmaier (1984)] serve as illustrative examples.

Dependent Pathways

According to this view, the CDC comprises a causal sequence of developmental events that eventually closes on itself (Figs. 3.5a, 3.6a), each of which is necessary before any later event can occur. These sequences are not necessarily linearly ordered, however, since branching networks (and even nested do-loops; Edmunds, 1975) may provide alternative pathways, some of which may operate concurrently (Fig. 3.5f,g,h).

In an elegant series of studies designed to distinguish between the fundamentally different dependent and independent pathway models, Hartwell and co-workers (1974) (see recent review by Pringle and Hartwell, 1981) dissected the circuitry of the CDC in the budding yeast *S. cerevisiae* (Fig. 3.7) by using temperature-sensitive, conditional cell- division-cycle (*cdc*) mutants in which development was arrested at the restrictive temperature at specific stages in the cell cycle, as determined by their cellular and nuclear morphology. These points [originally defined as "execution points" and later alternatively denoted as "block points" (Howell, 1974) or "transition points" (Nurse et al., 1976)] represent the stages of the CDC when the hypothesized defective gene products normally would have completed their functions in the wild-type cell. A shift to the restrictive temperature after (but not before) a given point in the CDC would have no effect on the succeeding cell division.

The phenotypes of these *cdc* mutants suggested that the landmarks of the cycle are ordered into two dependent pathways (Fig. 3.8). One pathway

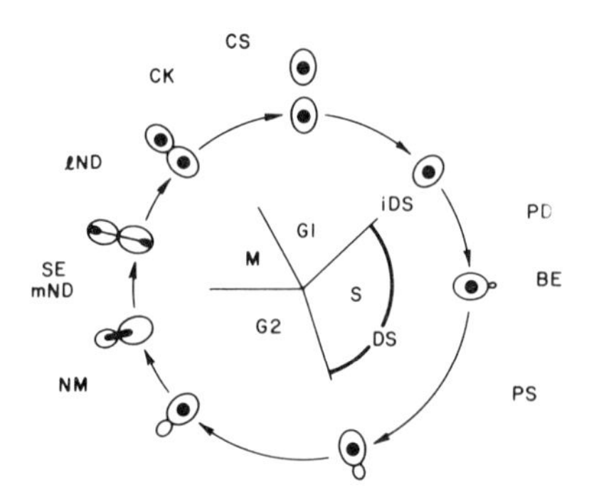

FIGURE 3.7. Landmarks in the cell division cycle of the budding yeast *Saccharomyces cerevisiae*. Distance between events does not necessarily reflect interval of time between events. *Key:* PD, nuclear plaque duplication; BE, bud emergence; iDS, initiation of DNA synthesis; DS, DNA synthesis; PS, nuclear plaque separation; NM, nuclear migration; mND, medial nuclear division; SE, spindle elongation; lND, late nuclear division; CK, cytokinesis; CS, cell separation. *Other abbreviations:* G_1, time interval between previous cytokinesis and initiation of DNA synthesis; S, period of DNA synthesis; G_2, time between DNA synthesis and onset of mitosis; M, period of mitosis. (Reprinted from Hartwell et al., 1974, with permission of American Association for the Advancement of Science.)

comprises plaque duplication (start), plaque separation, initiation of DNA synthesis, DNA synthesis, medial nuclear division, late nuclear division, cytokinesis, and cell separation. The other pathway was comprised of the following landmarks: bud emergence, nuclear migration, cytokinesis, and cell separation. No event in either pathway can occur without the prior occurence of preceding events, but the events in different pathways are independant of one another. For example, mutants defective in the initiation of DNA synthesis (*cdc* 4 and *cdc* 7) are unable to complete any subsequent events in this sequence but can undergo bud emergence and nuclear migration. Conversely, nuclear division is completed in mutants defective in bud emergence (*cdc* 24), but neither cytokinesis nor cell separation occurs. Only the inner pathway must be completed in one cycle in order for the cell to begin a second cycle. Since both pathways diverge from a common start (plaque duplication) and converge on a common event, cytokinesis, their integration is assured. Hartwell et al. (1974) found no evidence that an oscillator played a role in coordinating different events of the CDC. Rather, a variation of the dependent pathway model appeared to be sufficient to account for coordination of cell cycle events without recourse to the model of independent pathways and a central timing mechanism.

Provocatively, however, Hartwell et al. (1974) did discover evidence for the existence of a central timer that specifically controls the event of bud emergence. In an analysis of temperature-sensitive yeast mutants de-

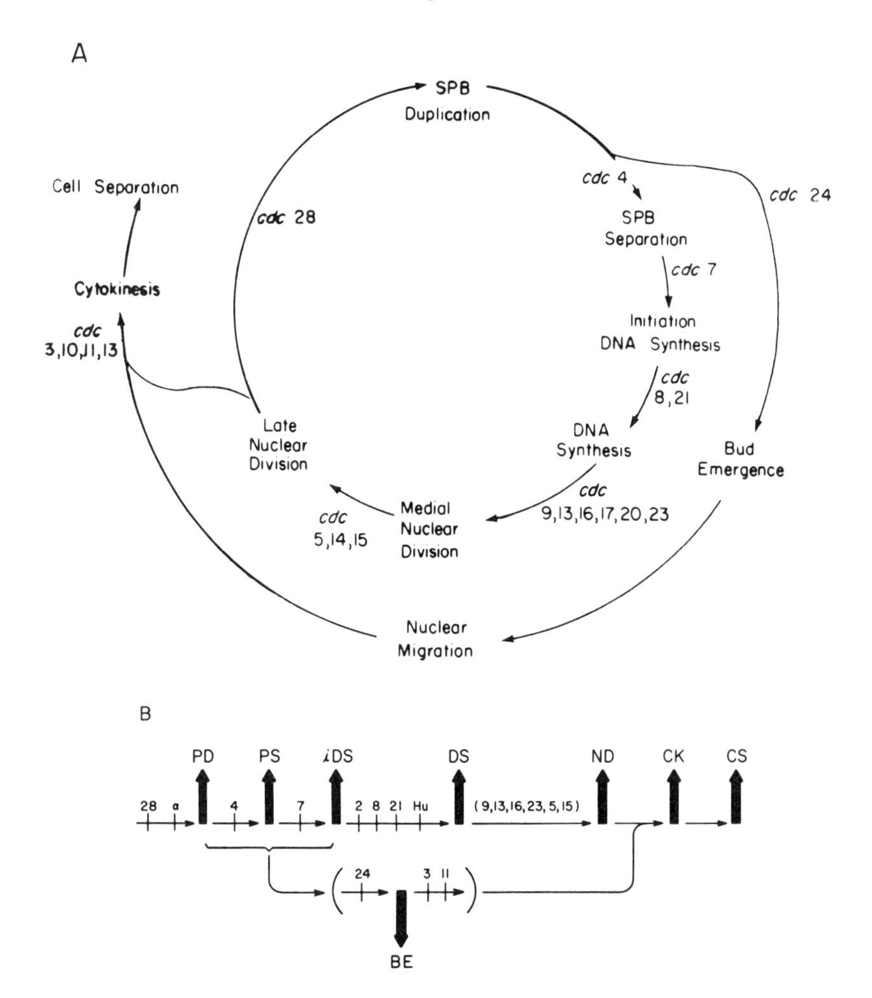

FIGURE 3.8. (A) The dissection of the yeast cell division cycle with temperature-sensitive mutants. Dependent pathways and landmarks derived from mutant phenotypes in *Saccharomyces cerevisiae*. Gene *(cdc)* designations are placed immediately preceding their diagnostic landmark. Upon a shift to the restrictive temperature, mutant cells arrest asynchronously at the position designated by the *cdc* number. All events connected by an arrow are proposed to be related such that the occurrence of any given event is dependent on the completion of the prior event. (Reproduced from Hartwell, 1978, with permission of Rockefeller University Press.) (B) Summary of dependent relationships in the CDC of *S. cerevisiae*. succeeding steps are related in a dependent sequence of given order. The relationships between arrows or symbols enclosed within parentheses are undefined. Completion of steps *cdc* 24, 3, and 11 are known to be dependent on the α-mating factor-sensitive step and independent of the hydoxyurea-sensitive step; their relationship, however, to steps mediated by *cdc* 4 and 7 is unknown. *Key:* numbers, *cdc* genes; α, mating factor; HU, hydroxyurea; other abbreviations as for Figure 3.7. (Reprinted from Hartwell, 1976, with permission of Academic Press.)

fective in the *cdc* 4 gene required for the initiation of DNA synthesis, Hartwell (1971) made the striking observation that buds were initiated repeatedly (remaining attached to the parent cell) in strain 314D5 at the restrictive temperature, in contrast to the complete cessation of bud initiation found in mutant strain 198D1. As many as five buds could be attained on a single mononucleate cell. The time interval (τ) between successive budding events in these *cdc* 4 mutants was roughly constant (between 100 and 150 min) and was about the same as the duration of the CDC g for this strain. Furthermore, these budding cycles continued in the absence of the initiation of DNA synthesis, DNA synthesis itself, nuclear division, cytokinesis and cell separation. These results, then, suggested the presence of a cellular clock controlling bud emergence that can express itself at only one discrete stage in the CDC, the stage of arrest in the *cdc* 4 mutant. The action of this timer might be a prerequisite for, be dependent upon, or even be part of the start event.

A similar mapping of the CDC using temperature-sensitive cell cycle mutants has been undertaken in several other microorganisms. Thus, a series of conditional, cell cycle-blocked *(cb)* mutants were isolated in *Chlamydomonas reinhardtii* (Howell and Naliboff, 1973; Howell, 1974) that grew normally at permissive temperature but were unable to complete a full cycle at the higher, restrictive temperature. There was a pronounced clustering of block points in the various *cb* mutants examined in the second and fourth quarters of the CDC, prompting speculation that there may be at least two major groups of processes required for cell division that terminate at about the same time (Howell, 1974).

A similar conclusion has been reached in *Schizosaccharomyces pombe*, the fission yeast, based on the analysis of a number of mutants that are conditionally defective in their progress through the cell cycle, which is about 2 h in duration with a G_2 phase about 180 min in length and with nuclear division and cell division's being about a quarter-cycle out-of-phase (Fig. 3.2c). The majority of these mutants are heat-sensitive and are defective in some aspect of DNA replication, mitosis, or septation at a restrictive temperature of 35°C (see reviews by Nurse and Fantes, 1981; Fantes, 1984a,b; Hayles and Nurse, 1986). Most notable among them are the *cdc* mutants, 27 of which have been isolated and characterized, and a large number of cell size *(wee)* mutants, whose most striking property is their reduced length and volume (cell width is unchanged) at division (Nurse et al., 1976). The *wee* mutants have yielded a particularly clear demonstration that entry into mitosis is subject to a cell size requirement (see next subsection). Two genes have been identified that are capable of mutation to give a *wee* phenotype: *wee* 1 and *cdc* 2. Both of these mutants have very pleiotropic phenotypes; the function of *cdc* 2^+, for example, is required for initiating both the S phase and mitosis (activity of the gene product is rate limiting for the latter), and the gene also controls a step in G_1 that is a switchpoint between the mitotic and conjugation-meiotic pathways (Fantes, 1984a).

Analysis of these two classes of mutant indicate that there are two parallel sequences of events in the CDC of *S. pombe* (Fig. 3.5j): (1) there appears to be a DNA division cycle, comprising DNA synthesis and nuclear and cell division as key events, and (2) the evidence also implicates a growth cycle, in which the synthesis of RNA and the synthesis of protein are the chief processes (Mitchison, 1974). These two sequences are only loosely coupled in that one can proceed if the other is blocked. Either can control the time of mitosis and cell size at that cycle stage (Fantes, 1984a). A summary of the relationships among the various events of the CDC and the genes that control them is given in Figure 3.9 (Nurse et al., 1976). The parallels between this scheme and that for *S. cerevisiae* are striking (Fig. 3.8). Indeed, Piggott et al. (1982) found that in the budding yeast the *cdc* 28 gene, once thought to control only a key event in G_1 (start), also codes for a function required for the completion of mitosis, just as the *cdc* 2 gene does in *S. pombe*.

A few words are in order concerning the regulation of growth and division of prokaryotes, whose division cycles might appear at first glance to be fundamentally different from those of eukaryotes (see general reviews and treatments by Slater and Schaechter, 1974; Helmstetter et al, 1979;

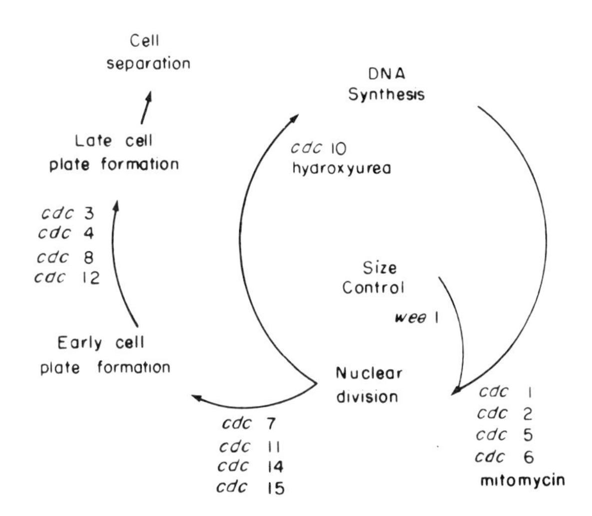

FIGURE 3.9. Summary of the relationships among the various events of the cell cycle and the genes that control them in the fission yeast *Schizosaccharomyces pombe*. The *cdc* genes and the cell cycle inhibitors are placed just before the event in which they are involved. The connecting arrows are not analogous to a biochemical pathway but only formally represent the interdependent relationships of the various cell cycle events. (Reprinted from Nurse et al., 1976.)

Lloyd et al., 1982b; Pritchard, 1984; Vicente, 1984). Although it is true that DNA replication occurs continuously in prokaryotes under certain conditions, in contrast to the eukaryotes, there are unifying analogies between the S period of the latter and the prokaryotic C period, when DNA is synthesized, and between the G_2 phase of eukaryotes and the prokaryotic D period, taken as the time between the termination of DNA synthesis and the onset of mitosis or cell division (Cooper, 1979). Indeed, the durations of these phases are relatively invariant, particularly in the case of slowly growing bacteria.

At the level of the genome, over 40 *E. coli* genes have been implicated in cell division and growth, and their positions in the genetic map have been determined (Bachmann and Low, 1980; Holland, 1987) (Fig. 3.10). If division were timed by the expression of several genes, their expression would have to be either continuous (as is the case for most major proteins in *E. coli*) or discontinuous at a low level. The latter alternative seems more plausible because it is difficult to explain how a pattern of continuous expression could time any discontinuous event (Vicente, 1984), unless one invoked a threshold mechanism.

The majority of models posited for the timing of bacterial cell division, the coordination among growth, DNA replication, and division, and the mode of surface growth, especially in intensively investigated *E. coli* and *Bacillus subtilis,* are deterministic (see reviews of Donachie, 1974, 1979, 1981; Donachie et al., 1973; Koch, 1977; Sargent, 1979; Vicente, 1984). For example, Cooper and Helmstetter (1968) assumed a coupling between chromosome replication and the division cycle and a step-by-step progression through the CDC, although one could also argue that cell division does not depend directly on the previous completion of chromosome replication but rather on the generation of signals that may independently trigger both processes (Koch, 1977). Similarly, Koch (1980) has proposed a model controlled by growth in *E. coli,* whereas Jones and Donachie (1973) reported the requirement for synthesis of "termination proteins."

This latter observation, in turn, was incorporated into a general model (Donachie et al., 1973) for the CDC of *E. coli,* which hypothesized a timer (reflected in mass doubling time), whose period varies with growth rate, and two parallel, separate sequences (chromosome replication and synthesis of termination protein; synthesis of division proteins), each probably a dependent pathway invariant with growth rate (within limits) and each of which must be completed for cell division to occur (see Fig. 3.13). That these two pathways are independent is supported by evidence that (1) if DNA synthesis is blocked for a short time, the normal delay between the end of a round of replication and the succeeding division does not occur, as if the sequence of protein synthesis were continuing to run unabatedly, and (2) in certain mutants of *E. coli,* cell division can occur in the absence of DNA replication, producing "anucleate" cells (Mitchison, 1974). Be-

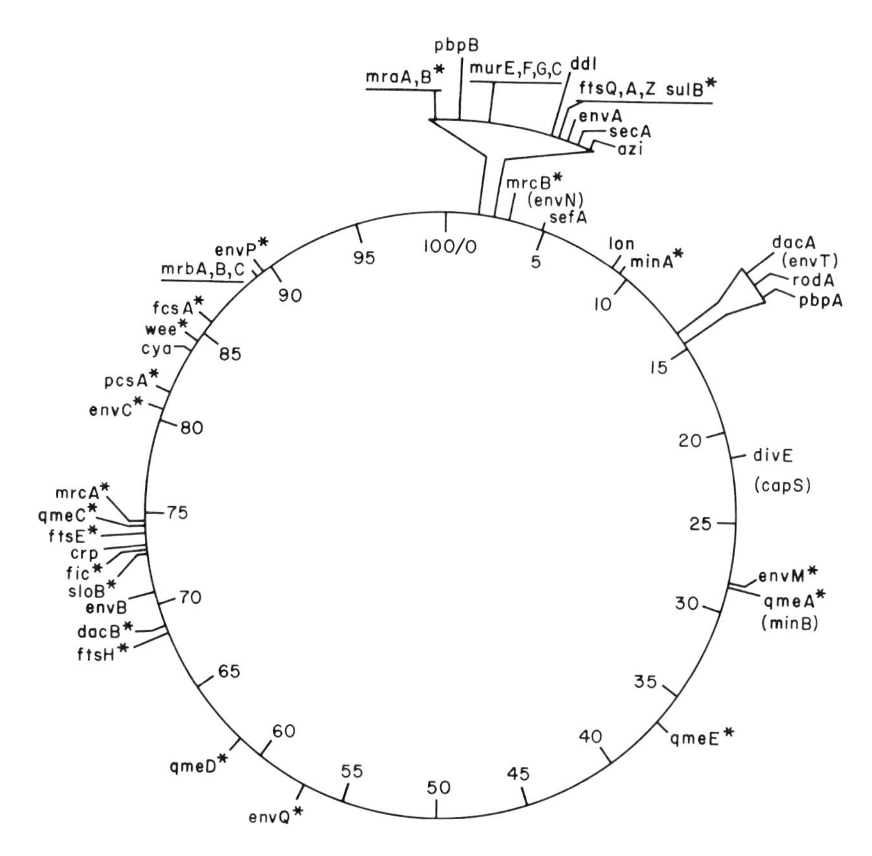

FIGURE 3.10. A map of the division genes found in the *Escherichia coli* genome. The map is that of Bachmann and Low (1980), in which the only positions indicated are those corresponding to genes related to cell division, morphology, or growth. Genes that are not found in the Bachman and Low map, in clockwise direction, are *murG, ftsQ, ftsZ, secA, fic,* and *wee*. In addition, *weeA* and *weeB* recently have been found to map, respectively, at minute 66.5 and between minutes 66 and 62 (M. Vicente, personal communication, 1985). Symbols are those used by Bachmann and Low; a star in a gene denotes that its position relative to adjacent genes is not precisely known. The position of a gene within parentheses is only approximated to a 5- to 10-min range. Expanded sections show the two main clusters of division genes. Although *sulB* is not a division gene, its approximate position within the 2.5-min cluster is indicated. (Reprinted from Vicente, 1984, with permission of CRC Press, Inc.)

cause this model incorporates the notion of timers and clocks underlying periodic cell division in bacterial populations, it is considered again in our later discussion of autonomous oscillators.

Cell Sizers

The central premise of models emphasizing cell size control of the CDC is that cells must attain a mimimum size before certain events (for example, DNA synthesis or nuclear division) can occur. Size itself could be measured by the cell in terms of cell volume, surface area, protein or RNA content, or other variables. If cell size were somehow thus monitored, cell growth could be coordinated with CDC events by the insertion of one or more size checkpoints into the CDC. For instance, a requirement for a certain cell size could be placed at the onset of DNA synthesis or cell division or at both points as long as appropriate causal links existed between secondary events of the cycle and the primary controlled event (Tyson, 1984, 1987; Tyson and Sachsenmaier, 1984).

There is considerable evidence for the operation of cell sizers in a variety of organisms (Fig. 3.11). Thus, Donachie (1968), assuming an exponential increase in cell volume and, therefore, mass in bacteria (reported by many workers), proposed that chromosome replication is initiated at a certain cell mass or multiples thereof (see Fig. 3.13). Similarly, in the budding yeast *S. cerevisiae,* the start event (plaque duplication; see Fig. 3.8) appears to be dependent on the cell's attaining a minimum size (Johnston et al., 1977), and in the fission yeast *S. pombe,* there seems to be a cell size requirement for both the onset of DNA synthesis and nuclear division (Nurse, 1975; Nurse and Thuriaux, 1977; reviewed by Fantes, 1984a,b; Hayles and Nurse, 1986). [Provocatively, Johnston and Singer (1985) have shown in *S. pombe* that mass accumulation, conversely, may be modulated by execution of the DNA division sequence: the rate of cell proliferation was decreased when the DNA division pathway was differentially slowed by low concentrations of the S phase inhibitor hydroxyurea in *wee* 1 mutants or by semipermissive temperatures in certain *wee* 1 *cdc* double mutants.] Size thresholds also may play an important role in the control of nuclear division in the acellular slime mold *Physarum polycephalum* (Sachsenmaier et al., 1972; Loidl and Sachsenmaier, 1982; reviewed by Tyson, 1984; Tyson and Sachsenmaier, 1984), of the number of daughter cells produced at the end of the CDCs of the green algae *Chlorella* spp. (Donnan and John, 1984; Donnan et al., 1985) and *C. reinhardtii* (Donnan and John, 1983, 1984; John, 1984; Donnan et al., 1985; McAteer et al., 1985), of cell division phasing in the marine diatom *Thalassiosira pseudonana* (Heath and Spencer, 1985), and of the onset of DNA synthesis in mammalian cells (Pardee et al., 1978).

Should the sizing process require a constant amount of time to be completed, its duration being independent of that of other events, it could be

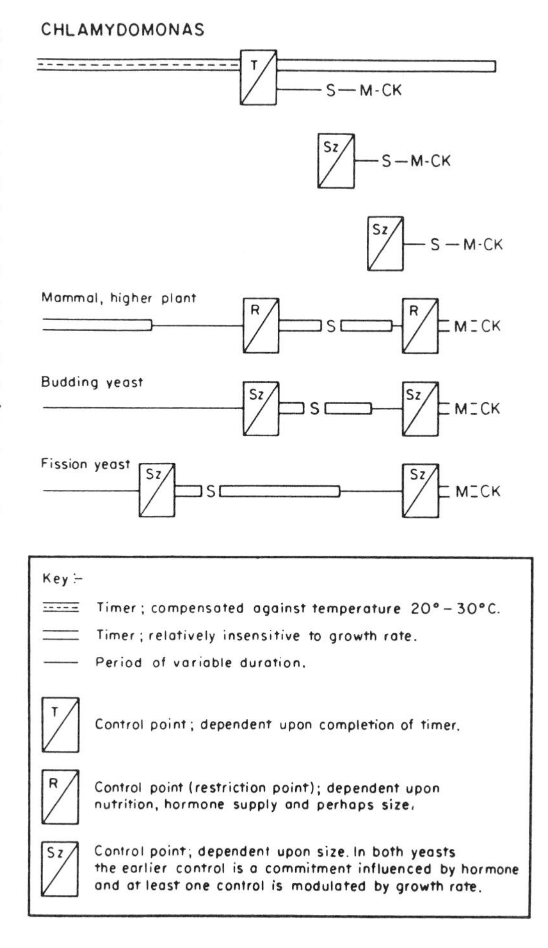

FIGURE 3.11. Control points in the CDCs of *Chlamydomonas* and several other eukaryotes. In the diagram for *Chlamydomonas*, the three commitments shown would result in eight daugter cells. Under slow growth conditions, which result in small mother cell sizes at the time of commitment, only the first commitment may be undertaken, resulting in two daughter cells and a cell cycle that closely resembles those of mammals, higher plants, and budding and fission yeasts. (Reprinted from John, 1984, with permission of Blackwell Scientific Publications Ltd.)

designated a "timer" or "clock" (Mitchison, 1974), of particular interest in this monograph. This type of CDC sizer control is considered in more detail in the next subsection and in Section 3.2). Of course, it is possible—even likely—that several types of mechanism in addition to cell sizers operate concurrently or in tandem to regulate the CDC in a single cell (Shymko et al., 1984). For example, both timer and sizer control function in cell cycle regulation in *Chlorella* and *Chlamydomonas* (Donnan and John, 1983, 1984; John, 1984; McAteer et al., 1985;) and possibly in marine diatoms (Heath and Spencer, 1985). Shymko and Klevecz (1981) have examined theoretically the consequences of superimposing a size threshold on a limit-cycle model of the cell cycle (see next subsection). In a similar fashion, Homma (1987) hypothesized that the CDC of *Gonyaulax polyedra* is divided into a cyclic part corresponding to G_1 and a noncyclic branch that constitutes the S + G_2 + M sequence, which has a fixed duration of approximately 11 h. The former constitutes a circadian clock and, therefore, it takes one circadian period to traverse the subcycle. According to this scheme, cells that satisfy a minimal volume requirement (between CT 12 and CT 18) exit probabilistically to the replication-segregation sequence culminating in division, and reenter the cyclic portion a fixed time interval (11 h) afterwards. Likewise, in order to explain the observed distributions of cell size and generation time (Tyson, 1985), Tyson and Hannsgen (1985) have posited a deterministic-probabilistic tandem model of the CDC, incorporating size control and random transition, which has been superseded by a model (Tyson and Diekmann, 1986) invoking sloppy size control and assuming that cell division is a random process with size-dependent probability (increasing with increasing cell size above a certain minimum). Finally, in an effort to reconcile theories for CDC regulation based on transition probability, size control, and inherited properties (pedigree), Webb (1986) has constructed a general model of cell population dynamics that can be viewed either probabilistically or deterministically, depending on the interpretation of the dynamic constituents in terms of probabilities or rates, with age and size taken as structure variables and the rate of cell growth assumed to be known.

Autonomous Timers: Cell Cycle Clocks and Oscillators

Inasmuch as mitosis in cell populations is a periodic event of short duration relative to the total length of the CDC, it is not surprising that various types of self-sustaining oscillatory systems, or biological clocks, have been proposed to underlie this and other events, or landmarks, comprising the CDC (Edmunds and Adams, 1981; Shymko et al., 1984). [The term clock is used as a generic term denoting a (real time) timekeeping device, without any bias about the underlying mechanism; this usage was adopted at the Dahlem Conference on The Molecular Basis of Circadian Rhythms in 1975 (Hastings and Schweiger, 1976).] These timers range from those of the relaxation type, such as a simple hourglass (Fig. 3.12A), to those exhibiting limit cycle behavior (Fig. 3.12B), in which two or more biochemical species oscillate, recovering their original form and period even after severe perturbations. Even if the CDC were merely a linear array of discrete metabolic states, each causing the next, and if it were controlled simply by the sequential transcription and subsequent translation of genes linearly ordered on the chromosomes (Masters and Pardee, 1965; Tauro et al., 1968), an oscillator could be formulated by invoking a recycling component that would initiate another time-metering, transcriptional cycle (Ehret and Trucco, 1967; see Section 5.3). A role for quantal cell cycles in the clustering of the generation times of cultured mammalian cells also has been proposed (Klevecz, 1976; Petrovic et al., 1984). Each of these types of cellular oscillator is considered in the following subsections; evidence for the interaction of circadian clocks and the CDC is detailed in Section 3.2).

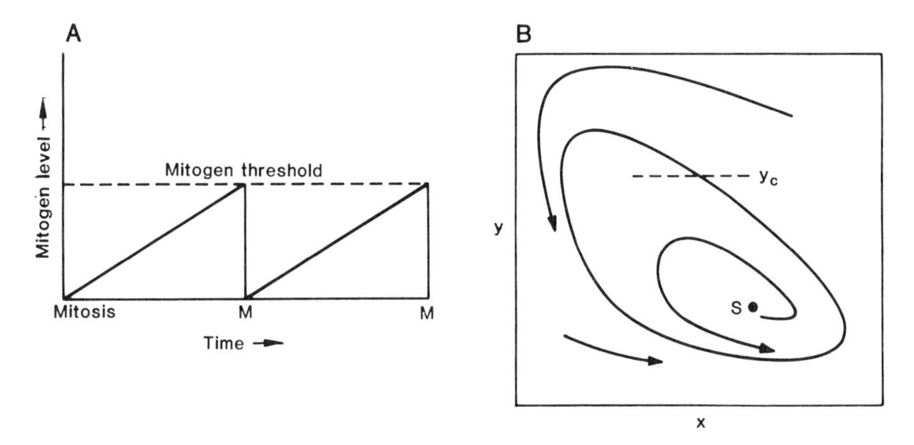

FIGURE 3.12. (A) Diagrammatic representation of a simple, discontinuous hourglass relaxation oscillator underlying mitosis. Mitogen accumulates linearly during the CDC and triggers mitosis at a critical threshold level. It is then destroyed, the clock is reset, and it reaccumulates again during the next cycle. (B) Diagrammatic representation of the phase-plane trajectory for a continuous biochemical oscillator exhibiting limit cycle behavior. Two interacting substances (X and Y) autonomously fluctuate; mitosis occurs when one of them (Y) attains a threshold level (Y_c). S denotes the point of singularity, which is unstable with respect to small perturbations. Trajectories spiral out to the closed curve, the limit cycle. (Reprinted from Edmunds and Adams, 1981, with permission of American Association for the Advancement of Science. See Tyson and Kaufmann, 1975; Tyson and Sachsenmaier, 1978.)

Relaxation Oscillators

Relaxation oscillations are one type of self-sustained oscillation observed in a system not subject to any noticeable periodic input, characterized by large energy exchanges, an oscillatory range relatively independent of external influence, and a frequency varying with environmental conditions (Klotter, 1960; Wever, 1965b). Typical physical relaxation oscillators are electric condensers and ordinary laboratory pipet washers, as well as sand-filled hourglasses that are turned over repetitively (Fig. 3.12A). At the other end of the scale are dedamped, pendulum-type oscillations, characterized by relatively small energy exchanges, with period virtually independent of external influences but range environment-dependent. Both of these classes of self-sustaining oscillations can be described by Van der Pol differential equations (Wever, 1965b), which have been used also to develop mathematical models for persisting circadian rhythms (Wever, 1965a; Edmunds, 1976). We now consider several hypotheses for mechanisms generating biological periodicities that have invoked relaxation-type oscillators (although one cannot necessarily determine the formal properties of the underlying oscillator from the shape or form of the overt rhythm).

Even though it is unlikely that bacterial cells are able to sense time, the discontinuity in their CDCs represented by cell division is, nevertheless, a well-timed event for individuals comprising the population, and the temporal sequence of the cell cycle can be conveniently described in terms of time-dependencies (Vicente, 1984; see Holland, 1987). Several clocklike processes have been described for *E. coli,* requiring a constant time to be completed at a given temperature and having a duration relatively independent of other events: the 40-min chromosome replication time (Cooper and Helmstetter, 1968; Helmstetter et al., 1968; Kubitschek and Freedman, 1971; Chandler et al., 1975), the 40-min timespan of protein synthesis required for cell division (Pierucci and Helmstetter, 1969), and the action of the *ftsA* gene product (Tormo et al., 1980), which may correspond to the termination protein postulated earlier by Jones and Donachie (1973). Other bacterial cell timers, whose rate depends on other events, such as growth rate [termed metronomes by Vicente (1984)], are represented by the accumulation of a theoretical initiator of DNA replication (Donachie, 1968), increase in cell mass (Dennis and Bremer, 1974), and cell elongation (Donachie et al., 1976; Grover et al., 1977).

Many of these observations were incorporated into a general model for the CDC of *E. coli* (Donachie et al., 1973), as diagrammed in Figure 3.13. This simplified description of the cyclic aspects of the CDC of strain B/rA (growing at 37°C at rates faster than one doubling per hour) embodied at least three empirical rules (Donachie, 1973; for further explanation, see Donachie, 1981): (1) replication is initiated at 2^n origins when cell mass equals 2^nM_i, where n is an integer and M_i is the initiation mass, (2) replication of the entire chromosome requires 40 min, and (3) the bacterial cell divides 60 min after the initiation of replication. The timer hypothesized to underlie the mass doubling time would vary with growth rate and, therefore, according to Vicente's terminology (1984), would constitute a metronome and not a true clock sensu stricto. In contrast, the two other parallel processes (each comprising a series of causally fixed, dependent events), whose rates would not vary with growth rate and which would be independent of each other even though both would have to be completed for division to occur, have been designated "clocks" (Vicente, 1984) or "dependent timer sequences" (Mitchison, 1974). One of these processes would be chromosome replication and the synthesis of the termination proteins, requiring 40 to 45 min to complete. The other would be a sequence of protein synthesis, followed by another process—perhaps, the assembly of a septum precursor—requiring approximately 60 min. At this point, the two independent sequences would converge through an interaction between the termination protein and the septum precursor(s), and septation and division would ensue a few minutes later.

This model has been further generalized by Cooper (1979, 1982a,b,

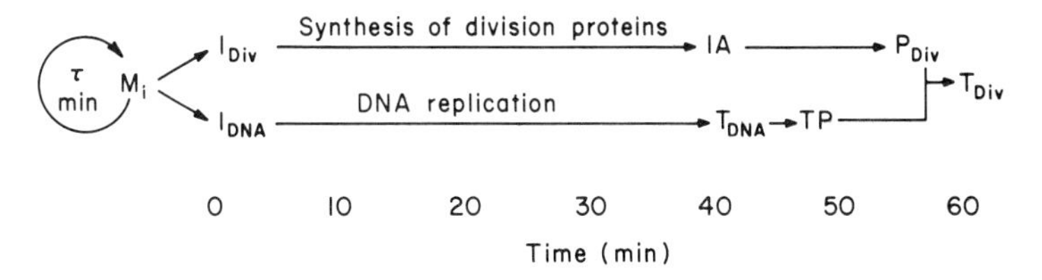

FIGURE 3.13. Model of the cell cycle in *Escherichia coli*. Doubling of the initiation mass (M_i) takes place every mass doubling time (τ min). At each doubling, two processes are initiated approximately simultaneously: DNA replication (I_{DNA}) and the sequence of events leading to division (I_{Div}). Termination of chromosome replication (T_{DNA}) at 40 min induces the synthesis of termination protein (TP). The first 40 min of the division sequence involves protein synthesis, which is then followed by the initiation of assembly (IA). (Assembly could actually start earlier.) After 15 to 20 min more, the cell has reached a stage (P_{Div}) where interaction between some septum primordium and termination protein leads to cell division (T_{Div}). In this view, the timing of the main events in the cycle is, therefore, periodic, occurring at intervals equal to the mass doubling time of the cell and at multiples (2^n) of a constant cell mass (M_i). This event triggers two parallel but separate sequences of events that take constant periods of time to complete, largely independent of the rate of cell growth. One of these processes is chromosome replication and the synthesis of the termination protein and requires 40 to 45 min to complete. The other is a sequence of protein synthesis, followed by another process, which may be assembly of some septum precursor. This sequence requires nearly 60 min, and, at the end of it, there is an interaction between the septum precursor(s) and the termination protein to give the final septum and cell division. The last event takes only a few minutes. (Reprinted from Donachie et al., 1973, with permission of Cambridge University Press.)

1984a,b), whose continuum model attempts to provide a unified description of the CDC of eukaryotes and prokaryotes and is antagonistic to the transition-probability model (Section 3.1.1) and to those that invoke restriction points in G_1 and G_0 and other CDC-specific controls. Its basic premise is that preparations for the initiation of DNA synthesis occur continuously during the CDC and are not confined to any one part of it. When the amount of initiator that has accumulated attains a certain value, it initiates the replication-segregation sequence, comprising the replication of the genetic material and the cell division that follows. Thus, the synthesis of a hypothetical initiator (Fig. 3.14) would occur independently of the events that are initiated—the synthesis of DNA and subsequent cell division. The frequency of division would be determined by the rate of initiator synthesis, whereas the time between the initiation of DNA synthesis and ensuing cell division would be relatively invariant. Division, therefore, would be the end of a sequence of events as opposed to being its start (Cooper, 1984b).

In terms of timer analogies, the continuum model postulates that reg-

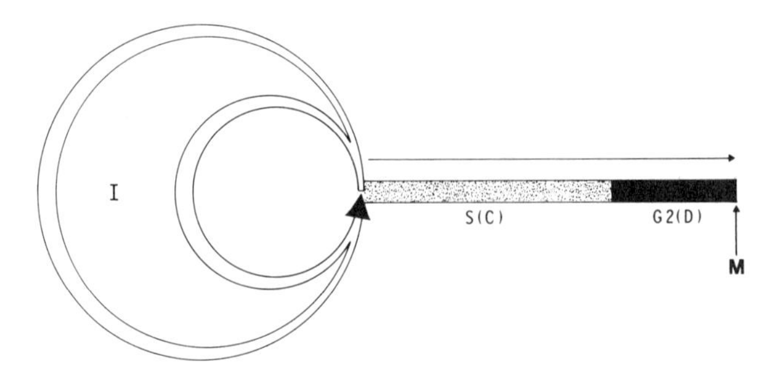

FIGURE 3.14. The continuum model, illustrating the continuous synthesis of initiator (I), which occurs independently of DNA synthesis (S) or cell division (M). (Reprinted from Cooper, 1979, with permission of Macmillan Journals Ltd.)

ulation of division does not occur by an event-determined classical clock, with the events that are rate-limiting for division occurring primarily during G_1, but rather by a continuous clock mechanism (e.g., an autonomous oscillator). Cooper (1984a) cites a cuckoo clock: the singing of the bird at each time interval would announce that sufficient initiator had accumulated to start a special event (e.g., the replication-segregation sequence) resulting in cell division. The frequency of the cuckooing would be determined by the speed of the underlying clock mechanism but would be independent of the duration of the song. The replication-segregation sequence itself would resemble an event-oriented clock (or dependent timer sequence). The manner by which this continuous synthesis of initiator could produce the observed CDCs with their discontinuous division events is illustrated in Figure 3.15. In effect, a smoothly running oscillator has been conjoined with a threshold requirement to generate a pipet washer-(relaxation-)type output.

A combination of timer and size monitoring (see previous subsection) controls also appears to regulate the CDC in *C. reinhardtii* and *Chlorella* spp., both green algae (Chlorophyceae) that divide by multiple fission (Donnan and John, 1983, 1984; John, 1984; McAteer et al., 1985). Both microorganisms show, in midcycle, a commitment to divide without further requirement for growth (Fig. 3.11). The time interval up to this commitment point for division in *C. reinhardtii* is characterized by immediate temperature compensation: a 10°C temperature change (in the range between 20 and 30°C) had only a small effect on the time between the beginning of daughter cell growth and the attainment of the first commitment to divide (Donnan and John, 1983, 1984), even though such a change halves or doubles the rate of most metabolic processes, including growth rate. This sort of temperature compensation is an important characteristic of circadian timers (Sweeney and Hastings, 1960; see Fig. 1.2), even those underlying circadian division rhythmicities in many unicells (Section 3.2.1).

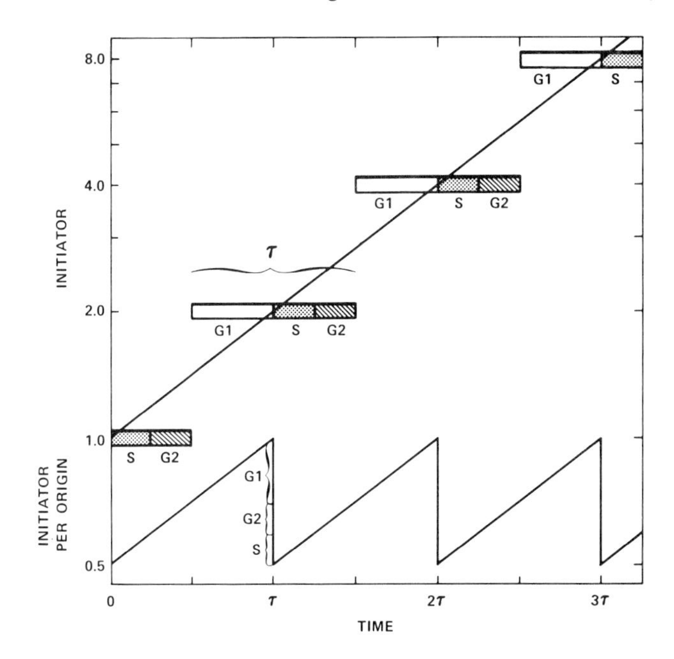

FIGURE 3.15. A formal description of the continuum model demonstrating how the continuous synthesis of initiator can produce the observed cell cycles. The lower sawtooth graph is the amount of initiator per origin, varying from 0.5 to 1.0. The upper boundary is the critical amount required for the initiation of DNA synthesis. At the start the cells have just entered an S period and the cells are progressing through S and G_2. The amount of initiator per origin is increasing from 0.5 to 1.0 over one doubling time (τ). When it reaches 1.0, another S period is initiated, but since at initiation there is a sudden doubling in the number of available origins of replication, the amount of initiator per origin decreases immediately to 0.5. If the rate of initiator synthesis was decreased and S and G_2 remained constant, G_1 would increase. There is no G_1-specific synthesis of initiator, but the synthesis is occurring continuously. As shown in the upper graph, this formal statement of the continuum model does not necessarily visualize any sudden appearance or disappearance of any particular molecules. (Reprinted from Cooper, 1982a, with permission of Plenum Press.)

The commitment timer formally implicated by these results, however, was only partly compensated for slow growth due to low energy supply (available light intensity), running more slowly. There was no evidence that commitment was triggered by attainment of a critical size (Donnan and John, 1984; John, 1984), in opposition to the hypothesis of Craigie and Cavalier-Smith (1982), although the data clearly showed that size itself (and not growth rate) at the time of commitment to divide did determine the division number (the number of rounds of commitment to doublings of cell number). In contrast, the duration of the postcommitment portion of the CDC (from commitment through S and M phases to completion of division) was more temperature-sensitive, exhibiting only slowly adaptive temperature compensation, and was unaffected by external energy supply or concurrent growth.

The circadian interval between divisions in these species of alga, therefore, appears to be determined by the sequential operation of at least two timers (with the sizer's beginning to function after the expiration of the commitment timer). As a consequence, the entire CDC would be stabilized against both temperature and growth rate. Furthermore, provided that minimum requirements of growth rate are met, a subsequent commitment to divide would be the inevitable consequence of a timer process that begins after division and runs through the G_1 phase (see Cooper, 1979). Because the operation of the commitment timer is interrupted by a post-commitment timer with quite different properties, Donnan and John (1984) conclude that both timers are of the hourglass type rather than the classic, autonomous circadian oscillator—involving specific, as yet unknown, timer processes, as opposed to more general metabolic ones—each commencing only after completion of the segment of the CDC that is controlled by the other. This conclusion is supported by the earlier finding by Spudich and Sager (1980), recently confirmed by McAteer et al. (1985), that circadian time was not maintained in newly divided *C. reinhardtii* held in darkness of varying duration (up to 48 h) and returned at intervals to light, the cells' revealing a constant capacity to resume CDCs of normal duration and of normal daughter cell yield.

A final example of possible control of the CDC by a relaxation oscillator, or even by a limit cycle clock, is afforded by the acellular slime mold (class Myxomycetes), *Physarum polycephalum*. In the diploid plasmodial stage of its life cycle, cell division does not occur. Rather, there is a remarkable degree of natural mitotic synchrony among the replicating nuclei (as many as 10^8) within the syncytium, which occurs every 8 to 12 h depending on growth conditions (Rusch, 1970; Sachsenmaier, 1981; Tyson, 1982). This ultradian rhythm of nuclear division is endogenous and relatively temperature-dependent. The CDC of *P. polycephalum* has no G_1 phase (not uncommon among lower eukaryotes), and daughter nuclei commence DNA synthesis immediately after separation. This fact indicates that there is a close coupling between nuclear division and DNA synthesis; indeed, DNA synthesis is obligatory for nuclear division, and the onset of the latter (but not its completion) is necessary before the former can take place. Further, the macroplasmodial protein/DNA ratio at mitosis seems to be under tight control, being quickly restored to normal levels following perturbation of the mitotic cycle by CHX, heat, or ultraviolet light (Tyson et al., 1979). These findings suggest that some sort of feedback exists between size (as measured by protein content) and DNA synthesis. Finally, plasmodia of different stages, or phases, of the CDC can be fused experimentally (as do microplasmodial fragments in nature), their nuclei and cytoplasm intermingle, and the fused pair then undergoes mitosis at some intermediate phase (Rusch et al., 1966; Sachsenmaier et al., 1972). These observations implicate an intracellular substance that accumulates as plasmodia progress through their CDCs and that is averaged out upon their out-of-phase fusion.

How can we account for the experimental data obtained with *P. polycephalum?* If one assumes that the onset of nuclear division and DNA synthesis is regulated by the plasmodia's monitoring the nucleocytoplasmic ratio—a type of cell size control (see preceding subsection)—one has several types of sizer model from which to choose (Fantes et al., 1975; Sudbery and Grant, 1975; Wheals and Silverman, 1982; Tyson, 1984, 1987; Tyson and Sachsenmaier, 1984). In one class of model, the cytoplasmic concentration of some substance is assumed to change monotonically until it reaches a critical threshold value, whereupon nuclear division and DNA synthesis would be triggered (the familiar mitogen-accumulation model). This type of relaxation oscillator includes several variants: (1) an activator is synthesized (or an inhibitor degraded) at a rate proportional to the cytoplasmic mass until a threshold is attained, division is triggered, and an aliquot of activator corresponding to the number of cellular genome equivalents is destroyed (Fantes et al., 1975), (2) an unstable inhibitor is synthesized at a rate proportional to the cellular DNA content, its total amount at all times being proportional to the number of genome equivalents in the steady state and its concentration decreasing during cell growth until it drops below a division-triggering threshold value (Sudbery and Grant, 1975), and (3) an unstable mitotic activator is synthesized at a mass-proportional rate and degraded by linear kinetics, with a rate-constant proportional to the nucleocytoplasmic ratio. The activator concentration (assuming relatively rapid turnover compared to CDC length) would always be inversely proportional to the nucleocytoplasmic ratio and would reach a critical level as growth diluted the nuclei; the nucleocytoplasmic ratio then would double as division occurred, and the activator concentration would abruptly halve (Wheals and Silverman, 1982; Tyson, 1983b).

A second class of model assumes that the subunits of a mitotic structure, either cytoplasmic or nuclear, are synthesized during the CDC at a rate proportional to the cytoplasmic mass until they eventually are assembled, leading to nuclear division and DNA synthesis, and then are either (1) inactivated or (2) disassembled for later reuse. In the former case, the concentration of subunits would be zero at the start of each cell cycle and would then reach a critical value that would effect the completion of the mitotic structure (Fantes et al., 1975). In the latter, the subunit concentration would be constant throughout the cell cycle, but as the cell grows, the number of subunits per genome equivalent would attain a critical value (Sudbery and Grant, 1975), perhaps binding to and gradually occupying corresponding genomic receptors.

Among these various models, the nuclear sites-titration model of Sach-

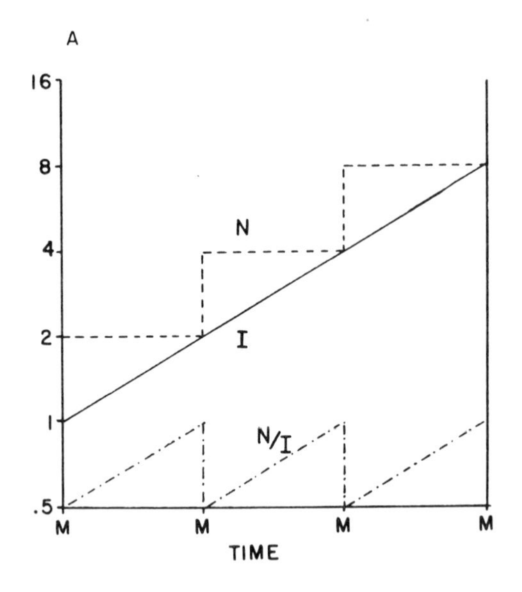

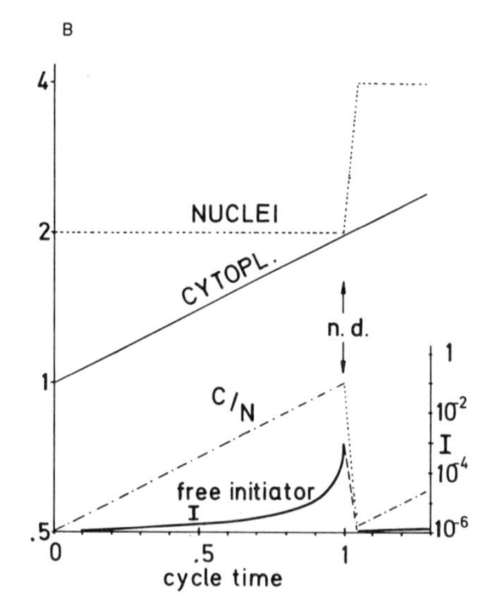

FIGURE 3.16. (A) Schematic diagram of the kinetics of the total mass (solid line) and the number of nuclei (dashed line) of exponentially growing synchronous plasmodia of *Physarum polycephalum*. The ratio of the total mass to the number of nuclei is shown by the dotted–dashed line. N, nuclear receptor sites; I, initiator; M, synchronous nuclear mitoses. (Reprinted from Sachsenmaier et al., 1972, with permission of Academic Press.) (B) Titration-model of the mitotic oscillator. Computer simulation of the expected changes of free mitotic initiator concentration (I) during nuclear division cycle. Total initiator accumulates proportional to the increase of cell mass (i.e., cytoplasm) and combines stoichiometrically with a given number of nuclear sites. The fraction of free initiator is calculated assuming a dissociation constant $K = 10^{-6}$ of the complex between initiator and nuclear

senmaier et al. (1972) is perhaps one of the more widely accepted (Fig. 3.16). This model characterizes the timing mechanism of nuclear mitosis in *P. polycephalum* as an hourglass, or discontinuous, extreme relaxation oscillator in which mitotic initiator molecules—an unstable activator, most likely a protein—are formed more or less continuously and proportionately to the increase in plasmodial mass during G_2 (Sachsenmaier, 1976). These molecules would be counted by combining tightly and stoichiometrically with a given number of nuclear receptor sites, the number of sites per nucleus being fixed and the total number proportional to the DNA content of the plasmodium. The concentration of free cytoplasmic activator is small until all the nuclear sites are occupied, at which time it would increase abruptly (as does the titration curve of a strong acid with a strong base) and uniformly throughout the plasmodium. Mitosis then would be triggered in all the nuclei simultaneously at this critical ratio of initiator to nuclei (i.e., nucleocytoplasmic ratio). The clock would be reset by each nuclear division (or by a related, obligatory event), at which time the number of nuclear sites would also double in a stepwise manner (Figs. 3.12A, 3.16). In this model, the event of mitosis is an essential feature of the oscillatory system (which distinguishes it from the limit cycle hypothesis discussed in the next subsection).

The nuclear sites-titration model is consistent with the data obtained from the experiments with plasmodial fusion alluded to previously. Thus, because the time until the following mitosis is determined in effect by the fraction of unfilled sites in a plasmodium, fused plasmodia should enter mitosis at an intermediate time reflecting the average of the fractions of unfilled sites in the individual plasmodia before fusion—as, in fact, they do (Rusch et al., 1966; Sachsenmaier et al., 1972; see Fig. 3.19).

Perturbations affecting mitotic timing in plasmodia have been used to distinguish between the nuclear sites-titration model and the others outlined earlier (Tyson, 1984; Tyson and Sachsenmaier, 1984). Thus, 30-min to 90-min pulses of CHX (10 μg/ml), a potent inhibitor of protein synthesis in *P. polycephalum*, given at progressively later times during the mitotic cycle and terminated by transferring the plasmodia to fresh medium, generated increasingly longer delays of mitosis (Scheffey and Wille, 1978;

---

sites. Note that most of the initiator is bound to the nuclei during interphase. Free initiator concentration increases steeply at the end of the cycle, analogous to the jump of pH at the endpoint of a base–acid titration. This signal may function as a trigger of mitosis. The number of nuclear sites doubles stepwise during or immediately after nuclear division. The concentration of free initiator drops by combining with the newly exposed nuclear sites and remains at a low level until the end of the next cycle. *Abscissa:* Cycle time; n.d.; nuclear division. *Left ordinate:* Relative amount of cytoplasm and number of nuclei; C/N, ratio of cytoplasm/nuclei. *Right ordinate:* Fraction of free initiator (I) in the cytoplasm. (Courtesy of Drs. W. Sachsenmaier and R. Exenberger, 1985.)

Tyson et al., 1979). The more-or-less linear dependence of the excess phase delay (mitotic delay in excess of the duration of the treatment itself) can be interpreted as evidence for the synthesis of an unstable mitotic activator-protein that accumulates gradually during the cycle. It presents difficulties, however, for the unstable inhibitor model, because if the inhibitor were a protein, it would be rapidly degraded during the CHX treatment, and the next mitosis would be advanced rather than delayed. Similarly, the initial delays produced by heat-shock (Brewer and Rush, 1968; Tyson et al., 1979) and by ultraviolet (UV) irradiation (Sachsenmaier et al., 1970; Sudbery and Grant, 1975; Tyson et al., 1979) presumably would interfere with heat-labile and UV irradiation-sensitive components of the nuclear division mechanism. It is noteworthy that perturbation with UV irradiation (as well as by heat-shock) caused a phase advance (shortening) of the next intermitotic time interval (probably because the dose-dependent destruction of a fraction of the nuclear population during the S phase resulted in an abnormally small nucleocytoplasmic ratio at the end of S, causing the critical level of mitotic effector substance to be reached more quickly than normally would be the case). Ensuing intermitotic intervals, however, displayed normal values. Whereas these results are consistent with the models invoking unstable inhibitors, unstable activators, or reusable subunits, they are inconsistent with simple concentration and inactive subunit models, both of which predict that this acceleration should be spread over several cell cycles (Fantes et al., 1975; Sudbery and Grant, 1975).

The assumption that the timing of mitosis in *P. polycephalum* is controlled by the accumulation of a mitotic activator is supported by experiments attempting to induce premature nuclear division in a plasmodium by treating it with extracts from another plasmodium further ahead in its own division cycle. For example, Oppenheim and Katzir (1971) found that mitosis in an early G_2 plasmodium could be advanced by 65 min relative to a control when extracts from a late G_2 plasmodium were applied to its surface. The phase-advancing activity of the extracts was nondialyzable and heat-labile, suggesting the presence of a protein. Further, proteins precipitated from a late G_2 phase extract with rabbit antibodies against proteins from the S phase dramatically accelerated mitosis in recipient plasmodia (Blessing and Lempp, 1978). Loidl and Gröbner (1982) recently have confirmed these earlier observations, finding that the mitosis-advancing capacity of extracts is heat- and ammonium sulfate-precipitable, nondialyzable, and pronase-sensitive.

These results, then, collectively implicate a mitotic activator-protein that is synthesized in G_2. Bradbury et al. (1974) have suggested histone kinase (or its activator) as a candidate for this cytoplasmic initiator. The application of H1 histone phosphokinase, prepared from Erlich ascites cells, to mid-G_2-phase plasmodia advanced mitosis by as much as 40 min. This effect could be correlated with fluctuations in the level of endogenous enzyme activity, which peaked about 2 h before to metaphase. Thus, histone phosphorylation might constitute a relaxation oscillator underlying the nuclear division cycle in *P. polycephalum* (Matthews et al., 1976), with histone kinase's and phosphorylated H1 histone's being formally analogous, respectively, to the mitotic activator and the fully titrated nuclear sites of the model of Sachsenmaier and co-workers. Other stimulators must also exist, however, because mitosis can be advanced in plasmodia treated with histone kinase before metaphase but some time after the peak in endogenous enzyme activity has occurred (Blessing and Lempp, 1978).

The fact that both the rate of tubulin synthesis (Laffler et al., 1981) and the level of tubulin mRNA (Schedl et al., 1984) increase 40-fold in the 2 to 3 h just before to mitosis in *P. polycephalum* suggests that microtubular proteins may be part of a mitotic oscillator. This hypothesis is supported by the findings that (1) microtubular inhibitors delay the timing of both mitosis and the premitotic accumulation of thymidine kinase (Eon-Gerhardt et al., 1981) and (2) heat-shock delays mitosis but not tubulin synthesis, the tubulin clock would continue to run independent of the heat-sensitive event (Carrino and Laffler, 1985). The α-tubulin and histone H4 genes of *Physarum* both constitute models for CDC-regulated gene expression (Laffler and Carrino, 1986). In fact, their transcription begins at the same point in the CDC (Carrino and Laffler, 1986).

Limit Cycle Oscillators

The relaxation oscillator model for nuclear division in acellular slime molds discussed in the preceding subsection is attractive in its relative simplicity and ability to account for many aspects of cell cycle behavior. Nevertheless, it has significant deficiencies, with perhaps the most serious problem's arising from the assumption that mitosis (or DNA replication) is a fundamental component of the underlying clock and is required for its cycling (Shymko et al., 1984). That this is not always the case is indicated by observations in a number of systems: the unicellular green alga *Chlorella* can undergo several rounds of cell division without DNA synthesis' occurring (Schmidt, 1966); the temperature-sensitive CDC mutant of *S. cerevisiae,* defective in the *cdc* 4 gene required for the initiation of DNA synthesis, undergoes bud emergence repeatedly in the absence of DNA synthesis, nuclear division, and cytokinesis (Hartwell et al., 1974), and cell-free extracts of sea urchin embryos may exhibit cyclic variations in protein synthesis, having a τ approximately equal to that of the CDC in the intact organism. Likewise, any model should take into account another general property of cell cycles—their stability—as reflected by the fact that cells return to their normal CDC even after large perturbations. This type of behavior can be described by the stable cycles found in nonlinear dynamic systems, the simplest form of which is the limit cycle (Pavlidis, 1973; Winfree, 1980).

The simple, pendulum-type oscillations (referred to in the previous subsection) represent conservative oscillations, whose amplitude and frequency characteristically depend on the initial conditions (displacement, velocity). Plots of the angular velocity as a function of the angle of deflection yield a phase plane filled with a nested continuum of periodic orbits, each curve of which describes the motion of the pendulum for a different set of initial conditions (at $t = 0$). Slight changes in the mechanistic assumptions often destroy the periodic nature of the oscillations (as illustrated, for example, by the damping observed in a pendulum if friction is taken into account), making it difficult to model mechanistically uncertain biological rhythms. The more robust limit cycle, however, is different from these conservative oscillations. It is an oscillation whose amplitude and period do not depend on initial conditions and is relatively stable to disturbing perturbations, with the system's eventually returning to its original amplitude and frequency (Fig. 3.12B). In response to light pulses of both very weak and strong intensity, most (if not all) circadian rhythms behave phenomenologically like a single, quickly recovering limit cycle (Pittendrigh, 1965), although it is possible to imagine nonlimit cycle models that could explain the data equally well (Winfree, 1975, 1980).

An oscillator (including the circadian oscillators discussed in Section 3.2) can be expressed mathematically as a set of differential equations comprising both state variables and parameters [see Tyson et al. (1976) for a succinct treatment]. The state variables characterize the state of the oscillation, with each set of values defining each phase of the oscillation. The parameters are constants constraining the manner in which the state variables change and determining the dynamics of the oscillation. A different set of parameter values gives a different solution of the rate equations. (It is important to realize, however, that this distinction among state variables, parameters, and hands is model-dependent and a function of its complexity.) For certain fixed parameter values, the differential rate equations have a limit cycle solution. Any transitory alteration or perturbation of either the state variables or the parameters can cause a permanent phase shift in an overt rhythm (i.e., the hands of the clock) but has no permanent effect on its period. In contrast, permanent changes in the parameter values can alter the steady-state period of the oscillation.

Limit cycle models have long been attractive to those analyzing the control of the CDC. More than 55 years ago, Rapkine (1931) hypothesized that autonomous, periodic, biochemical fluctuations might provide a signal for mitosis and cell division in proliferating cells. In brief, he believed that the appearance and subsequent disappearance of the spindle and asters of the mitotic apparatus resulted from a reversible polymerization and denaturation of proteins, involving oxidation-reduction cycles of glutathione (the reduced form of which was known to peak before division during cleavage of sea urchin eggs) and a protein. Polymerization of mitotic apparatus protein would result from the glutathione-mediated conversion of intramolecular sulfhydryl groups (—SH) to intermolecular disulfide linkages (—SS—). The breakdown of the spindle would then result from the reconversion of the latter by the intracellular reduced form of glutathione.

Subsequent work in various laboratories generally has confirmed Rapkine's original notion that these cyclic changes in thiol content could represent a periodic source of mitosis-initiating signals. Thus, Mano (1970, 1971, 1975) has isolated and partially reconstituted an autonomously oscillatory system of protein synthesis in cell-free extracts of sea urchin embryos, which brings about cyclic changes in —SH groups having a period corresponding to that of the normal CDC of the blastula. Further, theoretical analysis by Sel'kov (1970) of experimental data on the kinetics of thiol metabolism has shown that under certain conditions the —SH-controlling system has several alternative stationary states, some of whose surrounding trajectories are self-oscillatory in nature. Sel'kov (1970) was able to model the system using two differential equations, one for the concentration of —SH and another for that of —SS—. Certain solutions yielded limit cycles (Fig. 3.17) that could recover their original form and period even after severe perturbations. The transition of the system from

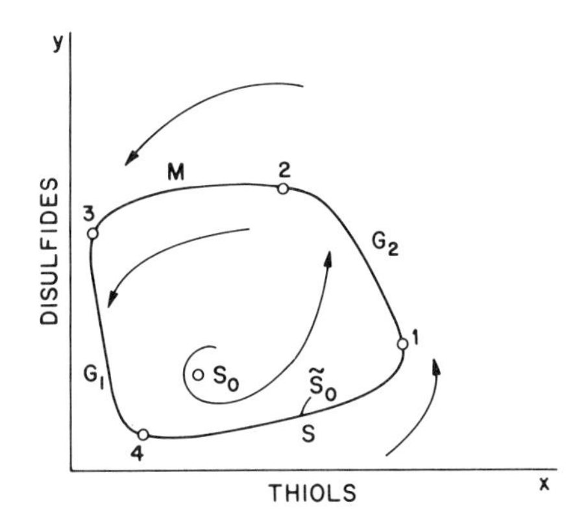

FIGURE 3.17. A limit cycle model of the cell cycle. The concentrations of thiol groups (—SH) and disulfide groups (—SS—) are imagined to oscillate indefinitely around a closed limit cycle (S_0) in the phase plane. Different portions of the limit cycle are identified with different phases of the cell cycle. Portion 1–2 represents preparatory phases of mitosis (G_2), portion 2–3 represents the mitotic phase (M), portion 3–4 represents the resting phase (G_1), and portion 4–1 represents the transient period of DNA replication (S). Perturbations away from the limit cycle are damped out as the system returns to the limit cycle along trajectories like the four curves with arrowheads. Inside the limit cycle is a resting point, or stationary state (S_0). If a perturbation lands the control system on S_0, oscillations in [SH] and [SS] are eliminated, and, presumably, the cell cycle is halted indefinitely. (Redrawn from Sel'kov, 1970, with permission of Vsesojuznoje Agentstvo po Avtorskim Pravam, Moscow.)

one stationary state to another, induced by some environmental perturbation, could lead to profound changes in cellular physiology. The CDC would be regulated by the normal functioning of the underlying oscillator (low division frequency and mean thiol level), quiescence would correspond to the system's resting on one of these stationary points, and malignant growth and carcinogenesis might be triggered if the control system switched to another oscillatory solution (high frequency of division, high mean thiol level)(Gilbert, 1974b, 1977, 1978a, 1980, 1981; Section 6.3.2). This system might represent a pacemaker for circadian biological clocks (see next subsection; Section 3.2; Section 5.2).

Given the same set of experimental facts for the mitotic synchrony observed in the plasmodia of *P. polycephalum* (see preceding subsection), Kauffman and Wille (1975) have suggested an alternative to the nuclear sites-titration, relaxation oscillator model proposed by Sachsenmaier et al. (1972)—namely, that the mechanism that controls the timing of nuclear division and DNA synthesis in this slime mold is a continuous limit cycle oscillator, analogous to those proposed for circadian rhythms, in which the time-dependent variables change smoothly. On this hypothesis two (or more) interacting components (X and Y) fluctuate autonomously, and mitosis would be triggered if one of them (an initiator) reached a threshold level (Fig. 3.12B). The model, analyzed in detail by Tyson and Kauffman (1975), postulates the synthesis of a mitogen precursor (X) at a constant rate throughout the cell cycle, which, in turn, is enzymatically converted at a constant rate to an active form (Y). This active form of mitogen autocatalyzes more of itself from X at another rate and decays (i.e., is degraded or inactivated) with first-order kinetics. The concentrations of X and Y both would exhibit sustained oscillations, and mitosis would be triggered when the level of Y exceeded a critical concentration (Y_c), as diagrammed in Figure 3.18A.

The critical distinctions between this model and the hourglass model of Sachsenmaier et al. (1972) are that in the limit cycle (1) the underlying oscillator runs independently of the events it is timing—mitosis does not function as an essential component of the oscillator system, and, thus, under certain conditions, the system may continue to oscillate at a subthreshold level even if mitosis is blocked and does not occur—and (2) although the amplitude and period do not depend on initial conditions (i.e., the limit cycle is stable, resisting, and recovering from most perturbations), a critical perturbation given at a singularity point (Winfree, 1970) results in a phaseless, timeless, motionless state (Fig. 3.18B).

These divergent predictions of the two types of model (difficult to examine directly as long as the identity of X and Y remains unknown) have been tested in detail by heat perturbation and plasmodial-fusion experiments (Wille, 1979; Tyson, 1982, 1984; Tyson and Sachsenmaier, 1984). It was assumed that heat-shocks destroyed both X and Y equally at a rate proportional to their concentrations and the duration of the perturbation

(Kauffman and Wille, 1975, 1976; Wille et al., 1977). A heat-shock, then, would displace the system off the limit cycle trajectory toward the origin ($X = Y = 0$), as shown in Figure 3.18B for five different phases. As is characteristic of limit cycles, phase shifts ($\Delta\phi$) in the overt rhythms (here, mitosis) presumably controlled by the oscillator subsequently are generated, whose sign and magnitude are a function of the time at which the pulse was applied and of the strength and duration of the pulse. Thus, a $+\Delta\phi$ of the next mitosis occurs if shocks are given shortly after mitosis, whereas progressively increasing phase delays ($-\Delta\phi$s) are produced for shocks given at 2 h after mitosis and thereafter until the late G_2 phase (driving the system in roughly the reverse direction to its normal progression around the cycle), at which time delays diminish up until mitosis. Perturbations of short duration applied just before mitosis displace the oscillation toward S (the nonoscillating, phaseless, steady-state, singularity point) and leave it in a state that is followed by a trajectory that crosses the critical threshold Y_c within the unperturbed limit cycle trajectory (Fig. 3.18A). Because this temporary trajectory leading to the next crossing of Y_c is shorter than the normal path, plasmodia receiving such a heat-shock should exhibit a shortened cycle (i.e., $+\Delta\phi$) immediately after the perturbation before resuming the limit cycle trajectory. In contrast, other trajectories within the limit cycle, such as those produced by longer heat shocks given late in the G_2 phase (but not in early G_2, whose oscillation is on the side of the limit cycle closest to the origin), may require two or three full cycles as they spiral outward before they cross Y_c again (see shaded area of Fig. 3.18A), leading to the prediction that the plasmodia may skip mitosis altogether. Finally, even stronger perturbations applied late in G_2 should drive the system past the singularity and completely across the limit cycle in the direction of the origin, where it will be near the normal limit cycle trajectory, and the $\Delta\phi$ will now become some fraction of a normal cycle duration. As the strength of the heat-shock increases, the delay before the trajectory next crosses Y_c should first jump from about 2 to 3 h to about 13 h, then to about 27 h (considerably more than one cycle length) as the system is driven well past S, and finally should drop to about 9 h (Wille, 1979).

Although the results of experiments using either short-duration (Kauffman and Wille, 1975) or long-duration (Wille et al., 1977) heat-shocks have confirmed many of the predictions of the limit cycle model, they do not demand it and are consistent with relaxation-oscillator models (such as the nuclear sites-titration scheme discussed in the previous subsection), which similarly postulate a heat-labile component of the nuclear division mechanism. Indeed, Tyson and Sachsenmaier (1978) have criticized these data on the grounds that prolonged exposures to high temperatures have a drastic effect on plasmodial growth. Immediately after a 1-h heat-shock, for example, the protein/DNA ratio is abnormally high and does not decline to normal levels until well after the first full postshock synchronous mi-

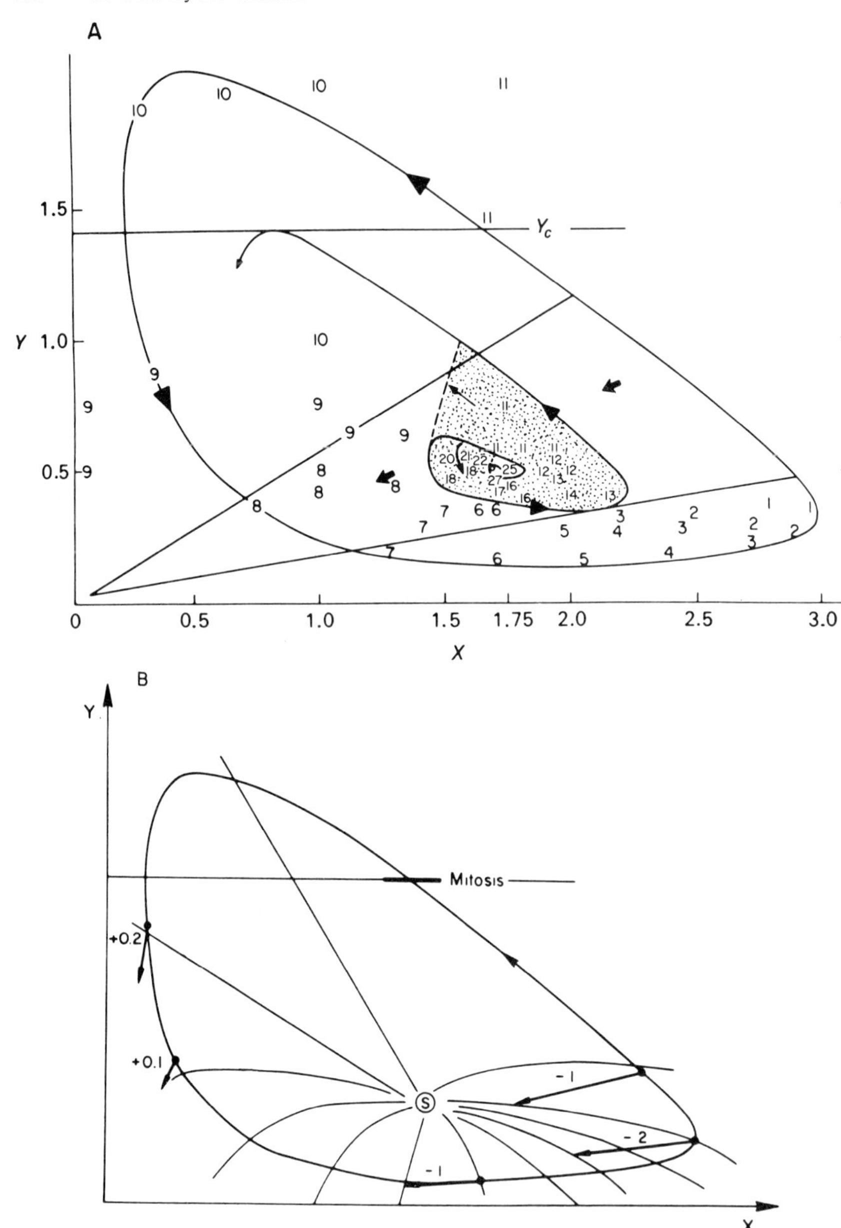

FIGURE 3.18. Limit cycle model for nuclear division in the acellular slime mold *Physarum polycephalum*. (A) Plot showing the time until the next crossing of Y_c from any state inside the XY limit cycle oscillation. The shaded region shows the concentric ring of states centered around the steady state, S, of the oscillations, which are more than one or two cycles away from their next crossing of Y_c. (Reprinted from Wille et al., 1977, with permission of Company of Biologists Limited.) (B) The effect of heat perturbations on the limit cycle and nuclear mitoses, assuming destruction of X and Y. A critical level of Y triggers mitosis. The nearly radial lines emanating from a point inside the cycle are isochrons, separating equal in-

tosis. Thus, one cannot conclude that a central, independent timer is the only mechanism involved; if it were, the perturbations would act directly and instantaneously on the control variables (Wille, 1979). Likewise, there is no clear-cut evidence that heat-shock actually causes skips in nuclear division or DNA synthesis. In fact, heat-shocks of any duration imposed shortly before metaphase appear to prevent nuclear separation at telophase, even though DNA synthesis proceeds normally, producing polyploid nuclei (Brewer and Rusch, 1968; Kauffman and Wille, 1975; Tyson and Sachsenmaier, 1978; Wolf et al., 1979).

Experimental fusions of out-of-phase plasmodia also have been used to test the limit cycle hypothesis in *P. polycephalum* (Kauffman and Wille, 1975; Wille et al., 1977; Loidl and Sachsenmaier, 1982), although, once again, the evidence obtained thus far is not definitive (Tyson and Sachsenmaier, 1978, 1984; Wille, 1979; Tyson, 1984; Shymko et al., 1984). Assuming that all pertinent variables mix freely, the simplest interpretation is that the fused plasmodium will adopt a single state, which then would spiral outward to the limit cycle trajectory. According to either the continuous limit cycle or the relaxation oscillator model, cells would be expected to synchronize to an intermediate, compromise phase biased toward the larger mass. These two models, however, have divergent predictions with regard to the details of the phase compromise behavior after fusion (Fig. 3.19).

For the limit cycle (in contrast to the hourglass), the fusion state generally would not be on the normal cycle trajectory but rather somewhere within the cycle, leading to the possibility of subthreshold oscillations and attendant skipping of mitosis or unusually long delays. For example, if a plasmodium just about to enter mitosis were fused with one that had just completed mitosis, the relaxation-oscillator model would predict that the fusion product should undergo nuclear division after about a half-cycle delay, whereas according to the relaxation oscillator model, the fused nuclei should divide either immediately or after a delay of full cycle. Experimental results seem to favor the former (Tyson and Sachsenmaier, 1978), that is, the relaxation-oscillator scheme, which has a discontinuous jump at some phase in the cycle, successfully predicts that if two plasmodia having phases on opposite sides of the jump—no matter how close—are

---

tervals of time along trajectories and along the limit cycle path and indicate hours before mitosis. All isochrons meet at the steady-state singularity, S, inside the limit cycle. The effect of 20% destruction of X and Y from five phases is shown as destruction vectors (arrows) that indicate the direction and magnitude in delay (−) or advance (+) of the next mitosis. Movement is counterclockwise in this model. Heat early in the cycle causes advances indicated by positive numbers, and delays are produced later in the cycle. Delay reaches a peak, then declines as mitosis is approached. (Reprinted from Wille, 1979, with permission of Academic Press; see Kauffman and Wille, 1975.)

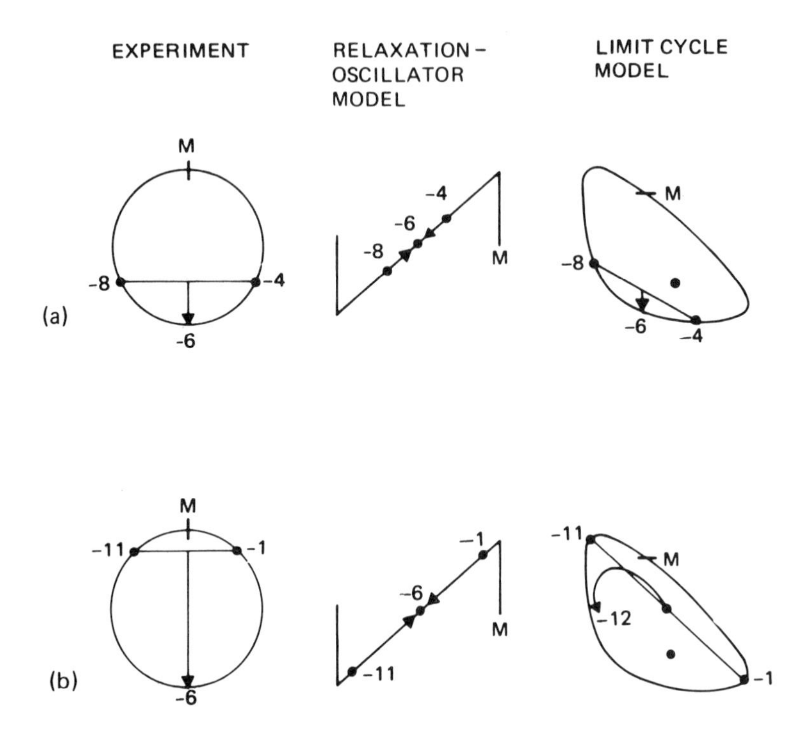

FIGURE 3.19. Divergent predictions of a continuous limit cycle model and a relaxation oscillator model. (a) If plasmodia 4 and 8 h from metaphase are fused, the mixed plasmodium undergoes synchronous nuclear division about 6 h after fusion. This result is predicted by both models of the mitotic control system. (b) If a plasmodium about 1 h before metaphase is fused with a plasmodium that has recently gone through metaphase, the mixed plasmodium undergoes nuclear division about one half to three quarters of a cycle after fusion. This is consistent with the relaxation oscillator model, but the limit cycle model predicts a full-cycle delay. (Reprinted from Tyson, 1982, with permission of Academic Press.)

fused, the phase of the fused plasmodial pair should be much different from the nearly common original phase of the constituent plasmodia, in contrast to the continuous limit cycle model (Sachsenmaier, 1981). There does not appear to be the unusual variability in compromise phases in reported data from fusion experiments (Chin et al., 1972; Sachsenmaier et al., 1972) that would be expected on the limit cycle hypothesis for pairs in which the averaged variables lie on or very near the steady-state singularity (S), in whose immediate vicinity are points that spiral out to all possible phases on the limit cycle trajectory (Tyson and Sachsenmaier, 1978).

In summary, although the experimental evidence obtained thus far does not rigorously exclude either the limit cycle hypothesis or the relaxation-oscillator model as an explanation for the mitotic synchrony observed in the plasmodia of *P. polycephalum*, the scales at present appear to be tipped in favor of the latter.

A biochemical, limit cycle oscillation structurally similar to the prokaryotic model of Helmstetter et al. for the bacterial CDC (see Section 3.1.2) has been proposed to underlie that of the yeasts *S. cerevisiae* and *Candida tropicalis* (Novák and László, 1986). This model, based on the periodic fixation of CO_2 observed during the CDC of synchronously proliferating yeast populations, posits that the start signal occurs when metabolic pools are filled as a result of the autocatalytic accumulation of tricarboxylic acid cycle intermediates produced by CO_2 fixation during catabolism. Thus, the correlation between critical cell size and the initiation of the CDC would be only indirect, and intracellular CO_2 concentration would be one of the mediators. The inhibition of diffusion by cell growth would result in an increase in CO_2 concentration, which in turn would enhance the filling of the metabolic pools and the start of another mitotic cycle. Simulation experiments supported this decrease in τ of the oscillator, agreeing with the observation that the bigger the cells are, the shorter their cycle time is. Finally, similar changes in the rate of CO_2 production have been reported in synchronous cultures of the fission yeast *S. pombe*. This periodic cell cycle event persisted even if the DNA division cycle (Novák and Mitchison, 1986) or protein synthesis (Novák and Mitchison, 1987) was blocked. It is interesting that the step changes in the activity of nucleoside diphosphokinase observed during the CDC of this species also continue after a block to the DNA division cycle (Creanor and Mitchison, 1986; see Section 5.2.4).

Quantal Cell Cycles

A somewhat different variant of the limit cycle oscillator is the quantal cell cycle, hypothesized by Klevecz (1976, 1978) to have a period of approximately 3 to 4 h and to be a reflection of a basic, central cellular clock. This model arose from time-lapse videotape microscopic observations of the generation times of mitotically selected cells taken from either synchronous or randomly dividing mammalian cell cultures. Rather than exhibiting a smooth, continuous distribution (with values of *g* perhaps being skewed toward longer values), generation times have a tendency to be clustered, or quantized, at intervals of roughly 4 h. For example, rapidly growing, Chinese hamster V79 lung fibroblasts often show a bimodal distribution of *g* values, with one peak's occurring at about 8.5 h and another, smaller one at 12 h (Fig. 3.20). At temperatures above the normal culture temperature of 37°C, V79 cells had a modal CDC about 7.5 to 8.5 h in duration, whereas at lower temperatures (33.5 to 36.5°C), the modal CDC was 12 h long.

In a survey of the literature, Klevecz (1976) found a similar quantizement of generation times for a variety of mammalian cell types (Fig. 3.21); the distribution of *g* formed a series with a greater range of values than that

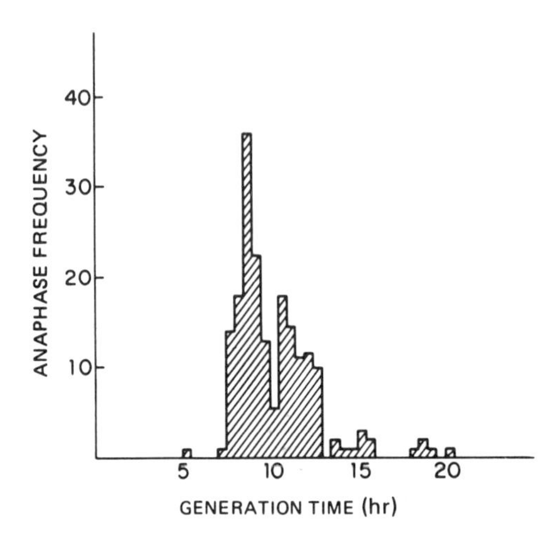

FIGURE 3.20. Generation time distribution for unperturbed V79 cells, showing clustering of generation times at intervals of about 4 h. (Reprinted from Klevecz, 1976, with permission of the author.)

observed for the V79 cell line alone. To satisfy this data, Klevecz proposed a quantal subcycle, G_q, which would have a traverse time equal to the period of an underlying limit cycle oscillator (apparently fixed at close to the same value in all mammalian somatic cells). This quantal cycle would be the fundamental cellular clock and, by Klevecz's scheme, could be visualized as a second cycle probably (but not necessarily) appended to the deterministic sequence, $S + G_2 + M$, of the classic CDC at a point *i* (Fig. 3.21, inset). In a sense, it would drive the CDC much as a small gear drives a larger one. Further, the timekeeping mechanism appears to be temperature-compensated, as a good clock should be (Fig. 1.2). The time required to traverse G_q is more or less constant over a temperature range of 6 to 7°C (Klevecz, 1976). In a more recent study, Klevecz and King (1982) found that the Q_{10} for cell division in the V79 line growing between 33 and 40°C was between 1.15 and 1.26, that the duration of the CDC increased more rapidly with decreasing temperature only at temperatures approaching the limits of reproductive viability, and that this increase in *g* tended to occur discretely in a quantal fashion (with $\tau \cong 4.6$ h). These data suggest that the increase in cell cycle time at lower temperatures (or lower serum concentration or high cell densities) is the result of an increase in the number of rounds of traverse of G_q.

Klevecz's quantal cell cycle model (1976) has the virtue of explaining the variability commonly observed in the duration of the G_1 phase of many cell types (and, hence, of the generation time of individual cells) as arising from the gated entry of cells into the S phase. It gains credence from two additional sets of observations. (1) Synchronous V79 cells perturbed by serum (Klevecz et al., 1978), heat-shock (Klevecz et al., 1980b), or ionizing irradiation (Klevecz et al., 1980a) at 0.5-h intervals across their 8.5-h CDC display a biphasic, 4-h PRC comprising both advances and delays in sub-

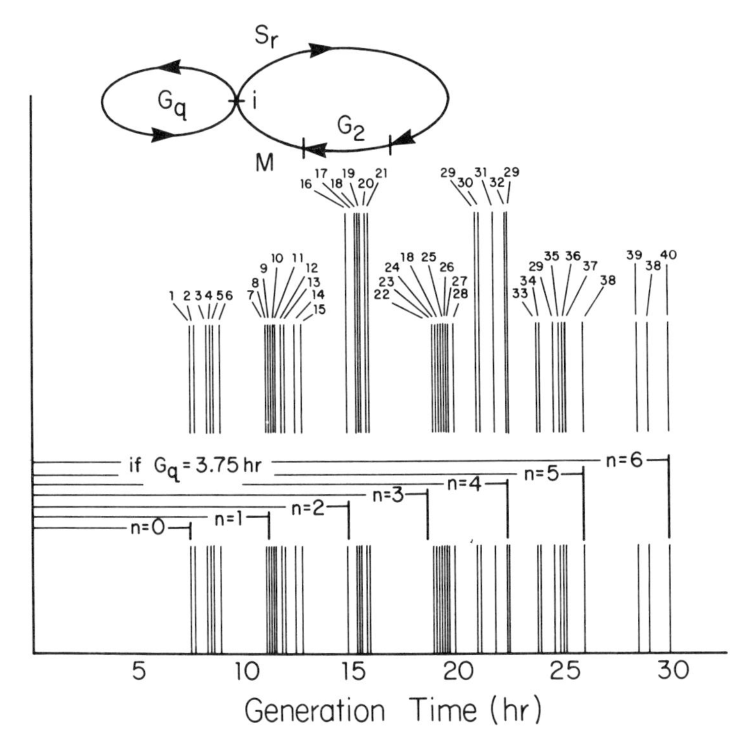

FIGURE 3.21. Quantized variation in generation times of mammalian cell lines. Generation times were determined from the published data on cells synchronized by mitotic selection or from time-lapse cinematography of random cultures. The list is not exhaustive but represents a sampling of papers published between 1961 and the present in which the stated generation time could be directly confirmed in the data. Wherever possible modal generation times were obtained, that is, the time of maximal mitotic index rather than the center of mass of the mitotic wave. Numbers above each line refer to individual published data. The calculation of possible generation times uses the simple expression $T_g = nG_q + 2G_q$. To obtain a fit to the data, values for the probability of emergence from G_q and the density function describing dispersion will have to be determined. T_q = generation time in hours. G_q (quantized G) = incremental increases in generation time in hours. (Reprinted from Klevecz, 1976, with permission of the author.)

sequent cell divisions (Fig. 3.22). These results suggest that the perturbing agent affected the clock directly, perhaps by the destruction of most of the macromolecules constituting the putative limit cycle oscillator. (2) The activity of a number of enzymes (e.g., lactate dehydrogenase) that have no obligatory relation with other periodic events, such as DNA synthesis, oscillates also with periods of 3 to 4 h, even if DNA and RNA synthesis are inhibited (Klevecz, 1969a,b). Further, these oscillations—presumably also manifestations of the underlying ultradian clock—appear to involve

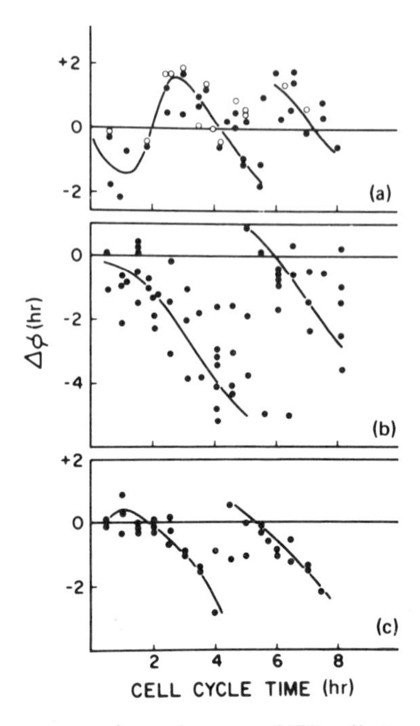

FIGURE 3.22. Phase-response of synchronous V79 cells to perturbation by serum, heat-shock, and ionizing irradiation. (a) Serum pulses: At intervals following mitotic selection, serum concentration in the medium was increased from 5 to 20%; midpoints of first (solid circles) and second (open circles) mitotic waves. (b) Heat-shock: Midpoints of the first mitotic wave after synchronization and a 10-min heat-shock (45°C) are compared for each pair of heat-shocked and control cultures as described in (a). (c) Ionizing irradiation: Synchronous V79 cells were exposed to 150 rad from a cobalt-60 source at 30-min intervals through the first synchronous cell cycle. Analyses of division advance or delay were determined by time-lapse videotape microscopy. (Reprinted from Klevecz et al., 1980a, with permission of Alan R. Liss, Inc.)

protein synthesis and degradation as well as modulation and are damped, but not obliterated, if DNA and RNA synthesis is inhibited.

Certain parallels can be drawn (Shymko et al., 1984) between the quantal cell cycle model and the transition-probability model (Smith and Martin, 1973; see Section 3.1.1). Although the latter makes no predictions about PRCs or the results of cell fusion (as with the acellular slime molds discussed in the preceding subsections), it does attribute the variability observed in generation times to the varying amounts of time that cells spend in the indeterminate A-state (Fig. 3.4). The G_q quantal limit cycle of Klevecz (1976) neatly accounts for the quantizement of this variation into discontinuous packets at nodes separated by about 3 to 4 h in many mammalian cell lines. If random fluctuations are added to the normal movement of the cells around the limit cycle and if a threshold crossing is required for gating into the next stage of the CDC, a part of the cloud representing a cell population might fail to cross the threshold and advance in the CDC. Further, if the random fluctuations are sufficiently large, a cell's position within the cloud during a given cycle will be uncorrelated with that one cycle later. Thus, the probability of a cell's attaining threshold would be constant per cycle. This system not only would generate the overall distribution of generation times that is characteristic of the transition probability model, as reflected in the right-skewed, negative-exponential tail of the α-curve (plot of the fraction of undivided cells as a function of cell age; Fig. 3.3), but also would reflect (Fig. 3.23) the clustering of cell cycle times at intervals corresponding to the period of the limit cycle (Klevecz et al., 1980a; Shymko and Klevecz, 1981). In principle, the random component (G_q) could be placed anywhere within the CDC and the limit cycle period could be of any duration.

The quantizement effect (Fig. 3.20) and the biphasic PRCs obtained from the phase perturbation experiments with V79 cells (Fig. 3.22) suggest a two-loop (or more) limit cycle model with a threshold crossing required for gating from one loop to the next (Klevecz et al., 1980a; Shymko and Klevecz, 1981; Shymko et al., 1984). (The greater variability associated with G_1 could be explained if the threshold associated with the first loop were higher than that for the second.) The existence of such a multiloop limit cycle nested within a single CDC has very different implications than do models invoking only one, such as the mitotic oscillator proposed for *P. polycephalum* (Kauffman and Wille, 1975—but see Wille, 1979, pp. 139–142). If a single oscillator cycled more than once each CDC, an additional marker would be required to determine the loop into which the

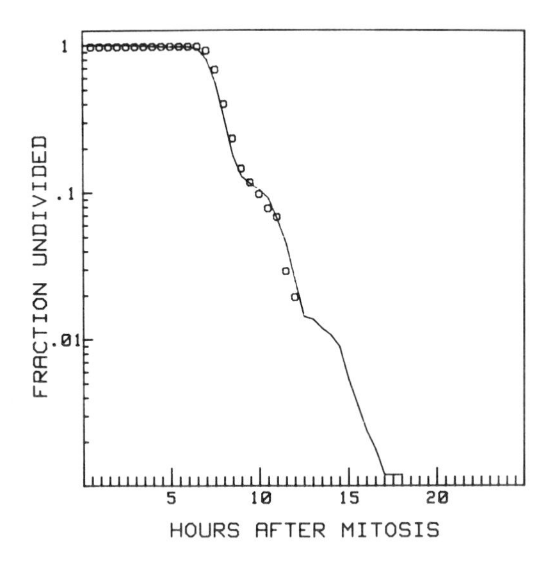

FIGURE 3.23. α-Curve for V79 cells along with a simulation (solid line) using a two-loop limit cycle model with random noise added, resulting in subthreshold oscillations. Limit cycle period was 4.25 h. The undulations in the tail of the curve reflect the quantizement of generation times. (Reprinted from Klevecz et al, 1980a, with permission of Alan R. Liss, Inc.)

cell had entered and to ascertain the stage of the CDC to which it had progressed. Indeed, cell cycle progress could be measured if each limit cycle loop were assumed to trigger passage of the cell into a subsequent CDC stage and if these loops were counted. Initiation or completion of these causal sequences would provide such a counting mechanism, whereas the independent oscillator, or central timer (Figs. 3.5, 3.6), would control the timing of individual events (Shymko et al., 1984).

It is possible that cell size is an appropriate general parameter for representing the causally related events in the CDC—a recurrent theme in cell cycle investigations. Shymko and Klevecz (1981) have analyzed the theoretical consequences of superimposing a size threshold on the two-loop limit cycle model (and associated threshold crossing required for transit through successive loops) of the CDC discussed previously. This further constraint requires that a critical size be reached before gating is implemented to exit from the second loop (Fig. 3.24). If the rate of cell growth were sufficiently rapid, cell size would have no effect on the timing of mitosis. Otherwise, the cell would remain on the second trajectory until the critical size was attained. This refinement would retain the global behavior of cycling cells and additionally would predict that the *g* values of sibling cells would be positively correlated (those of mother and daughter pairs would be negatively related). Similar blends of deterministic and probabilistic models for the CDC have been proposed by Tyson and

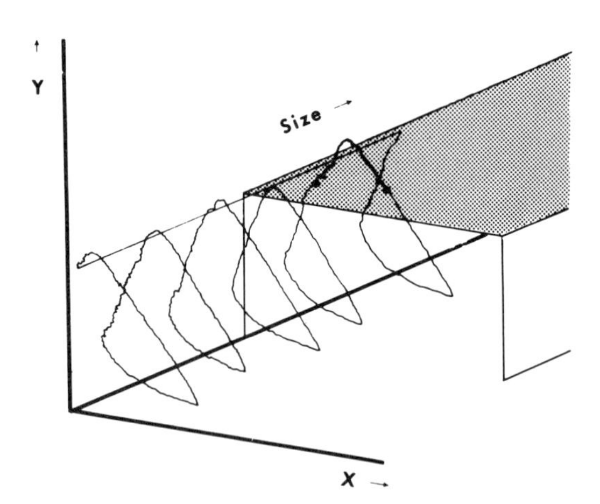

FIGURE 3.24. Schematic representation of a limit cycle model with a size threshold added. Limit cycle oscillations occur in X and Y; size increases in the third dimension. Crossing of the Y threshold (stippled top of the box) is ineffective in triggering mitosis until the threshold size (open face of the box) is reached. (Reprinted from Shymko and Klevecz, 1981, with permission of Elsevier, Amsterdam.)

Hannsgen (1985) and Tyson and Diekmann (1986), who posit a random-exiting phase of the CDC and a minimum size requirement for entry into this phase, and by Webb (1986), who attempts to reconcile models for proliferating cell populations based on transition probability, size control, and inherited properties (pedigree). No doubt, nature has achieved even more complex mixes.

The quantizement of cell generation time has been reported in single *Paramecium tetraurelia* (Kippert, 1985b) and *T. pyriformis* (Kippert, personal communication, 1986; Readey and Groh, 1986) cells in the ultradian growth mode under free-running conditions (see Section 3.2.2). In more slowly growing cultures ($g > 24$ h) of *G. polyedra*, Homma (1987), using flow cytometric techniques, also has found that the values of g in LL are variable, but are quantized to integral multiples of the circadian period. He hypothesized that the CDC of this unicell is divided into a cyclic part corresponding to G_1 and a noncyclic branch that constitutes the $S + G_2 + M$ sequence, which has a fixed duration of approximately 11 h. The former constitutes a circadian clock and, therefore, it takes one circadian period to traverse the subcycle. According to this scheme, cells that satisfy a minimal volume requirement (between CT 12 and CT 18) exit probabilistically to the replication-segregation sequence culminating in division, and reenter the cyclic portion a fixed time interval (11 h) afterwards.

Cell generation time also is quantized in single, cultured human bone and cartilage cells in the infradian growth mode (reviewed by Petrovic et al., 1984). These skeletal cells display circadian periodicity of mitosis both in situ and in organ culture (see Section 3.2.1), which results not only from the g of individual cells but also from the temporal organization of the cell populations (Stutzmann and Petrovic, 1979). Only in the single, isolated skeletal cell, however, is g itself approximately 24 h; in situ it is considerably longer, suggesting that cellular interaction masks the endogenous circadian periodicity of g in populations. As soon as g increases significantly in individual cells, the lengthening is quantized in a stepwise pattern. The increase of the intermitotic intervals appears to be discontinuous but not stochastic. Indeed, g of skeletal cells was consistently quantized toward multiples of 24 h by the recurrence of 6-h to 8-h increments (Fig. 3.25). Occasionally a quantum of 3 or 4 h was found, corresponding to the G_q described by Klevecz (1976). The values of g for sibling cells in a progeny of skeletal cells were always tightly correlated (perhaps reflecting an aspect of gene-coded phenomena), although there appeared to be cell and tissue specificity for the temporal pattern within an animal species. Only ultradian rhythmicities were detected in a population of malignant skeletoblasts. Strikingly, the period of these fluctuations in mitotic index ($\tau = 6$ to 9 h) corresponded to the quantal component of normal skeletal cells.

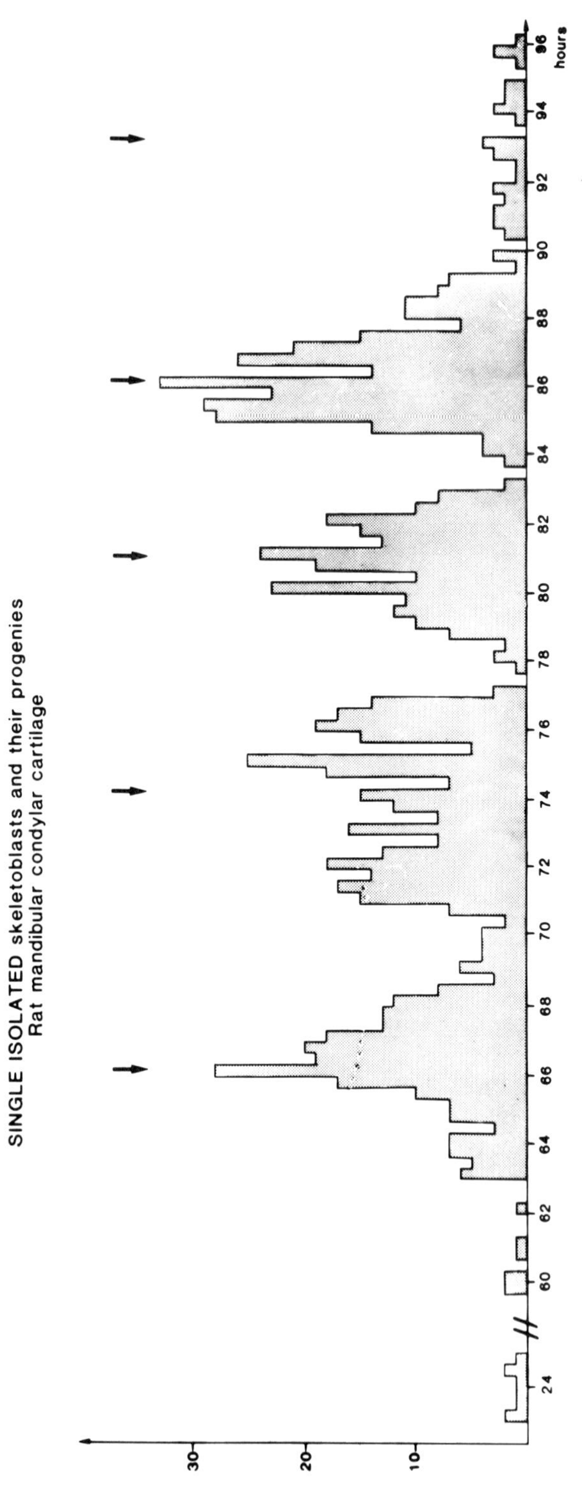

FIGURE 3.25. Polymodal distribution of the generation times of progenies issued from a single isolated skeletoblast. The skeletoblast originated from the condylar cartilage of the rat mandible. Prevailing values occur every 6 to 7 h. The number of generation times measured: $n = 987$. (Reprinted from Petrovic et al., 1984, p. 334, with permission of Marcel Dekker, Inc.)

3.2 Cell Division Cycles and Circadian Oscillators

In exponentially increasing cultures of microorganisms, the phase points of the individual cell cycles of the cell population are distributed randomly; they are developmentally synchronous. A striking contrast is afforded by developmentally synchronous populations (Edmunds, 1978): cell division (as well as at least some of the other events comprising the CDC) in cultures of numerous protists, algal unicells, and cells dissociated from plant or animal tissues can be synchronized by a variety of inductive treatments so that there is a one-to-one mapping of similar phase points of the cell cycles throughout the culture with respect to time (Cameron and Padilla, 1966; Mitchison, 1971; Padilla et al., 1969, 1974; Zeuthen, 1964). These techniques include the addition of inhibitors (e.g., thymidine, fluorodeoxyuridine, amethopterin, hydroxyurea, or 5-aminouracil—all inhibitors of DNA synthesis—or colcemid, nitrous oxide, or vinblastine—metaphase blocking agents) and selective lethal agents to asynchronous cultures. The operational principle here is that a particular stage(s) of the CDC is differentially sensitive to the agent; the cells accumulate at this stage, therefore, and after release from the inhibitor (usually by washing it out), they divide synchronously for several rounds until the synchrony decays due to extrinsic, microenvironmental fluctuations and to intrinsic, karyotypic heterogeneity (Engelberg, 1964). This concept of cell cycle blockage points can be easily extended to embrace those synchronization methods in which the culture is starved or allowed to enter the stationary phase and then rejuvenated by the addition of fresh medium or even to the use of single or repetitive heat-shocks or cold-shocks, as has been documented, for example, in *Euglena* sp. (Terry and Edmunds, 1970a) and *Tetrahymena* sp. (Zeuthen, 1971).

A general model for such synchronization by shifts in environmental conditions has been proposed by Campbell (1957, 1964). It was assumed that under a given set of extrinsic conditions, the cell progresses through a sequence of stages and that under a different set of conditions, the cell cycle consists of the same sequence but that the relative time spent between such stages is different. If a series of shifts is performed between these two sets of conditions of such duration that one period of the regimen corresponds to the doubling of cell number, the model predicts a gradual attainment of complete synchronization. In the well-synchronized culture, therefore, a majority of the cells pass through the same developmental stage at the same time and that which is determined for the entire culture can be assumed to obtain as a first approximation for the individual cells. It is for this reason that synchronously dividing cultures of microorganisms have proved such useful tools for elucidating numerous biochemical and physiological problems associated with the cell developmental cycle.

In addition to the techniques just described, it is well known that appropriately chosen light cycles (LD) can synchronize cultures of microorganisms, and it is the method of choice for most photosynthetic unicellular algae (Edmunds, 1975), including *Euglena* (Cook and James, 1960; Edmunds, 1965a,b). Although it would seem possible to ascribe this light-induced division synchrony to the same mechanism responsible for that obtained by inhibitor blocks and heat-shocks, invoking repetitive shifts between two sets of environmental conditions (i.e., light and darkness) to which specific stages of the cell cycle are differentially sensitive, it will be seen in Section 3.2.1 that this type of model is not sufficient to explain the observed facts and, indeed, may not even be relevant at all under certain experimental conditions.

3.2.1 Interaction of Circadian Oscillators with the Cell Cycle

As a working hypothesis, we have assumed that an endogenous, autonomously oscillatory (see Section 3.1.2), light-entrainable, circadian clock—with all those properties (see Fig. 1.2) that characterize the timing mechanism(s) underlying the persisting 24-h rhythms that have been documented in virtually every higher organism examined—may underlie diurnal division rhythmicity in light-synchronized cultures. This oscillator would gate cell division to restricted intervals of time during successive 24-h time spans. Further, it is postulated that it is this same putative oscillation and its coupled pathways that generate and regulate such other overt circadian rhythms as phototaxis, motility, photosynthetic capacity, bioluminescence, and amino acid incorporation not only in dividing cultures of microorganisms but also in nondividing (infradian phase), cultures. The *Euglena* system is discussed in detail, since it provides some of the most compelling evidence supporting this hypothesis (Edmunds, 1984a), but other relevant studies in both unicellular (Edmunds, 1984b) and mammalian cell (reviewed by Scheving, 1984) populations are mentioned also.

Circadian Oscillators and the CDC of Phytoplankton and Other Unicells

There is abundant evidence that the CDCs of many unicellular algae, fungi, and protozoans in culture do exhibit persisting circadian rhythms of cell division, or hatching (Edmunds, 1975, 1978). A number of these protistan systems are itemized in Table 3.1, including some recent experimental entries.

As is evident from this tabulation, the division of a number of algal cells occurs only at a certain phase of the circadian cycle, very often the times (subjective nights) in DD or LL corresponding to dark intervals in an environmentally synchronizing LD cycle. This phenomenon, commonly referred to as "gating," has been documented for a number of years in *Chlamydomonas* (Bruce, 1970; Bruce and Bruce, 1981), *Chlorella* (Pirson and Lorenzen, 1958; Hesse, 1972), *Euglena* (Edmunds, 1966, 1971; to be

TABLE 3.1. Some unicellular, eukaryotic microorganisms for which persisting circadian rhythms of cell division (or hatching) have been documented.

Organism	Reference
Protozoans	
Paramecium bursaria	Volm (1964)
Paramecium multimicronucleatum	Barnett (1969)
Tetrahymena pyriformis	
Strain W	Wille and Ehret (1968a); Edmunds (1974)
Strain GL	Edmunds (1974)
Algae	
Ceratium furca	Weiler and Eppley (1979); Adams et al. (1984)
Chlamydomonas reinhardi	Bruce (1970); Bruce and Bruce (1981)
Chlorella fusca var. *vacuolata*	Wu et al. (1986)
Chlorella pyrenoidosa Chick (strains 211-8b, 211-8k)	Pirson and Lorenzen (1958); Hesse (1972); Lorenzen (1980); Lorenzen and Albrodt (1981); Lorenzen et al. (1985)
Emeliania huxleyi (MCH, 451B)	Chisholm and Brand (1981)
Euglena gracilis Klebs (Z strain)	Edmunds (1966, 1971); Edmunds and Laval-Martin (1984); Edmunds and Funch (1969a,b)
P_4ZUL mutant	Jarrett and Edmunds (1970)
P_7ZNgL mutant	Mitchell (1971)
W_6ZHL mutant	Edmunds et al. (1976)
W_nZUL mutant	Mitchell (1971)
Y_9ZNa1L mutant	Edmunds et al. (1976)
Gonyaulax polyedra	Sweeney and Hastings (1958)
Gymnodinium splendens	Hastings and Sweeney (1964)
Hymenomonas carterae (Cocco 2)	Chisholm and Brand (1981)
Olisthodiscus sp.	Chisholm and Brand (1981)
Prorocentrum sp.	Chisholm and Brand (1981)
Pyramimonas sp.	Chisholm and Brand (1981)
Pyrocystis fusiformis Murray	Sweeney (1982)
Skeletonema costatum	Østgaard and Jensen (1982)
Fungi	
Candida utilis	Wille (1974)
Saccharomyces cerevisiae (Y185 *rho*$^+$)	Edmunds (1980c); Edmunds et al. (1979)

Updated from Edmunds, 1984b, with permission of CRC Press.

considered in detail later), and *Gonyaulax* (Sweeney and Hastings, 1958; Hastings and Sweeney, 1964) and more recently in a number of marine phytoplankton (Chisholm, 1981; Chisholm et al., 1980, 1984; Weiler and Chisholm, 1976). For example, Chisholm and Brand (1981) have examined the cell division patterns of 11 species of phytoplankton maintained in an LD: 14,10 cycle followed by 3 days of LL. All species except two diatom clones showed typical phased or synchronized division in the LD regimen, although the timing of division relative to the LD cycle was variable. In six of the species (representing the algal classes Prasinophyceae, Xanthophyceae, Haptophyceae, and Dinophyceae), the division rhythm persisted throughout the 3 days of LL, suggesting endogenous control, although there was some evidence of damping (Fig. 3.26). In others, the

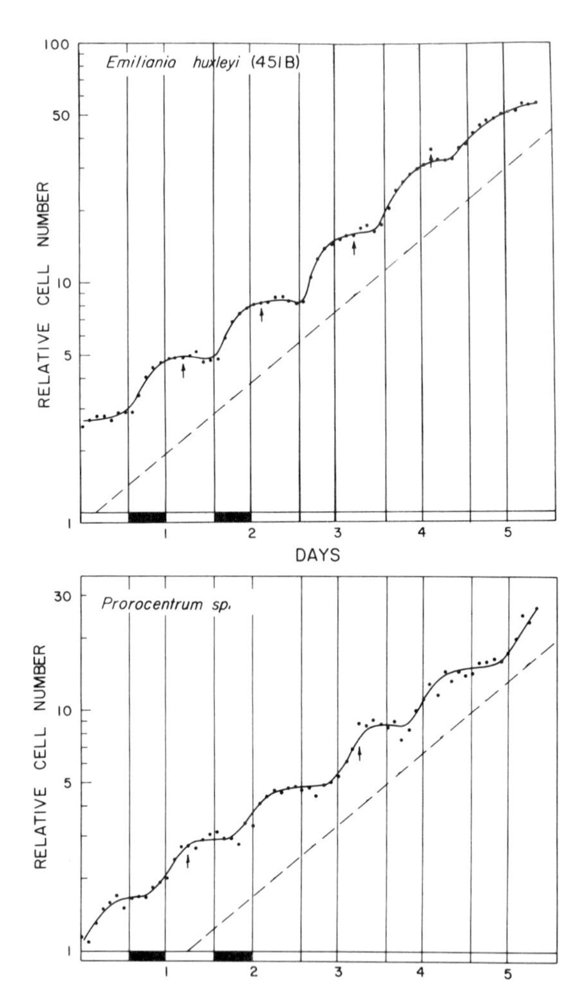

FIGURE 3.26. Photocycle-induced cell division patterns of two species of marine phytoplankton grown in semicontinuous batch culture. Arrows indicate times when the cultures were diluted, and dashed line has a slope of one population doubling per day. Note that division phasing persists for at least 3 days in LL and that the average population growth rate does not change. (Reprinted from Chisholm et al., 1984, p. 368, with permission of Marcel Dekker, Inc.; see Chisholm and Brand, 1981.)

rhythm completely damped out after 1 day or, in the case of the two diatoms, was never evident in the first place.

Although the circadian clock model of cell division control appears to be the only viable explanation of the observed growth pattern of some species of phytoplankton (Heath and Spencer, 1985), particularly the dinoflagellates (which display some of the tightest cell cycle phasing), other conceptual models also are possible. Spudich and Sager (1980) have given evidence that the regulation of the CDC in *C. reinhardtii*—a so-called clock-controlled species—is more adequately explained in terms of sequentially timed intervals operating independently within the cycle. Phasing of division in this species would result from the CDCs being driven directly by the imposed LD cycle (as opposed to the interposition of an autonomously oscillatory, light-entrainable circadian clock) and thus would reflect a forced oscillation. By monitoring the progress of this photosynthetic algal cell through its CDC (by measuring cell volume and DNA content), these workers demonstrated a light-dependent segment demarcated by a primary arrest point (A), where the CDC is blocked in the absence of light, and a transition point (T), where a cell is committed to progress through the cycle to division, even in the absence of illumination (Fig. 3.27a). The inhibitor of electron transfer from PS II, DCMU, was able to mimic the effect of darkness (Spudich and Sager, 1980). Further, circadian time was not maintained in newly divided cells held in darkness of varying duration (up to 48 h) and returned at intervals to light. The cells revealed a constant capacity to resume CDCs of normal duration and of normal daughter cell yield (an observation recently confirmed and extended by McAteer et al., 1985).

The relative simplicity and testability of this block-point model have been its attractiveness for many workers. The hypothesis easily could be extended by postulating other block-points in the CDC, such as for cell size, mitogen levels, and protein and RNA synthesis (see Section 3.1.2). Thus, Donnan and John (1983, 1984) propose on the basis of their experiments that division in *C. reinhardtii* and *Chlorella* sp. is determined by the sequential operation of a commitment timer and a postcommitment sizer, each measuring time from an initiating stimulus and hourglass in nature (see Section 3.1.2). Synchrony would arise from the differing dependence of these two timers on concurrent growth. All cells would initiate the commitment timer when reillumination (in LD or after prolonged darkness) allows growth to begin and could occur even in LL. Thus, circadian rhythmicity in cell division patterns during LL (cells cannot divide in DD) does not necessarily demand an autonomous oscillator but would result simply because a new CDC follows the completion of a previous one, and the two endogenous hourglass timers together would specify a circadian period (McAteer et al., 1985), presumably through natural selection. McAteer et al., (1985) reason that this type of hourglass timing in multiple-fission algae allows division to be synchronized over a wider

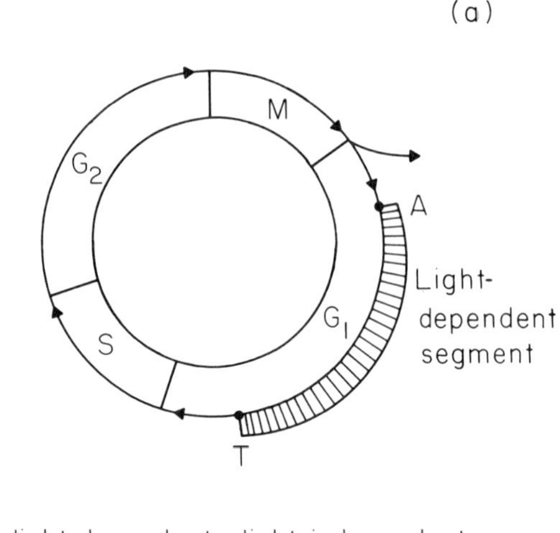

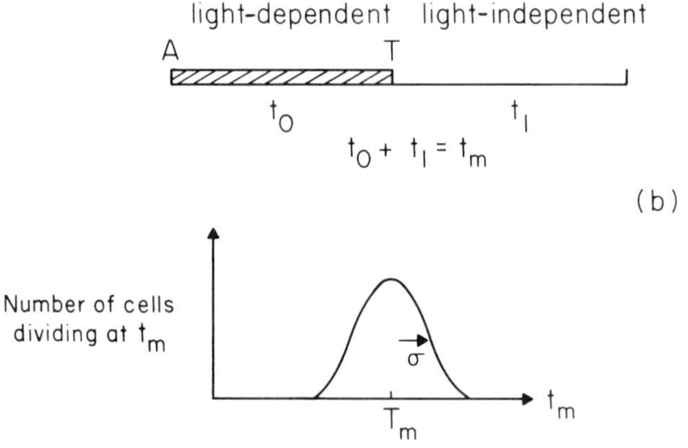

FIGURE 3.27. (a) The cell cycle with a light-dependent segment according to Spudich and Sager (1980). The cell cycle is viewed as a succession of four segments: G_1, S, G_2, and M. S corresponds to DNA synthesis and M to mitosis. Spudich and Sager postulated the existence of a light-dependent segment in G_1 from A to T (see text). (b) Stochastic model for entrained populations adapted from the Spudich and Sager model. Individual cell cycles are composed of two segments: one that is light dependent of length t_0 (the arrest point has been placed at the beginning of G_1 for simplicity), one that is light independent of length t_1. The total length of the cycle is t_m ($= t_0 + t_1$); t_0 and t_1 are random variables normally distributed among the population [$N(T_0\sigma_0)$], implying a normal distribution for t_m: $N(T_{m,\sigma})$. For simplicity the variability has been concentrated in the light-independent segment ($\sigma = 0$, $t_0 = T_0$). Thus $\sigma = \sigma_1$ and $T_m = T_0 + T_1$. T_0 is the mean light requirement and T_m the mean generation time. The time evolution of such a stochastic population has been simulated numerically with a technique similar to the age transfer method used by Bronk et al. (1968). (Reprinted from Chisholm et al., 1984, p. 377, with permission of Marcel Dekker, Inc.)

range of growth rates than oscillator control might (which allows phasing to the diurnal cycle at typically slower growth rates by deferring division of some cells to the next cycle or gate), since the larger mother cell sizes attained at high growth rates can be balanced by higher division numbers. [For further discussion of the optimization of resource use, see Chisholm et al. (1984).] An analogous model to that envisaged for *C. reinhardtii* (Donnan and John, 1983, 1984) also has been presented by Heath and Spencer (1985) for the marine diatom *Thalassiosira pseudonana* Hasle and Heimdal (clone 3H).

A similar tack has been taken by Chisholm et al. (1980), who have been impressed by the high degree of variability in diatom generation times and their patterns of cell division phasing by LD cycles. They conclude that not only are these patterns not characteristic of species in which cell cycles are coupled to a circadian clock but also that they cannot be explained on the basis of the simple block-point theory of Spudich and Sager (1980). Rather, they adapt the block-point model with its light-dependent and light-independent segments (Fig. 3.27a) inserting a stochastic component into the light-independent segment consisting of a closed loop with a probability distribution governing exit once the energy requirements for division have been fulfilled) (Fig. 3.27b). A simple gaussian distribution was used, although other forms for the variability could be invoked (Smith and Martin, 1973; Klevecz, 1976; Cooper, 1982b). The evolution of the population was simulated by following the progression of individual cohorts of cells through the CDC. By choosing and varying realistic parameters (Chisholm et al., 1980, 1984; Vaulot and Chisholm, 1987), several of the qualitative properties (entrainment, phase shifting, transients, damping, and decay during free-run) of population growth that have been attributed to circadian regulation have been reproduced using the block-point model (Fig. 3.28). This type of modeling also has proved to be particularly useful for examining cultures in which the average population growth rate is greater than one doubling per day or in which only a portion of the cells divide during each 24-h period, as, for example, in light-limited or nitrogen-limited populations of the marine dinoflagellate *Amphidinium carteri* (Olson and Chisholm, 1986).

The block-point model, however, does not seem to have much relevance for the persisting circadian rhythms of cell division described for protozoans and fungi (Table 3.1) or, for that matter, for organotrophically cultured algal mutants incapable of photosynthesis (see next subsection), in which rhythmicities are exhibited in both LL and DD following entrainment by an appropriate LD cycle. Thus, a sudden increase or decrease in irradiance (such as that afforded by either a DD/LL or an LL/DD transition) could initiate a circadian rhythm of division (τ = 19 to 23 h) in culture of *T. pyriformis* (strains W and GL) in the infradian growth mode, and an LD: 10,14 cycle was able to photoentrain the cells (Wille and Ehret, 1968a; see Section 2.1.1). In an extension of these studies, Edmunds (1974) was also able to obtain long trains of 24-h oscillations in cell number in

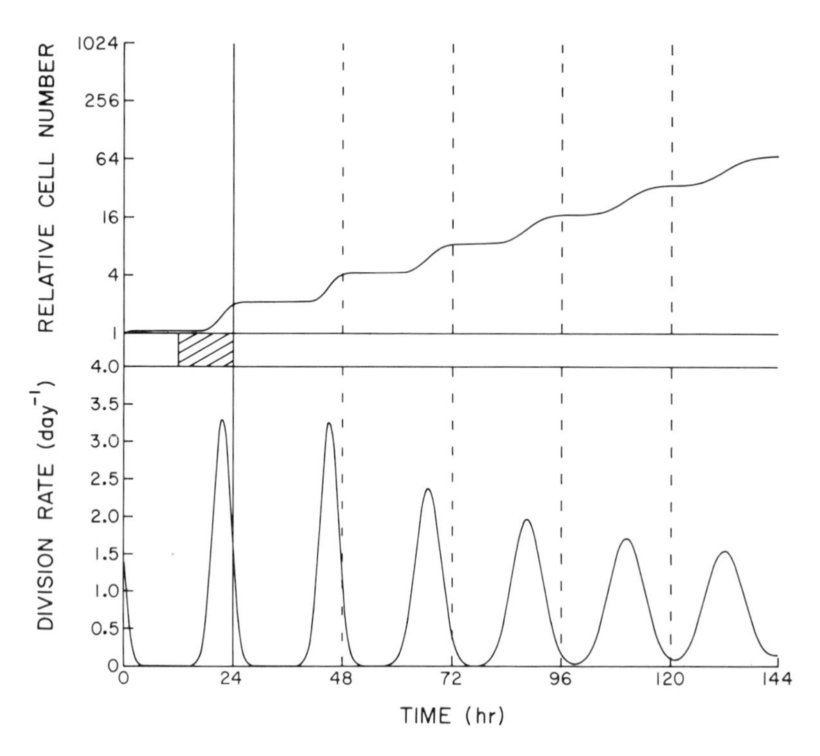

FIGURE 3.28. Synchronized population allowed to free-run showing slowly decaying oscillations. Model parameters: $T_0 = 7$ h, $T_1 = 15$ h, $T_m = 22$ h, and $\sigma = 2$ h (coefficient of variation = 0.09). The oscillations have a period close to 22 h (T_m). See Figure 3.27. (Reprinted from Chisholm et al., 1984, p. 382, with permission of Marcel Dekker, Inc.)

semicontinuous, infradian cultures phased by LD: 12,12 or LD: 6,18 (see Section 2.1.1), and the entrained rhythm persisted (τ = 23.8 to 24.4 h) for at least 6 days in DD (Fig. 3.29). These observations would appear to exclude any hypothesis for light-induced cell division phasing based simply on photoinhibition of growth. Similar circadian division rhythmicities have been reported in *Paramecium* spp. (Volm, 1964; Barnett, 1969; see Section 2.1.2). Finally, a light-entrainable, persisting circadian rhythm ($\tau \cong 26$ h) in bud release has been discovered (Edmunds et al., 1979 see Section 2.3.2) in cultures of *S. cerevisiae* grown at low temperatures ($g > 24$ h) in LD: 14.10 and released into DD (Fig. 3.30).

Circadian Clock Control of the CDC in *Euglena*

The relationship between the CDC and circadian oscillators perhaps has been most intensively investigated in the algal flagellate, *Euglena gracilis*

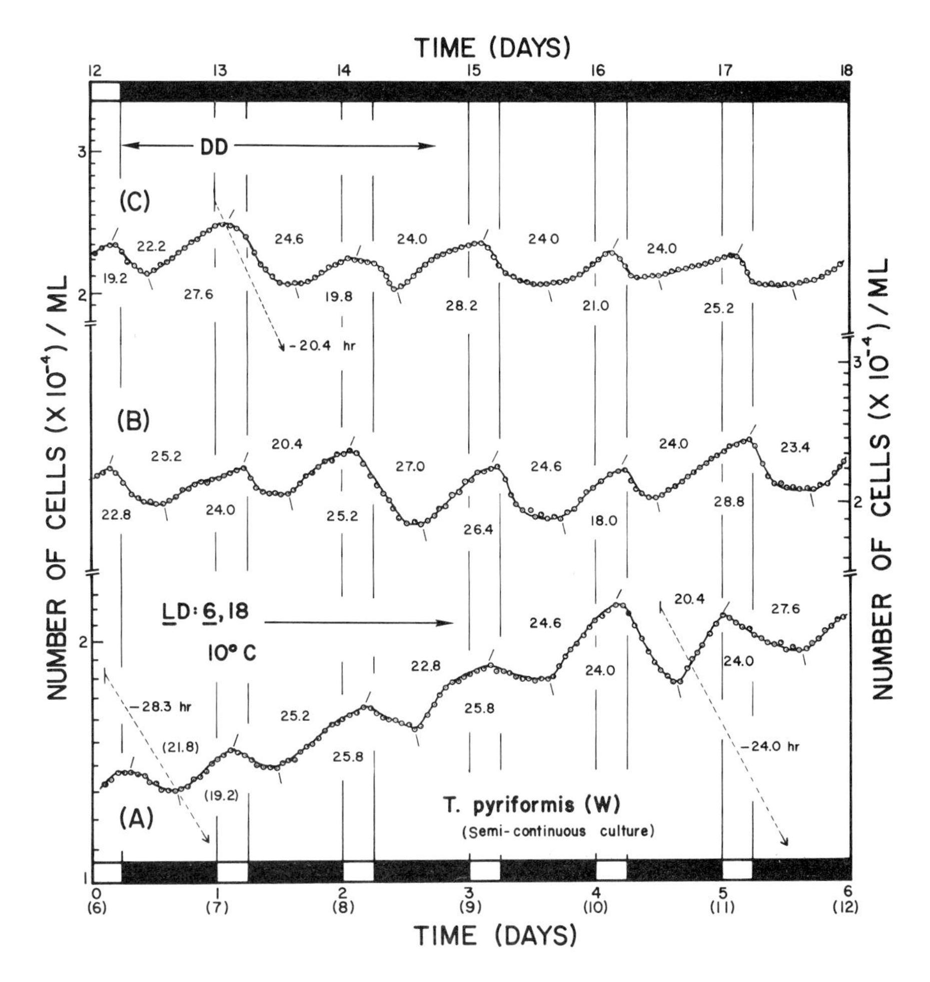

FIGURE 3.29. Entrainment and persistence of division rhythmicity in a semicontinuous culture of *Tetrahymena pyriformis* (W) on PPL medium at 10°C ± 0.05°C. The culture was diluted continuously with fresh medium supplied by a peristaltic pump at a predetermined rate designed to exactly replenish the aliquots withdrawn automatically at intervals of 72 min for determinations of cell concentration and to balance the rate of increase in cell titer. The three segments of the curve shown (A, B, C) represent a continuous monitoring of cell number over a timespan of 18 days and should be read from left to right (starting with curve A) against the corresponding ordinate and abscissa scales. For the first 12 days the culture was maintained in LD: 6,18, which phased divisions as indicated. On day 4 the effective equivalent dilution, or washout rate (indicated by the dashed, diagonal lines), was adjusted from − 28.3 h to − 24.0 h to balance exactly the calculated generation time of the culture (see curve B). On the 12th day of entrainment, the culture was placed into continuous darkness (DD) for 6 days. This resulted in a decrease in the overall generation time to 20.4 h and required a matching adjustment in the dilution rate as shown. Period lengths between inflection points (either troughs or peaks) of successive oscillations are given (Reprinted from Edmunds, 1974, with permission of Plenum Press; see Wille and Ehret, 1968a.)

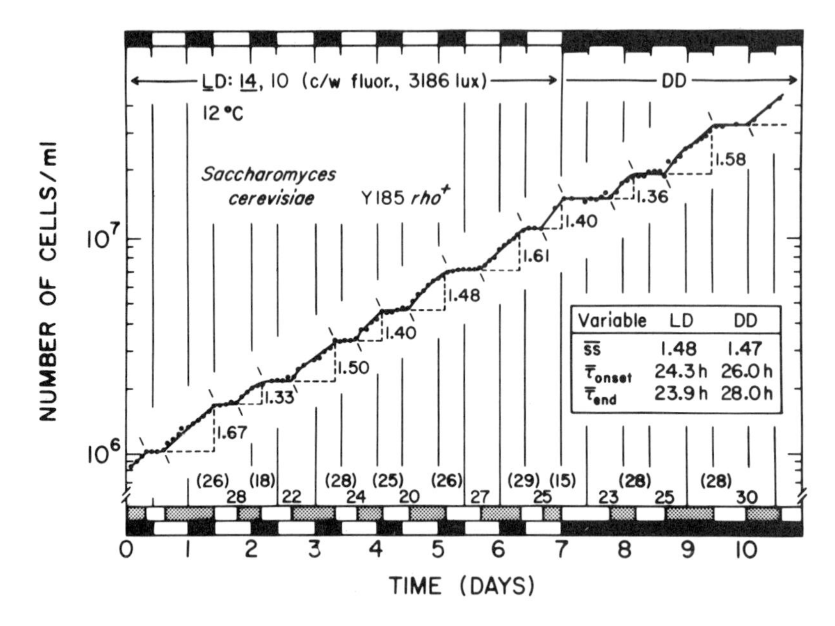

FIGURE 3.30. Entrainment of the rhythm of cell division (bud release) by LD:14,10 (L = 3180 lx) and its persistence in constant darkness (DD) in a population of *Saccharomyces cerevisiae* batch-cultured at 12°C. Dark intervals are indicated by heavy black bars. Stippled bars denote times during which an increase in cell number was actually observed. Beside each fission burst on the curve, the step size *(ss)*, or factorial increase, is indicated. Time intervals (hours) between successive onsets of division, or between endpoints (in parentheses) are shown at the bottom of the figure; average stepsize *($\overline{ss}$)*, period of onsets ($\bar{\tau}_{onset}$), and period of endpoints ($\bar{\tau}_{end}$) are shown in the box. In the LD cycle, increases in cell number occur at intervals approximating 24 h (matching the period of the imposed light regimen), whereas for the segment of the curve in DD, the period (τ) is only circadian, averaging 26 to 28 h. Note that although not every cell divides during a given 24-h timespan, those that do do so during intervals restricted primarily to (subjective) night. (Reprinted from Edmunds et al., 1979, with permission of Pergamon Press.)

Klebs, which will be used as a basis for discussion (Edmunds, 1978, 1987; Edmunds and Adams, 1981; Edmunds et al., 1987). This organism can be grown on a variety of different, completely defined media, either photoautotrophically in the presence of CO_2 or organotrophically in the light or dark on carbon sources ranging from acetate and ethanol to lactic, glycolic, glutamic, and malic acids over a wide pH range. This versatility in growth mode, in conjunction with the fact that cell division can easily be synchronized by appropriate 24-h (and other) lighting schedules (Cook and James, 1960; Edmunds, 1965a) and temperature cycles (Terry and Edmunds, 1970a), has made *Euglena* an important experimental organism for a variety of physiological and biochemical investigations (Buetow, 1968a,b, 1982). In addition, the temporal organization of *Euglena* has been extensively examined (Edmunds and Halberg, 1981; Edmunds, 1982; see Section 2.2.3), and a large number of persisting circadian rhythms have been reported in both dividing and nondividing (or very slowly dividing) cultures (see Table 2.4, Fig. 2.4). We now review the major lines of evidence that implicate and endogenous, self-sustaining circadian oscillator

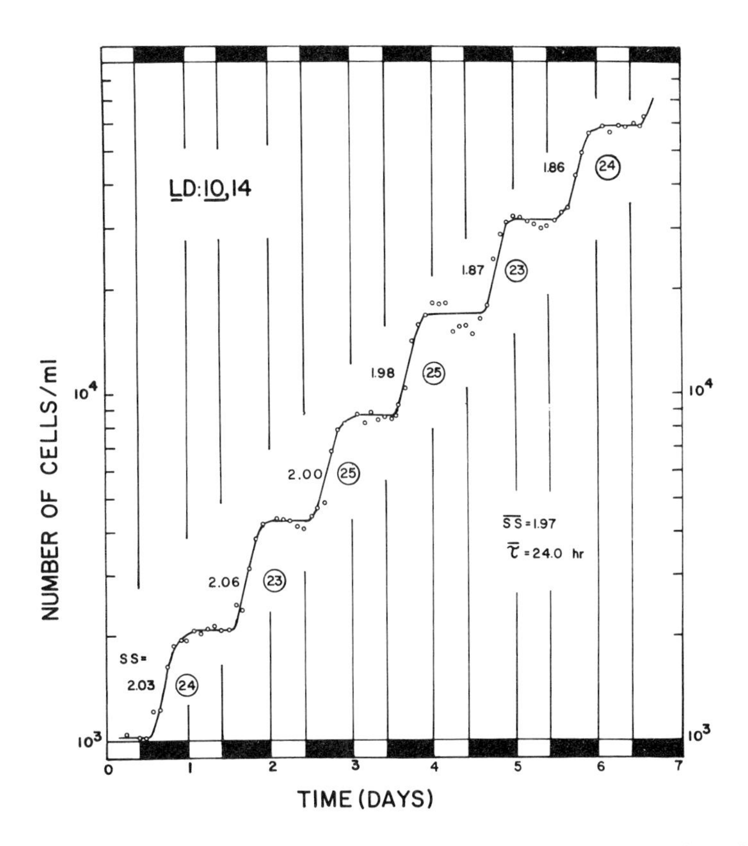

FIGURE 3.31. Entrainment of the cell division rhythm in populations of *Euglena gracilis* (Z strain) batch-cultured photoautotrophically at 25°C by a full-photoperiod, diurnal LD: 10,14 light cycle. Step sizes (*ss*, ratio of number of cells per milliliter after a division burst to that just before the onsets of divisions) are given for successive steps; the estimated period (τ) of each oscillation (intervals between successive onsets of division) is indicated by the encircled numbers (hours). The average period ($\bar{\tau}$) of the rhythm in the culture was almost identical to that (T) of the synchronizing LD cycle. Divisions were confined primarily to the main dark intervals, commencing at their onsets. A doubling of cell number ($ss \cong 2.00$) usually occurred every 24 h in this full-photoperiod LD cycle. (Reprinted from Edmunds and Funch, 1969a, with permission of American Association for the Advancement of Science.)

(see Fig. 1.2) in the control of the CDC in this unicell and, by generalization, those microorganisms listed in Table 3.1.

Entrainability

Photoautotrophically grown cultures of wild-type *Euglena* (Z strain) can be routinely synchronized, or entrained—a key property of circadian rhythms (see Fig. 1.2)—by empirically chosen repetitive, 24-h LD cycles so that cell division synchrony is confined entirely to the dark intervals (Cook and James, 1960: Edmunds, 1965a). For example, in LD: 10,14 (Fig. 3.31) a batch-cultured population doubles (factorial increase, or *ss* $\cong$ 2.0) at each step, the period (τ) of the rhythm in the population exactly matching that (T) of the imposed LD cycle, and, by inference, the length of the individual CDCs also must average 24 h (the rate of cell death is insignificant). Long-term synchronous cultures can be obtained (Fig. 3.32) using continuous or semicontinuous culture techniques by which the cell titer is maintained at a constant level without reaching limiting conditions (Terry and Edmunds, 1969), as occurs in batch cultures (due primarily to reduced light intensity caused by the mutual shading of the cells).

If one reduces the total duration of the light interval (for example, LD: 8,16) within a 24-h framework (Fig. 3.33), the amplitude of the cell division

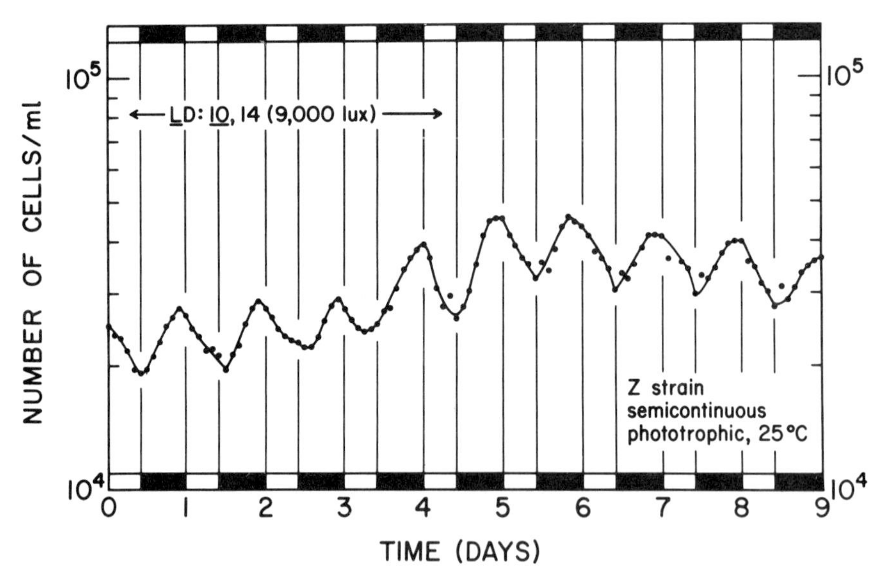

FIGURE 3.32. Entrained rhythm of cell division in *Euglena gracilis* (Z strain) grown photoautotrophically at 25°C in LD: 10,14 in semicontinuous culture. Note that the rate of dilution with fresh medium approximately balances the overall growth rate of the culture. A doubling of cell number occurs about every 24 h, with cell division's being confined to the dark intervals. (Reprinted from Edmunds and Halberg, 1981, with permission of Williams & Wilkins Co.)

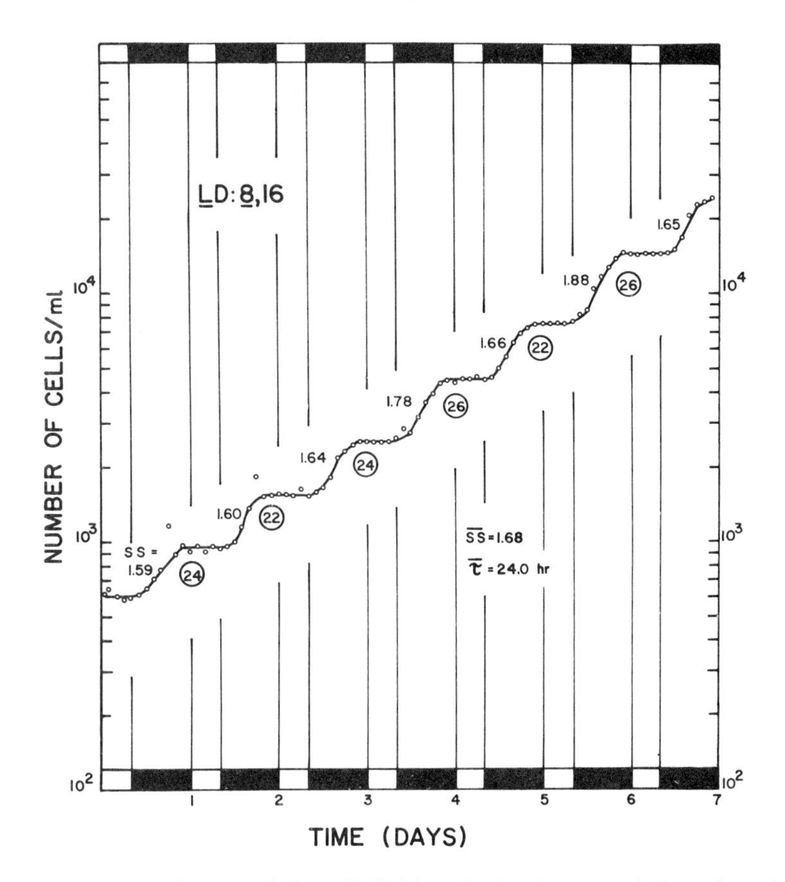

FIGURE 3.33. Entrainment of the cell division rhythm in a population of *Euglena gracilis* grown photoautotrophically in LD: 8,16. Labels are as for Figure 3.31. Although τ of the rhythm is precisely that of the synchronizing LD cycle, *ss* of the successive fission bursts is substantially less than 2.0, indicating that not all of the cells divide during any one cycle. (Reprinted from Edmunds and Funch, 1969a, with permission of American Association for the Advancement of Science.)

rhythm in the population is proportionately reduced (*ss* = 1.68), and the average doubling time *(g)* of the culture is lengthened to about 36 h (Edmunds and Funch, 1969b). Not every cell divides during each division burst, and the culture is now no longer developmentally synchronous to the extent of a one-to-one mapping between the stages of the CDC in all constituent cells (Edmunds, 1978). Nevertheless, cell division, when they do occur, do so during the dark intervals only at intervals of 24 h, and the culture continues to be synchronized in the sense of event simultaneity (Edmunds, 1978). This rhythmicity observed in LD: 8,16 and in similar LD cycles stand in sharp contrast to the asynchronous, exponential growth curve obtained in LL, having a minimum doubling time of 12 to 14 h (Edmunds, 1965a).

Similarly, so-called skeleton photoperiods (PP_s) comprising the framework of a normal, full-photoperiod cycle (for example, *LD*: *3*,6,*3*,12) will also entrain the rhythm to a precise 24-h period (Fig. 3.34). In this case, division are confined to the main dark intervals, commencing at their onsets (Edmunds and Funch, 1969b). Entrainment by LD cycles having $T \neq 24.0$ h (e.g., LD: 10,10) may also occur but only within certain limits (Edmunds and Funch, 1969b; Ledoigt and Calvayrac, 1979). Additionally, appropriate diurnal temperature cycles (for example, 18°, 25°C: 12,12 or 28°C, 35°C:

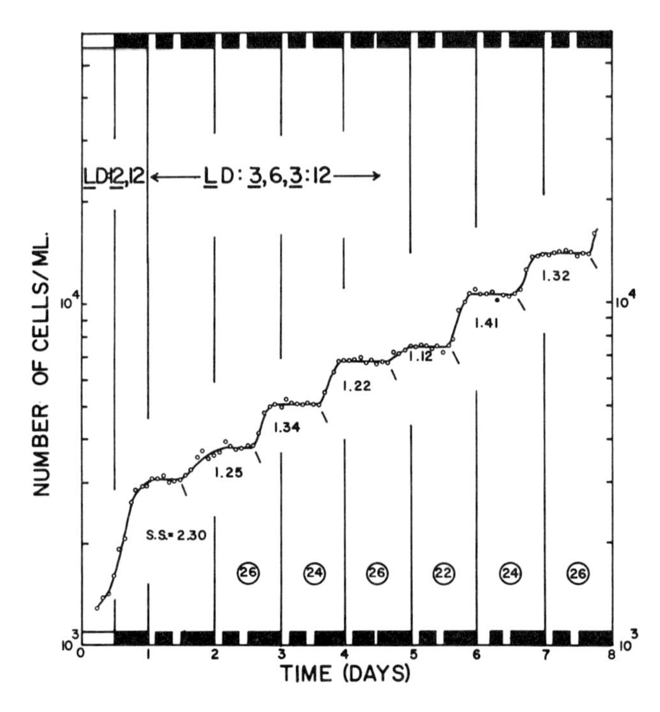

FIGURE 3.34. Entrainment of the cell division rhythm in populations of *Euglena gracilis* (Z strain) batch-cultured photoautotrophically at 25°C by a skeleton *LD*: *3*,6,*3*,12 regimen comprising the framework of an LD: 12,12 cycle. Labels are as for Figure 3.31. Note the effects of the accidental reduction in the length of the second light pulse on the fourth day of the skeleton test cycle. The average τ of the rhythm in the culture was almost identical to that (T) of the imposed LD regimen; divisions were confined primarily to the main dark intervals, commencing at their onsets. Whereas a doubling of cell number (ss = 2.00) usually occurred every 24 h in the full-photoperiod regimen (see Fig. 3.31), the amplitude of the rhythm was necessarily diminished in the skeleton LD cycle, corresponding to the reduction of the photofraction. (Reprinted from Edmunds and Funch, 1969b, with permission of Springer-Verlag.)

12,12) will synchronize the cell division rhythm in cultures maintained in LL (Terry and Edmunds, 1970a).

Even more conclusive evidence for the role of a basic circadian oscillator in the control of the CDC has been obtained by using photosynthetic mutants (all obligate heterotrophs) of *E. gracilis* (Edmunds, 1975, 1978). These studies have circumvented effectively the problem of the dual use of imposed light cycles and signals: as an energy source, or substrate, for growth, on the one hand, and as a timing cue *(Zeitgeber)* for the underlying clock on the other. Thus, the UV light-induced P_4ZUL mutant (Jarrett and Edmunds, 1970), although not entrainable to a 24-h period by diurnal

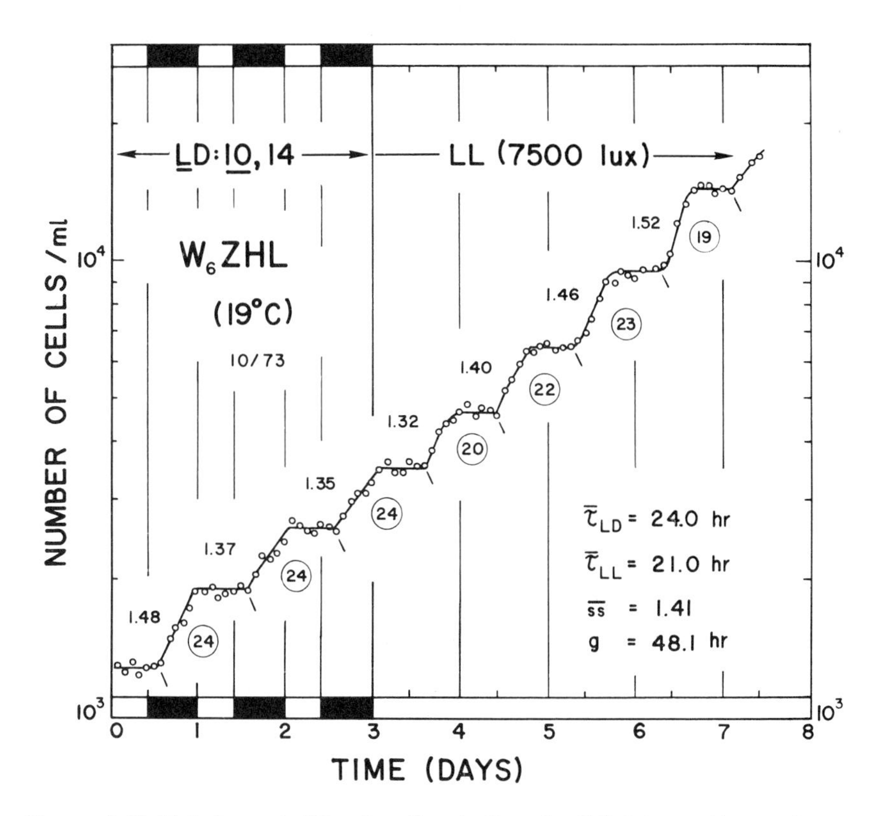

FIGURE 3.35. Entrainment of the circadian rhythm of cell division and its persistence for at least 5 days under continuous illumination in the aplastidic, heat-bleached W_6ZHL mutant of *Euglena gracilis* (Z strain) batch-cultured photoautotrophically at 19°C in an LD: 10,14 cycle and subsequently released into LL (7500 lx). Labels and notations are as for Figure 3.31. Note that although the period of the population rhythm was about 24 h (i.e., circadian), the period of the average individual CDC approximated 48 h, since less than half of the cells divided during any one 24-h timespan. (Reprinted from Edmunds, 1975, with permission of Editions du Centre National de la Recherche Scientifique, Paris; see Edmunds and Halberg, 1981.)

LD cycles when growing in the ultradian mode ($g < 24$ h) at 25°C where a doubling of the population occurs every 10 h, could be synchronized by an LD: 10,14 cycle if the temperature were lowered to 19°C, (yielding an exponential curve if the temperature were lowered to 19°C, (yielding an exponential curve with a g of $\cong$ 24 to 26 h in DD or LL). In this infradian ($g > 24$ h) growth mode (Ehret et al., 1977), divisions were set back or delayed 8 to 10 h at 24-h intervals (Fig. 3.50). These results have been extended both to the naladixic acid-induced Y_9ZNa1L photosynthetic mutant and to the white, heat-bleached W_6ZHL strain (Fig. 3.35) that totally lacks chloroplasts (Edmunds, 1975: Edmunds et al., 1976) and are consistent with the date of Mitchell (1971) for the pale green, nitrosoguandine-mutagenized P_7ZNgL strain and the white, ultraviolet-bleached W_nZUL mutant.

Persistence

Although the synchronization of the cell division rhythm in *E. gracilis* by diurnal, full-photoperiod LD cycles is consistent with the notion that a putative circadian clock is entrained by the imposed light regimen and, in turn, phases or gates cell division to the dark intervals every 24 h [by acting, perhaps, on one or more key control points (Spudich and Sager, 1980)] of the CDC, it does not demand it. Light (or darkness) could be acting by directly inhibiting (or promoting) division, and periodic shifts between light and dark would synchronize the culture (Campbell, 1957). A number of other observations, however, render this seemingly straightforward hypothesis untenable.

One of the most basic tests for the existence of a circadian rhythm (sensu stricto) is to determine whether it will continue to free-run for a number of cycles after transfer of the organism to conditions held constant with respect to the major environmental *Zeitgeber* (light and temperature). Characteristically, the period (τ) under such conditions only approximates 24 h, as might be expected of an imperfect biological clock (see Fig. 1.2).

Indeed, rhythmic cell division has been found to persist for a number of days (τ = 24.2 h) in the autotrophically grown Z strain, batch-cultured under dim LL (Edmunds, 1966). This series of experimental results (Fig. 3.36) was perhaps the most definitive but was restricted by the low light intensities (800 lx) that had to be used. (The rhythm damped out in more intense LL and, or course, could not be observed in DD.) Nevertheless, although the division bursts were relatively small ($g \cong 5$ days), those cells that did divide did so during their subjective night at the times when they would have done had the entraining LD cycle previously imposed been continued.

We have also observed free-running circadian rhythms of cell division in a variety of higher-frequency (e.g., LD: 1,3 or LD: 1/3,1/3) light cycles (Edmunds and Funch, 1969b; Laval-Martin et al., 1979) and even random illumination regimens (Edmunds and Funch, 1969a), although we do not

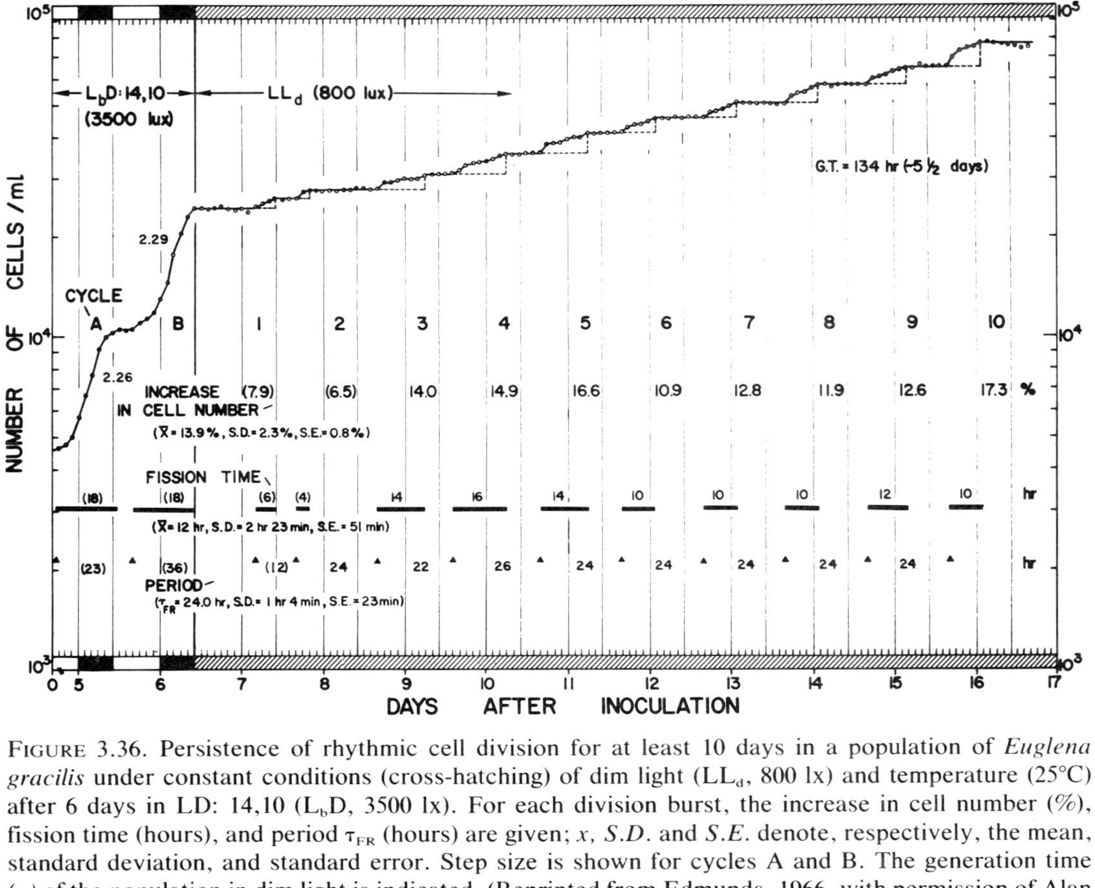

FIGURE 3.36. Persistence of rhythmic cell division for at least 10 days in a population of *Euglena gracilis* under constant conditions (cross-hatching) of dim light (LL_d, 800 lx) and temperature (25°C) after 6 days in LD: 14,10 (L_bD, 3500 lx). For each division burst, the increase in cell number (%), fission time (hours), and period τ_{FR} (hours) are given; *x*, *S.D.* and *S.E.* denote, respectively, the mean, standard deviation, and standard error. Step size is shown for cycles A and B. The generation time (*g*) of the population in dim light is indicated. (Reprinted from Edmunds, 1966, with permission of Alan R. Liss, Inc.)

exclude the possibility that such short-period light regimens (which provide no information to the cells with regard to 24-h periodicities) may modulate the period to some extent (Adams et al., 1984). Those high-frequency cycles (Fig. 3.37B; see Fig. 3.41) with symmetric photophase and scotophase (e.g., LD: 1,1 or LD: 3,3) and properly constructed random regimens (Fig. 3.37A) have proved particularly useful in that they afford an amount of light during a 24-h timespan identical to that received in a full-photoperiod LD: 12,12 entraining reference cycle ($T = \tau = 24$ h) yet elicit free-running circadian rhythms ($T << \tau \cong 24$ h) not only of cell division (fig. 3.37B; see Fig. 3.41) but also of random motility (Schnabel, 1968) and of photosynthetic capacity and chlorophyll content (Laval-Martin et al., 1979; Edmunds and Laval-Martin, 1981). They have been used (Edmunds et al., 1982) to advantage in the derivation of a PRC for light perturbations and in studies on temperature-compensation. With these exotic light cycles, therefore, the total duration of light afforded for growth is held constant (reflected in the comparable step sizes of the observed rhythmic growth curves), whereas the signaling information inherent in the light perturbations can be varied to manipulate the light-sensitive circadian oscillator.

Persisting circadian rhythms of cell division have been observed routinely in the P_4ZUL (Jarrett and Edmunds, 1970) and W_6ZHL (Edmunds, 1975; Edmunds et al., 1976) photosynthetic mutants batch-cultured in the infradian growth mode at temperatures less than 19°C in DD or LL (Fig. 3.35). Similar observations have been made in semicontinuous culture of the P_4ZUL mutant (Fig. 3.38): long trains of circadian oscillations were exhibited in DD or LL over a span of 10 to 14 days (Edmunds et al., 1976) that both of these mutants gradually may lose both their capacity to exhibit division synchrony in LL or DD and to be entrained by LD cycles, reverting to random exponential growth ($g < 24$ h). These properties could

FIGURE 3.37. Persistence of the free-running, circadian rhythm of cell division in cultures of *Euglena gracilis* (Z strain) grown photoautotrophically at 25°C under two exotic LD regimens, each of which furnished a total duration of 12 h of illumination (cool-white fluorescent, 7500 lx) and 12 h of darkness during a 24-h timespan but which provided no 24-h time cue to the cells. The growth curve (A) at the top was obtained in a random illumination regimen, the duration of whose signals varied between 15 min and 180 min according to the distribution indicated in the inset. That at the bottom (curve B) was obtained from a culture exposed to a high-frequency LD: 1,1 cycle (T = 2 h). The step sizes *(ss)* and periods for the oscillations are indicated as in Figure 3.31. The mean value of the free-running period (τ) for the rhythm obtained in the random LD regimen was 28.5 h as read from the growth curve, whereas that in LD: 1,1 was 27.3 h. These values corresponded, respectively, to those computed by time series analysis of 28.2 h and 26.6 h. (Reprinted from Edmunds and Laval-Martin, 1984, p. 306, with permission of Marcel Dekker, Inc.)

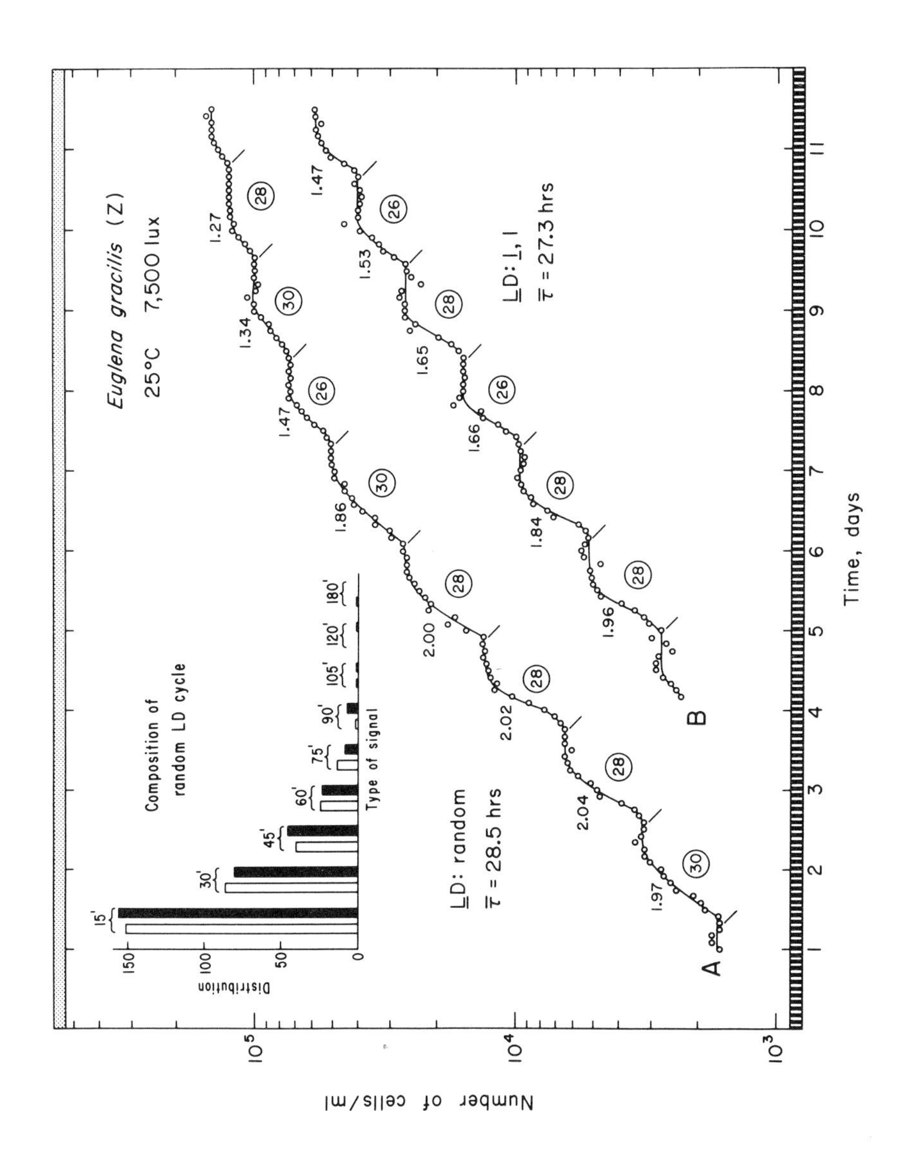
Euglena gracilis (Z)
25°C 7,500 lux
Composition of random LD cycle
Distribution
150
100
50
0
15'
30'
45'
60'
75'
90'
105'
120'
180'
Type of signal
LD: random
τ̄ = 28.5 hrs
LD: 1, 1
τ̄ = 27.3 hrs
A
B
1.97
2.04
2.02
2.00
1.86
1.47
1.34
1.27
1.96
1.84
1.66
1.65
1.53
1.47
30
28
26
Number of cells/ml
10^5
10^4
10^3
1
2
3
4
5
6
7
8
9
10
11
Time, days

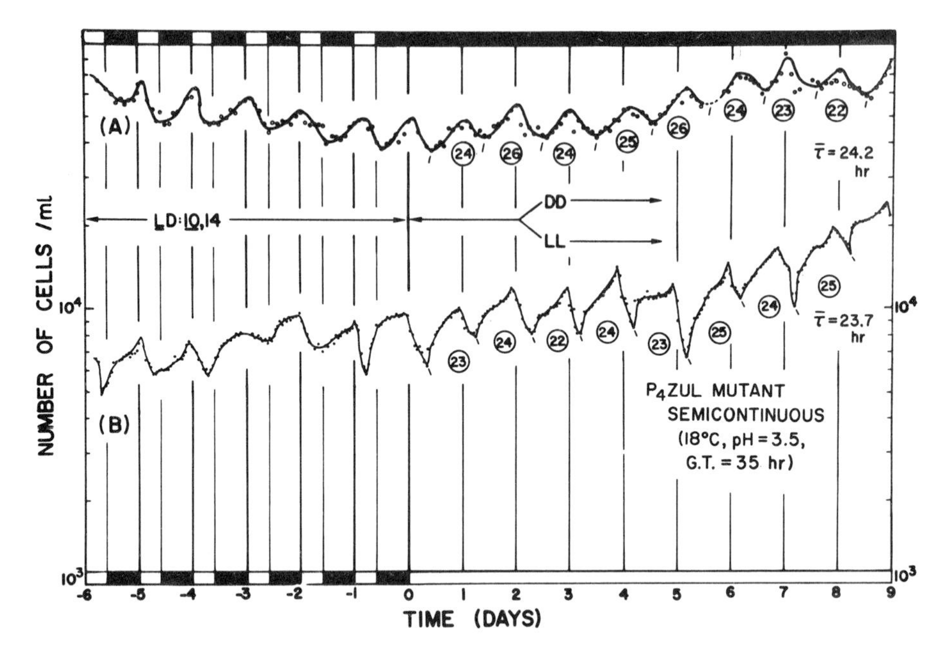

FIGURE 3.38. Long-term, persisting circadian rhythm of cell division in two different semicontinuous cultures of the P_4ZUL photosynthetic mutant of *Euglena gracilis* grown on low-pH glutamate-malate medium at 18°C. The cultures were first synchronized by LD: 10,14 (6 monitored days shown) and then placed either in DD (curve A) or in LL (curve B); the first nine cycles under constant conditions are indicated. The overall generation time *(g)* of both cultures was calculated from the known dilution rate to be about 35 h. Successive period lengths are encircled just below each free-running cycle, and the average period (τ) is given to the right of each curve. (Reprinted from Edmunds et al., 1974, with permission of Igaku-Shoin.)

be restored, however, by the addition of certain sulfur-containing compounds (such as cysteine or methionine) to the medium at the onset of the experiment (Fig. 3.39). If these substances were added at various times to an arrhythmic P_4ZUL culture in LL (after prior exposure to LD: 14,10), periodic division was induced whose phase was that predicted on the assumption that the underlying clock had been running undistrubed (but unexpressed) throughout the experiment, merely having been uncoupled from division itself until the sulfur compounds were added (Fig. 3.40). This hypothesis could be tested further by monitoring simultaneously another rhythm (such as motility), presumably a manifestation of the same oscillator, but whose coupling to the clock would not be affected by these compounds (or lack thereof).

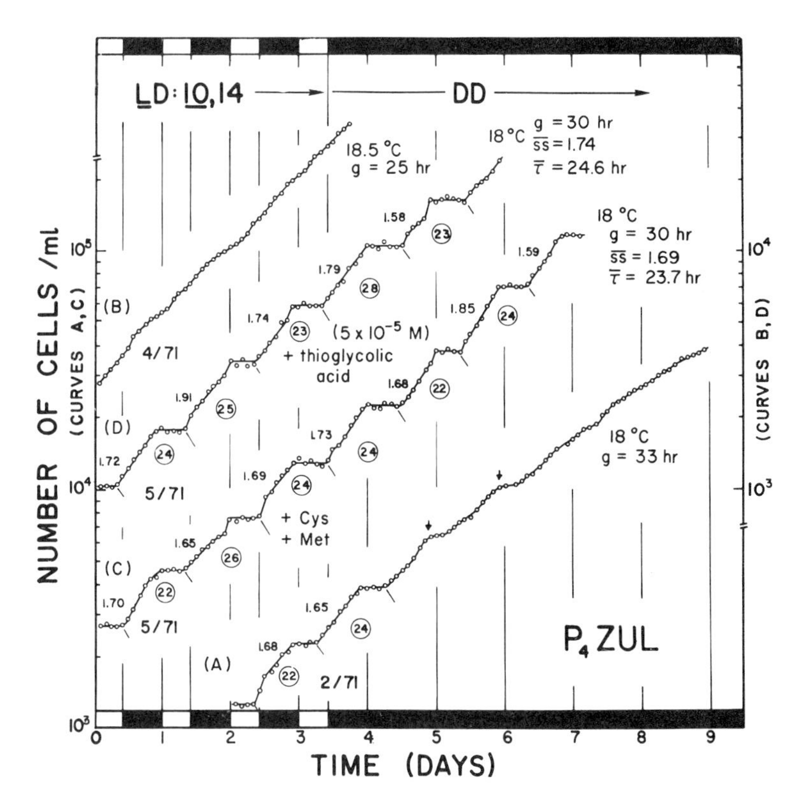

FIGURE 3.39. Gradual loss and subsequent restoration of circadian division rhythmicity in organotrophically batch-cultured populations of the UV light-induced P_4ZUL photosynthetic mutant of *Euglena gracilis*. (A) Initial loss of persistence of the rhythm in DD. (B) Loss of capacity for entrainment of the rhythm by an LD: 10,14 cycle a few months later. (C) Restoration of the capacity for both entrainment and a DD free-run by the addition of cysteine and methionine (1×10^{5} *M*) to the medium. (D) Restoration of synchrony by the addition of thioglycolic acid (5×10^{-5} *M*) to the medium. The generation time *(g)*, the mean step size *(ss)*, or fractional population increase for the successive division bursts, and the mean free-running period (τ) in DD for the growth curves are indicated. (Reprinted from Edmunds et al., 1976, with permission of Springer-Verlag.)

Initiation

It is a common feature of circadian rhythms that they damp out or are not observed in continuous bright illumination (in contrast to dim LL or DD). This also holds true for the cell division rhythm in *E. gracilis:* is more intense LL (e.g., 3500 to 7500 lx), photoautotrophic cultures of the wild type (z strain) at 25°C (Edmunds, 1965a) revert to asynchronous, exponential growth (g = 11 to 15 h). A single switch-down in irradiance

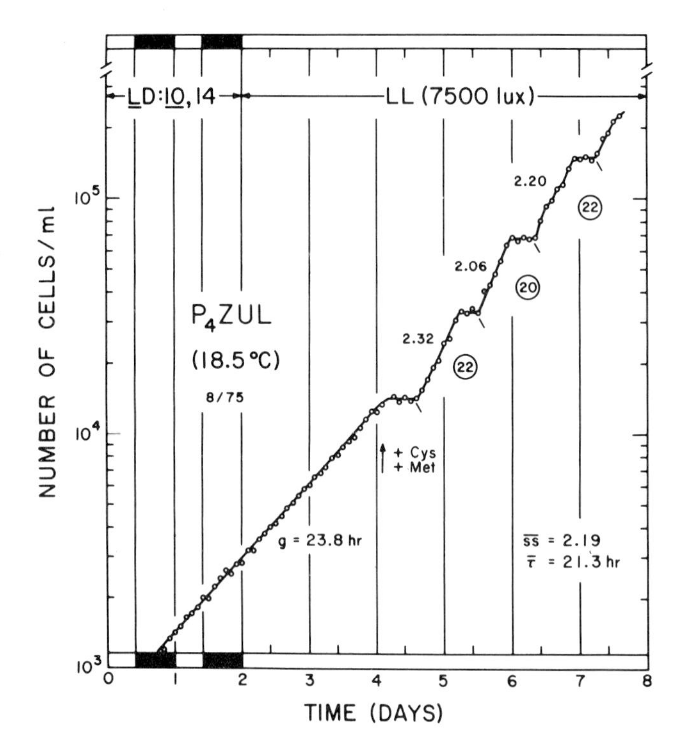

FIGURE 3.40. Induction of circadian division rhythmicity in an asynchronously dividing (exponentially increasing) population of the P_4ZUL mutant of *Euglena gracilis* organotrophically batch-cultured in LD: 10,14 and subsequently released into LL by the addition of cysteine (1×10^{-5} *M*) and methionine (5×10^{-5} *M*) at the time indicated by the arrow (during the third day in LL). The phase of the elicited rhythms is comparable to that usually found in LD, with divisions occurring in the subjective night. Other notations are as for Figure 3.31. (Reprinted from Edmunds et al., 1976, with permission of Springer-Verlag.)

to 800 lx was sufficient to elicit a free-running circadian rhythm that persisted for at least two or three cycles, with the first ϕ_rs being observed about 14 h after the switch-down (Edmunds, 1966). Two hypotheses could explain these observations. Either (1) the endogenous oscillators in the individual cells comprising the population were absent or arrested at some specific phase point and subsequently initiated or released by the transition (primary arrhythmicity), or (2) the individual oscillators were all running in bright LL but were out-of-phase with each other (secondary arrhythmicity). The first explanation appears to be more likely (Edmunds, 1966), although this may not be the case for the initiations of rhythmicity (Jarrett and Edmunds, 1970) in arrhythmic cultures of the P_4ZUL mutant growing in DD by a single switch-up in irradiance to LL (5000 lx).

More recently, we explored this problem further in cultures of *E. gracilis*

(Z strain) exposed to the higher-frequency cycles discussed in the previous subsection. It is not necessary for the cultures to have been exposed to a prior entraining diurnal light cycle for rhythmicity to be exhibited. Cells precultured in LL and then maintained in LD: 1,1 (Fig. 3.37B) or LD: 3,3 (see Fig. 3.41) or random illumination regimens (Fig. 3.37A) from the moment of inoculation all were rhythmic. Furthermore, if arrhythmic cultures growing exponentially in LL (7500 lx, 25°C) with a value of *g* of 11 to 15 h were placed in any of these cycles, circadian periodicity was quickly induced, with values of τ varying from 26.6 to 30.6 h depending on the substrain and regimen used. For example, in some 15 different cultures transferred from LL to LD: 3,3 the ϕ_rs of the elicited rhythms were found to occur at intervals after the onset of the first dark pulse of the imposed cycle given by the expression 19.1 h CT + $n\tau$ (where n is an integer and all times have been normalized to 24 h). Since the ϕ_r is known to occur at about CT 12 with reference to an LD: 12,12 cycle, one can infer from this empirical finding that all the oscillators comprising the population were stopped at approximately CT 16.9 in LL and were then set into motion from this phase point by the first transition (or, alternatively, that all the clocks were running in LL and were subsequently reset to CT 16.9 by the dark pulse). This value is somewhat larger than that of CT 12 found as an LL arrest point for many circadian systems. We note, however, that the LD: 3,3 regimen itself may not be without effect (Adams et al., 1984).

Phase Shiftability

Another basic property of circadian rhythms (see Fig. 1.2) is that their phase can be reset (or shifted)—by the lengthening or shortening of the period of one or more oscillations—by single light (or dark) or temperature signals. Indeed, the twin notions of phase and period control are key elements of Pittendrigh's theory (1965) for the entrainment of circadian clocks by light cycles. It is also characteristic of circadian rhythms that both the sign and magnitude of phase shifts engendered by the light signals are predictably dependent on the subjective circadian time at which the perturbations are applied. That light can reset circadian division rhythms also has been demonstrated in a general sort of way for several unicellular organisms, such as the dinoflagellate *Gonyaulax polyedra* (Sweeney and Hastings, 1958; Hastings and Sweeney, 1964; McMurry and Hastings, 1972a).

Similarly, the phenomenon of phase perturbation is inherent in most if not all models for cell cycle oscillators. The response of cells to external influences is often strongly dependent on the time of the CDC at which the agent is imposed. Thus, synchronous cultures of eukaryotic unicells, such as the ciliate *T. pyriformis* (Zeuthen, 1974) or the fission yeast *S. pombe* (Zeuthen, 1974; Polanshek, 1977), and of animal cells, such as mouse fibroblast L cells (Miyamoto et al., 1973), all exhibit increased

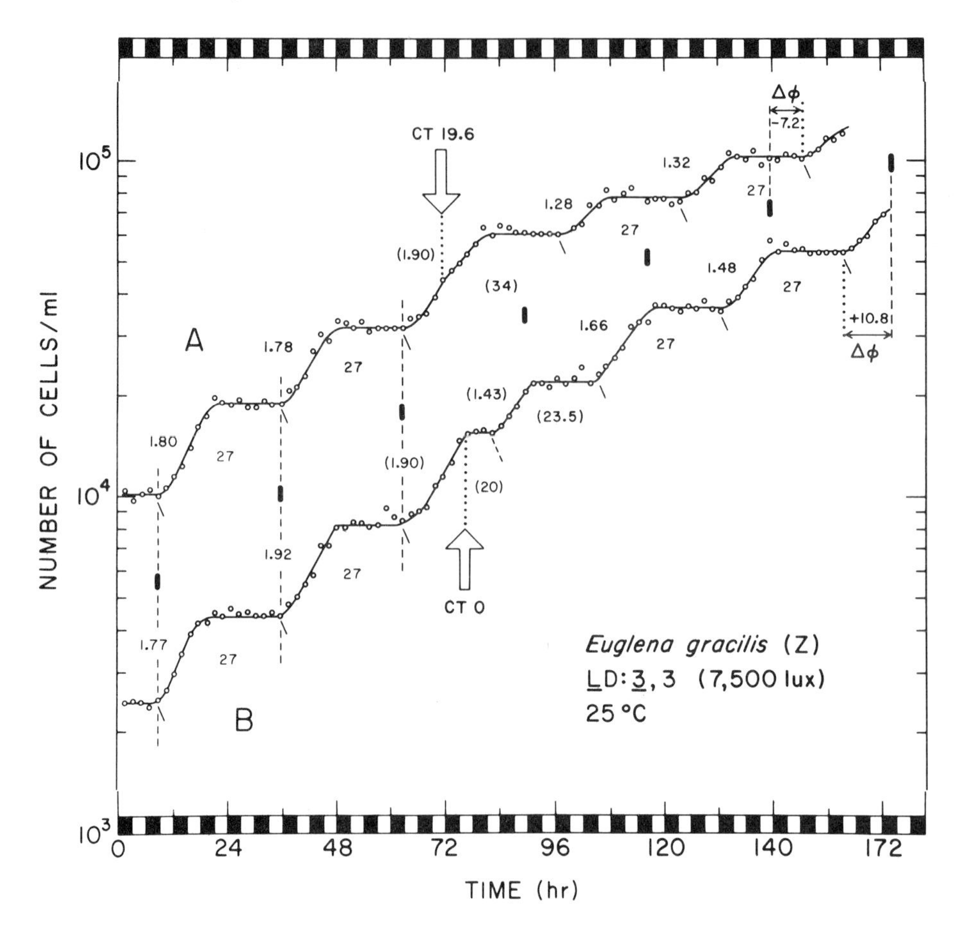

FIGURE 3.41. Examples and methodology used in calculation of phase-shift delays ($-\Delta\phi$) and advances ($+\Delta\phi$) of the free-running rhythm of cell division obtained in cultures of *Euglena gracilis* maintained in LD: 3,3 at 25°C and perturbed by 3-h light (7500 lx) signals given at different circadian times. Two in-phase, synchronously dividing cultures (A, B), at approximately the same titer (ordinate) and exhibiting a stable period of 27 h (vertical dashed lines) over a timespan of at least 72 h (abscissa), were subjected to a light pulse whose midpoint occurred at either hour 70.5 or hour 76.5, corresponding to circadian times (modulo 24) CT 19.6 or CT 24.0 (= CT 0), respectively. Onsets of cell division were taken as phase-reference points (ϕ_r) and considered to fall at CT 12.0, corresponding to onset of darkness in LD: 12,12. After transients had subsided and free-running τ of 27 h had been reestablished, the difference in phase ($\Delta\phi$, here given in real time) between the shifted rhythm (vertical dotted lines) and that projected (solid markers) for the rhythm before it was perturbed was measured and taken to be the steady-state phase shift engendered by the light signal. Other notation are as for Figure 3.31. (Courtesy of Edmunds and Laval-Martin, unpublished; see Edmunds et al., 1982.)

division setback (or excess division delay) to heat-shock and to a variety of chemical agents applied to progressively later times during their CDC. Perhaps more germane is the finding in *S. pombe* (Smith and Mitchison, 1976) that the same agent can produce delays when imposed at one point of the CDC and advances when given at another. Indeed, Klevecz and co-workers have discovered that synchronous mammalian V79 cells perturbed by serum (Klevecz et al., 1978), heat-shock (Klevecz et al., 1980b), or ionizing radiation (Klevecz et al., 1980a) at 0.5-h intervals across their 8.5-h CDC display a biphasic, 4-h PRC comprising both advances and delays in subsequent cell divisions (see Fig. 3.22).

Despite these investigations, however, a detailed PRC for a circadian mitotic clock has been lacking. Recently, we filled this void by using photoautotrophic cultures of *E. gracilis* free-running in a high-frequency LD: 3,3 cycle and displaying, under the experimental conditions used (Edmunds et al., 1982), a stable circadian period ($\tau \cong 27$ h). At different times throughout the 27-h division cycle, 3-h light perturbations were imposed systematically on free-running cell populations by giving light during one of the intervals when dark would have fallen in the LD: 3,3 regimen (Fig. 3.41). Using the onset of division as the ϕ_r, the net steady-state phase advance or delay ($\pm\Delta\phi$) of the rhythm was determined after transients, if any, had subsided (usually in 1 or 2 days) relative to an unperturbed control culture. Both $+\Delta\phi$ and $-\Delta\phi$ were found with maximum values of 11 to 12 h being obtained at about CT 23 (the breakpoint). Little if any phase shift occurred if the light signal was given between CT 6 and CT 12 (Fig. 3.42, top panel). The phase-shifting curve (Fig. 3.42, bottom panel) obtained by plotting new phase (ϕ') versus old phase (ϕ) was of the type 0 (strong) variety (Winfree, 1980). Light pulses, no matter when imposed, engendered new phases that mapped to a relatively restricted portion (CT 6 to CT 13) of the circadian cycle (Edmunds et al.,, 1982).

Several other interesting features of the PRC for *E. gracilis* emerge. In the first place, the CT at which all delay or maximum advance phase shifts were achieved corresponded to the approximate position of the CDC during which division occurred (commencing about CT 12) and to the first few hours of the G_1 phase. These times, moreover, were exactly those when the free-running rhythm of photosynthetic capacity (Laval-Martin et al., 1979; Edmunds and Laval-Martin, 1981) in LD: 3,3 displays the lowest values. Maximum values (CO_2 fixed/cell/hour) occur at the very time (CT 6 to 12) that 3-h light signals were virtually ineffective in phase-shifting the division rhythm. This observation demonstrates that although light is needed as a substrate for photosynthesis and the progression of the CDC in photoautotrophically cultured *E. gracilis*, it serves a distinct and separable function in phase shifting and entraining the circadian oscillator(s) hypothesized to underlie the rhythm of cell division, whose rhythmic sensitivity to light is reflected in the PRC. The phase shifts ob-

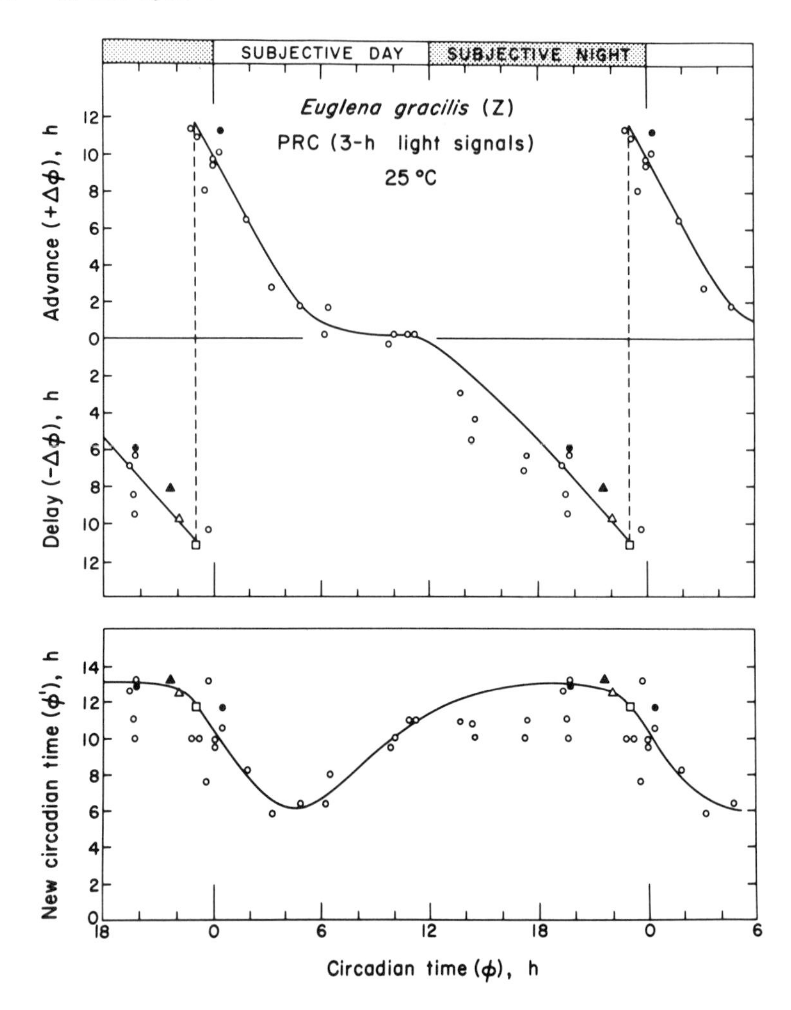

FIGURE 3.42. Top panel: Phase-response curve for the action of 3-h light signals (○, 7500 lx; □, 2800 lx; △, 1500 lx; ●, 700 lx; ▲, 300 lx) on the free-running rhythm of cell division in photoautotrophic cultures of *Euglena gracilis* maintained in LD: 3,3. The steady-state phase shift ($\pm \Delta\phi$), determined as described in Figure 4.41 but normalized to 24 h, is double-plotted as a function of the circadian time (modulo 24) at which the midpoint of the signal was given. The breakpoint inherent in this type of plot, indicated by the dashed line at CT 23, would disappear if the curve were drawn in monotonic form with all the phase shifts being treated as delays. Lower panel: The resultant new CT ($\phi' = \phi \pm \Delta\phi$) following the perturbation is plotted as a function of the CT (ϕ) at which the signal was given. The smooth sinusoidal curve thus obtained indicates that the PRC is of the type 0 (strong) variety. (Reprinted from Edmunds and Laval-Martin, 1984, p. 310, with permission of Marcel Dekker, Inc; see Edmunds et al., 1982; Malinowski et al., 1985.)

served appeared to constitute true developmental advances and delays. In the case of the former, a higher cell titer was attained before the same cell concentration was reached in the unperturbed control, whereas the converse held for phase delays.

The PRC derived for the effects of white light pulses on the circadian rhythm of cell division in *E. gracilis* (Fig. 3.42) is qualitatively similar to those for many other organisms (Pittendrigh, 1965; Winfree, 1980). We note, however, that the breakpoint is displaced somewhat to the right; if the onset (as opposed to the midpoint) of the 3-h light pulse or the onset of the resultant 9-h light span were used in the calculations, the PRC would be shifted to the left and the discontinuity would lie at approximately CT 19 to 20, close to that found for the eclosion rhythm in *Drosophila pseudoobscura,* for example (Pittendrigh, 1965). Finally, we caution that we have taken what is perhaps the simplest interpretation for the action of these light signals and have disregarded possible complications due to the interaction of the imposed LD: 3,3 cycle (free-running conditions) with the circadian oscillator [see Adams et al. (1984) for a more detailed analysis].

In contrast to the phase-shifting effects of light pulses, the action of dark pulses on free-running circadian rhythms (usually in LL) has received far less attention. In animals, such as the tropical nocturnal bat *Taphozus melanopogon* (Subbaraj and Chandrashekaran, 1978) or the golden hamster, *Mesocricetus auratus* (Boulos and Rusak, 1982), the PRCs for light or for dark perturbations have been reported to be mirror images of each other (i.e., 180° out-of-phase), although Ellis et al. (1982) find that dark signals in LL perturb the hamster circadian system in a somewhat more complex manner. Similarly, light and dark pulses seem to have opposite and mutually complementary influences in plants, such as *Bryophyllum fedtschenkoi* (Wilkins, 1960), and in the unicells *G. polyedra* and *Acetabularia mediterranea* (Karakashian and Schweiger, 1976a). Another generalization from these studies is that PRCs for dark pulses seem to have smaller amplitudes than those of PRCs for light signals of the same duration. The PRC for 3-h dark perturbations (replacing a normal 3-h light interval in a manner analogous to that employed for the derivation of the light-pulse PRC described in Fig. 3.42) for the circadian rhythm of cell division in *E. gracilis* free-running in LD: 3,3 also seems to follow these generalizations, with very little, if any phase advance being observed (Edmunds and Laval-Martin, 1984).

Singularity Point

Biological pacemakers may exhibit limit cycle dynamics (Kauffman and Wille, 1975; Shymko et al., 1984; see Section 3.1.2). Pavlidis (1968), among others, reformulated in terms of space-state topology the analytical model of Pittendrigh (1965) for the entrainment of circadian oscillators by light cycles whose dynamic behavior can be described by a set of two first-order differential equations involving two variables (which, if known at any given time, allow prediction of the behavior of the system at some later time). These two variables, termed the "state of the systems," can be represented as coordinates in a plane, and the behavior of the system can be described in terms of plane curves or trajectories. The oscillatory motion of a biological clock, in turn, can be represented as a stable periodic trajectory that closes in on itself and is termed a "stable limit cycle." Such a system, if distributed (as, e.g., by a light pulse or an inhibitor pulse), will always tend to return to an equilibrium configuration.

Theoretical and mathematical studies have predicted that a circadian oscillator might be rendered arrhythmic—characterized by a phaseless, motionless state—by a critical pulse of a certain strength and duration given at a specific time (termed the "singularity point") in the circadian cycle (Winfree, 1970; for extensive treatment, see Winfree, 1980). As Winfree (1980) has pointed out, as stimulus strength is increased, the transition from type 1 (weak pulse) to type 0 (strong pulse) resetting is necessarily discontinuous at one special phase point, the breakpoint, corresponding to this unique singularity. This prediction has now been demonstrated for light perturbations in several circadian systems, including *D. pseudoobscura* (Winfree, 1970), *Kalanchoë blossfeldiana* (Engelmann et al., 1978), *Sarcophaga argyrostoma* (Saunders, 1978), and *Culex pipiens quinquefasciatus* (Peterson, 1980) and most recently for critical pulses of anisomycin (an inhibitor of protein synthesis on 80S ribosomes) in the dinoflagellate *G. polyedra* (Taylor et al., 1982b).

The threshold intensity of illumination for phase-shifting the circadian

FIGURE 3.43. Localization of a singularity point in the behavior of the circadian oscillator underlying the free-running rhythm of cell division in cultures of photoautotrophically grown *Euglena gracilis* (Z) perturbed by a critical 3-h light signal. Three initially in-phase, synchronously dividing cultures were used, each displaying a stable τ of approximately 30 h in an LD: 3,3 regimen at 25°C. At circadian time (CT) 0.4, culture A was perturbed by a 3-h light signal of 700 lx, whereas culture B was perturbed by a 3-h light signal of only 300 lx (each replacing one of the normal light intervals in the imposed LD cycle having an intensity of 7500 lx, as represented by the hatched arrows). Whereas the former signal elicited a phase shift ($\Delta\phi$) of + 11.2 h (modulo 24) CT (equivalent to 14.0 h real time) comparable to that seen for 7500-lx light signals (Fig. 3.42), the latter, weaker light pulse (curve B) induced total arrhythmicity, although cell division continued in the population. If a third culture (curve C) were perturbed by a 3-h signal with an identical intensity of 300 lx but imposed at a slightly different circadian time (CT 21.5), the cell division rhythm continued although its phase was delayed ($\Delta\phi$ = −8.0 h CT, equivalent to −10.0 h real time) in a manner observed for 7500-lx pulses (Fig. 3.42). (Reprinted from Malinowski et al., 1985, with permisssion of Springer-Verlag.)

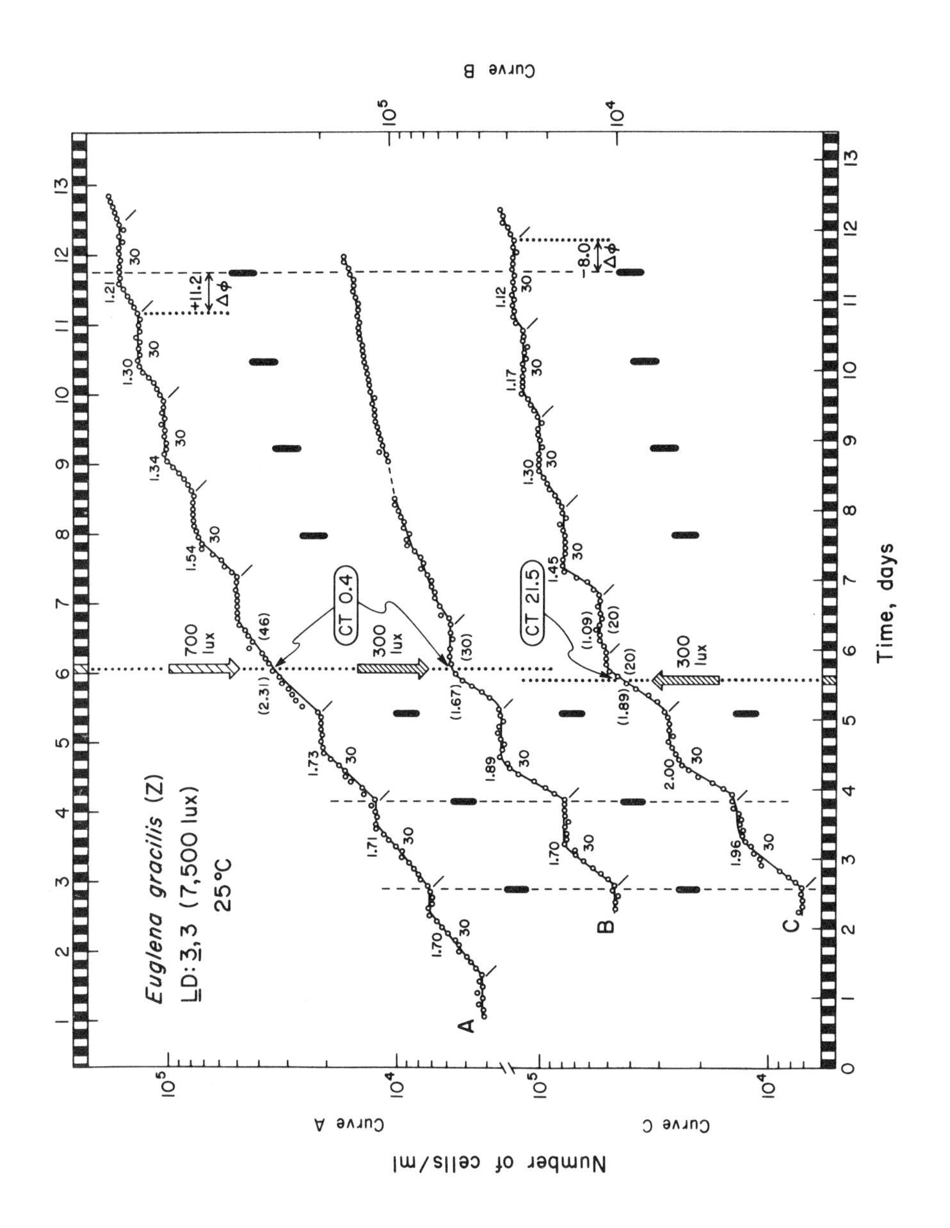
Euglena gracilis (Z)
LD: 3,3 (7,500 lux)
25°C
700 lux
300 lux
300 lux
CT 0.4
CT 21.5
Time, days
Number of cells/ml
Curve A
Curve B
Curve C

rhythm of cell division in *E. gracilis* by 3-h light signals has been examined (Malinowski et al., 1985). Perturbations with an intensity within the range of 700 to 7500 lx generated the same phase shift when imposed at a given CT (Fig. 3.43A). Light signals of lower intensity, however, elicited different responses, some quite dramatic. Thus, a 40-lx to 400-lx pulse given at CT 0.4 (the approximate location of the breakpoint at about CT 23) induced arrhythmicity (Fig. 3.34B), the population reverting to asychronous, exponential growth. The intensity of this annihilating pulse and the CT at which it was imposed were found to be very specific: a 300-lx stimulus given at CT 21.5 (Fig. 3.43C) merely generated a phase delay of the same magnitude found for 7500-lx signals (Fig. 3.41). Different degrees of asynchrony were observed as one approached the boundaries (lx and CT) of the critical pulse. The location of the singularity point at approximately CT 24 (CT 0) corresponds to the steepest part of the type 0 resetting curve (Fig. 3.42, bottom panel) and probably lies on the axis the three-dimensional phase-resetting surface of *E. gracilis*.

The existence of this critical pulse and its corresponding singularity point not only further supports the hypothesis that a circadian oscillator regulates the CDC in *E. gracilis* but also suggests (although it does not demand) that the pacemaker may have limit cycle dynamics. Inasmuch as the results were obtained with populations of cells, we cannot deduce the state of the oscillator(s) in individual cells. On the one hand, arrhythmicity in the culture may be the result of a dispersion of the phases of individual CDCs (cellular incoherence), entailing that the pulse differentially phase shift the cells that were synchronous at the time of the pulse. On the other hand, the critical pulse may have stopped each circadian clock comprising the population (but not the CDCs, which continue to run). To distinguish between these two alternatives for the eclosion rhythm of *D. pseudoobscura*, Winfree (1970) gave a second light perturbation at different times after the first critical pulse. His finding of a unimodal distribution of phases suggested that reinitiation rather than resynchronization had occurred. A similar situation probably exists for the cell division rhythm in *E. gracilis*, although this second-pulse test is difficult to perform and proper controls are not feasible in this system.

Temperature Compensation

A final and rather remarkable property of circadian rhythms (see Fig. 1.2) is that their period, but not their amplitude, is only slightly affected by the ambient steady-state temperature over the physiological range (Sweeney and Hastings, 1960). In fact, this is just what one would anticipate in a functional biological clock measuring astronomical time. In contrast, the duration of the CDC commonly is thought to be highly dependent on temperature, and, indeed, this is true for *E. gracilis* also (Terry and Edmunds, 1970a). A study (Anderson et al., 1985) of the effects of different constant temperatures ranging between 16 and 32°C on the generation

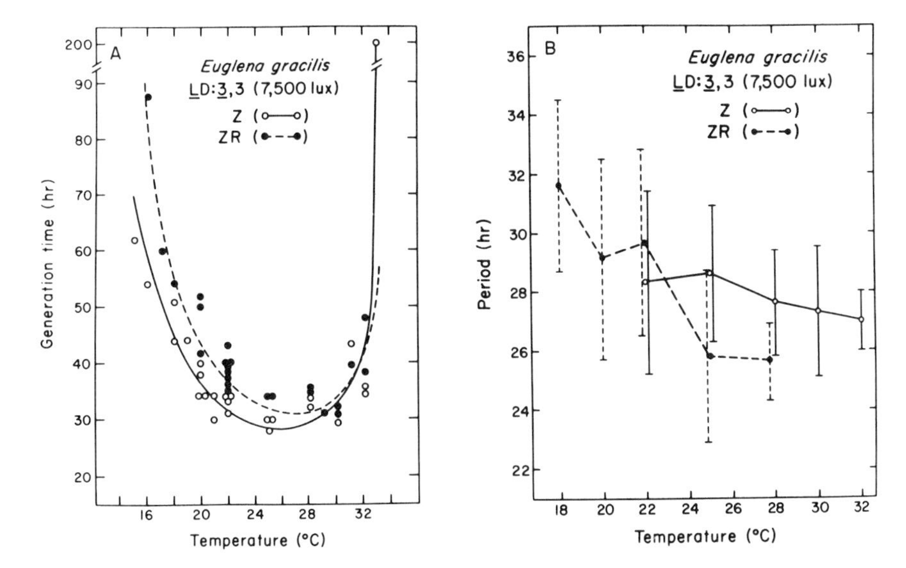

FIGURE 3.44. Contrasting effects of temperature on generation time *(g)* and the period (τ) of the circadian rhythm of cell division in *E. gracilis*. (A) Influence of temperature on the average *g* of two different strains (Z, wild type; ZR, DCMU-resistant) maintained in LD: 3,3 (7500 lx). Growth curves obtained at each temperature during the exponential growth phase where light intensity had not become limiting (usually between 10,000 and 60,000 cells/ml) were fitted with straight lines by linear regression (least squares) to obtain their slopes. (B) Temperature compensation of τ of the free-running circadian rhythm of cell division in the Z and ZR strains cultured in LD: 3,3. Periods were measured as the time between successive onsets of division (ϕ_r) until the stationary phase of growth was reached (usually at a cell titer approximating 100,000 cells/ml). Note that in strain ZR there was some lengthening of the period at lower temperatures. Although growth of the Z strain could be obtained at temperatures less than 22°C (18°C for ZR), rhythmicity disappeared. Errors bars denote ± 1 standard deviation of the mean. (Reprinted from Anderson et al., 1985, with permission of Academic Press.)

time of the wild-type Z strain and DCMU-resistant line (ZR) photoautotrophically batch-cultured in LD: 3,3 clearly illustrates this dependence (Fig. 3.44A).

Despite this apparent paradox, several lines of evidence suggest that the period of the circadian oscillator hypothesized to underlie rhythmic cell division and to produce the gating effect observed in a population of cells is conserved. For example, Klevecz and King (1982) found that the Q_{10} for cell division of the V79 line of Chinese hamster lung fibroblasts growing between 34 and 40°C was between 1.15 and 1.26, thus indicating that the mammalian CDC is temperature-compensated over a span of 6 to 7°C. We have observed previously (Edmunds and Adams, 1981) a similar compensation of the period of the free-running rhythm of cell division in cultures of the P_4ZUL photosynthetic mutant of *E. gracilis* grown in DD over a temperature range of about 7°C (14 to 21°C).

In a more extensive and rigorous comparative study of temperature compensation of the free-running period in the Z and ZR strains of *E. gracilis* maintained in LD: 3,3 at different steady-state temperatures within the physiological range (Anderson et al., 1985) a Q_{10} of 1.05 in the former was found (Fig. 3.34B), indicating that it was virtually unaffected by changes in temperature over a 10°C range (22 to 32°C). The circadian clock was not as well compensated in ZR (Fig. 3.44B), in which an average value for the Q_{10} of 1.23 was observed over a temperature range of 18 to 28°C. The difference between the two strains was reflected also in the distribution of their period lengths over the range of temperatures used the spread was considerably greater in ZR. Even the circadian oscillator of ZR was affected only modestly by temperature as compared to typical biochemical reaction rates and other biological processes, such as membrane transport. Finally, a critical permissive temperature of 22°C was found for the Z strain (18°C for ZR). Although slow exponential growth occurred at lower temperatures, synchrony was not seen.

Circadian Rhythms of Mammalian Cell Proliferation

Circadian rhythmicity in the CDC is not limited to unicellular organisms. Many data on synchronized cell cycles (with particular emphasis on the S and M phases) in rats and mice are available, in which many different tissues have been analyzed. Although many of these rhythmicities have been examined only under LD cycles (typically, LD: 12,12), in most instances where the incorporation of [^{3}H]TdR into DNA or the mitotic index have been monitored in DD or LL, the oscillations persist with a circadian period (see reviews by Scheving and Pauly, 1973; Rensing and Goedeke, 1976; Scheving, 1981, 1984; Scheving et al., 1983). The results of an intensive chronobiological study of the intestinal tract of the mouse are shown in Figure 3.45 (Scheving, 1981).

The findings from a number of earlier investigations are summarized in

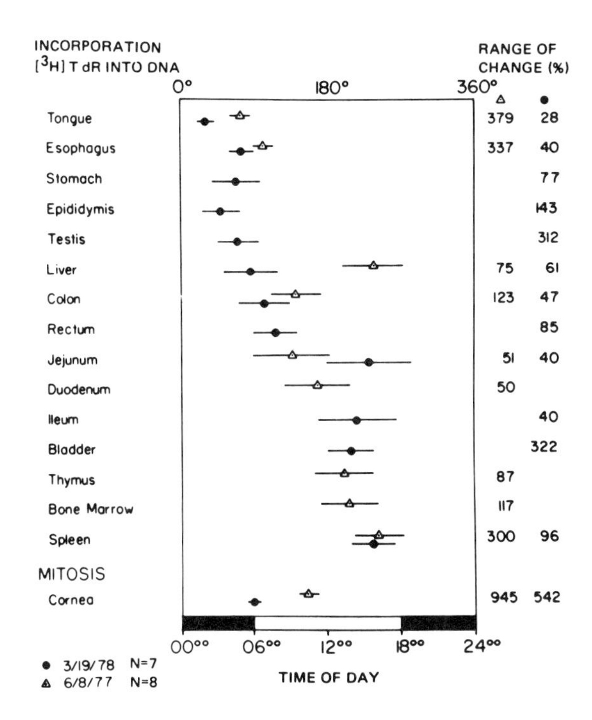

FIGURE 3.45. An acrophase map, showing the the approximate peak of the circadian rhythm in either the incorporation of [³H]TdR into DNA or in the mitotic index of the corneal epithelium of mice. The acrophase is represented by a dot or triangle, and the bars extending out from the dots or triangles represent the confidence limits. The dark bar on the abscissa portion represents the 24-h dark span of the environmental LD cycle. The percentage range of change reflects the change from the lowest to the highest time point means, when the lowest means equal to 100%. Values are rounded to the nearest integer. (Reprinted from Scheving, 1981, with permission of Alan R. Liss, Inc.)

Figure 3.46 (Rensing and Goedeke, 1976). In most tissues, DNA synthesis occurs around (subjective) dusk, followed by mitosis at approximately (subjective) dawn. In many cases, diminished rates of both mitosis and DNA synthesis have been recorded at other times also throughout a 24-h time span. More recent studies (Thorud et al, 1984) have confirmed the existence of statistically significant diurnal and circadian rhythms in cell population kinetics (rate of cell flux, as measured by the rate of mitotic or S phase influx and efflux) and in the distribution and duration of CDC phases in self-renewing mammalian tissues (epidermis, corneal epithelium, hemopoietic tissue, gastrointestinal tract, liver). Similarly, Keiding et al. (1984) have found a diurnal variation in the influx and transitions in the S phase of hamster cheek-pouch epithelial cells.

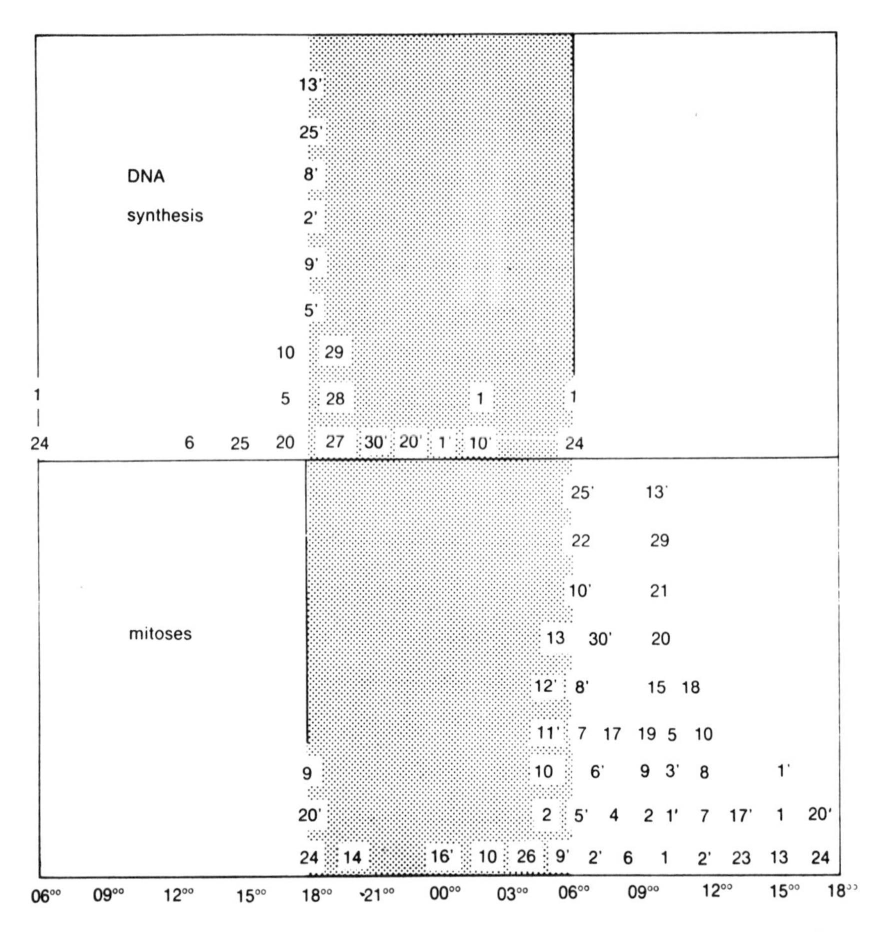

FIGURE 3.46. Maxima of DNA synthesis and mitoses in different tissues of rat and mice (primed numbers) maintained in LD: 12,12. Many of these diurnal rhythms have been shown to persist in DD or LL. Individual numbers refer to studies referenced by Rensing and Goedeke (1976) and reviewed by Rensing (1969) and Scheving and Pauly (1973). (Reprinted from Rensing and Goedeke, 1976, with permission of Publishing House Il Ponte.)

The results from studies of unicells and mammalian tissues do not necessarily imply that circadian modulations of the CDC will occur in isolated cells of metazoans (see Section 4.1). Rhythmicities in tissues, for example, might be generated or forced by synchronizing signals emanating from other organs in the body and transmitted by the blood or interstitial fluid system. Even observations of circadian oscillations in tissues cultured in vitro do not demand self-sustaining cellular oscillations because of the possibility of chemical or electrical cross-coupling among the constituent cells. Rensing and Goedeke (1976) have addressed this problem by determining the frequency of the different phases of the CDC in various cell cultures using impulse cytophotometry. Yoshida-ascites hepatoma cells, human embryonic fabroblasts, and rat liver and rat hepatoma cells all exhibited persisting circadian changes in the percentage of cells in the G_1, S, and G_2+M phases, although in certain of these cell types, the amplitude of the rhythm decreased markedly after 74 h in DD.

3.2.2 Ultradian, Circadian, and Infradian Interfaces

The preceding overview has attempted to present the evidence for the key role of a circadian oscillator(s) in the control of the CDC in the algal flagellate *E. gracilis,* taken to be representative of other unicellular (and, perhaps, multicellular) systems that display persisting circadian rhythms of cell division (see Table 3.1). Several other questions arise concerning the nature of the interaction of this putative oscillator with the CDC and with other overt periodicities in rapidly (ultradian) and slowly (infradian) cycling cells.

The Characteristics of the Oscillator

The formal properties of circadian clocks—entrainability, persistence, initiation, phase shiftability, and temperature compensation (see Fig. 1.2)—were found also to characterize the circadian rhythm of cell division in *E. gracilis* (Edmunds and Laval-Martin, 1984). The recent discovery of a singularity point at which the imposition of a critical light pulse generated an arrhythmic population (we do not yet know if the oscillators are stopped in individual cells of the culture or are merely dispersed in their phases) suggests that the underlying oscillator may be of the limit cycle type, although it does not demand it (Malinowski et al., 1985). Such oscillators are stable, under fixed conditions, and are independent of the starting concentrations of the components, whose levels vary rhythmically with time (see Section 3.1.2).

According to the working hypothesis, mitosis would not be an essential part of the oscillator but would lie downstream from it. Blockage of cell division should not stop the system from oscillating, at least at a subthreshold level. In this sense, cell division would be a hand of the underlying clock. This hypothesis has been tested recently in two ways (Edmunds and Laval-Martin, 1984: (1) Vitamin B_{12} deprivation has been shown to block completely the CDC in *E. gracilis* (Bré et al., 1981). If the division rhythm (free-running in LD: 3,3) was temporarily arrested due to low initial levels of vitamin B_{12}, and if this inhibition subsequently was released by readdition of vitamin B_{12} to the medium, the division rhythm started up again in phase with an unperturbed control (or in phase with an extrapolation of the stable period of the culture before vitamin B_{12} deprivation), thereby suggesting that the putative underlying oscillator had continued to run unabatedly throughout the timespan during which the CDC had stopped cycling. (2) If a pulse of lactate was given to a free-running culture, thereby temporarily accelerating the CDC (values of *g* approached 10 to 12 h) and overriding circadian oscillator controls (Jarrett and Edmunds, 1970), the phase of the rhythm when it was finally restored after the supplemental substrate had been depleted appeared to be in phase with that of an unperturbed control, leading to a similar interpretation. These latter results are consistent with those found for the in-phase restoration of rhythmicity in the P_4ZUL mutant free-running in LL by the addition of sulfur-containing compounds to the medium (Edmunds et al., 1976; Fig. 3.40). The circadian clock, therefore, phases the CDC, and not vice versa.

The question arises whether other events comprising the CDC besides division itself can be controlled by a circadian oscillator. To approach this question experimentally, one needs an organism with a value of *g* greater than 1 day and several clearly distinguishable morphological landmarks. Recently, Sweeney (1982), working with the large (900-μm long) dinoflagellate *P. fusiformis* Murray whose CDC has a minimum duration of 5 or 6 days, was able to isolate single cells or dividing pairs in capillary tubes kept in LD or DD and to determine five discrete morphological stages that could be correlated with the classic phases (G_1, S, G_2, M and C) of the CDC. Cells changed from one stage to the next only during the night phase (subjective) of the circadian cycle (Fig. 3.47), and cells in all stages displayed a circadian rhythm of bioluminescence. All morphological

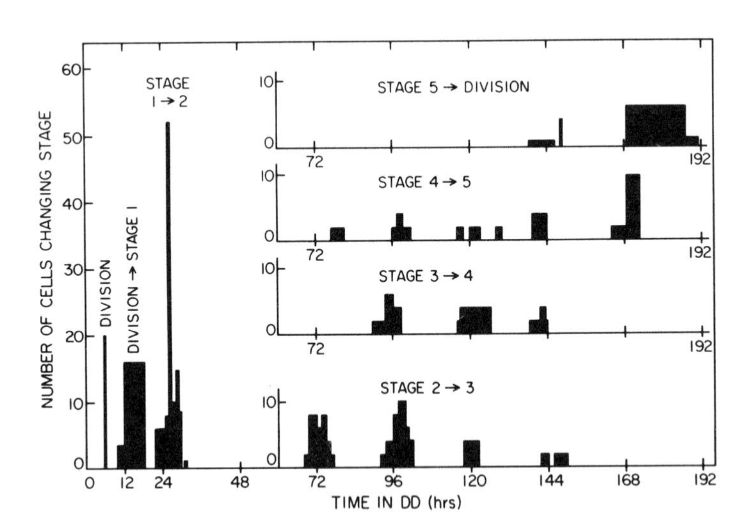

FIGURE 3.47. The timing of transitions between morphological stages of the cell cycle in the dinoflagellate *Pyrocystis fusiformis* in DD following an LD: 12,12 cycle. Zero time on the abscissa corresponds to the onset of an environmental light interval. Note that all changes are distributed discontinuously in time, at intervals of about 24 h, and hence are quantized to $n \cdot 24$ h. (Reprinted from Sweeney, 1982, with permission of American Society of Plant Physiologists.)

stages, therefore, and not only division, appear to be phased by the circadian oscillator(s). Sweeney (1982) interpreted these findings as being incompatible with a mechanism for circadian oscillations that invokes cycling in G_q, the quantal cycle that Klevecz (1976) has hypothesized to be appended to the G_1 phase (see Fig. 3.21). In contrast, Honma (1987) has found using flow cytometric techniques that the values for *g* of *G. polyedra* in LL are variable, but are quantized to integral multiples of the circadian period. He hypothesizes that the CDC of this unicell is divided into a cyclic part corresponding to G_1 and a noncyclic branch that constitutes the $S + G_2 + M$ sequence, which has a fixed duration of approximately 11 h. The former constitutes a circadian clock and, therefore, it takes one circadian period to traverse the subcycle. According to this scheme, cells that satisfy a minimal volume requirement (between CT 12 and CT 18) exit probabilistically to the replication-segregation sequence culminating in division, and reenter the cyclic portion a fixed time interval (11 h) afterwards.

Insertion and Deletion of Time Segments in Cell Cycles

The evidence reviewed in earlier sections (Edmunds and Laval-Martin, 1984) formally demands that a clock of some sort predictably insert time segments into, or delete them from, the CDC of *E. gracilis*. Some sort of mechanism, coupled in some manner to the LD cycle, seems to be necessary to stretch the CDC successively day by day to account for the fact that in slowly growing cultures (where $g > 24$ h and $ss < 2.0$) some cells divide during each subjective night but not in the intervals between (Figs. 3.33, 3.36). Similarly, in the growth curves at 19°C for the P_4ZUL mutant (Figs. 3.40; see Fig. 3.50), although divisions were seemingly set back by 8 to 12 h (plateaus), the increases in cell number, when they did occur, took place at a rate ($g \cong 14$ to 16 h) greater than that found for asynchronous, exponential growth in LL or DD ($g \cong 25$ to 26 h). The net result of this compensatory process was that a doubling of cell number still occurred every 24 h. The phase-shifting data reflected in the experimentally delivered PRC (Fig. 3.42) for light signals require an explanation for the observed phase advances and delays in the rhythm of cell division.

How, then, does a master (circadian) clock generate at the biochemical or molecular level the observed shortenings and lengthenings of individual CDCs? Assuming that this time dilation or contraction has an immediate molecular basis, one can envisage at least two ways by which such time segments could be added to or subtracted from the CDC (Edmunds and Adams, 1981). An indeterminate, variable number of traverses of Klevecz's g_q subcycle (1976) could generate the necessary variance in *g* values (see Section 3.1.2). The fundamental period of 4 h hypothesized for mammalian cell cultures in the ultradian growth mode would either have its analog in a circadian oscillator, which would then generate CDCs whose lengths would be integer multiples of 24 h or would be transformed somehow by a circadian clock into a longer subcycle. The former notion implicates a multiplicity of clocks (a clockshop; see Section 6.2.1), and the latter requires frequency demultiplication by another control system or by an internal modification of a versatile, pliable oscillator. Unfortunately, the quantal cycle thus far is only a deceriptive notion without molecular basis, although concomitant oscillations in enzyme activity have been observed in mammalian cell cultures (Klevecz, 1969a,b).

Alternatively, there is no reason to suppose that only one G_q subcycle is possible and that the duration of G_1 is to be accounted for solely by summation of a variable number of G_q. Other subcycles of different lengths and functional roles might exist from which the cell could choose. For example, one way that the CDC might be programmed would be for a collection of timing loops of different lengths to couple together in various combinations to form a flexible timer, or cytochron (Edmunds and Adams, 1981) (Fig. 3.48). This scheme is sufficiently generalized to apply to any eukaryotic cell cycle (both its strength and its weakness), although it was originally devised to incorporate experimental findings in *E. gracilis* (Edmunds and Adams, 1981; Adams et al., 1984). Temporal loci, or control

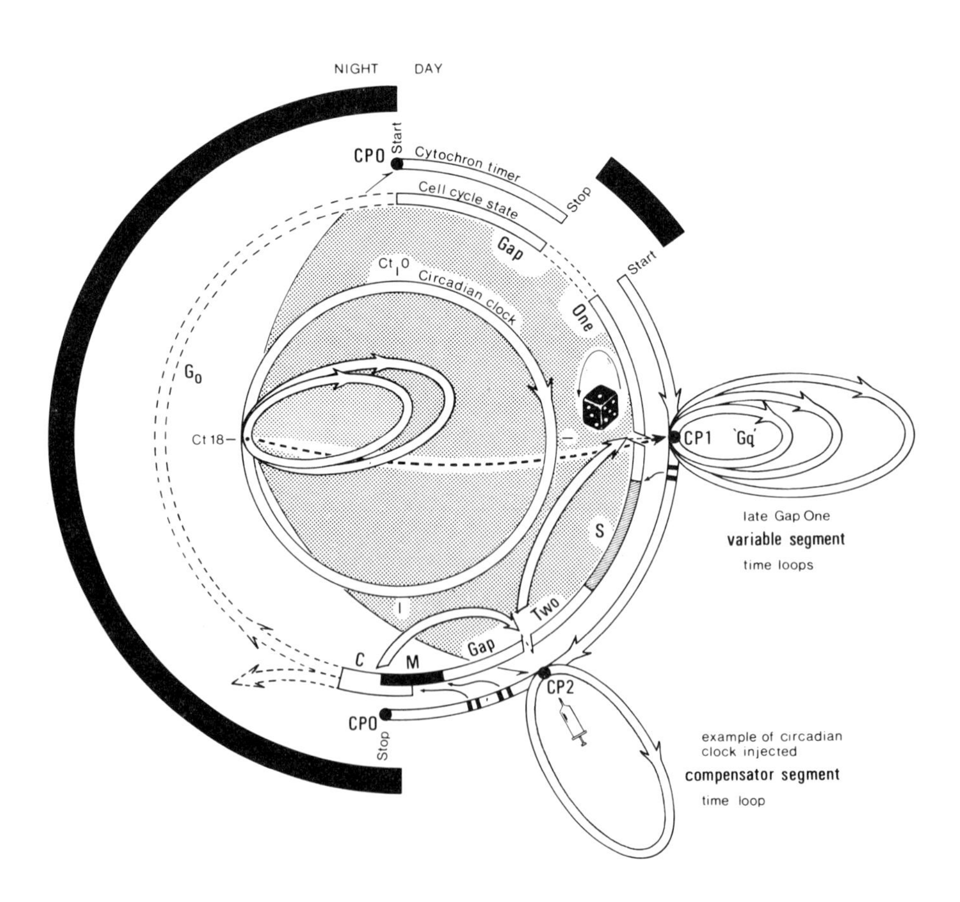

points, would exist along the cytochron track at which decisions would be made about the addition or deletion of time loops and would correspond to the G_1 (start) and G_2 potential arrest points of *S. cerevisiae* and *S. pombe* (see Section 3.1.2). We have hypothesized (Edmunds and Adams, 1981) the existence of chronogenes (the formal time segments) whose transcription would meter time and that would be inserted (or deleted) in varying numbers into the programmable cytochron by interaction with a circadian oscillator (see Section 5.3.3). Thus, the cytochron and the circadian clock are posited to be functionally independent, although not necessarily entirely separate as to mechanism.

Further analysis of the phase-shifting experiments described in Section 3.2.1, Fig. 3.41) has led to some modification of the cytochron concept (Adams et al., 1984). The width of the division steps in the growth curves appears to be expanded or contracted by advanding or delaying the circadian clock. Thus, the time gaps represented by the steps actually would be metered by the circadian oscillators themselves, which would stop and restart the cytochrons at specific certain circadian times. Similarly, the variation in the length of the CDC that results in the typical generation

FIGURE 3.48. A generalized model for the insertion and deletion of time segments in the track of the cytochron (cell cycle clock) hypothesized to program the events of the cell division cycle in *E. gracilis*. Cytochrons in each cell start up synchronously at dawn and meter time (unless interrupted by a dark pulse) around to the control point (CP0), triggering sequentially (black bars) the events leading to chromatin replication (S), mitosis (M), and cytokinesis (C). In photoautotrophic cultures in LD regimens, they stop (noncyclic mode) at CP0 in the dark, having triggered M and C, leaving cells in an untimed G_0 state until dawn restarts the cycle. In LL, or on certain organic substrates, however, the cytochron is cyclic and runs on through CP0 (i.e., there is no G_0). At CP1, the addition of one or more variable-segment time loops by a random selector (dice symbol) generates variability in the duration of G_1 and disperses (desynchronizes) subsequent cell divisions across the next dark interval. A circadian clock, entrainable by LD *Zeitgeber*, can couple to the cytochron at a unique circadian time (CT 18) and inject (syringe symbol) a determinate compensator-segment time loop into the cytochron track anywhere within the arc CP0 to CP2 (stippled), so that the cell cycle is extended to a subsequent circadian cycle. Thus, divisions are phased in bursts or clusters at 24-h intervals, the scatter within each cluster being generated by the G_1 variable-segment time loops, dawn synchronization of the cytochrons ensuring that each burst is confined to the dark interval in each LD cycle. The circadian clock can also apparently delete G_1 variable loops and reduce cell cycle variation under some regimens (dashed arrow). Division is suppressed in cells approaching the infradian (stationary) phase of population increase, but multiple rounds of S raise the ploidy level. When conditions improve, ploidy is reduced by successive rounds of M and C. Anticlockwise arrows represent these temporary loop closures. (Reprinted from Edmunds and Adams, 1981, with permission of American Association for the Advancement of Science.)

time distribution (skewed toward shorter values of *g*) seems to be generated by a separate timer–clock mechanism (the varichron) that runs in parallel rather than in series with the time-metering track of the cytochron.

The behavior of this circadian clock that stops and starts the cytochrons in *E. gracilis*—termed the "circadachron" to distinguish it from the other types of timer in the same cell having cycle times of about 24 h—appears to be governed by the following formal rules (Adams et al., 1984): (1) The circadachron stops when it reaches CT 12 during a light interval or pulse and remains stopped until light-off. (2) If darkness prevails as the circadachron meters its time track through CT 12 of the circadian cycle, it will

FIGURE 3.49. Characterization of the *Euglena gracilis* circadachron in response to light signals. (a) Phase response curve (PRC) for the resetting of the circadian rhythm of mobility and cell division by light pulses and photoperiods. Combined data from Schnabel (1968) (●), Feldman (1967) (▲), for mobility; from Adams et al. (1984) (■), and from the LD: 3,3 data of Figure 3.41 (□) for cell division. Available values for *Ceratium furca* (△) are plotted for comparison. Data points plotted at the time planes of light-on for advances, and at light-off for delays. All delays assumed to result from stopping at CT 12 in the light or resetting back to CT 12, the clock remaining at CT 12 in both cases until darkness ensues. Note the satisfactory fit to a -1 slope consistent with this asumption. Examples of resetting with 6-h light pulses, shown as narrow bars: a 6-hour pulse starting at CT 9 gives only a 3-h delay because the clock does not stop until it reaches CT 12. A 6-h pulse just starting before CT 18 will reset the clock back to CT 12 and hold it there until the end of the pulse. A 6-h pulse just starting after CT 18 will advance the clock at a light-on of the pulse as defined by the advanced segment of the PRC. (b) Resetting map for the light-induced resetting of the *Euglena* circadachron. Concentric arrows, clockwise for the advancing segment and anticlockwise for the reset to CT 12 segment (delay), link the times that the circadachron is challenged by a light pulse with the times to which it resets. Note the double cusp in the advancing segment reset pattern caused by two nodes, or chronostats, centered on the -1 slope. (c) The relationship between PRC slope and resetting map spacing ratio. The times at which successive equally spaced pulses (a, b, c, d) fall on a segment of a given slope and are reset (a′, b′, c′, d′) are linked by arrows. For a given slope (S), there is a simple relationship between the spacing of the time of occurrence of the light signals (set at unity) and the spacing of the times they reset the clock to. The spacing ratio is defined as 1:S + 1. For slopes within the segment 0 to -2, differences in timing between clocks in a population are reduced after being reset. All clocks are set to the same time at a slope -1, and on either side the pattern opens out until the initial and final spacings are equal at slopes of 0 (spacing ratio 1:1) and -2 (spacing ratio 1:$-$1), and any timing differences are conserved. Note that for slopes of < -1, the sequence a, b, c, d is inverted to d′, c′, b′, a′, hence the negative ratio. For slopes of > 0 or < -2 the differences in timing between clocks are compounded by the resets. Thus for slopes of $+2$ and -4, differences are magnified by a factor of 3. (Reprinted from Adams et al., 1984, pp. 416-417, with permission of Marcel Dekker, Inc.)

(**a**)

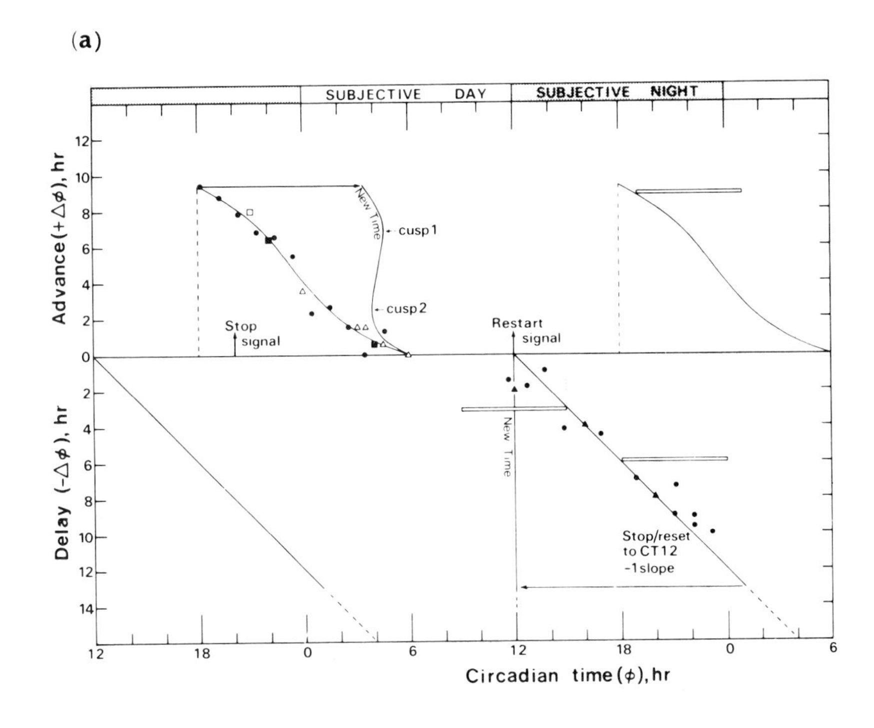
SUBJECTIVE DAY
SUBJECTIVE NIGHT
Advance (+Δφ), hr
Delay (−Δφ), hr
New Time
cusp 1
cusp 2
Stop signal
Restart signal
Stop/reset to CT12 −1 slope
Circadian time (φ), hr

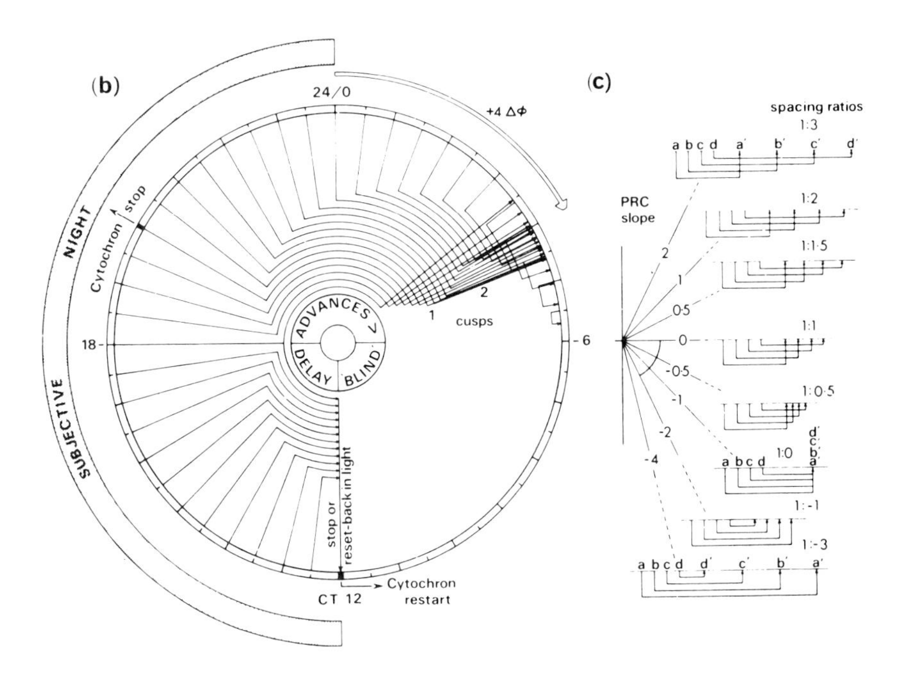
(b)
24/0
+4 Δφ
NIGHT
SUBJECTIVE
Cytochron stop
18
6
2
1
cusps
ADVANCES
DELAY
BLIND
stop or reset-back in light
CT 12
Cytochron restart
(c)
spacing ratios
1:3
a b c d a' b' c' d'
PRC slope
1:2
1:1·5
1:1
1:0·5
1:0
1:-1
1:-3
2
1
0·5
0
-0·5
-1
-2
-4

run on, but should a light pulse intervene during its traverse of the CT 12–18 segment, it will be set back to CT 12 and held there until darkness again ensues. (3) If the circadachron reaches CT 12 in the light (or is set back to CT 12) and is stopped and held there until dusk, it cannot be set back again by a subsequent light pulse impinging of the CT 12–18 segment during the same circadian cycle. (4) The segment between CT 6 and about CT 12 is insensitive, or blind, to light signals. (5) The circadachron is sensitive to light-on (D/L transition, but blind to light-off (L/D), in the segment from about CT 18 to about CT 6, the responses in this segment being + $\Delta\phi$s whose magnitudes are nonlinear and dependent on the CT of the D/L transition. (6) The phase advance expected of a given light-on signal is negated unless it is confirmed by a light pulse in excess of about 4 to 5 h. These formal features are summarized in the light-response resetting map and the PRC shown in Figure 3.49.

Thus, the posited cytochron–varichron–circadachron complex represents several linked timing devices that would work in conjunction with each other to ensure that as day length increases, a greater and greater proportion of the population receives two or more daily inputs of solar energy before being committed to cell division. In this way, the cells would have a greater chance of fixing enough carbon to reach an optimal size before fission. Simple sizer models (see Section 3.1.2) would be inadequate. The cells divide when their clocks program them to do so, and unbalanced growth in successive generations would be due to the natural selection on a microcomputerlike clockshop that on average maintains a balance between growth and cell division by adjusting the timing of division in relation to photoperiod, temperature, and nutrient status (Adams et al., 1984).

The Circadian–Infradian Rule

Our survey of circadian organization in eukaryotic microorganisms (see Chapter 2) has documented amply the point that in cells not proceeding through their normal CDC (as with cells in the plateau, or infradian, stage where little, if any, cell division is occurring, i.e., where the developmental sequence culminating in mitosis has been blocked or arrested), the circadian clock continues to operate, as evidenced by the manifestation of numerous overt circadian rhythms (see, e.g., Table 2.4 for the *E. gracilis* system, Section 2.2.3). These overt rhythmicities typically can be abolished by changing the environmental conditions so that the overall g of the culture is less than 24 h (ultradian growth mode), as, for example, by raising the temperature (Fig. 3.50A), increasing the intensity or duration of illumination (photosynthetic systems), or introducing usable carbon sources (Fig. 3.51) into an autotrophic system (Jarrett and Edmunds, 1970; Edmunds, 1978, 1981). Presumably, the clock either is operating at a higher frequency matching that of the fast-cycling CDC (with a lower limit equal

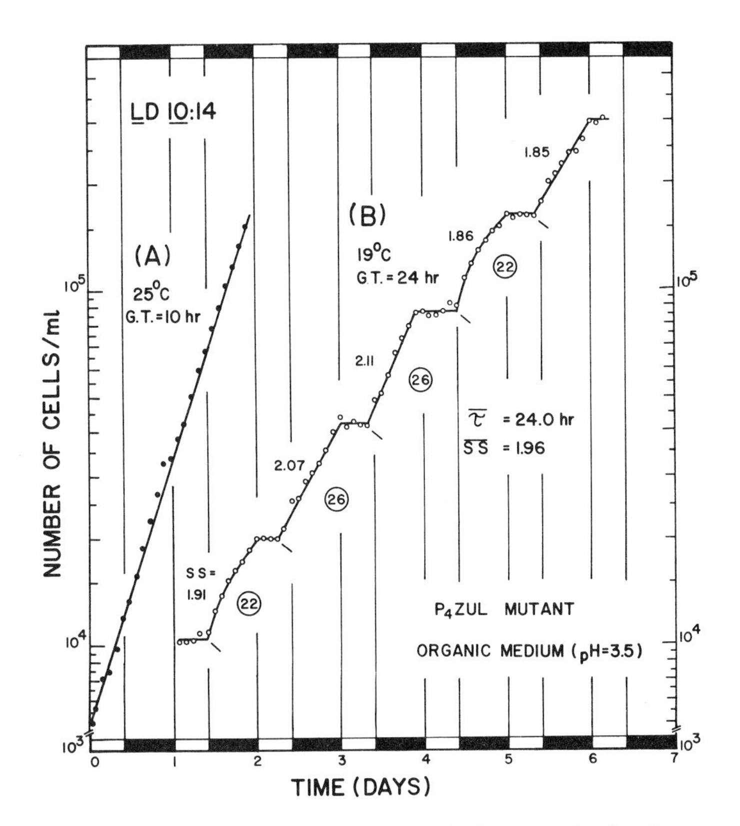

FIGURE 3.50. The circadian–infradian rule. Population growth of a photosynthetic mutant (P_4ZUL) of *Euglena gracilis* var. *bacillaris* strain Z (Pringsheim) grown on defined organotrophic medium in LD: 10,14. Curve A: exponential increase in cell number (g = 10 h) at 25°C. Curve B: Entrainment of cell division rhythm at 19°C. Other labels as for Figure 3.31. The average period ($\bar{\tau}$) of the rhythm in the population was 24.0 h, and the average step size ($\overline{ss}$) was 1.96, yielding a generation time (g) of about 24 h. (Reprinted, with permission, from Jarrett and Edmunds, 1970; copyright the American Association for the Advancement of Science.)

to that of the minimum possible g for a species—about 8 to 10 h in *E. gracilis*) or is running with a circadian period but is uncoupled from the CDC or perhaps in even stopped or absent (Ehret et al., 1977; Edmunds, 1978).

In any case, circadian rhythms have rarely, if ever, been observed in cultures of microorganisms in the ultradian growth mode (Ehret and Wille, 1970). However, there are some recent challenges to this generalization. Some phytoplankton species appear to maintain a persisting diel rhythm under constant conditions while reproducing at a rate greater than one

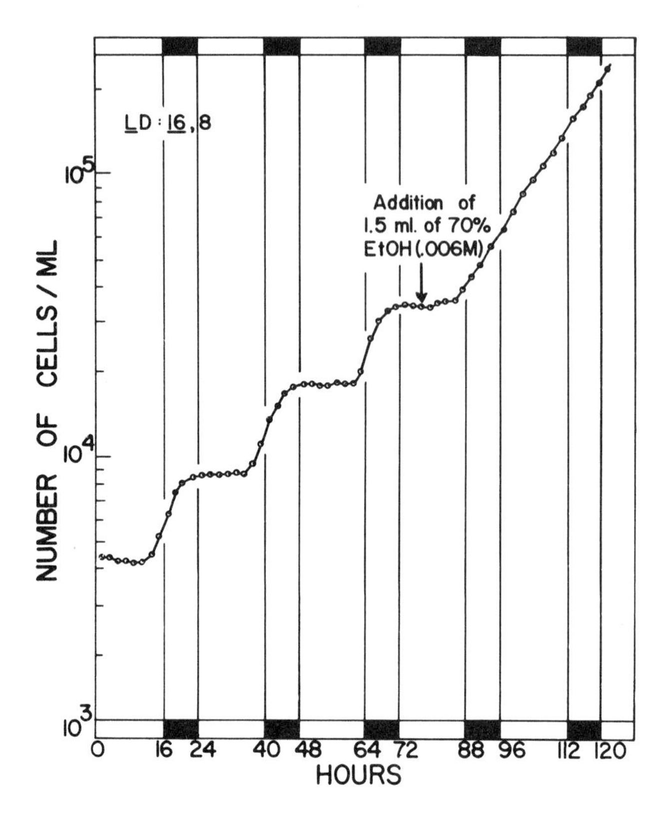

FIGURE 3.51. Disruption of synchronized cell division in cultures of *Euglena gracilis* cultured autotrophically in LD: 16,8 by the addition of ethanol to the vessel at the time indicated by the arrow. The final concentration of ethanol was 0.006 *M*. (Reprinted from Edmunds, 1965a, with permission of Alan R. Liss, Inc.)

division per day (Chisholm et al., 1980; Brand, 1982). The mycelial mass of *Neurospora crassa* sometimes may be doubling at a rate greater than every 24 h, yet the circadian conidiation rhythm continues to be expressed (J. Feldman, personal communication; Perlman et al., 1981), and in photoautotrophically grown cultures of *E. gracilis* induced to enter the ultradian growth mode by a pulse of lactate, the phase of the temporarily suppressed rhythm of cell division when it was finally restored (after the supplemental substrate had been depleted) appeared to be that of an unperturbed control (Edmunds and Laval-Martin, 1984).

Cultures in the ultradian growth mode, however, can be phased (if not truly synchronized) by LD cycles whose period is also less than 24 h. Using LD: 6,6 (T = 12 h), Ledoigt and Calvayrac (1979) have phased *E. gracilis* cultures growing on lactate in DD so that a doubling occurred every 12 h (g = 12 h), and Kämmerer and Hardeland (1982) have syn-

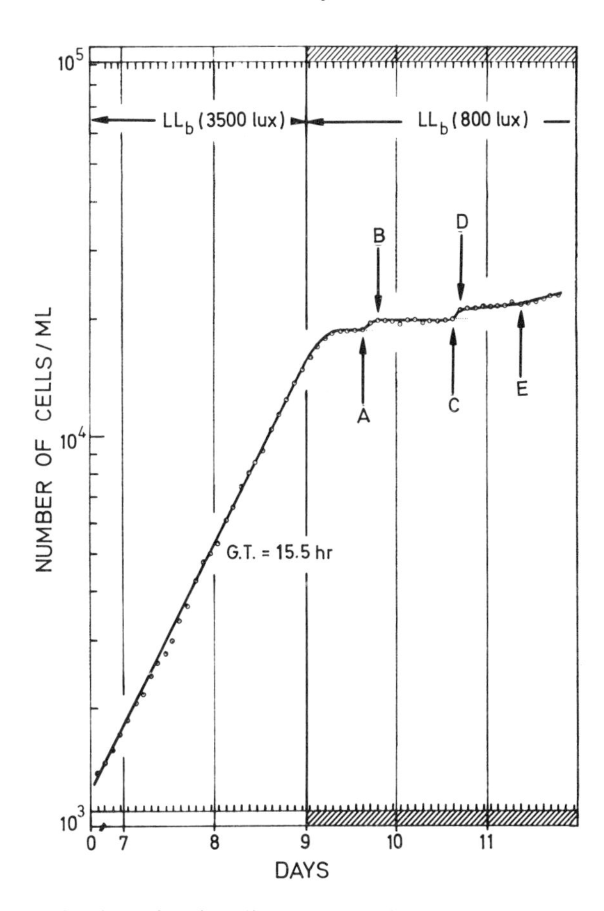

FIGURE 3.52. Induction of a circadian rhythm of cell division in *Euglena gracilis* by a single transition from continuous bright illumination (3500 lx, cool-white fluorescent) to a continuous dim illumination (800 lx). The generation time of the exponentially increasing culture before the transition was 15.5 h. Afterward, it stabilized at an overall value of 118 h. (Reprinted from Edmunds, 1966, with permission of Alan R. Liss, Inc.)

chronized ultradian cultures of the ciliate *T. pyriformis* by a 5-h temperature cycle (see Section 2.1.1). Similarly, Readey (1987) has demonstrated that ultradian LD cycles can entrain cultures of *T. pyriformis* (strain GLC) growing in the ultradian mode at low cell titers in a nephelostat as long as the period of the synchronizer does not exceed the nearest modal value of g observed in free-running single cells. Thus, LD cycles (in fact, bright L alternating with dim L) having T values of 3, 5, or 6 h (i.e., LD: 1,2; LD: 1.5,3.5; LD: 2,4) induced synchronous division (as well as changes in mean cell volume) in populations with mean cell generation times of 5.63 to 5.98—values corresponding closely to the value (5.4 h) found for individual cells in the same growth medium at the same temperature. Under subsequent dim LL, the cultures free-ran for several cycles with τ values $\sim$ 3 or $\sim$ 6h. These findings are consistent with the observation (Readey and Groh, 1986) that single cells in constant conditions display a polymodal distribution of generation times (see Section 3.1.2), with *g* values' being clustered around either 3.25 or 6.25 h.

The duration of the CDC in individual cells of *P. tetraurelia* in the ultradian growth mode appears to be a discrete multiple of the ultradian cycle of motility, whose τ of about 1 h is temperature-compensated over the range 18 to 33°C (Kippert, 1985b). The number of motility subcycles varied in stepwise fashion, from 3 (for short *g*s at 30°C) to 12 (for long *g*s at 18°C). Divisions occurred at the same phase of the motility rhythm, a fact that suggests that ultradian rhythms function in gating the CDC in a manner similar to their circadian counterparts (see Section 3.2.1). The CDC in *Tetrahymena* spp. also appears to be a discrete multiple of subcycles of the temperature-compensated ultradian rhythm of 0_2 uptake (Kippert, personal communications, 1986).

Conversely, as *g* exceeds 24 h (e.g., by lowering the temperature or by nutritional limitation), the periods of the basic oscillator and that of the

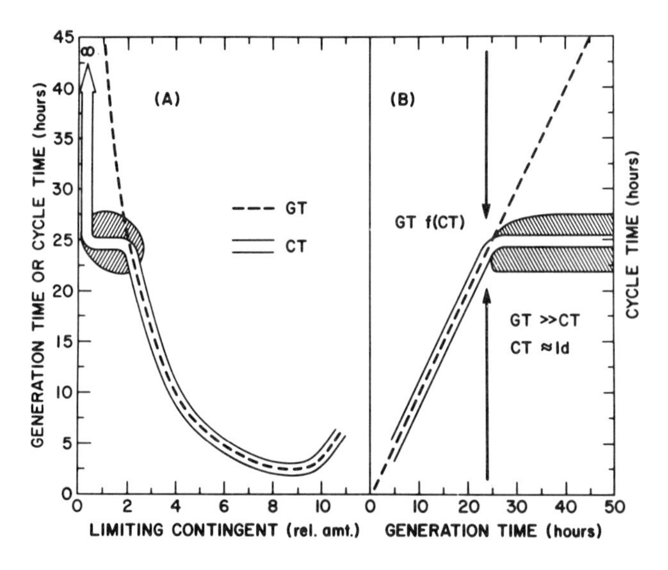

FIGURE 3.53. The circadian–infradian rule. (A) Diagrammatic representation of generation time (GT, broken line) and cell cycle time (open path) to an environmental factor (limiting contingent). Only in lethal environments does the cycle time (CT) exceed circadian values (approaching infinity, arrow). Hatched areas represent the range for CT. (B) Generation time is a function of cell cycle time only during ultradian growth. As GT approaches infinity, CT remains approximately constant (circadian). (Reprinted from Ehret and Dobra, 1977, with permission of Publishing House Il Ponte.)

CDC seem to start to diverge in the other direction (Ehret et al., 1977; Edmunds, 1978). In the limiting case, where *g* approaches infinity (as in very slowly dividing, stationary-phase cultures), low- amplitude division bursts occur at circadian intervals (Fig. 3.52; see Figs, 3.33, 3.36) in the population (but CDC > 24 h), along with numerous other cyclic physiological and biochemical events that are not necessarily related to the CDC.

This distinction between *g* and the circadian cycle in slowly dividing cell populations has been formalized in the circadian–infradian rule (Ehret and Wille, 1970; Ehret and Dobra, 1977; Ehret et al., 1977) and is diagrammatically represented in Figure 3.53. As *g* (or GT) tends toward infinity, the cell cycle time remains approximately constant (circadian) and is nearly temperature-independent. Thus, although CDCs, like circadian rhythms, are endogenous, self-sustaining oscillations, they differ in that the period (τ) of the circadian oscillation would not be a function of either the period (*g*) of the CDC or of the temperature (it arrives at an apparently genetically determined limit value of about 1 day, circadian). In contrast, the length of the CDC can take on all values from 24 h to infinity. (Only in lethal environments would CT exceed circadian values, surely a null case.)

But what is happening to the cellular circadian clock during its presumed replication at intervals of less than 24 h? Ehret et al. (1977) believe that during ultradian growth cell, CT would be a function of *g* (as well as temperature). That is to say, CT equals *g*, which equals the length of the interdivisional period. According to this scheme, the basic iteration period (CT) could take on all values between the minimum *g* possible under optimal growth conditions for a given cell and approximately 24 h (see also Homma, 1987). More precise answers await further elucidation of biological clock mechanisms.

4
Experimental Approaches to Circadian Clock Mechanisms

Several approaches to elucidating the nature of biological clocks, particularly circadian oscillators, have emerged over the years (Edmunds, 1983; Johnson and Hastings, 1986). These include the attempt to locate the anatomical loci responsible for generating these periodicities, efforts to trace the entrainment pathway for light signals (and other *Zeitgeber*) from the photoreceptor(s) to the clock itself, the experimental dissection of the clock using chemicals and metabolic inhibitors and employing the exciting new techniques of molecular genetics, and the characterization of the coupling pathways and the transducing mechanisms between the clock and the overt rhythmicities (hands) it drives. Each of these strategies are discussed, and the key experimental results obtained to date are summarized. It should become readily apparent that what constitutes mechanism for one investigator is merely descriptive phenomenology for another working at a lower level of organization! Because of the immense amount of work done in these areas, no attempt is made to provide an exhaustive review; rather, several major experimental systems (not surprisingly, most at the cellular level) have been chosen for illustrative purposes, with an effort being made to integrate the results obtained with those of other systems. Consequently, cross-referencing is provided to the more general review of circadian organization in eukaryotic microorganisms presented earlier (see Chapter 2). Finally, this survey sets the stage for the discussion of various biochemical and molecular models for circadian clocks in Chapter 5.

4.1 Quest for an Anatomical Locus: Autonomous Oscillators in Isolated Organs, Tissues, and Cells

One obvious approach to discovering the clock is to attempt to localize it within the organism, assuming, of course, that it has an anatomically defined locus. We now briefly examine some illustrative in vivo and in vitro examples of this line of attack.

4.1.1 Circadian Pacemakers at the Organ and Tissue Levels

At higher levels of vertebrate and invertebrate organization, several circadian pacemakers, or master oscillators, have been localized in the nervous system and the endocrine system (for recent reviews, see Block and Page, 1978; Jacklet, 1981, 1984; Rusak and Zucker, 1979; Suda et al., 1979; Takahashi and Zatz, 1982; Turek, 1985). Discrete centers of the nervous system are responsible for the generation of circadian rhythms of higher animals, apparently becoming localized to specialized tissues and acquiring controlling status during the course of evolution, just as other tissues and organs have assumed specialized functions (Jacklet, 1985). Brain centers known to contain such circadian pacemakers are the paired suprachiasmatic nuclei (SCN) of the vertebrate hypothalamus, the optic lobes or the median neurosecretory cells of the brain of certain insects (e.g., the cockroach, or the housefly and the domestic cricket, respectively), and the eyes of certain marine gastropods (*Aplysia* spp., *Bulla* spp.). In some instances, these brain centers control overt rhythms by hormonal secretion; the brain of the silk moth can be transplanted into the abdomen of a debrained host (Truman's loose-brain experiment, 1972) and determine the phase of the host's eclosion rhythm via the eclosion hormone. Other pacemakers are coupled to the overt rhythms by direct neural connections. They constitute the driving oscillators in a hierarchically ordered system that provides temporal coordination of physiological function through widespread neural projections. The cell bodies of ocular neurones appear to be required for circadian pacemaker activity, and only a few neurones are necessary for rhythmic expression (Jacklet, 1984). Finally, the avian pineal gland, a so-called neuroendocrine transducer, contains an important circadian pacemaker.

Circadian oscillators are not confined to animals. The classic rhythmic sleep (nyctinastic) movements of leaves observed by Darwin, especially common in leguminous plants, have an anatomical basis in the leaf pulvini and are caused by the shuttling of ions between opposing sides of these organs (reviewed by Satter and Galston, 1981; see Section 4.3.2). Analogously, the circadian periodicity of stomatal opening and closure seems to be attributable to clock-controlled ion movements within the stomatal apparatus that generate rhythmic changes in osmotic potential and turgor pressure in the constituent epidermal guard cells (Raschke, 1975; Snaith and Mansfield, 1985, 1986). Finally, recent evidence suggests that Darwin's hypothesis that the rhythmic movements of twining shoots and tendrils is under endogenous, nastic (as opposed to tropistic) control well may be correct. Rhythmic circumnutation (τ = 100 min) in shoots of *Phaseolus vulgaris* L. appears to be due to a wave of turgescence that moves around the climbing apex, which in turn generates a rhythm of growth in the epidermal cells comprising the bending zone (Millet et al., 1984). If so, a tissue-based ultradian oscillator would have been localized.

The Hypothalamus

Some years ago, ablation experiments identified the bilaterally symmetrical nuclei of the hypothalamus of the mammalian brain—each one in the rat containing about 10,000 neurons (Moore, 1979) and lying just above the optic chiasm and receiving light via a retinohypothalamic tract (RHT)—as the site of an important circadian oscillator. Lesions of these SCN in the rat resulted in a loss of the circadian rhythm of adrenal corticosterone (Moore and Eichler, 1972) and of locomotor activity and drinking behavior (Stephan and Zucker, 1972). Each of the SCN can function independently—hamsters sustaining complete or partial unilateral lesions exhibit normal activity rhythms and even show identical increases in τ ($\cong$ 30 min) compared to intact animals as the result of treatment with 10% deuterium oxide (see Section 4.3.1), although the variability in τ also increased (Pickard and Zucker, 1986). On the other hand, some circadian rhythmicities, such as those of temperature and feeding, still persist after bilateral SCN ablation, suggesting that other pacemakers must also exist. An organizational scheme of this pacemaker, based primarily on anatomical evidence and results from experimentally induced lesions in the rat, is given in Figure 4.1 (Moore, 1983). Rusak (1982) has presented a similar organizational scheme for the SCN circadian system in the golden hamster.

This scheme provides for entrainment pathways (with light information's

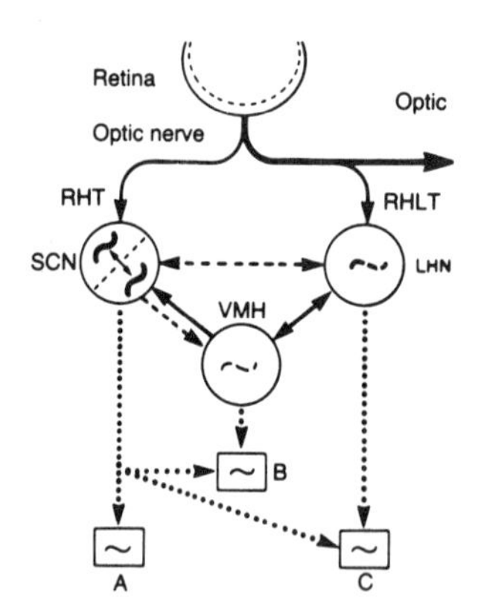

FIGURE 4.1. Organizational scheme of the SCN circadian system in the rat. The principal circadian pacemakers are the paired interacting suprachiasmatic nuclei (SCN). Putative pacemakers in the ventral medial hypothalamus (VMH) and lateral hypothalamic nuclei (LHN) may interact with the SCN. A, B, and C represent secondary, passive systems driven by the main pacemakers. Entrainment by external light cycles requires an intact retinohypothalamic tract (RHT) to the SCN and RHLT to the LHN for information received in the retina to reach the circadian pacemakers. Solid lines indicate established connections; interrupted lines indicate uncertain or proposed connections. (Reprinted from Moore, 1983, with permission of Federation of American Societies for Experimental Biology.)

reaching the SCN and lateral hypothalamus from the eyes via the RHT or, in the hamster, via the primary optic tract and the lateral geniculate, for multiple interacting pacemakers (in the SCN and in the ventral, medial, and lateral hypothalamus) arranged hierarchically within a multioscillator system, and for neural and possibly neurohormonal pathways by which the pacemakers can drive passive, secondary (overt) rhythms. For example, the circadian rhythm in melatonin in the pineal gland (see Fig. 4.4), known to be generated by the rhythmic activity of the pineal enzyme *N*-acetyltransferase (NAT), is controlled in turn by the SCN through the release of norepinephrine from superior cervical neurons and the stimulation of β-receptors on the pineal organ itself. In addition, vasoactive intestinal polypeptide (contained within the cell bodies and terminals of the SCN), can stimulate NAT activity independently of the β-receptors (Yuwiler, 1983). It is important to note, however, that the chicken pineal is different from that of the rat. It is a relatively independent pacemaker capable of generating rhythms of NAT activity and melatonin secretion in vitro (see Fig. 4.5), although it is not excluded that the SCN plays an important, even crucial, role in vivo in the avian circadian system (Takahashi and Menaker, 1979a,b; 1982b).

Melatonin secretion by the pineal gland is known to mediate photoperiodic effects on the reproductive system of seasonally breeding mammals, such as the hamster (Goldman and Darrow, 1983) and sheep (Bittman et al., 1983). Because lesions of the SCN, of the periventricular area dorsal to the SCN, or of the paraventricular nucleus (PVN) eliminate the normal nocturnal increase in pineal melatonin synthesis or prevent testicular regression or both in hamsters exposed to short-day photoperiods, it has been suggested that the SCN relays relevant light information to the pineal gland via a dorsal periventricular projection to the PVN (Eskes et al., 1984; Klein et al., 1983; Lehman et al., 1984; Pickard and Turek, 1983; Rusak, 1980; Rusak and Morin, 1976). Such lesions, however, also disrupt periventricular cell groups and rostrocaudal projections in the area, thereby complicating the description of the route by which the SCN efferents influence photoperiodic responses. Accordingly, Eskes and Rusak (1985) used horizontal knife cuts in the SCN area of the hamster (presumably interrupting only the connections between the SCN and PVN) and found that hamster gonadal responses to photoperiod were prevented. In contrast, these knife cuts did not systematically influence the circadian rhythm of wheel-running activity. These results confirmed the earlier report of Pickard and Turek (1983) that PVN ablation did not affect circadian parameters and indicate that the projections regulating circadian rhythms do not run dorsally from the SCN.

Further support, albeit indirect, for the proposition that the SCN of the hypothalamus is an autonomous circadian pacemaker comes from several other sources. Thus, Rusak and Groos (1982) found that electrical stimulation of the SCN caused phase shifts and period changes in the free-running feeding rhythms of rats and activity rhythms of hamsters. The PRC for SCN stimulation appears to parallel that for light pulses. Uhl and Reppert (1986) have implicated vasopressinergic SCN neurons (as opposed to those in the PVN and supraoptic—SON—hypothalamic zones) in generating the circadian rhythm in the concentration of the neuropeptide vasopressin in the cerebrospinal fluid of rats; SCN vasopressin mRNA exhibited a circadian variation also. In a different approach in which the metabolism (as opposed to the anatomy) of brain structures was examined by an autoradiographic method, the incorporation and utilization of labeled 2-deoxy[^{14}C]glucose (2-DG) by the intact SCN of normal and blinded rats was found to be rhythmic in LD and to persist in DD (Schwartz et al., 1980; Newman and Hospod, 1986). As might be expected, the pattern of glucose utilization reflected the functional activity of the SCN. A peak occurred during the subjective day, corresponding to the time at which the rate of neuronal firing was highest. Finally, Sawaki et al. (1984) demonstrated that if hypothalamic tissue containing the SCN and taken from neonatal (day 1) rats was transplanted into the third ventricle of young adult rats that previously had received bilateral electrolytic lesions of the SCN (and whose rhythm of wheel-running activity thus had been disrupted), circadian rhythmicity reappeared from 2 to 8 weeks after transplantation. Vasoactive intestinal polypeptide (VIP)-containing neurons were found in the ventrolateral portion of the transplanted SCN by use of immunocytochemical techniques, and vasopressin (VP)-containing neurons were identified in the dorsomedial portion (Kawamura and Nihonmatsu, 1985), further confirming that neural communication between the graft and the host brain had been established. DeCoursey and Buggy (1986) reported a similar restoration of circadian locomotor rhythmicity in SCN-lesioned golden hamsters by transplantation of fetal (14-day) hypothalamic tissue containing SCN anlagen to the third ventricle of the host. Such grafts contain an organization of peptidergic cells similar to the intact SCN and appear to establish both afferent and efferent connections with the host brain (Lehman et al., 1987a). Relatively few peptidergic cells and fibers need be present, however, for the restoration of circadian rhythmicity. Lehman et al. (1987b) found that mechanically dispersed cell suspensions prepared from trypsin-digested, fetal hamster SCN, labeled with [^{3}H]thymidine and injected (80,000 cells per 5-μl injection) into the medial hypothalamus of arrhythmic, SCN-lesioned adults, restored free-running locomotor periodicity in DD of 6 of 12 recipients (although light entrainability and gonadal photoperiodic responses were lacking). Only isolated or small groups of VIP cells were found in the medial hypothalamus, along with small plexuses of VIP and VP fibers. Interestingly, the $\bar{\tau}$ of the restored rhythms after injections was shorter that that of lesioned hamsters whose CRs recovered after whole-tissue grafts.

The ablation approach has been refined with the so-called Halasz-knife technique by which islands of hypothalamic tissue containing the SCN can be created in situ. Neural, multiple-unit activity can be recorded simultaneously with extracellular electrodes both within the island and at other brain locations outside the island. Using this methodology, Inouye and Kawamura (1979) found that diurnal rhythmicities (in LD: 12,12) were present only within the island in rats (Fig. 4.2A). Similarly, free-running (DD), circadian rhythms of neuronal activity in rats blinded by bilateral ocular enucleation were lost at all brain locations recorded outside the island (raphe, substantia nigra, reticular formation) but persisted within the island that contained the SCN (Fig. 4.2B) for up to 35 days (Inouye and Kawamura, 1982) with τs slightly longer than 24 h. The rhythmicity of the island was thus not dependent on afferent inputs from elsewhere in the brain. When the RHT was spared, re-entrainment to a shifted LD cycle was attained after a period of transients lasting 3 or 4 days (Inouye and Kawamura, 1982). One-hour light signals given during a DD free-run at various times of the subjective day under similar circumstances, although also capable of generating phase-dependent shifts of the SCN rhythm, produced only constant increases in SCN multiple-unit discharge rates (Inouye, 1984). Whereas the direct effect of a light stimulus at the input side of the SCN did not vary within a circadian period, the outcome of the processing of the light information within the SCN did show circadian phase-dependence in keeping with the notion that the SCN is a generator of circadian rhythmicity.

These findings do not exclude the possibility that the rhythmic activity of the SCN is driven by circadian temperature or humoral rhythms generated outside the island. A more stringent requirement for designation of the SCN as an autonomous master oscillator would be the demonstration in vitro of endogenous rhythmicity in SCN preparations. Recently, several investigators have developed techniques allowing for electrical recording from single cells in SCN explants (blocks or slices), which in some cases at the present state-of-the-art have remained functional for 3 days or more. Thus, Green and Gillette (1982) found that the firing rates of single neurons in SCN-containing brain slices removed from rats at various times of the day maintained a circadian rhythm in phase with the multiple-unit activity rhythm recorded in vivo in the SCN of the donor animal. Periodogram analysis of the composite curve obtained from the in vitro recordings of single-unit activities of a population of individual neurons for up to 36 h showed a peak at about 24 h and minor peaks at 6, 12, and 18 h (Gillette, personal communication, 1985; Shibata and Oomura, 1985). A similar finding of circadian rhythms in electrical discharge of rat SCN neurons (but not those in explants of the retrochiasmatic area or the arcuate nucleus) recorded in vitro has been reported by Groos and Hendriks (1982). Thus, the circadian clock appears to continue to run in SCN slices removed from the brain, assuming, of course, that the mechanical stimulation inherent in the sacrifice and dissection procedures does not introduce artifacts. In fact, Gillette (1986) has demonstrated that when slices were

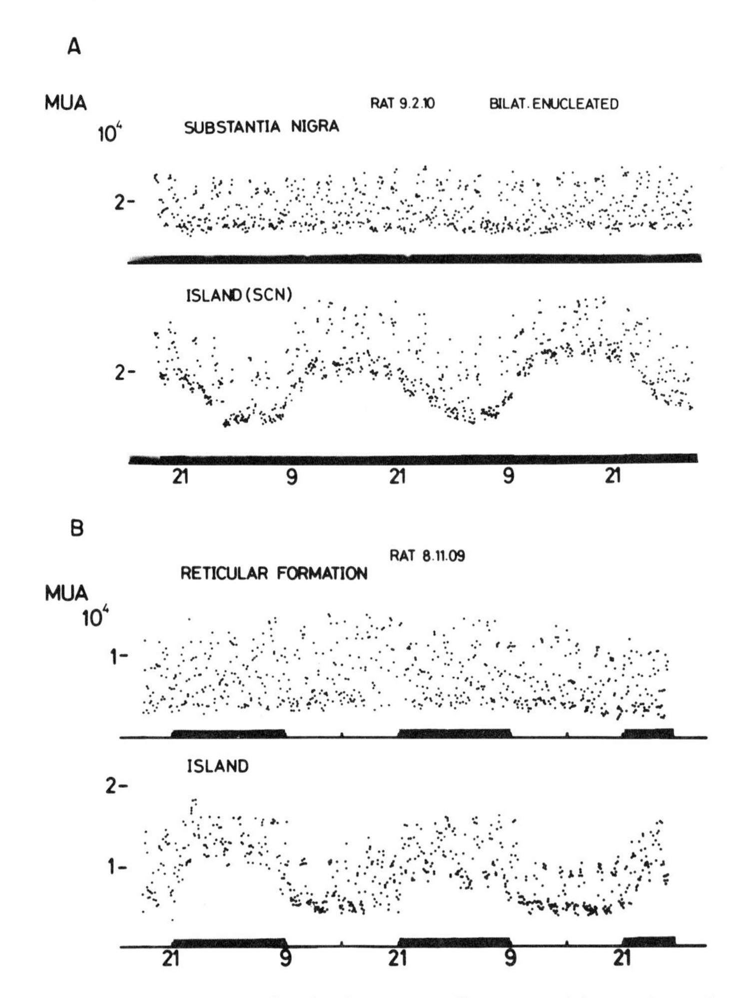

FIGURE 4.2. Circadian pacemaker in the mammalian suprachiasmatic nucleus. (A) Persistence of a daily rhythm within the hypothalamic island of an unblinded rat. No rhythm persisted in the midbrain reticular formation. (B) Persistence of a circadian rhythm within the hypothalamic island of a blinded (bilateral enucleation) rat in DD. Record started 3 days after blinding. Free-running circadian rhythm persisted only in the SCN, with maximum activity during the projected daytime (09:00 to 21:00). (Reprinted from Inouye and Kawamura, 1979, with permission of the authors.)

prepared during the donor's dark period (necessitating a light exposure of 20 to 30 sec during handling, decapitation, and surgery until the optic tracts were severed), the rhythm in neuronal firing rate was phase-shifted in a manner similar to that for intact animals, and this $\Delta\phi$ was preserved in vitro. Barbacka-Surowiak (1986) demonstrated a circadian activity rhythm in single anterior hypothalamic neurons grown in vitro and measured every 3 h over a 24-h timespan (or longer) by means of a fluorescent dye method (3,3′-dihexyloxacarbocyanine iodide, 1.17×10^{-6} *M*).

The autonomy of the SCN oscillator has been further supported by the results of several recent lines of experimental work. For example, circadian rhythms of vasopressin (VP) release have been reported for perfused rat SCN explants in vitro (Earnest and Sladek; 1985; Reppert and Gillette, 1985), the SCN neurons retained their capacity to release VP in response to membrane depolarization even on day 4 in culture, and mediation by a calcium-dependent mechanism has been implicated (Earnest and Sladek, 1986). Newman and Hospod (1986) reported a rhythm of labeled 2-DG in slice preparations of rat hypothalamus: after 1 h, 2-DG uptake into the SCN was proportional to the rate of glucose utilization present in vivo at the corresponding time of day (Schwartz et al., 1980), and during an 8-h incubation in vitro, the SCN was capable of spontaneously changing its metabolic rate. The concentration of calcium ion may be important for maintaining and regulating this rhythm of metabolic activity in the SCN (Shibata et al., 1986). One-hour pulses of cAMP analogs (8-benzylamino-cAMP, 8-*p*-chlorophenylthio-cAMP), which are potent stimulators of cAMP-dependent protein kinases (see Section 4.2.2), can reset the phase of the circadian rhythm of electrical activity in rat hypothalamic brain slices. The PRC (comprising both $+\Delta\phi$ and $-\Delta\phi$) for 1-h pulses given at 8 different CTs was 12 h out-of-phase with that for light pulses administered in vivo (Prosser and Gillette, 1986). Pulses of CHX and of the calmodulin antagonists W7, CPZ and TFP (Section 4.3.3) cause similar $\Delta\phi$s (R. Y. Moore, unpublished results, 1987). Gillette et al. (1986) demonstrated that changes in proteins and phosphoproteins accompany the circadian oscillation in firing rate in the isolated SCN of rat hypothalamic brain slices.

We note that the hypothalamus is also the anatomical site of an ultradian oscillator. In all vertebrates examined thus far, including man and other higher primates, the secretion of the gonadotropic hormones is a pulsatile phenomenon. Each pulse of luteinizing hormone (LH) and follicle-stimulating hormone (FSH) released from the pituitary is the consequence of a cascade of events initiated by a pulse generator located in the arcuate region of the mediobasal hypothalamus, which causes the synchronous activation of neuronal elements whose activity, in turn, leads to the release of packets of LH-releasing hormone (LHRH) from their terminals into the pituitary portal circulation (reviewed by Knobil, 1980). In the absence of gonadal hormones and of inputs from higher centers, the frequency of this LHRH pulse generator is about one discharge per hour, and hence, it has been termed a "circhoral" clock. The period can be increased by progesterone (as during the luteal phase of the menstrual cycle) or testosterone, and barbiturate anesthesia, opiates, and α-adrenergic and dopaminergic blocking agents either reduce the frequency or inhibit activity completely. Direct electrical stimulation of the hypothalamic region can modulate the period of the oscillator and can stimulate the pulsatile release of LHRH. Finally, the proper operation of this neuroendocrine oscillator at an appropriate frequency is essential for normal gonadal function of both sexes: (1) The normal, 28-day ovulatory menstrual cycle can be restored in rhesus monkeys in which LHRH production has been abolished by hypothalamic lesions by the administration of a pulse of LHRH each hour with a suitably programmed infusion pump, (2) The continuous delivery of LHRH or the administration of LHRH to monkeys bearing hypothalamic lesions at frequencies greater than one pulse per hour leads to a complete suppression of gonadotropic secretion (in essence, a chemical hypophysectomy), and (3) delivery of LHRH to lesioned monkeys at lower frequencies (e.g., a pulse every 2 h) markedly alters the ratio of LH to FSH and causes abnormal follicular development and ovulation (Knobil, 1980).

Gastropod Eyes

Gastropod eyes, particular those of the sea hare *Aplysia californica* and of *Bulla gouldiana*, the cloudy bubble snail, have proved excellent material for analysis of the cellular and biochemical (see Sections 4.2 and 4.3) bases of circadian pacemakers (reviewed by Jacklet, 1984). The isolated eyes of both molluscs (as well as those of *Navanax* and *Bursatella*) display circadian rhythms of large amplitude in the frequency of the compound action potentials (CAP) of optic nerve impulses (Fig. 4.3D). The rhythm in CAP has been recorded for up to 2 weeks in DD in *Aplysia* (Jacklet, 1969; Block and Wallace, 1982).

Each of the paired eyes of *Aplysia* is about 0.7 mm in diameter and comprises a central lens surrounded by a retina composed of some 4000 photoreceptors and 1000 secondary neurons. The long (1 cm) optic nerves facilitate extracellular electrical recording. Both light and high potassium concentrations shift the phase of the ocular rhythm during subjective night, whereas serotonin and inhibitors of protein synthesis phase-shift the rhythm during the subjective day, suggesting that two entrainment pathways may exist (Jacklet, 1984, 1985; Section 4.2).

The eye of *Aplysia* has been surgically reduced by trimming away the distal (corneal) portions in an effort to identify the retinal cells responsible for the rhythm in CAP. Jacklet and Geronimo (1971) found that the lens and 80% of the eye could be removed without affecting the free-running period, although the amplitude of the rhythm was progressively reduced. Further reduction abolished circadian rhythmicity, although CAP pace-

---

FIGURE 4.3. Circadian pacemaker in the eye of *Bulla*. The photoreceptors surrounding the central lens and a cluster of basal neurons giving rise to optic nerve fibers are shown in (A) (Jacklet and Colquhoun, 1983). A cut (line) between the photoreceptor and neurons does not disrupt the circadian rhythm generated by the remaining basal neurons. The spikes of basal pacemaker neurons occur in 1:1 relationship with the optic nerve (ON) CAP, as shown in (B) during direct de-

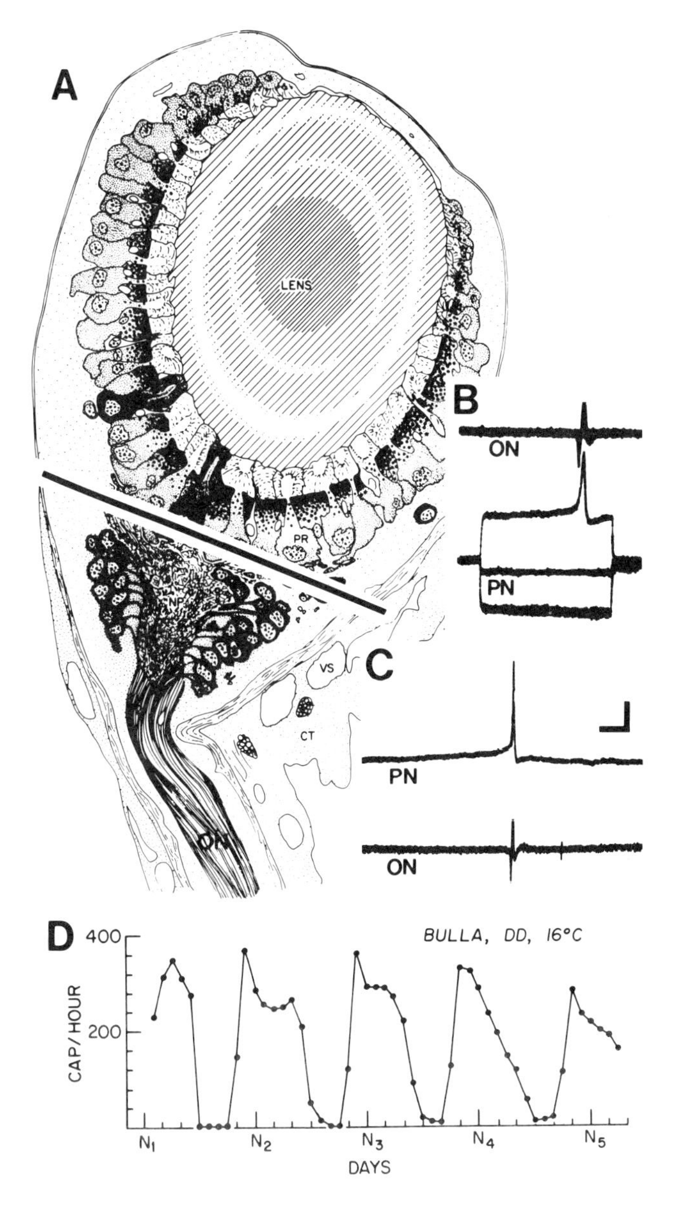

polarization of a single pacemaker neuron (PN) and in (C) during spontaneous firing. The frequency of optic nerve CAP activity exhibits a free-running circadian rhythm (D) when the isolated eye is maintained in constant darkness and temperature. The rise in CAP frequency normally precedes dawn, when the rhythm is entrained. *Scales:* (A) Eye has a diameter of 0.4 mm. (B) Scale bar represents 200 msec. 50 μV (ON), and 10 mV (PN). (C) Scale bar represents 500 msec, 10 mV (PN), and 50 μV (ON). (D) N_1, N_2, and so on are the successive noons of day 1, 2, and so on. (Reproduced from Jacklet, 1985, with permission of Elsevier Biomedical Press.)

maker activity continued at a constant rate, and shorter and shorter periods were expressed. Indeed, ultradian periodicities (τ = 1 to 12 hours) of very low amplitude sometimes were noted in these preparations, suggesting that the accuracy of this isolated clock system might depend on the number of cells it contained. At least three alternative hypotheses for the organization of the endogenously active neuronal (oscillator) population might explain these provocative results (Jacklet and Geronimo, 1971): (1) a population driven by a master circadian pacemaker cell to which the other cells were enslaved (but then τ should have remained constant until this master clock were suddenly cut away), (2) a population of circadian oscillators (but then a circadian period should have been displayed up until the very last moment and not have shown a change at the 20% level), or (3) a population of noncircadian, ultradian oscillators that together display a circadian periodicity, perhaps due to their inhibitory, cross-coupling interaction (see Section 5.2.3). Unfortunately, the possible effects of surgical trauma, desynchronization, and unequal perturbation of the constituent cells by the trimming procedure make it difficult to distinguish among these alternatives (Jacklet, 1984).

The retina of *Bulla* (Fig. 4.3) is particularly favorable for electrophysiological study of the circadian pacemaker system, since many retinal cells are relatively large, and there is good spatial separation between morphologically distinct cell types (Block and Wallace, 1982; Jacklet and Colquhoun, 1983). It contains approximately 1000 large photoreceptors with distinct villous-bearing distal segments, which form a layer around a solid lens. There is also a population of about 100 neurons that surround a neuropil at the base of the retina. Continuous intracellular recording from these basal retinal cells (BRN) reveals neurons that are spontaneously active and fire action potentials in precise synchrony with the compound impulses in the optic nerve. They are coupled tightly electrically and are responsible for the CAP (Block et al., 1984; McMahon et al., 1984).

Normally, if the CNS is intact, both eyes of *Aplysia* also are mutually coupled, and their interaction can be observed in vitro (Roberts and Block, 1985). Thus, if intact animals were placed in DD for up to 72 days, the two pacemakers did not desynchronize, and the observed τ (24.4 h) of the intact system was longer than that (23.7 h) of the CAP rhythm recorded from isolated eyes. Additionally, if one eye was stimulated electrically, a signal was immediately generated in the other. If the rhythm in one of the two interconnected eyes were experimentally desynchronized in vitro, by applying seawater in which Ca^{2+} was replaced partially by manganese (Eskin and Corrent, 1977) or by giving cold pulses, resynchronization occurred if the pacemakers were allowed to interact for 48 h.

Two entrainment pathways appear to exist in *Bulla* also, one for light and one for serotonin (see Section 4.2.2). Light signals cause depolarization, which can be mimicked by direct electrical stimulation of a single BRN (Fig. 4.3B). Pulses of serotonin hyperpolarize the membrane and generate a PRC that is 180° out-of-phase with that derived for light signals. Depolarization may be causing changes in the ionic flux of Ca^{2+}, allowing for its entry, perhaps by opening and closing Ca^{2+} channels. Indeed, the calmodulin antagonist W7 appears to induce phase shifts (Block, personal communication, 1985). If the extracellular Ca^{2+} concentration was reduced from 10 m*M* to 10^{-4} with the chelator EGTA, light-induced phase shifts of the *Bulla* ocular rhythm were blocked (McMahon and Block, 1986). This was not the case for the phase delays generated by the convulsant agent, pentylenetetrazole, which is known to directly modulate intracellular Ca^{2+} levels, indicating that $\Delta\phi$s can be produced despite reduced extracellular Ca^{2+} (Khalsa and Block, 1986).

As with *Aplysia*, one can surgically reduce the *Bulla* eye in order to localize further anatomically the circadian pacemaker. Thus, surgical removal of the entire photoreceptor layer did not alter the circadian period of the CAP rhythm and did not prevent phase-shifting by light pulses (Block and Wallace, 1982). Only a small proportion of the 100 or so BRNs were required for the expression of the circadian rhythm in optic nerve frequency. Ocular fragments with as few as 6 BRN somata remained rhythmic and exhibited a τ indistinguishable from intact eyes, a finding that suggests that it is most likely that individual neurons are autonomous circadian pacemakers that are responsible for modulating an extrasomatic action potential (Block and McMahon, 1984), an assertion that could be tested by recording from individual eye neurons in cell culture. If so, BRN somata that are severed from their connections with the neuropil should express a circadian rhythm in membrane potential but not be able to generate spontaneous action potentials. Membrane potential itself could serve two functions: as an input pathway to transfer environmental entraining signals to the cell and, as an output pathway by means of which the BRN soma would modulate impulse frequency (Block and McMahon, 1984). The critical question, of course, is whether such changes in membrane potential actually are part of the causal loop generating the circadian oscillation, that is, are they gears or merely hands of the circadian clock (see Sections 5.2.5 and 5.4).

The Pineal Gland

The pineal gland—if not the seat of the rational soul, as Descartes thought in 1677—also has been central to the study of the physiology and biochemistry of vertebrate circadian rhythms. Arising from evaginations of the diencephalon of the brain and constituting one class of circumventricular organ, pineal structure and innervation differ in lower and higher vertebrates (Wurtman et al., 1968). Further, whereas the pineal acts as a circadian pacemaker in birds, in mammals it merely expresses a driven rhythm (reviewed by Takahashi and Menaker, 1979a,b; Takahashi and Zatz, 1982). The regulation of pineal biochemistry also differs in these two classes, although, in both, the pineal expresses an overt circadian rhythm of indoleamine metabolism that results in the nocturnal synthesis and secretion of melatonin.

Pineal Biochemistry

Melatonin is synthesized from serotonin by N-acetylation, catalyzed by serotonin NAT, and O-methylation, catalyzed by hydroxyindole-*O*-methyltransferase (HIOMT).Daily rhythms in the melatonin content of the pineal, in serotonin content, and in the activities of both NAT and HIOMT have been studied in both chicks and rats kept in LD cycles (Binkley et al., 1973). In the former, the 10-fold nocturnal rise in melatonin was phased identically with the 27-fold increase in NAT activity (Fig. 4.4), whereas

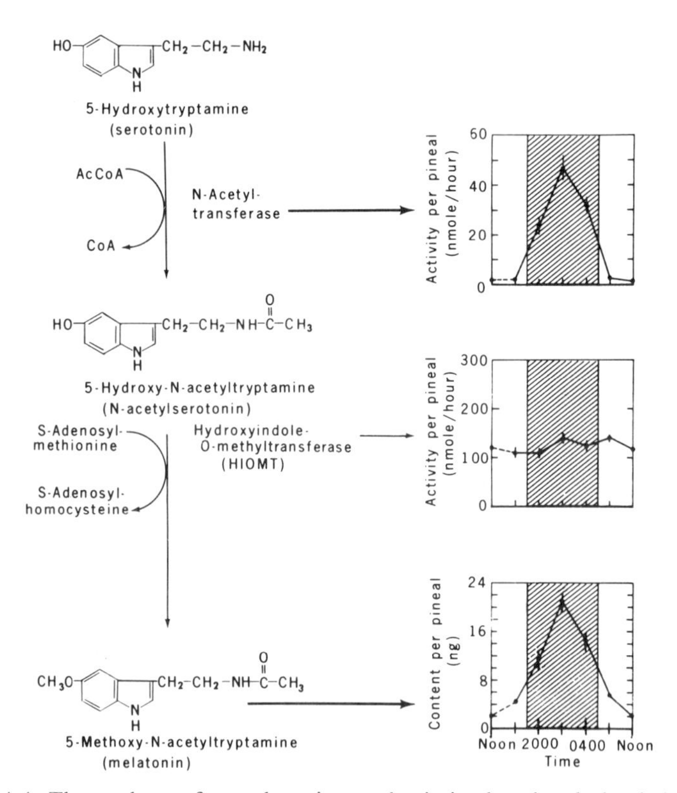

FIGURE 4.4. The pathway for melatonin synthesis in the pineal gland (left) and graphs of *N*-acetyltransferase activity, hydroxyindole-*O*-methyltransferase activity, and melatonin content measured in aliquots of a single set of 36 chicken pineal glands (right, 6 glands per time point). The standard errors are indicated when they exceed the size of the symbol. The hatched area represents the dark portion of the LD cycle. AcCoA, acetyl coenzyme A; CoA, coenzyme A. (Reprinted from Binkley et al., 1973, with permission of American Association for the Advancement of Science.)

the relatively small changes (20%) in that of HIOMT did not appear important in causing the large changes in melatonin. Furthermore, the phase of the rhythms of NAT activity and melatonin content relative to the LD cycle were quantitatively similar to that of rats, whereas the sleep–wake cycle of chickens was about 180° out-of-phase with that of rats. In contrast, the low-amplitude changes in HIOMT activity were out-of-phase in the two species. Thus, the melatonin rhythm appears to be regulated primarily by the production of *N*-acetylserotinin at the NAT step in chicks and rats. Klein and Weller (1970) reached a similar conclusion for the melatonin and serotonin rhythms in the rat pineal gland. A circadian rhythm in the activity of NAT in chickens persists in vivo for at least two cycles in DD (Binkley and Geller, 1975) and, as we shall see, also has been reported in vitro.

Pineal Physiology and Circadian Organization

Melatonin, rhythmically synthesized in the pineal gland and secreted into the bloodstream, permits hormonal coupling between this organ and the circadian system at remote sites. Thus, when melatonin was administered continuously in low doses to the house sparrow, *Passer domesticus* (a passerine bird), τ of the circadian activity rhythm was shortened, but at high doses continuous activity was induced (Turek et al., 1976). Furthermore, daily injection of melatonin synchronized the activity rhythms of pinealectomized starlings (Gwinner and Benzinger, 1978). Similarly, daily injections of melatonin could entrain rat circadian rhythms, although this was dependent on the presence of the SCN (Cassone et al., 1986).

That the pineal gland plays a major role in the circadian organization of passerine birds has been supported by many lines of evidence (Takahashi and Menaker, 1979a,b). Removal of the pineal gland eliminates the persistence of free-running locomotor activity rhythms (Gaston and Menaker, 1968; McMillan, 1972), transplantation of pineal tissue into the anterior chamber of the eye restores rhythmicity in arrhythmic pinealectomized hosts (Zimmerman and Menaker, 1975), and the restored rhythm bears the phase of the donor. Presumably the recipient bird's activity rhythm became quickly coupled to the transplanted pineal clock by the melatonin secreted by the latter (Zimmerman and Menaker, 1979).

Although the pineal thus is necessary for the persistence of rhythmicity in constant conditions in the house sparrow, not all birds are alike. Pinealectomy disrupts the circadian system but does not cause permanent arrhythmia in starlings (Gwinner, 1978), and gallinaceous birds, such as the chicken *(Gallus domesticus)* and the Japanese quail *(Coturnix coturnix japonica),* are not affected (Takahashi and Menaker, 1979a). Even in the sparrow, oscillatory components remain after pinealectomy. The locomotor activity rhythm can be synchronized by LD cycles, and the entrained rhythm only gradually decays and approaches arrhythmicity upon transfer of the bird to DD (Gaston and Menaker, 1968; Takahashi and

Menaker, 1982a). These findings suggest that structures other than the pineal gland are involved in the circadian system of the sparrow. Takahashi and Menaker (1982b) have shown that if the SCN of *P. domesticus* (homologs of the mammalian SCN) were lesioned, circadian rhythmicity in DD was completely abolished. Although entrainment of the lesioned birds by LD: 12,12 was still possible, synchronization was disrupted in a short-photoperiod LD:1,24 cycle. These results demonstrate, therefore, that the SCN are important for the generation of overt circadian rhythmicity in birds, and they are in accord with those obtained with mammals.

The relationship between the avian pineal gland and the SCN, nevertheless, is still ambiguous. There are at least three hypotheses that could account for the behavior of sparrows having SCN lesions (Takahashi and Menaker, 1982b): (1) The SCN may be components of the output pathway between a circadian pacemaker located elsewhere (perhaps in the pineal) and the overt locomotor rhythm. Arrhythmicity in SCN-lesioned birds would result from the uncoupling of the pacemaker from its output by the interruption of this pathway. On the assumption that the SCN are the target for the pineal, a persisting rhythm of melatonin production still might be expected to occur. (2) The SCN could play a permissive role and be necessary for sustaining pineal rhythmicity in vivo under constant conditions. It must be noted, however, that chicken pineals can oscillate when isolated in vitro (see next subsection). (3) The SCN and the pineal gland may interact and function en ensemble as a complex pacemaker. Neither structure alone would be able to sustain the normal overt rhythmicity. On the assumption that the two organs behave as damped oscillators (suggested by the results observed both in organ culture and in pinealectomized sparrows), the self-sustained rhythmicity exhibited by intact birds would result from mutual entrainment. The population of oscillators in the SCN would be coupled by a rhythmic hormonal (melatonin?) signal from the pineal gland, even in the pineal-transplant experiments in which the resultant phase was that of the donor, and the SCN would then generate a feedback signal (either hormonally or via the sympathetic fibers that innervate both pineal and iris) that would maintain the amplitude of the pineal oscillation. Vertebrate circadian organization among different species would be based on differentially weighted interactions among the pineal, the SCN, and perhaps other brain regions (Takahashi and Menaker, 1979a, 1982b).

In mammals, the rhythm of melatonin synthesis is driven by a circadian pacemaker in the brain. The pineal gland is coupled to the oscillator by sympathetic nerve fibers whose cell bodies reside in the superior cervical ganglia and that transmit oscillatory information from the hypothalamus to the pineal gland via changes in the release of the neurotransmitter norepinephrine (Takahashi and Zatz, 1982; Moore, 1983). If the connection between the pineal gland and the spinal cord is severed, or, in fact, if a lesion is placed anywhere between the SCN and the pineal gland, the rhythm in indoleamine metabolism is abolished.

In both mammals and birds, light acts in two ways on the pathway of melatonin synthesis. On the one hand, LD cycles entrain the rhythms of NAT activity and melatonin content; on the other, acute light exposure at night results in a rapid reduction in both enzyme activity and the level of melatonin. This photosensitivity occurs only in vivo in the rat, requiring that all components of the pathway between the eye and the pineal gland be intact, whereas neither the eyes nor sympathetic innervation are required in the avian pineal gland—light is directly inhibitory in vitro (see next subsection). Constant illumination also is inhibitory, and its action is paralleled in rats by a concomitant decrease in the weight of the pineal glands and an increase in the weight of the ovaries and an accelerated estrous cycle. Since melatonin is known to have antigonadal effects in both birds and mammals, it is obvious that daylength, mediated by the SCN and pineal gland (and acting in coincidence with the light-sensitive circadian clock), must play a crucial role in the control of reproduction in photoperiodic species. Thus, it is not surprising that lesions of the SCN block testicular responses to photoperiod and prevent antigonadal effects of melatonin in mammals and that pinealectomy causes similar disruptions (Goldman and Darrow, 1983).

Pineal Pacemakers In Vitro

Compelling evidence that the avian pineal gland is the anatomical locus of a circadian pacemaker also has been afforded by studies on pineals in organ culture and on reduced pineal glands in vitro (Takahashi and Zatz, 1982). Kasal et al. (1979) demonstrated that the rhythm in NAT activity in cultured chick pineal glands persisted for two cycles in DD. Similar observations were made by Binkley et al. (1978), Deguchi (1979a) and Wainwright and Wainwright (1979). When pineal glands were cultured in LD: 12,12, the rhythm continued to oscillate in phase with the LD cycle for 3 days, indicating that the glands were photosensitive.

In these studies, however, populations of pineal glands were used (necessitated by the relatively low levels of NAT activity). Takahashi et al. (1980) subsequently developed a flow-through culture system for the individually isolated pineal gland by which the release of melatonin could be measured continuously from superfused glands over long periods of time. The highly rhythmic release of melatonin observed in chick pineal glands in vivo also occurred in vitro in LD cycles and persisted with a circadian period in DD (Fig. 4.5), although in constant conditions the amplitude was lower and appeared to be damping. If a 1-h or 2-h light exposure were given during the DD free-run at the peak of the first oscillation in melatonin content, melatonin release into the perfusate was quickly and strongly inhibited. Thus, light had at least two effects: cyclic light input synchronized the isolated pineal, whereas acute light exposure during the subjective night rapidly inhibited the release of melatonin. Organ-cultured lizard (*Anolis carolinensis*) pineal glands also have an LD-entrainable cir-

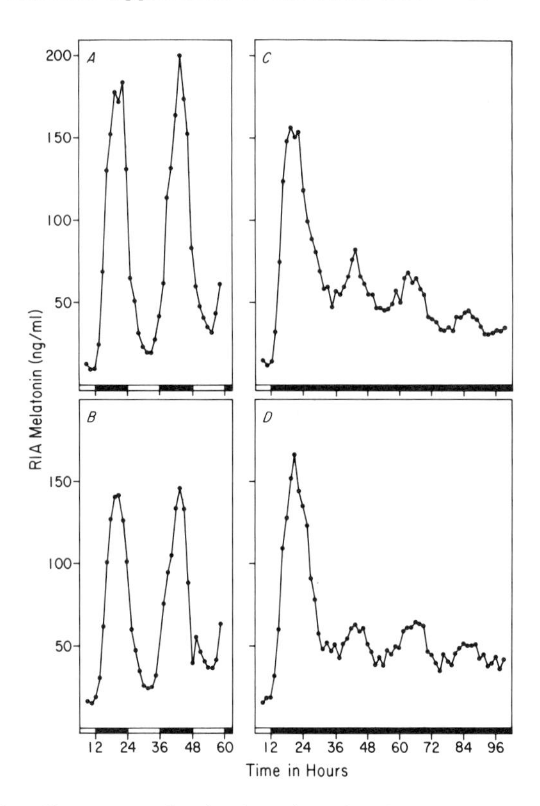

FIGURE 4.5. Circadian pacemaker in the avian pineal gland. Rhythms of melatonin release from individual superfused chicken pineal glands in vitro in an LD cycle (A and B) and in DD (C and D). The light treatments are indicated at the bottom of each panel. (A and B) Two individual pineal glands from a 60-h experiment with a 12-h light (350 to 500 lx cool-white fluorescent) and 12-h dark (1 to 5 lx red light) cycle. (C and D) Two individual pineal glands from a 96-h experiment in constant dim red illumination (1 to 5 lx). Pineal glands in both experiments came from the same group of chickens. (Reprinted from Takahashi et al., 1980, with permission of the authors.)

cadian rhythm of melatonin secretion, which persists for up to 10 days in DD and which is temperature-compensated ($Q_{10} \cong 1.14$) over the range 22 to 37°C (Menaker and Wisner, 1983).

Further work (Takahashi and Menaker, 1984) demonstrated that the capacities for circadian oscillation and photoreception are distributed throughout the chicken pineal gland and that tissue reduction has no detectable effects on τ of the melatonin rhythm. Pineals were surgically reduced progressively into lobules that were cultured individually. Multiple pieces (up to eight) from a single pineal gland all were capable of circadian oscillation (Fig. 4.6), establishing directly that a gland contains at least eight oscillators and thereby extending the earlier results of Kasal and

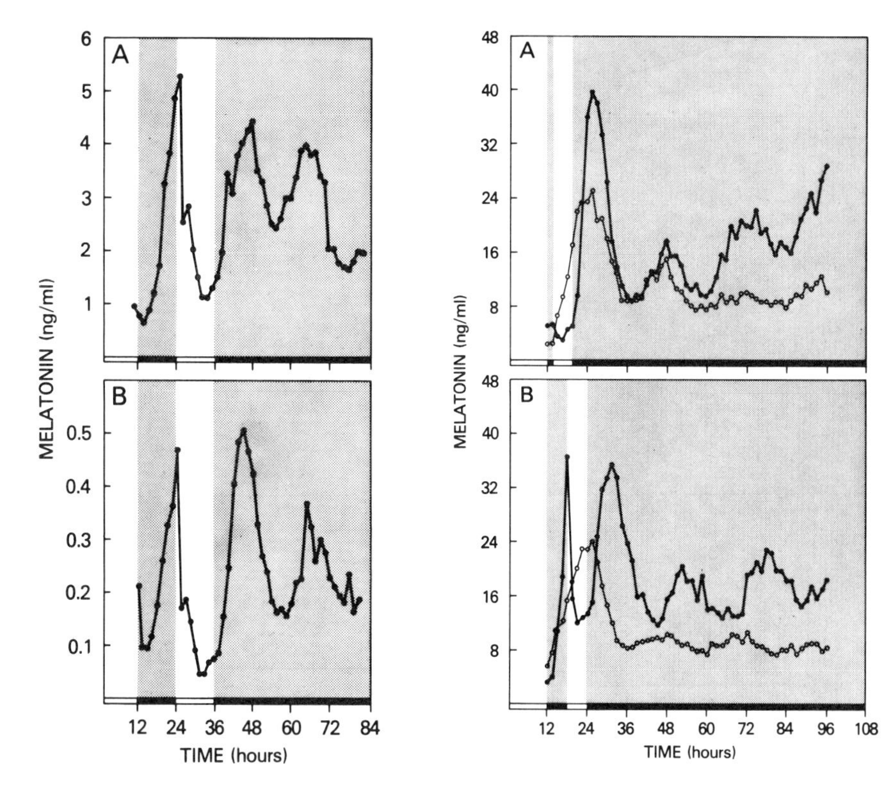

FIGURE 4.6. Circadian pacemaker in the surgically reduced avian pineal gland. *Left:* Circadian rhythms of melatonin release in vitro from an eighth chicken pineal gland (A) and from a lobule (B). All eight pieces from this gland were rhythmic. Note that the rising phase of the melatonin peak on the first day in culture was delayed and that the light cycle appeared to truncate the peak by inhibiting melatonin. *Right:* The effect of single 6-h light pulses on the phase of the melatonin rhythm. Each panel shows melatonin release from two quarter-pineals from the same gland. (A) One pineal piece (closed circles) was exposed to a 6-h pulse (1500 lx) beginning at hour 13.5, whereas the other piece (open circles) was maintained in DD beginning at hour 12. (B) One pineal piece (closed circles) was exposed to a 6-h pulse beginning at hour 18, whereas the other piece (open circles) was maintained in DD. (Reprinted from Takahashi and Menaker, 1984, with permission of Springer-Verlag.)

Perez-Polo (1980) that provided evidence that a population of half-pineal glands could oscillate in culture. All pineal pieces were responsive to light, which appeared both to inhibit melatonin production during the 12-h photoperiod and to maintain the amplitude of the first peak that occurred after the light-to-dark transition into DD. Further, single light pulses shifted the phase of the rhythm (Fig. 4.6). Thus, because pieces equivalent to less than 1% of the whole pineal gland were rhythmic, an individual gland

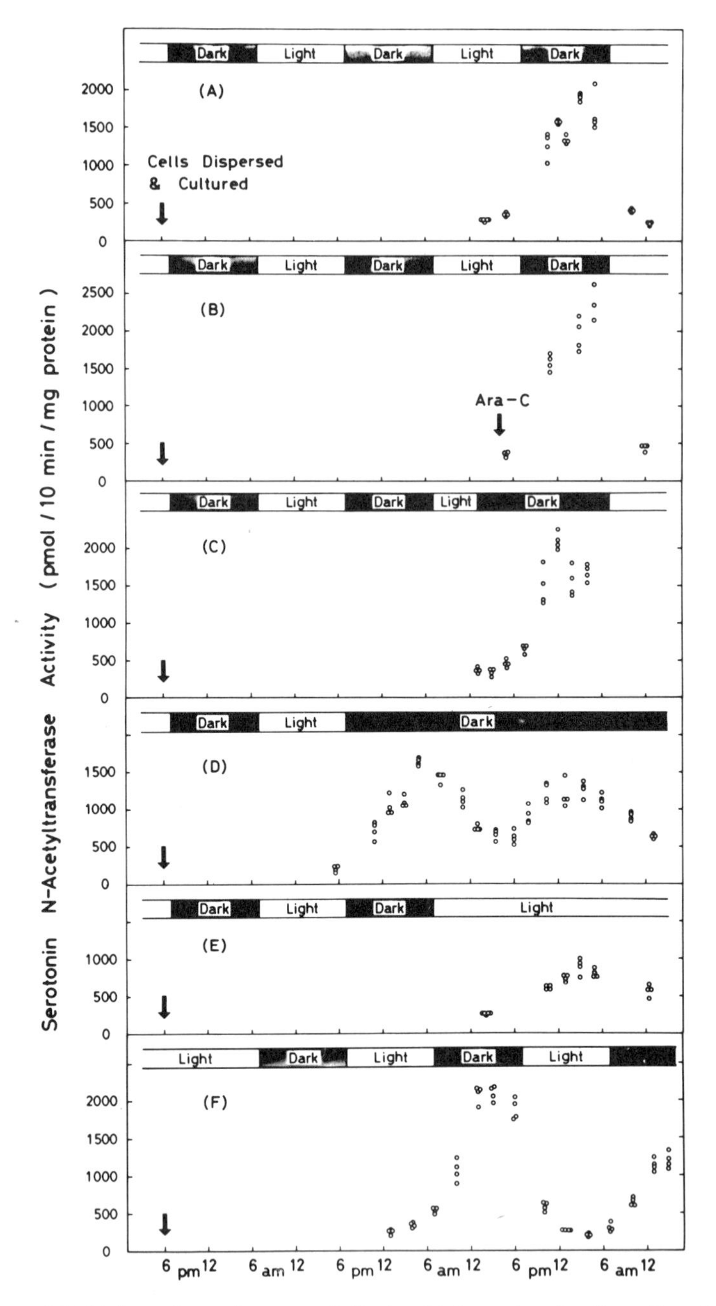

FIGURE 4.7. *N*-Acetyltransferase activity in chicken pineal cells cultured in vitro under various lighting schedules. Each point represents the specific enzyme activity on an individual dish. Ara-C indicates the addition of cytosine-1-β-arabinofuranoside (1×10^{-5} *M*) to the medium. (Reprinted from Deguchi, 1979b, with permission of Macmillan Journals Limited.)

appears to be composed of multiply redundant circadian oscillators. This population of clocks might be coupled and interact (see Section 5.2.3) although if there is such coupling it is probably weak, given the strong damping of the rhythm in DD. Such damping suggests a progressive asynchrony in an oscillator population in vitro. The fact that it does not occur in the intact chicken during prolonged darkness (at least 2 weeks) could be due to inputs from the SCN or from elsewhere in the organism (Takahashi and Menaker, 1984).

Deguchi (1979b) has shown that dispersed cell cultures of chicken pineal glands express circadian rhythms of NAT activity in DD for 2 or 3 days and respond to environmental lighting in the same manner as in the organ culture (Fig. 4.7). There was an approximate 10-fold increase in activity at night as compared to the daytime value (similar to the values found in cultured glands). Although the cells in these cultures could be in physical or chemical contact with each other, the evidence suggests that both the circadian oscillation and the response to light (inhibition of NAT activity) might be properties of individual pineal cells. Recent work in which a PRC was derived for the phase-shifting effects of 6-h light pulses on the rhythm of melatonin release by cultured, collagenase-dispersed cells from chick pineal glands supports this view (Robertson and Takahashi, 1986). Similarly, a circadian rhythm of ^{3}H-labeled melatonin secretion by dispersed pineal cells in primary cell culture has been found to persist for at least 2 weeks in LD: 12,12 and for three cycles under constant red illumination (Zatz and Moskal, 1986).

Other Animal Organs and Tissues

Certain other anatomical loci for circadian oscillators have been discovered. Some portions of the nervous system of insects, for example, serve as pacemakers for circadian rhythmicity (Block and Page, 1978; Jacklet, 1984). When these areas of the CNS are isolated surgically, they can be cultured and then transplanted into other individual insects (in the same manner that was done for the avian pineal gland). By this procedure, it has been found in silk moths, *Drosophila*, and cockroaches that the phase of the rhythm in the recipient insect is changed to that of the tissue donor. Particularly convincing recent evidence that the optic lobes of cockroaches are the sites of circadian pacemakers has been furnished by Page (1982), who showed that transplanted optic lobes regenerate neural connections to the brain and that circadian rhythmicity is reinstated by 4 to 6 weeks after transplantation. Isolated optic lobes in situ also reconnect to the brain within a similar time course, followed by the reappearance of circadian rhythms. In these experiments, the host cockroach always adopted the free-running period of the donor (whose τ differed by as much as 1.5 h) after transplantation of the lobes and neural regeneration had occurred.

Circadian rhythms have been observed in the adrenal gland in culture. For example, Andrews and Folk (1964) found that O_2 consumption in cultured hamster adrenals varied from 60% above the mean at midnight to 60% below the mean some 12 h later at midday (Andrews, 1971). The rate of uptake of [^{14}C]acetate and its incorporation into steroids was 6 to 8 times greater at midnight than at midday (Andrews, 1968). A similar increase in the secretion of corticosterone by cultured adrenal glands was observed by Shiotsuka et al. (1974). In contrast, O'Hare and Hornsby (1975) reported an absence of a circadian rhythm of corticosterone secretion in monolayer cultures of adult rat adrenocortical cells that had been either stimulated with ACTH or unstimulated; this could have been due to desynchronization of the cells.

Circadian and ultradian (12 h) rhythms in the rate of cardiac contraction in the isolated rat heart have been reported, as well as in dissociated heart cells and heart cell networks (Tharp and Folk, 1965). Since the networks were formed from scattered heart cells, the rhythmicity in the intact organ cannot be due to a particular arrangement of cells. Rather, it appears to reside in individual cells that then become synchronized in the network. The organization of cardiac pacemakers represents both a hierarchical system, in which there is a dominant, driving pacemaker that imposes its period and phase upon subordinate, slave oscillators, and a system of mutually coupled oscillators, wherein no single oscillator serves as a dominant master but in which redundant oscillators share the pacemaker role (Pittendrigh, 1974; Pittendrigh and Daan, 1976; Takahashi and Zatz, 1982). Although the sinoatrial node usually plays the pacemaker role and imparts its phase to the rhythm of cardiac activity, in its absence the atrioventricular node takes over control of heart rate. If both nodes are absent, other groups of cardiac cells continue to pace the heart, beating in unison in cell culture with a period dependent on the number of constituent cells and the strength of their mutual coupling (Goshima, 1979). Thus, surgical removal of a putative pacemaker does not necessarily abolish the overt rhythmicity nor fully reveal its normal role (Takahashi and Zatz, 1982).

Plant Pulvini and Stomata

The leaves of many leguminous plants *(Albizzia julibrissin, Mimosa pudica, Phaseolus coccineus (multiflorus)* and *P. vulgaris, Samanea saman, Trifolium repens)* fold together into a vertical position during the night and reopen horizontally during the next day. The movements commence even before the arrival of dusk or dawn, respectively. This sleep (nyctinastic) behavior was described more than 2300 years ago by Androsthenes during the march of the army of Alexander the Great, and more than 250 years ago by de Mairan (1729), who reported that leaf movements in *Mimosa* persisted in the absence of environmental perturbations. These circadian systems have proved useful in the dissection of clock mechanisms (reviewed by Satter and Galston, 1981), particularly since cellular membranes and ion transport may play a key role in the pacemaker mechanism (see

Section 4.3.3), and perhaps serve more generally as the Rosetta stone of plant behavior (Satter and Galston, 1973).

Most nyctinastic movements are generated by changes in the size and shape of cells in the pulvinus, a motor organ at the base of each leaf. Plants with compound leaves have secondary or tertiary pulvini associated with each leaflet. The pulvinus itself is a flexible cylinder, straight during the day and curved at night, which consists of a central vascular core containing xylem and phloem that is surrounded by several concentric layers of collenchyma cells and numerous layers of cortical cells. The cortical cells in the outer few layers (the motor cells) have elastic walls that permit turgor changes to be translated into leaf movements. Leaflet movement is accompanied by potassium flux into and out of these pulvinule motor cells (Satter and Galston, 1971a). Pulvinar movement is accomplished by the interaction between groups of motor cells located on opposite sides of the pulvinus that alternately swell or shrink out-of-phase with each other. Opposing regions of the pulvinus are defined functionally: extensor cells gain turgor during leaf opening, and flexor cells gain turgor during leaf closure (Satter et al., 1974b).

Pulvini are self-contained systems for the generation of circadian, turgor-based leaf movements capable of being phased by light (Satter and Galston, 1973, 1981; Gorton and Satter, 1983). In addition to the motor cells and the ions required for the regulation of turgor (Satter et al., 1970, 1974a), each pulvinus contains a circadian oscillator (Satter and Galston, 1971b; Satter et al., 1974b), the photoreceptor pigments that interact with the clock (Koukkari and Hillman, 1968; Satter and Galston, 1971b; Satter et al., 1974b; Watanabe and Sibaoka, 1973), and the metabolic machinery and substrate that keep the clock running for at least one circadian cycle in excised organs (Simon et al., 1976a,b). Indeed, the persistence of free-running, light-sensitive oscillations after the selective dissection of either extensor or flexor cortical cells clearly shows that the clockwork and associated pigments are contained in each half of the pulvinus (Palmer and Asprey, 1958). Thus, the pulvinus represents another anatomically defined locus for a circadian pacemaker, this time in plant material.

It is striking that the behavior of the pulvinule motor cells resembles that of guard cells controlling stomatal opening and closure (Evans and Allaway, 1972). The mechanisms underlying these rhythmic changes in turgor pressure as well as those underlying ultradian rhythms of circumnutation (Millet et al., 1984) are discussed further in Section 4.3.3.

4.1.2 Circadian Rhythms in Isolated Cells

One can conclude from the precending section that isolated organs and tissues can oscillate independently, but what of lower levels of anatomical complexity? That populations of eukaryotic microorganisms exhibit circadian organization is quite clear (see Chapter 2), although there is always the formal (but probably unlikely) possibility that rhythmicity is generated by a network of intercommunicating cells and that no individual cell is capable of circadian outputs (see Section 6.2.2). Mammalian cell cultures also display circadian periodicity (Kadle and Folk, 1983). For example, cultured liver cells from 3- to 7-day-old rat pups have been shown to have a circadian rhythm in the activity and the inducibility of tyrosine aminotransferase (TAT) and in [^{3}H]leucine incorporation (Hardeland, 1972; 1973a,b). The highest activity in TAT was observed during the light interval in LD: 12,12, and the lowest occurred during darkness. The rhythm persisted in LL for several days but eventually disappeared after 2 weeks (Hardeland, 1973a). This damping was probably due to desynchronization among the cells, indicating the importance in vivo of cell-to-cell contact and communication via gap junctions. Rhythmicity could be reinduced by a single 1-h dark pulse (Hardeland, 1973b). We have discussed in the previous section the circadian rhythms reported in dissociated heart cells (Tharp and Folk, 1965) and in cultures of dispersed cells from chicken pineal glands (Deguchi, 1979b).

The situation for isolated single eukaryotic microorganisms is less ambiguous. Oxygen evolution has been recorded by flow-through techniques (see Section 2.2.1) in a single cell of *Acetabularia mediterranea* in LL for long timespans, even up to 7 weeks (Schweiger et al., 1964a; Mergenhagen and Schweiger, 1973; Karakashian and Schweiger, 1976a). A clear-cut circadian rhythm was evidenced, with no signs of damping (Fig. 4.8A). Damping of the rhythm in groups of cells (Karakashian and Schweiger, 1976a) reflects the variability of the average period in individual cells and the absence of a mutual effect among them (Mergenhagen and Schweiger, 1974). Similarly, the rhythm of photosynthesis in single cells of *Gonyaulax polyedra* (see Section 2.2.4) has been followed under constant conditions with a cartesian diver (Sweeney, 1960), and the rhythm of glow bioluminescence in an isolated cell of this dinoflagellate (see Section 2.2.4) also displays circadian periodicity in DD (Hastings et al., 1985). The wave form of the first glow peak after transition from an LD cycle had a shape similar to that of a population of cells (Fig. 4.8B). The broadening of the glow peak in cell cultures during ensuing days and consequent damping of the rhythm (Hastings et al., 1985) was indicative of the expected desynchronization among individual cells (Njus et al., 1981). Rhythms of plastid movement (Töpperwien and Hardeland, 1980), spontaneous bioluminescence (Hardeland, 1982), and chloroplast expansion and contraction and the intracellular distribution of bioluminescent microsources (Hardeland and Nord, 1984) have been reported in various species of the dinoflagellate genus *Pyrocystis*. A rhythm of mating reactivity has been demonstrated within single cells of *P. bursaria,* even in individuals taken from arrhythmic populations in which circadian rhythmicity had damped out after 2 weeks in dim LL of 1000 lx (Miwa et al., 1987). Thus, population arrhythmicity appears to be the result of desynchronization among the

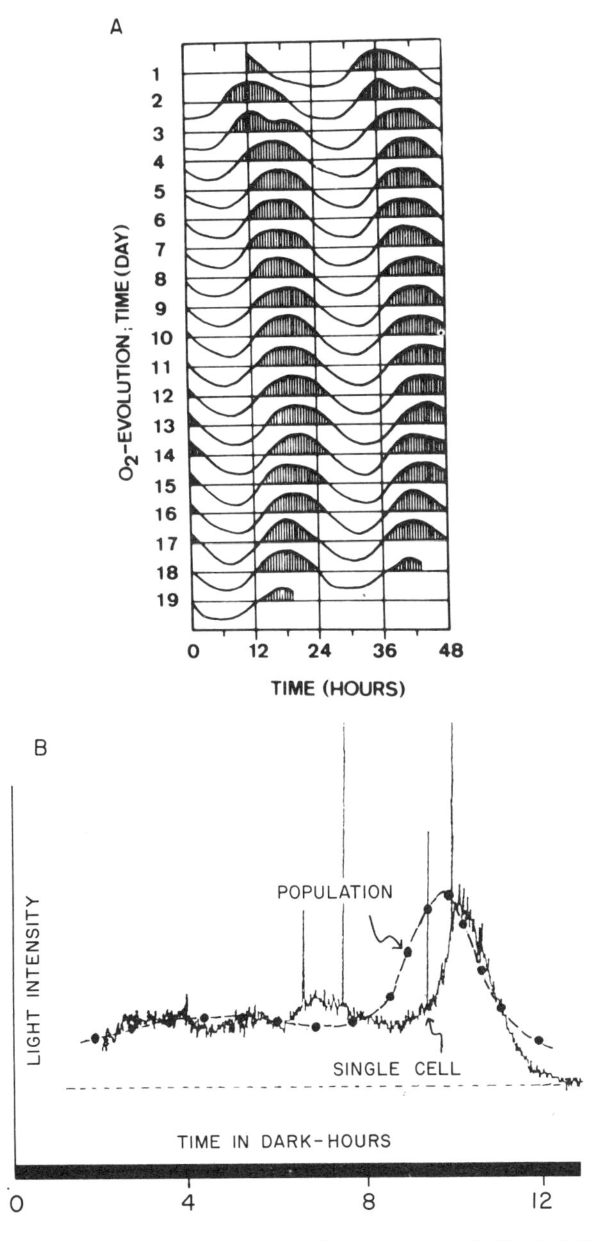

FIGURE 4.8. Circadian rhythmicity in isolated eukaryotic unicells. (A) Computerized data for oxygen evolution by a single cell of *Acetabularia mediterranea*. (Reprinted from Schweiger and Schweiger, 1977, with permission of Academic Press, Inc.) (B) Comparison of time courses of glow peaks measured in DD from a single *Gonyaulax polyedra* cell (jagged line with four flashes superimposed) and a population (10,000 cells; solid dots, flashes not shown). *Ordinate:* Light intensity, arbitrary units. *Abscissa:* Time in hours from beginning of dark. Dashed line, base line. (Reprinted from Hastings et al., 1985, with permission of Springer-Verlag; see Krasnow et al., 1981.)

constituent oscillators. The fact that a 9-h dark pulse was able to restore rhythmicity in a population could be explained by its differential phase-shifting action (a dark-pulse PRC was derived) on the individual cells, the phases of which were randomly distributed at the time of the signal.

4.1.3 Subcellular Circadian Rhythmicity

We now turn to the possibility that circadian organization may reside at the subcellular level. Perhaps the entire cell is not required for the generation of circadian periodicities. Indeed, we have already seen this to be the case for *Acetabularia crenulata* and *A. major* (Sweeney and Haxo, 1961) and for *A. mediterranea* (Schweiger et al., 1964a,b), in which a circadian rhythm of photosynthetic activity was demonstrated in enucleated single cells (see Section 2.2.1). The nucleus—although not necessary for entrainment, phase shifting, or the maintenance of the persisting rhythm—did determine its phase (see Fig. 2.3). A PRC for dark pulses has been constructed for individual, anucleate *A. mediterranea* cells (Karakashian and Schweiger, 1976a) that was almost identical to that found for nucleate cells (Fig. 4.9).

Requirement of a nucleus for the expression of circadian rhythmicity has been investigated in human erythrocytes cultured in vitro. This minimal system is attractive because the cells lack nuclei, mitochondria, ribosomes, endoplasmic reticula, and vacuoles. Ashkenazi et al. (1975) first reported 12-h and 24-h rhythms in the activities of a number of enzymes in such incubated erythrocyte suspensions. Similarly, Cornelius and Rensing (1976) detected daily rhythmic changes in the activity of membrane-bound Mg^{2+}-ATPase in an LD cycle but with quite small amplitudes. Hartman et al. (1976) labeled human red blood cells (RBC) with a fluorescent cyanine dye (a membrane probe) and considered the circadian changes in the intensity of fluorescence of the suspension to be due to a varying binding capacity of the membrane for the dye. Mabood et al. (1978) reported significant circadian changes in the activity of acetylcholinesterase in incubated human erythrocytes. On the other hand, attempts by several other laboratories to confirm the earlier findings of Ashkenazi and coworkers have been unsuccessful (Cornelius, 1980). Thus, Peleg et al. (1979) found ultradian, but not circadian, oscillations in enzyme activity and calcium-binding capacity of suspensions of human RBC and RBC ghosts in vitro. Cornelius (1980) measured the activities of the tightly membrane-bound enzyme acetylcholinesterase and of partly bound pyruvate kinase during a 24-h timespan in incubated suspensions of human erythrocytes, as well as membrane fluidity (by a fluorescence-depolarization method) and valinomycin-mediated K^+ transport. Although fluctuations were found in all of these measurements of enzyme activity and dynamic membrane structure and function, some of which appeared to have a 9 h-, 12 h-, or 24 h-period, statistically significant periodicities were not obtained after av-

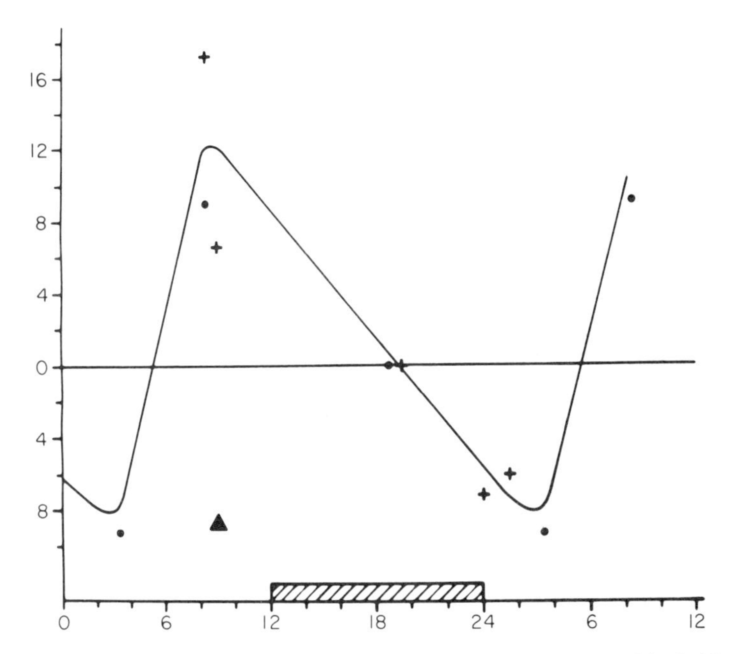

FIGURE 4.9. Approximate phase response curve (O_2 production) of individual nucleated and enucleated *Acetabularia mediterranea* cells subjected to single, 8-h dark pulses at various times in the circadian cycle. Phase changes are plotted (advances above, delays below the horizontal line) with reference to the midpoint of the dark pulse. Both nucleated (+) and enucleated (●) cells were phase-sensitive to darkening during the first 12 h of the circadian cycle. Dark pulses given early in this interval induced phase delays, whereas pulses a few hours later caused phase advances. Similar pulses given late during the second half of the circadian cycle (hatched bar) effected no change in phase. For orientation in the circadian cycle, the maximum of O_2 production (▲) detected by the electrode is shown at hour 9. The results for nucleated cells were determined with cells maintained at a constant temperature of 25°C, whereas those for enucleated cells were derived from cells monitored at 20°C. Two of the points obtained for enucleated cells are reproduced on both sides of the profile for clarity of presentation. (Reprinted from Karakashian and Schweiger, 1976a, with permission of Academic Press.)

eraging the individual experiments. Thus, at best one can speculate that the RBC clock (assuming that it exists) represents a rudiment of the full-fledged circadian clock found in other cell types, especially given the fact that the mammalian erythrocyte has lost its nucleus, organelles, and capacity for protein synthesis during ontogeny (Cornelius, 1980).

It is interesting to note, however, that Radha et al. (1985) reported that glutathione (GSH) levels in human platelets, but not in erythrocytes, display a circadian rhythm in vitro. GSH was measured at 4-h intervals over a 24- to 48-h timespan in LL, DD, or LD: 16,8 in platelet-rich plasma and concentrates stored (pH = 7.4) in sterile Falcon flasks, which were shaken gently at the times of sample withdrawal. The phasing of the rhythm (peak in subjective night) was approximately the same under all illumination regimens, and platelets from night-shift workers displayed a similar periodicity to that of day workers, observations that indicated that the rhythm was independent of both the LD cycle and the sleep–wake cycle of the donor. More than 90% of the total GSH in platelets was found to be in the reduced form. Incubation of platelets with buthionine sulfoximine (1 μ*M*), an inhibitor of glutamyl cysteine synthetase, abolished the periodicity in GSH, a finding that suggested that the rise in GSH levels was due to de novo synthesis. The stored platelets actively incorporated [^{14}C]glutamine into a small peptide that coeluted with GSH on a sulfhydril affinity column.

Mergenhagen and Schweiger (1975a) have shown that the circadian rhythm of oxygen evolution in *A. mediterranea* is manifested also in a variety of cell fragments. Isolated caps, basal nucleate and anucleate fragments, apical fragments (some only 10 mm in length) with and without stalks, and small segments taken from the middle of the stalk all displayed circadian rhythmicity (Fig. 4.10). Adamich and Sweeney (1976) have demonstrated that the circadian rhythm of luminescence continues in protoplasts of *G. polyedra,* which lack cell walls. It would be most provocative if circadian periodicity could be demonstrated in an isolated subcellular component, such as an in vitro chloroplastic or mitochondrial preparation (Edmunds, 1980b). In such a case, an anatomical locus for a circadian oscillator would have been demonstrated at the organellar level—perhaps not impossible if one considers these structures to have had an ancestry

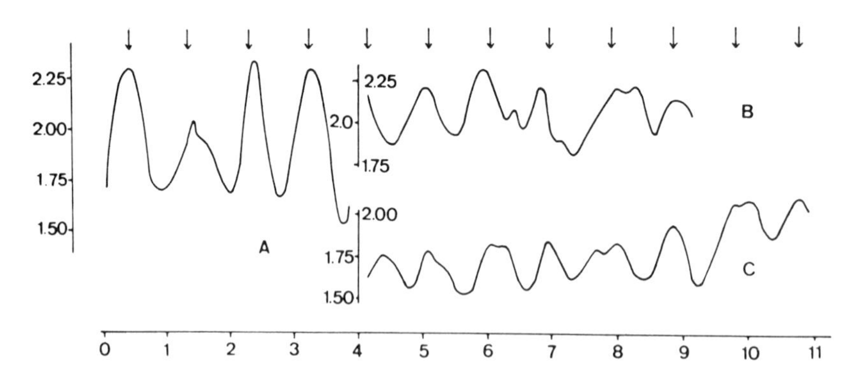

FIGURE 4.10. Circadian rhythm of O_2 evolution in cell fragments of *Acetabularia mediterranea. Abscissa:* Time (days). *Ordinate.* O_2 evolution in relative units. The arrows indicate the circadian time. Recordings made (A) in the whole cell, (B) in the anucleate capless apical fragment, and (C) in the nucleate basal fragment. (Reprinted from Mergenhagen and Schweiger, 1975a, with permission of Academic Press.)

as an invading endosymbiont, replete with genome and membranes (see Sections 5.2.5 and 6.1).

4.2 Tracing the Entrainment Pathway for Light Signals

In the preceding section, several anatomical loci for circadian pacemakers in higher organisms were identified (although the situation is less clear in isolated single cells and in unicells). A conceptual model of a circadian system is given in Figure 4.11 (Eskin, 1979a). One strategy for discovering the mechanism of a circadian clock is to delineate the pathway of entrainment by light. We have seen, for example, that the isolated gastropod eye contains a photoentrainable pacemaker. The morphological and physiological routes by which environmental information flows to the clock, therefore, must end on the pacemaker mechanism itself. Characterization of the entrainment pathway—a classic problem in information processing by the animal nervous system—is of particular interest because the time course of major events is relatively long and because the transduced information ultimately must affect some sort of intracellular metabolic process that is the oscillator.

There are several steps in the entrainment of a circadian system (Eskin, 1979b). A photopigment in an organized receptor or receptor cell must first absorb the photons of the light signal. This information must then be transduced, coded and propagated to the pacemaker. Afterward, a decoding process takes place, resulting ultimately in a phase shift of some overt rhythmic output of the clock.

4.2.1 Nature and Localization of the Photoreceptor

Light influences circadian rhythms in various ways (Bruce, 1960; Ninnemann, 1979): light perturbations initiate rhythmicity in arrhythmic populations, LD cycles and repetitive pulses entrain rhythms, single steps-up and steps-down in light intensity, as well as discrete light or dark pulses, phase-shift rhythms, light signals cause photoperiodic induction in higher plants and animals, which is correlated with seasonal rhythmicity under natural conditions, higher light intensities inhibit the expression of rhyth-

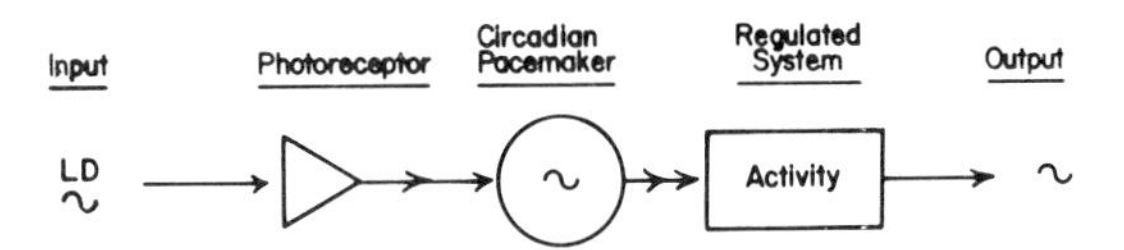

Figure 4.11. A conceptual model of the circadian system. (Reprinted from Eskin, 1979a, with permission of Federation of American Societies for Experimental Biology.)

micity, perhaps arresting the clock itself, and constant illumination of various intensities modulates the free-running period, shortening or lengthening it in a consistent manner (codified as Aschoff's circadian rules: see Hoffmann, 1965; Wever, 1965a). Thus, we are confronted with a number of types of light effect and with the attendant possibility that a different photoreceptor system may underlie each action, even in the same species.

Even for a given light action on a particular organism, there is the possibility that different circadian rhythms may be associated with different photoreceptors. For example, both the oscillator and the photoreceptor controlling the eclosion rhythm in *Drosophila* seem to be different from those controlling adult activity (Engelmann and Mack, 1978). The matter may be even more complicated: different photoreceptor pigments might mediate the effects of light on a given rhythm in the same species at different circadian times. Thus, Chandrashekaran and Engelmann (1973) found that the early subjective-night phase of the free-running circadian rhythm of eclosion in *Drosophila pseudoobscura* required 10 times less energy of blue light (442 nm) for phase-shifting than did the late subjective night, a finding that could suggest that two different pigments or primary processes may be responsible for light absorption at the two times. Likewise, Winfree (1970) has reported that this same circadian system exhibits two qualitatively distinct types of resetting behavior (type 0 and type 1) in response to two quantitatively different resetting stimuli (>50 sec and <50 sec, respectively).

A classic approach to the identification of a photoreceptor pigment is the determination of an action spectrum for a given process in the hope that it will match the absorption spectrum of some putative substance present in the cell or tissue. In her comprehensive and critical review of photoreceptors for circadian rhythms, Ninnemann (1979) compiled the results of such experiments in both plants and animals. Given the complexity of the problem as just discussed, it perhaps is not surprising that Ninnemann concludes that there is no specific light-absorbing pigment in organisms that is effective for initiating, phase-shifting, or inhibiting circadian rhythmicity. Action spectra for phase-shifting circadian rhythms in *Neurospora, Gonyaulax, Drosophila,* the moth *Pectinophora,* and the hamster *Mesocricetus* are all different. Just as no specific universal structure has been developed as the anatomical locus for circadian oscillators (see Section 4.1), no universal photoreceptor or pigment has evolved. In retrospect, this is reasonable given the likelihood that during their evolutionary history organisms selected and used photoreceptors from the plethora of pigments they already possessed, rather than creating them de novo (Bünning, 1973). Nevertheless, some generalizations can be made (Ninnemann, 1979).

Plant Photoreceptors

Lower plants appear to use blue light-absorbing photoreceptors, such as flavoproteins and carotenoids (Briggs, 1976; Senger, 1980), whereas many higher plants use phytochrome and perhaps chlorophyll as photoreceptor pigments for circadian rhythms. Thus, the action spectrum for the inhibition by LL of the conidiation rhythm of *Neurospora crassa* (see Section 2.3.1) had a maximum effect at 465 nm and other peaks at 485, 415 and 375 nm (Sargent and Briggs, 1967). This spectrum resembles that for both flavins and carotenoids, although the latter appear to be excluded by the finding that LL-induced damping was unaffected in an albino double mutant *(al-1, al-2)* having blocks in the pathway of carotenoid biosynthesis. Blue light also is most effective (and red light unaffective) in phase shifting the conidiation rhythm (Sargent et al., 1966), and the action spectrum is similar to that for the suppression of conidial banding (Feldman and Dunlap, 1983). For these and other reasons, the flavoprotein-mediated reduction of a *b*-type cytochrome was considered as a possible primary event of the induction and phase-shifting of rhythmic conidiation in *Neurospora* (Muñoz and Butler, 1975).

Additional evidence for a role in photoreception of this flavin-cytochrome *b* complex, which may be localized in the plasma membrane (Brain et al., 1977a,b), has been afforded by studies on several other mutants of *Neurospora* that affect light sensitivity (see Section 4.4.3, Table 4.2). Klemm and Ninnemann (1979) have suggested that nitrate reductase, a molybdoflavoprotein containing a *b*-type cytochrome serves as the photoreceptor for blue light-stimulated coniditation, but this enzyme does not appear to be important for the LL suppression and the light-induced phase-shifting effects, since mutants *(nit)* deficient in nitrate reductase activity (or cultures grown under conditions in which enzyme activity was repressed) nevertheless exhibited normal photoresponses (Paietta and Sargent, 1982).

The involvement of cytochromes or flavins or both has been suggested by results obtained with *Saccharomyces cerevisiae* (Section 2.3.2). Blue light ($\cong$ 410 nm) was the most effective wavelength in the visible spectrum for the inhibition of growth and transport. Petite mutants *(rho$^-$)* deficient in cytochromes *b*, a/a_3, or both exhibited decreasing photosensitivity (Ułaszewski et al., 1979; Edmunds, 1980a). Further, the light-entrainable, circadian rhythm of amino acid transport exhibited by the wild type could not be synchronized by LD cycles in cultures of the Y185 *rho$^-$* mutant (Edmunds et al., 1979), a finding consistent with the hypothesis that flavin–cytochrome complexes may serve as photoreceptors for both the direct inhibitory action of visible light and the entrainment of biological clocks by light in this microorganism.

Red and blue light seem to be more effective in inducing photoresponses in algae. Thus, the action spectrum for the photoinhibition of the circadian rhythm of luminescence capacity in *G. polyedra* (see Section 2.2.4) showed a maximal effect at 440 nm and a lesser effect at 660 to 735 nm, whereas minimal suppression occurred at 600 nm (Sweeney et al., 1959). These results suggested that chlorophyll and other photosynthetic pigments (chl *a* and chl *c*, peridinin, and other carotenoids) may serve as photoreceptors for this response. Phytochrome was not implicated because far-red light failed to reverse photoinhibition. The action spectrum for phase-shifting the rhythm of luminescence showed maximal effectiveness at 475 and 650 nm; red light was considerably less effective than blue (Hastings and Sweeney, 1960).

Phytochrome and chlorophyll appear to serve as photoreceptors for the circadian system in many higher plants as well. Continuous red light lengthened τ of the leaf movement rhythm in *P. coccineus (multiflorus)*, whereas blue, green, and infrared light were without effect (Lörcher, 1958). Entrainment of the rhythm by a red light:dark cycle was reversed by far-red irradiation applied together with the red light (or immediately thereafter), and in an alternating red and far-red light cycle, the wavelength of the last exposure determined entrainment or its suppression (Bünning and Lörcher, 1957). The wavelength of the perturbing light pulse determined the shape of the PRC for the leaf-movement rhythm. Red light generated both $+\Delta\phi$s and $-\Delta\phi$s, but blue and far-red light yielded only $+\Delta\phi$s or no phase shift at all (Bünning and Moser, 1966). The authors suggested different primary processes for the advancing and delaying effects of light.

In the short-day plant *Coleus*, red and blue light affected τ and ϕ of the circadian rhythm of leaf movement in opposite ways (far-red light showed no reversal effects), leading Halaban (1969) to conclude that photoinduction of two different photochemical processes occurred by red or blue light to which the circadian system was periodically responsive and that the photosynthetic apparatus was involved in photoreception. Rhythmic petal movement in the CAM (Crassulacean Acid Metabolism) plant *Kalanchoe blossfeldiana* also is regulated by red light. Isoquantal blue light red-orange light were equally effective in reinitiating a damped rhythm (Karvé et al., 1961); continuous red light was more effective than blue in suppressing rhythmicity (and was antagonized by far-red light), and the action spectrum for phase-shifting the petal movement rhythm closely resembled the absorption spectrum for the red light-absorbing (λ_{max} = 660 nm) species of phytochrome (Schrempf, 1975).

Phytochrome also seems to be a key photoreceptor for the entrainment and phase-shifting of the circadian rhythm of CO_2 evolution in *Bryophyllum fedtschenkoi* (Wilkins, 1973; Harris and Wilkins, 1978), for entrainment of rhythmic CO_2 output in the duckweed *Lemna perpusilla* (Hillman, 1971), and for photoresponses of the nyctinastic leaf movements (see Section 4.1.1) of *A. julibrissin* (Hillman and Koukkari, 1967; Jaffe and Galston, 1967) and of *S. saman* (Satter et al., 1974a).

All of the plant pigments invoked as photoreceptors are membrane-bound or are at least associated with membranes. The far red-absorbing form of phytochrome (P_{fr}) appears to influence the level of K^+ ion in different parts of the pulvinus in plants exhibiting leaf nyctinasty and, therefore, to regulate the circadian rhythm of K^+ flux observed in these species (Satter and Galston, 1971a,b; Satter et al., 1974a,b). This rhythmic ionic flux, in turn, is responsible for the changes in turgor pressure that generate the overt leaf movement rhythm. Phytochrome also regulates changes in transmembrane potential in *Samanea* pulvini (Racusen and Satter, 1975). Indeed, membrane-bound phytochrome has been incorporated into models for membrane oscillators (Wagner et al., 1974a,b) and has been proposed as an integral part of the circadian clock (Heide, 1977). The role of membranes in circadian oscillators is considered further in Sections 4.3.1 and 5.4.

Photoreceptors in Animals

Less is known about the identity of absorbing pigments for circadian systems in animals. If one had to generalize, the insects appear to use flavoproteins (or carotenoids) as photoreceptors, whereas rhodopsin appears to be the clock photopigment in mammals (Ninnemann, 1979). Thus, blue light (360 to 520 nm) perturbations generate both $+\Delta\phi$s and $-\Delta\phi$s for the eclosion rhythm of *D. pseudoobscura* (Frank and Zimmerman, 1969). A detailed action spectrum for phase delays (Klemm and Ninnemann, 1976) revealed 457 nm to be most effective, with secondary maxima at 473, 435, 407, and 375 nm. Wavelengths longer than 520 nm were ineffective. As noted earlier, the sensitivity of the system to such blue light pulses was a function of the CT at which the signal was imposed (Chandrashekaran and Engelmann, 1973). This result may indicate participation of more than one photoreceptor pigment. Similarly, Bruce and Minis (1969) have derived an action spectrum for the initiation of the circadian rhythm of egg hatching in the pink bollworm *Pectinophora gossypiella* by a single blue light pulse (λ = 390 to 520 nm), given after the fifth day of embryogenesis. All blue wavelengths were approximately equal in their inductive effect; wavelengths longer than 520 nm were not effective. The rhythms of pupal eclosion and of oviposition by the adult moth also can be entrained and phase-shifted by blue (λ = 480 nm) light but not by red light (Pittendrigh et al., 1970). The action spectra for these two insects are similar to that for *Neurospora* (Sargent and Briggs, 1967), and all three resemble the absorption spectra of many flavins, flavoproteins, or carotenoids. The last species of pigment seems to be excluded: *D. melanogaster* grown on a diet depleted of carotenoids showed no change in photosensitivity of the circadian rhythm of eclosion, whereas the sensitivity of the visual receptor decreased by three orders of magnitude (Zimmerman and Goldsmith, 1971).

In vertebrates, there is a paucity of action spectra for the primary photobiological processes involved in the light-mediated steps of chronobiology (Ninnemann, 1979). The data that exist, however, suggest that retinal rhodopsin is the primary photopigment for the circadian system. For example, Cardinali et al. (1972) found that green light had the greatest effect in inhibiting the activity of HIOMT in the rat pineal gland (see Section 4.1.1), blue or yellow light had less effect, and red light had no effect at all. These results correspond to the absorption spectrum of rhodopsin or a pigment with similar absorption characteristics. The structures within the mammalian eye that participate in perceiving entraining light signals, nevertheless, may not be the visual rods or cones. There is a great body of evidence for light reception by extraretinal photoreceptors (ERP) in addition to the eyes in birds and mammals (as well as the lower animal phyla), such as the pineal gland (see Section 4.1.1) and the harderian glands (containing red protoporphyrins) (Wetterberg et al., 1970), which would transfer environmental information to the hypothalamus and SCN.

In fact, Menaker and Eskin (1967) were able to test the Bünning hypothesis for the role of circadian clocks in photoperiodic time measurement by making use of the different spectral sensitivities (and presumably different photoreceptors) for entrainment and for photoperiodic induction in the house sparrow, *P. domesticus*. Noninductive green light, capable of entraining circadian rhythms even in blinded birds (Menaker, 1968) but without effect on testicular recrudescence, was used to phase the rhythm of motor activity and that of sensitivity to light signals so that perturbations by potentially inductive white light fell at different circadian times. As anticipated, the white light exposures were either inductive or noninductive depending on the CT at which they were imposed. A provocative study of the clock receptor in hamster retina (Takahashi et al., 1984) showed that this rhodopsin-containing photoreceptor had a very high threshold (approximately 1 million times higher than that for rod vision), but the reciprocity between intensity and duration of light stimuli held for pulses as long as 45 min. This system, therefore, appears to serve as a photon integrator. Whether it is characteristic of the photoreceptors of other circadian pacemakers is unknown.

4.2.2 Coupling Links Between Receptor and Clock

In Section 4.2.1, we began to trace the entrainment pathway for light signals to the clock by attempting to identify via action spectra the photoreceptor pigments themselves, which, in higher organisms, reside in the anatomical loci for circadian pacemakers (see Section 4.1). One must proceed with caution. The cellular structure that contains the photopigment for light input into the circadian system may not be part of the clock mechanism or pacemaker itself. We have seen in the isolated eye of *Bulla* (see Section 4.1) that surgical removal of the entire photoreceptor layer (Fig. 4.3) did not alter the circadian period of the rhythm in CAP. All that remained were the 100 or so basal retinal neurons (Block and Wallace, 1982). Thus, in this case the information from the photoreceptor must be propagated electronically (passively) or via action potentials and by either secretory or electrical junctions between cells (Eskin, 1979b). To further complicate matters, in many animals the retina and the brain form a neural loop. The sensitivity of the lateral eye of the horseshoe crab *Limulus polyphemus,* for example, is modulated by efferent optic nerve impulses transmitted from a circadian clock located in the brain—a direct effect of the CNS on photoreceptor function (Barlow et al., 1977, 1987). Should the circadian oscillator be associated with the photoreceptor cells themselves, the flow of entrainment information might or might not involve an electrical membrane event. If not, only intracellular processes would transmit information from the primary photochemical reactions.

These next steps in the entrainment pathway have been dissected pharmacologically in *Aplysia*—an experimental approach that has proved particularly fruitful in this organism and that has been aided by the fact that one isolated eye can serve as a control for the other treated one (Eskin, 1977, 1979b; Jacklet, 1984). On the one hand, drugs that do not themselves perturb the clock and that act on some specific target can be used to try to block the phase-shifting action of light signals on the CAP rhythm; on the other hand, pulses of various chemical agents can be used in an attempt to mimic the effects of light and produce similar PRCs. Both types of experiment should help elucidate the physiological and biochemical processes that occur during and after a phase-resetting light pulse.

Light-Blocking Experiments

If an isolated *Aplysia* eye was exposed to a bathing solution that was low in sodium and calcium and high in magnesium (LoNaHiMgLoCa) during the time that a normally phase-shifting ($\Delta\phi = +3$ h), 6-h light-pulse was given, no phase shift occurred (Eskin, 1977). This treatment by itself did not generate a $\Delta\phi$. This observation suggested that a change in membrane conductance or membrane potential (photoreceptor, action, or synaptic) was a required step in the entrainment pathway. The LoNaHiMgLoCa solution inhibited optic nerve impulses during the light and had large effects on photoreceptor potential. In fact, solutions that decreased the electroretinogram by 90% or more also blocked light-induced phase-shifting. Action potentials and those resulting from secretory processes, however, did not appear to be necessary, since treatment of the eye with tetrodotoxin and a HiMgLoCa solution (which should inhibit such potentials) had no effect on light-induced $\Delta\phi$s (Eskin, 1977). These findings suggested that passively (electrotonically) propagated changes in photoreceptor membrane potential couple the reception of light signals to the pacemaker mechanism, which would reside either in the photoreceptor cells of *Aplysia* themselves or in adjacent neuronal cells coupled to them by electrical gap junctions. A critical test to distinguish between these two possibilities would be to attempt to generate a $\Delta\phi$ by light in the presence of an agent that would disrupt gap junctions but that alone would not produce a $\Delta\phi$ of the CAP rhythm when administered to the eye.

Light-Mimicking Experiments

In order to test the hypothesis based on the results of the light-blocking experiments, attempts were made to mimic the effects of light pulses on the CAP rhythm of *Aplysia* by treatments that caused changes in membrane potential of the neurons in the isolated eye (Eskin, 1979b). Membrane-depolarizing agents had phase-shifting effects that were similar to those produced by light. Both elevated external potassium concentration (HiK) and incubation in the presence of strophanthidin (an inhibitor of Na^+–K^+-pump ATPase) had PRCs that were relatively similar and resembled derived for light pulses (Eskin, 1972, 1979b; see Fig. 4.27). Phase shifts were dose-dependent, and the sharp transition (breakpoint) from delays to advances for all three agents was centered at CT 18 (Jacklet and Lotshaw, 1981). Neither agent seemed to be acting indirectly by causing secretion, since the $\Delta\phi$s generated were not affected by simultaneous treatment with a HiMgLoCa solution. Further, the $\Delta\phi$'s generated by strophanthidin were negated if the eye was treated with LoNa, supporting the notion that the inhibitor acted by decreasing transmembranal ionic and potential gradients. The effects of both HiK and strophanthidin seemed to be due primarily to changes in potential and not to the concomitant changes in the intracellular concentration of Na^+ and K^+ ions, which would be expected to be opposite to one another in the two treatments.

Entrainment Pathway for Serotonin

During the course of experiments attempting to trace the entrainment pathway for light, serotonin (5-hydroxytryptamine, 5-HT), a neurotransmitter present in the *Aplysia* eye, was found to produce significant advance and delay phase shifts (Corrent et al., 1978, 1982). Its PRC (Fig. 4.12A; see Fig. 4.27), however, was about 180° out-of-phase with that obtained for light pulses, a finding that suggests that 5-HT is used in a different entrainment pathway. Indole analogs of 5-HT, such as bufotenine and LSD, also generated $\Delta\phi$s, whereas tryptophan and 5-hydroxytryptophan, the immediate precursors of 5-HT, did not (Eskin, 1979b, 1982a). Other putative transmitter substances (dopamine, acetylcholine, and neostigmine), although producing $\Delta\phi$s, did not do so at the same circadian phases as those at which 5-HT was effective, indicating a degree of specificity for the latter (Eskin, 1979b). Accordingly, Corrent et al., (1978, 1982) hypothesized that 5-HT plays a role in transmitting circadian information from the CNS to the eye via efferent activity in the optic nerve. Both the photic entrainment pathway contained within the eye and this serotonergic pathway would converge on the ocular circadian pacemaker to regulate its phase.

The question then arose about the mediation of the effects produced by 5-HT. Serotonin did not seem to be acting on the pacemaker by changes in membrane conductance to Na^+, Cl^-, or Ca^{2+} because large reductions

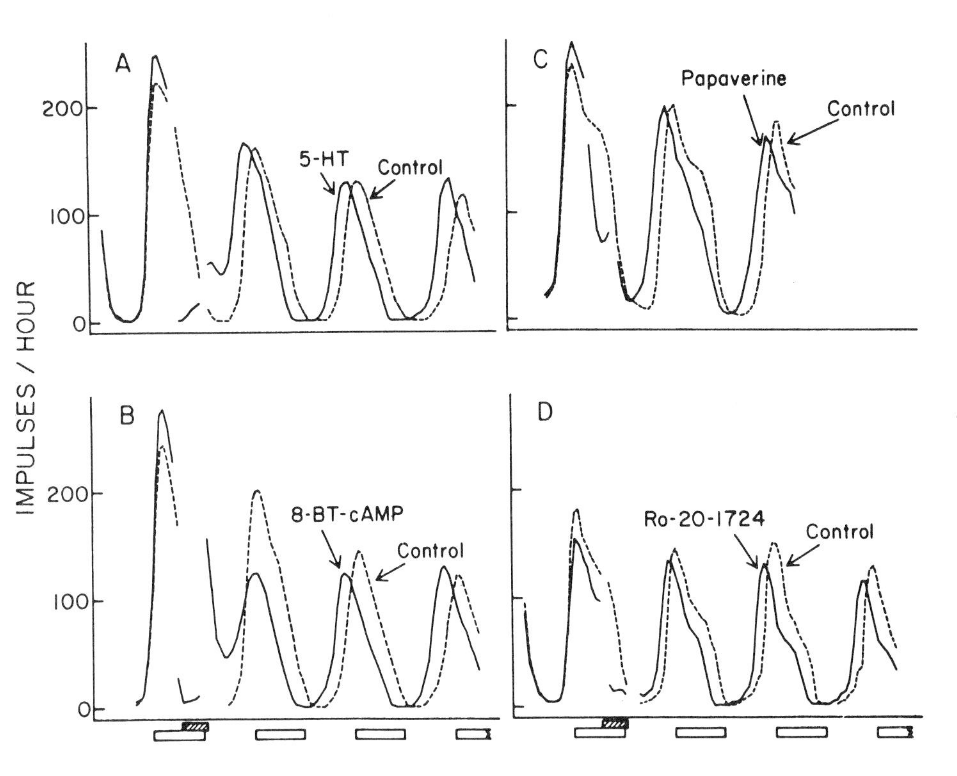

FIGURE 4.12. Advance phase shifts of the circadian rhythm of nerve impulses produced by serotonin (5-hydroxytryptamine, 5-HT) or treatments capable of increasing intracellular cAMP in the isolated eye of *Aplysia californica*. The frequency of spontaneous optic nerve impulses is plotted as a function of the time that the isolated eyes were in DD. Each graph contains rhythms for eyes of the same animal. Experimental eyes received 6-h treatments during the time (CT 7-13) shown by the hatched bars on the bottom of the graphs. The open bars at the bottom of the graph are 12 h in length and represent the projected light portion of the LD cycle that was used to entrain the intact animals. The concentrations used in the treatments were: (A) 10 μ*M*, (B) 2 m*M*, (C) 200 μ*M*, (D) 60 μ*M*. (Reprinted from Eskin et al., 1982, with permission of the authors.)

in the extracellular concentrations of these ions had no effect on the magnitude of the $\Delta\phi$s generated by 5-HT (Corrent et al., 1978). In striking contrast, treatments that should increase intracellular levels of cAMP mimicked the action of 5-HT (Eskin et al., 1982). For example, 8-benzylthio-cAMP, an analog of cAMP, advanced and delayed the rhythm at phases in which 5-HT had similar effects (Fig. 4.12B). In addition, two phosphodiesterase inhibitors, papaverine and Ro-20-1724, generated phase advances at the same time (CT 7-13) that 5-HT did (Fig. 4.12C,D). The phosphodiesterase inhibitors Ro-20-1724 and 3-isobutyl-1-methylxanthine each potentiated the effect of subthreshold doses of 5-HT on the rhythm. The effects of 5-HT and 8-benzylthio-cAMP were nonadditive, indicating that these two agents affect the rhythm through a common pathway. In addition, 5-HT produced large changes (13-fold) in the levels of cAMP in the eye.

Additional evidence that cAMP mediates the phase-shifting effects of 5-HT on the rhythm in the *Aplysia* eye has been obtained with forskolin, a highly specific activator of adenylate cyclase (Eskin and Takahashi, 1983). This agent produced both advance and delay $\Delta\phi$s (Fig. 4.13A,B), whose circadian phase-dependence was identical to that for 5-HT (Fig. 4.13C), known also to stimulate adenylate cyclase in the eye. The ability of both treatments to activate adenylate cyclase in homogenates of the eye was correlated with their ability to shift the phase of the circadian oscillator. There appears to be a circadian rhythm in the level of cAMP in DD in the eye, whose phase corresponds to that of the rhythm of spike discharge (Eskin, 1985, personal communication).

The foregoing results are consistent with two roles for cAMP (Eskin et al., 1982): it may be part of an input pathway by which serotonin entrains the oscillator [in which case the molecule would be acting as an intracellular second messenger for the effects of 5-HT, the next steps of the entrainment pathway entailing, perhaps, protein phosphorylation through activation of a kinase by cAMP; see Lotshaw and Jacklet (1983, 1987)], or it may be part of the clock mechanism itself, in which the cAMP generating and degrading system would form the basis of the oscillator in the eye (see Section 5.2.5). It is interesting to note that cAMP is involved in mediating the output of the circadian pacemaker in the avian pineal gland (Takahashi and Zatz, 1982). Norepinephrine (or its analog isoproterenol) interacts with the β-adrenergic receptors on cultured pinealocytes, resulting in the synthesis of cAMP, which, in turn, causes the induction of serotonin NAT activity. Although the avian pineal gland and the gastropod eye constitute

FIGURE 4.13. Effects of forskolin on the phase of the circadian rhythm of the isolated eye of *Aplysia californica*. (A) Delay phase shift produced by 6-h treatment with forskolin (10^{-6} *M*) during the time shown by the hatched bar on the bottom of the graph. The frequency of spontaneous optic nerve impulses is plotted as a

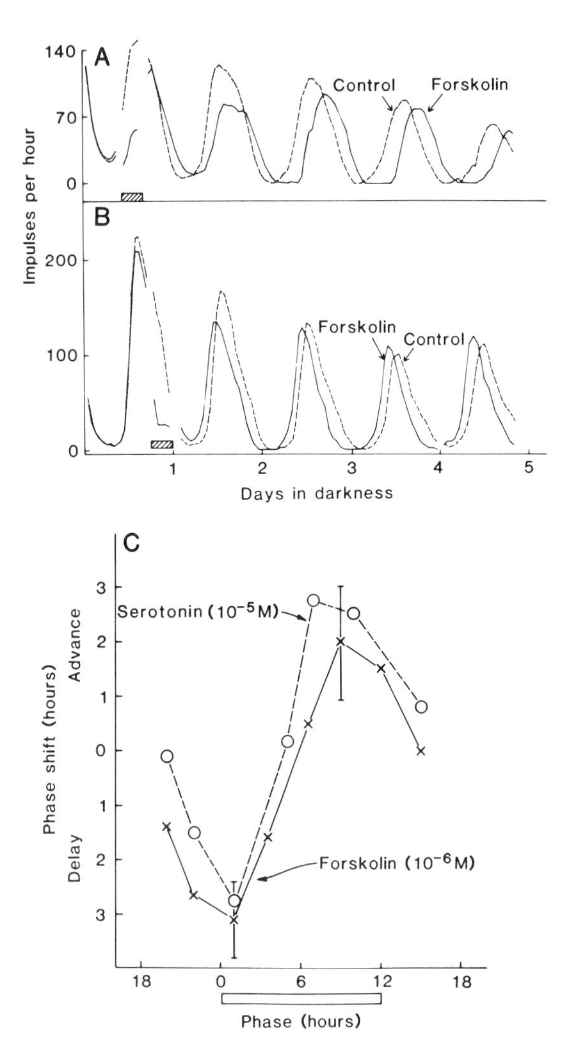

function of the time isolated eyes were in DD. Rhythms shown are from eyes of the same animal. The abscissa begins at projected dusk (CT 12) of the LD: 12,12 cycle used to entrain the intact animal. (B) Advance phase shift produced by forskolin (10^{-6} *M*) given during a later time than the treatment in (A). (C) Comparison of 6-h treatments of forskolin and serotonin. The graph shows phase shifts produced in the rhythm as a function of the time (phase) of exposure of the eyes to the treatments. Treatments were given during the first day of isolation of the eyes, and phase 0 is the time of projected dawn. Data are plotted with respect to the midpoint time of the treatment. In the forskolin experiment, each point is the mean phase shift for three or four pairs of eyes. A total of 28 pairs of eyes was used in this experiment. The 95% confidence intervals are shown for the maximum delay and advance ($n = 4$). Data for serotonin are from Corrent et al. (1982). (Reprinted from Eskin and Takahashi, 1983, with permission of American Association for the Advancement of Science.)

markedly different systems, if cAMP were part of a common mechanism that generates circadian oscillations, it might provide a point of convergence for both clock inputs and outputs (see Section 4.3.3).

In contrast, the photic entrainment pathway in *Aplysia* discussed previously does not appear to involve mediation by cAMP: light has no detectable effects on cAMP (Eskin et al., 1984a). Rather, cyclic guanosine 3′,5′-monophosphate (cyclic GMP, cGMP) has been implicated. Light increased the level of cGMP in isolated eyes. Six-hour exposures to an analog of cGMP, 8-bromoguanosine 3′,5′-cyclic monophosphate (8-bromo-cGMP), generated both advance and delay $\Delta\phi$s, depending on the CT that the pulse was imposed (Fig. 4.14A). The PRCs for light and for 8-bromo-cGMP were indistinguishable (Fig. 4.14B), and the kinetics of phase-shifting were similar, indicating that the latter compound mimics the action of the former on the circadian pacemaker. Light and 8-bromo- cGMP appear to use convergent mechanisms for entrainment, since the effects of these two treatments were nonadditive, and LoNa solutions antagonized the effects of both (Eskin et al., 1984a). In addition to perturbing circadian rhythmicity, both light and cGMP increased the frequency of spontaneous optic nerve impulses. These results suggest that membrane depolarization mediates the effects of 8-bromo-cGMP on the rhythm, although the cellular site of action has not yet been identified. It may be involved in the transduction of non-R-type photoreceptors (there was no effect on the membrane potential of R-type photoreceptors). Alternatively, cGMP may not be involved in transduction but may be elevated in cells as a result of the photoreceptor potential (Eskin et al., 1984a).

There is a requirement for protein synthesis in the regulation of the CAP rhythm by the serotonergic pathway (Eskin et al., 1984b; Lotshaw and Jacklet, 1986, 1987; Yeung and Eskin, 1987). Exposure of eyes to anisomycin (ANISO), an inhibitor of protein synthesis that binds to the ribosomal subunit and prevents peptide bond formation, completely blocked the advance phase shift produced by 5-HT. Although ANISO alone can produce $\Delta\phi$s (see Fig. 4.27), it did not affect the rhythm at the phases where the blocking experiments were performed. The action was specific. Deacetylanisomycin, an analog of ANISO that is not effective in inhibiting protein synthesis, did not affect the phase shift produced by 5-HT. Furthermore, ANISO did not inhibit spontaneous optic nerve ac-

FIGURE 4.14. Cyclic guanosine 3′,5′-monophosphate mimics the effects of light on a circadian pacemaker in the eye of *Aplysia californica*. (A) Advance (top) and delay (bottom) phase shifts produced by a 6-h treatment with 8-bromo-cGMP (2×10^{-3} *M*). The frequency of spontaneous optic nerve impulses is plotted as a function of the time isolated eyes were in DD. The two rhythms shown in each graph are from eyes of the same animal. The time of exposure to 8-bromo-cGMP is shown by the hatched bars at the bottom of each graph. (B) Comparison of PRCs for 6-h treatments with light and 8-bromo-cGMP (2×10^{-3} *M*). The graph

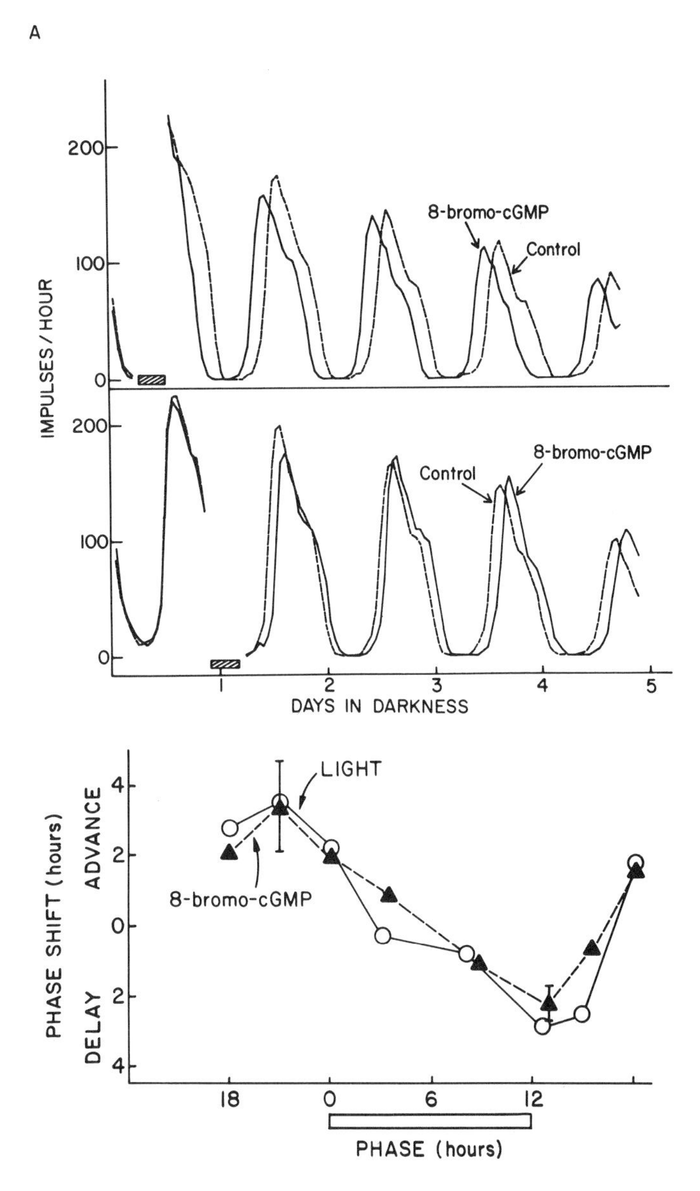

shows phase shifts produced in the rhythm as a function of the time (phase) of exposure of the eyes to the treatments. Treatments were given during the first day of isolation of the eyes, and phase 0 is the time of projected dawn. Data are plotted with respect to the midpoint time of the treatment. The error bars are 95% confidence intervals with $n = 9$ for the maximum advance phase shift and $n = 7$ for the maximum delay phase shift. All other data points for the 8-bromo-GMP curve are averages of data from four pairs of eyes. Data for the light PRC are from Corrent et al., (1982). (Reprinted from Eskin et al., 1984a, with permission of Society for Neuroscience.)

tivity or prevent an increase in photosensitivity of the eye—two other effects of 5-HT that also appear to be mediated by cAMP. ANISO also inhibited the ability of 8-benzylthio-cAMP to shift the phase of the rhythm, indicating that the step in the serotonin phase-shifting pathway that is sensitive to ANISO probably occurs after the cAMP step. Finally, 5-HT appears to regulate directly the expression of at least one eye protein: two-dimensional gel electrophoresis indicated an increase in the synthesis of a protein (as measured by [^{3}H]leucine incorporation) with a molecular weight of 34,000 (Eskin et al., 1984b). Indeed, 6-h pulses of 5-HT significantly increased the synthesis of the 34-kDa protein at phases where 5-HT delayed or advanced the ocular rhythm but did not do so at a phase where 5-HT did not generate $\Delta\phi$s, and similar treatments with forskolin and 8-benzylthio-cAMP, both of which mimic the phase-shifting action of 5-HT by increasing the concentration of cAMP (see Figs. 4.12A,B, 4.13), likewise mimicked the effect of 5-HT on the 34-kDa protein (Yeung and Eskin, 1987). The possible role of protein synthesis in circadian clock mechanisms is considered further in Section 4.3.2 and Chapter 5.

4.3 Dissection of the Clock: Perturbation by Chemicals

Another approach toward the elucidation of circadian clock mechanisms is to perturb the oscillator by drug treatment and then to determine the biochemical site of action. Thus, one could administer a variety of chemicals or inhibitors (such as valinomycin, an ionophore, or cycloheximide, an inhibitor of 80S protein synthesis) to a rhythmic system to determine if they affect the oscillator, as reflected by a shift in the phase of the overt rhythm or by an alteration in its period length. In this case, the rationale is that the known specific cellular target of the drug may be a part of the clock or at least intimately associated with it. The chemical agent could be present continuously (a step experiment, as in much of the earlier literature) or, more desirably, could be given as a pulse (the treatment being terminated by destroying or metabolizing the agent or diluting it below its threshold of effectiveness) in a manner analogous to the use of different intensities of continuous light to modulate τ or to the imposition of light signals at various CTs to generate $\Delta\phi$s and derive a PRC. For drug pulses, one must also consider the parameters of duration and concentration as well as the circadian phase of application. For each drug, ideally a dose-response curve (DRC) would have been obtained, in which the drug concentration was varied but ϕ and pulse duration were held constant.

We have seen (see Chapter 3) that the phenomenon of phase perturbation (by light, temperature, or drugs) is also inherent in the various models of oscillators that have been hypothesized to underlie CDCs in microorganisms and mammalian cells (Edmunds and Adams, 1981; Edmunds, 1984a;

Shymko et al., 1984). Cellular response to external influences often is strongly dependent on the time of the CDC at which the agent is imposed. Thus, synchronous cultures of eukaryotic unicells, such as the ciliate *Tetrahymena pyriformis* (Zeuthen, 1974) or the fission yeast *Schizosaccharomyces pombe* (Polanshek, 1977; Zeuthen, 1974), and of animal cells, such as mouse fibroblast L cells (Miyamoto et al., 1973), all exhibit increased division setback (excess division delay) to heat-shock and a wide variety of chemical agents applied at progressively later times during their CDC. Perhaps more provocative is the finding in *Schizosaccharomyces* (Smith and Mitchison, 1976) that the same agent can produce delays when imposed at one point of the CDC but advances when given at another. Similarly, Klevecz et al. (1980a,b) have discovered that synchronous mammalian V79 cells perturbed by heat shock, serum pulses, or ionizing radiation at 0.5-h intervals across their 8.5-h CDC display biphasic, 4-h PRCs comprising both advances and delays in subsequent cell division.

With a few exceptions, however, most attempts to perturb the biological oscillators specifically and to elucidate their biochemistry have been confounded by the difficulties inherent in distinguishing between the so-called hands of the clock (clock mechanism-irrelevant events) and the gears themselves (clock-relevant processes) due to possible side effects or chains of secondary effects induced by the perturbing agent (Sargent et al., 1976). Thus, a drug pulse might affect the clock only indirectly through some unknown pathway rather than directly via its known target. Similarly, rather than being the direct result of drug inhibition, the observed effects of a perturbation on an overt rhythm might occur merely as a secondary consequence of inhibition. Some important criteria for ascertaining the role of a given biochemical component in circadian clock function have been discussed (Engelmann and Schrempf, 1980; Roeder et al., 1982). In a further attempt to circumvent the problem of secondary effects of drug pulses, Goto et al. (1985) have introduced a more restrictive rationale for designating the target of a perturbing drug or other agent as an integral element of a circadian clock.

An oscillator can be expressed mathematically as a set of differential equations comprising both state variables and parameters (Tyson et al., 1976). The state variables characterize the state of the oscillation, with each set of values defining each phase of the oscillation. The parameters are constants constraining the manner in which the state variables change and determining the dynamics of the oscillations. A different set of parameter values gives a different solution of the rate equations. Any transitory alteration or perturbation of either the state variables or of the parameters can cause a permanent $\Delta\phi$ in an overt rhythm but has no permanent effect on its τ. In contrast, permanent changes in parameter values can alter τ of the oscillation.

As we noted earlier, one of the most important drawbacks inherent in the phase-shift experiments widely used in determining a state variable or a parameter has been that we do not know whether the $\Delta\phi$ occurred as a result of an effect on the postulated primary target or rather on some other site secondarily affected by the drug. Goto et al. (1985) have operationally designated any element as a gear (G) of a circadian clock if it can be expressed as a state variable or a parameter. If not, it is a nongear (~G), that is, a hand or other mechanism-irrelevant element. Together, the ensemble of gears would constitute a closed control loop, or oscillator. [A note of caution: The distinction among state variables, parameters, and hands is model-dependent; a target initially designated as a ~G may later have to be included among the Gs of a more complicated model as more subtle properties of the oscillation come under investigation (Tyson et al., 1976).]

Because an unperturbed ~G in its normal or physiological oscillatory range would not be expected to regulate the operation of the Gs themselves, its artificial perturbation within this range should not perturb circadian timekeeping (no steady-state $\Delta\phi$ in an overt rhythm should be observed). [A $\Delta\phi$ might occur if the level of the input of ~G were so high or low (i.e., outside of the normal range of its oscillation) that it rate-limited or otherwise affected the normal operation of the Gs comprising the oscillator. In this latter case, even if both the high-level and low-level inputs of ~G perturbed the overt circadian rhythm, they would probably be affecting different gear elements of the clock (Goto et al., 1985).] Consequently, if an experimental alteration in the level of a target within its normal range perturbs the clock and generates steady-state $\Delta\phi$s, then that target is most likely a G (criterion A). It is conceivable, however, that the activation and resulting increase in the level of a ~G might perturb timekeeping, whereas its inhibition would not (or vice versa). For example, inhibition of protein synthesis might cause a $\Delta\phi$ in an organism but its activation might not, or experimentally increasing (but not decreasing) the level of cyclic AMP might alter τ of an overt rhythm. To differentiate more stringently between G and ~G, therefore, a further requirement for a target to be classified as a G (Goto et al., 1985) is that both the direct activation and the direct inhibition of the target must perturb the clock (criterion B).

Sometimes, it will be impossible to attack a presumed target directly. Phase shifts generated by several drugs that only secondarily activate and inhibit a target do not demonstrate necessarily that such a target is a G. Nevertheless, if a given target (B) is regulated by another target (A) and, in turn, regulates yet a third target (C), and if it is probable that A and C are gears, it is likely that B is one also. In contrast, if both targets A and C are known to be hands (~G) of the oscillator, target B, of course, is also a ~G. Only in the case where target A is shown to be a G and target C is proven to be a ~G can target B not be classified (Goto et al., 1985).

Despite many experiments using chemical perturbations, few targets that can be classified as a G on this rationale have been reported thus far (see Section 5.2.5). Nevertheless, let us now proceed to briefly review what evidence exists in the various circadian systems surveyed in Chapter 2 and Sections 4.1 and 4.2 (Edmunds, 1983, 1984c; Engelmann and Schrempf, 1980; Hastings and Schweiger, 1976; Jacklet, 1984; Jerebzoff, 1986; Schweiger and Schweiger, 1977; Schweiger et al., 1986; Sweeney, 1983; Wille, 1979).

4.3.1 Nonspecific Compounds and Respiratory Inhibitors

Early Studies: The Chemical Shelf

Earlier work with *G. polyedra* and several other unicellular systems perhaps supported the old adage, "only the uninhibited use inhibitors." Practically the entire chemical shelf was thrown at the glow luminescence rhythm (see Section 2.2.4): EDTA (a chelator), $AgNO_3$, $CaCl_2$, KCl, and $FeCl_3$ (all metabolic poisons), the growth factors, gibberellin and kinetin, the mitotic inhibitors, urethane and fluorodeoxyuridine (FUdR), respiratory inhibitors, such as arsenate, cyanide, and *p*-chloromercuribenzoate (PCMB), 2,4-dinitrophenol (DNP), an uncoupler of oxidative phosphorylation; and the herbicides DCMU (diuron) and CMU (chlorophenyl-1,1-dimethylurea), which specifically inhibit photosynthesis (Hastings, 1960; Hastings and Bode, 1962). In most cases, the compounds were added to vials of culture and then washed out later by centrifugation and resuspension of the cells into fresh medium. In this sense, they constituted chemical analogs of the light pulses used for phase-shifting the overt glow rhythm. Unfortunately, however, little could be concluded from these studies except that the clock seemed to be remarkably insensitive to perturbations by ordinary agents. In some cases, $\Delta\phi$s of sorts occurred, whereas in others, the rhythm damped out. The problem remained of differentiating between merely affecting the expression of the rhythm (by uncoupling the hands of the clock) and actually perturbing the underlying oscillator itself. One point did emerge: neither photosynthesis nor cell division was needed for clock function inasmuch as the glow rhythm continued to run unabated when these other processes were blocked (on the assumption that one clock controlled all three processes, but see Section 6.2).

Early attempts to affect the period of the rhythm of phototaxis in *Euglena gracilis* (see Section 2.2.3) were similarly unrewarding. KCN, phenylurethane, the adenine growth factor analog 2,6-diaminopurine sulfate, the pyrimidine and nucleic acid analog 2-amino-4-methylpyrimidine, the nucleic acid bases, and gibberellic acid and kinetin were all without consistent effect on period and phase (Bruce and Pittendrigh, 1960). More specific (but still not well understood) perturbations have been obtained by altering the nutritional conditions. Feldman and Bruce (1972) reported that the addition of acetate (100 m*M*) to autotrophic cultures of *E. gracilis* lengthened the phototaxis rhythm to 27 h and that 10 m*M* pulses of acetate administered at different phases of the free-running rhythm induced phase shifts. Pulses of other carbon sources (succinate, lactate, pyruvate), however, caused a temporary cessation of the rhythm and then a resumption with variable phase shifts. Feldman and Bruce argue that the changes were probably caused by a general metabolic switch (i.e., from autotrophy to mixotrophy) rather than by any specific effect. The nutritional mode also has been shown to play a role in the rhythm of dark motility (Brinkmann, 1966, 1971; see Section 2.2.3), and Sweeney and Folli (1984) have demonstrated that nitrate deficiency shortens τ of the rhythms in stimulated luminescence and photosynthesis in *G. polyedra* suspended in unsupplemented sea water.

Deuterium Oxide

Deuterium oxide (^2H_2O), or heavy water, which has been shown to reversibly lengthen τ of circadian rhythms in a variety of organisms, including sand-beach isopods, fruit flies, pigeons, mice, and higher plants, and which slows down several ultradian biological rhythms having τs in the millisecond range (reviewed by Enright, 1971), similarly alters both the period and phase of the phototaxis rhythm in *E. gracilis*. Thus, a culture adapted to 95% ^2H_2O increased its τ to about 27 h but returned to a τ of about 24 h when it was readapted to H_2O again (Bruce and Pittendrigh, 1960). As Enright (1971) has speculated, ^2H_2O may retard biological rhythms by influencing the ionic balance across the cellular membrane, altering the nerve activity of higher animals and related membrane processes of simpler organisms. Inasmuch as high-frequency rhythms originate in pacemakers dependent on diffusion processes involving ion exchange, it is possible that biological clocks with longer (circadian) periods are also based on diffusion-dependent pacemakers (see Section 5.4).

The problem with these studies, of course, is the fact that the action of ^2H_2O is general and nonspecific, and the period-lengthening effects are equally compatible with kinetic, transcriptional, and membrane models for circadian oscillators (see Chapter 5). Kreuels and Brinkmann (1979) have found in a comparative study of the effects of ^2H_2O on the cell-bound circadian oscillator underlying the motility rhythm in *E. gracilis*, on the cell-free glycolytic oscillator of yeast, and on the Belousov-Zhabontinsky chemical reaction having a known network structure (Kreuels et al., 1979) that although the circadian and glycolytic oscillations were slowed down to an extent depending on the ^2H_2O concentration, τ of the chemical reaction was either lengthened or shortened, respectively, at high or low catalyst concentrations. They concluded, therefore, that the generalized period-lengthening effect of ^2H_2O can be explained only by a more complex network approach.

The period-lengthening effect of ^2H_2O has been formally compared (Pittendrigh et al., 1973) to a diminishing of the apparent temperature (the so-called low-temperature equivalence hypothesis). It was, therefore, of interest to test this idea using the rhythm of glow bioluminescence in *G. polyedra,* which, as we have seen (see Section 2.2.4; Hastings and Sweeney, 1959), exhibits over-compensation (i.e., τ is shorter at lower temperatures). Accordingly, McDaniel et al. (1974) assayed glow luminescence in deuterated (6 or 12%) cultures of this dinoflagellate at either 22 or 16°C in dim LL (1180 lx) following prior entrainment by LD: 12,12. The period of the rhythm was lengthened in a dose-dependent manner at either constant temperature. Furthermore, in contrast to the phase angle delay with respect to the LD cycle induced by low temperature and ^2H_2O in *D. pseudoobscura* (Pittendrigh et al., 1973), the glow rhythm of deuterated *G. polyedra* cultures displayed a phase angle advance at the lower temperature but a phase angle delay at the higher temperature. These results, then, did not support the low-temperature equivalence hypothesis.

Respiratory Inhibitors

In 1973, Bünning proposed that the circadian clock comprises an energy-requiring, physiological night phase and an energy-independent day phase. Results from a number of systems have given support to this general notion: oxygen deprivation at specific CTs phase-shifts the rhythm in the growth rate of the *Avena* coleoptile (Ball and Dyke, 1957; Wilkins and Warren, 1963), and 4- to 6-h treatments with respiratory inhibitors (such as cyanide, DNP, and azide) generate CT phase-specific $\Delta\phi$s of the rhythms of leaf movement in *Phaseolus* (Bünning et al., 1965; Mayer, 1981), of potassium uptake in *Lemna* (Kondo, 1983), of glow luminescence in *Gonyaulax* (Hastings, 1960), and of compound action potential in the isolated eye of *Aplysia* (Eskin and Corrent, 1977). Usually, the magnitude of the $\Delta\phi$ approximately equalled the duration of the treatment. In most of these experiments, however, the levels of neither respiratory activity nor ATP were measured directly—minimal correlations essential to the conclusion that energy depletion caused by the inhibitors resulted in the phase-shifting of the clock.

Nakashima (1984b) examined the effects of 3-h pulses of several respiratory inhibitors on the conidiation rhythm of *N. crassa* (see Section 2.3.1) and concomitantly measured respiratory (O_2 consumption) activity and ATP content. All inhibitors [KCN, NaN_3, antimycin A, and carbonyl cyanide *m*-chlorophenyl hydrazone (CCCP)] shifted the phase by as much as 10 h during the subjective day (CT 0–12) but not during the subjective night (CT 12–24), suggesting that processes other than respiration may mediate the effects of metabolic inhibitors on the clock. All the PRCs derived for these agents were similar (with both advances and delays) and resembled that for cycloheximide but were different from that for 5-min light pulses (Fig. 4.15). Although the magnitude of the $\Delta\phi$ induced by azide and CCCP was proportional to the lowering of respiratory activity and

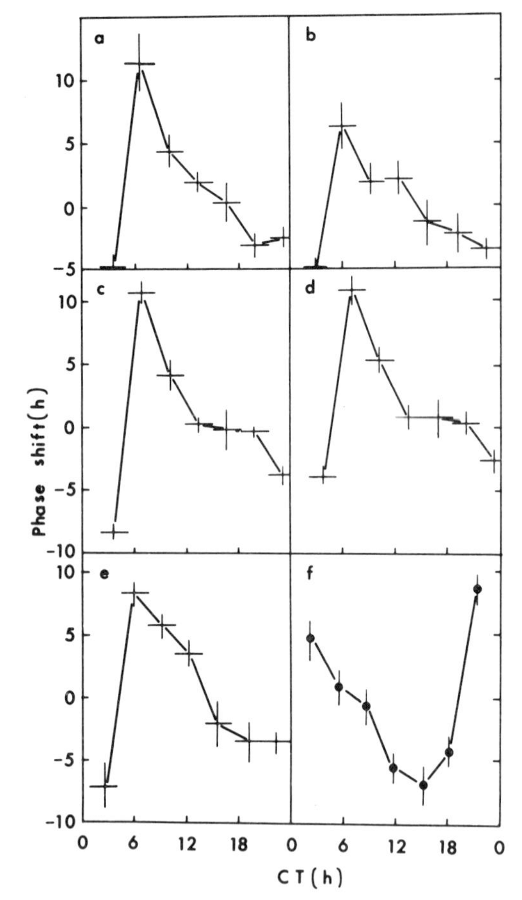

FIGURE 4.15. Phase-response curves for respiratory inhibitors, cycloheximide, and light in *Neurospora crassa*. Mycelial discs were treated with 30 m*M* CCCP (a), 2 m*M* potassium cyanide (b), 4 m*M* NaN_3 (c), 1 mg/ml cycloheximide (d), and 0.7 mg/ml antimycin A (e) for 3 h and transferred individually to race tubes, or they were irradiated with white light (1000 lx) (f) for 5 min at different circadian times, cultured for 3 h, and then transferred to race tubes. Error bars are ± SD. (Reprinted from Nakashima, 1984b, with permission of American Society of Plant Physiologists.)

ATP content, such a correlation was not observed for cyanide and antimycin A. Indeed, cyanide (0.5 m*M*) completely depleted the ATP of the cultures but did not significantly shift their phase. It was impossible to exclude the possibility that the clock stopped immediately when ATP content was lowered but then ran normally when ATP levels recovered 3 h later (thereby generating a small $\Delta\phi$ equal to the duration of the pulse). It is clear, nevertheless, that the large phase shifts caused by these inhibitors cannot be due solely to a decrease in ATP levels and energy supplied by respiratory activities. They could be attributed to an affected mitochondrial function other than respiratory metabolism during the sensitive phase, since these chemicals act on mitochondria. This lack of phase-shifting effect by ATP depletion does not support the possibility that the $\Delta\phi$s induced by pulses of diethylstilbestrol (DES) and related compounds (see Section 4.3.3, Fig. 4.32) are generated by the indirect inhibition of plasma membrane ATPase activity resulting from the lowering of ATP levels by DES (Nakashima, 1982b, 1984b).

The analysis of the effect of respiratory inhibitors on the circadian clock of *N. crassa* has been extended by Schulz et al. (1985), who also examined the phase-shifting action of 3-h pulses of various inhibitors of ATPases, phosphatases, and the mitochondrial ATP synthetases, such as vanadate (Na_3VO_4), molybdate (Na_2MoO_4), N-ethylmaleimide (NEM), oligomycin, and N,N′-dicyclohexylcarbodiimide (DCCD). As Nakashima (1984b) had found earlier, pulses of azide and cyanide were effective in phase-shifting the conidiation rhythm, with maximal $+\Delta\phi$s being observed during the subjective day at about CT 6 and maximal $-\Delta\phi$s at CT 15–21 (although there were some differences in dose-dependency between the two studies). Treatments with vanadate, molybdate, oligomycin, and NEM yielded similar PRCs. If two out-of-phase rhythms of energy charge in the cytoplasm and mitochondria were decisive for clock function (see next subsection), as postulated by Wagner (1976a,b), one might expect the PRCs for inhibitors of ATP synthesis to differ from those for inhibitors of ATP breakdown, such as vanadate (Schulz et al., 1985). A common effect of these drugs, however, may be to block phosphate transfer reactions, either by inhibiting transferring enzymes, such as ATPases and phosphatases (vanadate, molybdate, NEM, azide, DES), or by other means (azide, cyanide, oligomycin, CCCP, antimycin A). Some of these agents also induce HSP synthesis and repress the synthesis rate of normal proteins, suggesting a link between energy metabolism and protein synthesis. Pulses of DCCD and light, however, yielded PRCs that were about 180° out-of-phase with those of the other drugs (maximal $+\Delta\phi$ at CT 18–24) and that were significantly greater in amplitude than those obtained with the cytochrome-deficient mitochondrial mutant *poky*. (Such differences in PRCs between the wild type and *poky* were not found for cycloheximide.) These results suggest that DCCD and light have a common target and that only the transmittance of the perturbing pulses and not the oscillator itself is changed in *poky*. On the basis of their results, Schulz et al. (1985) concluded that energy metabolism, despite having an intimate relation with protein synthesis, is not involved directly in clock function in *Neurospora* but is controlled by it.

Energy Charge

The overall adenylate energy charge (AEC), defined as a ratio between the adenine nucleotides [(ATP + 0.5 ADP)/(ATP + ADP + AMP)], has been used as a measure of metabolic energy stored in the adenine nucleotide pool (Atkinson, 1968). The AEC reflects the direct and indirect inhibition of ATPases, phosphatases, mitochondrial ATP synthetase, and electron transport (all of whose action involves phosphate transfer) and would be expected to be perturbed by most of the respiratory and metabolic inhibitors discussed in the previous subsection. Were this energy variable part of the clock mechanism, it should itself cycle. Circadian oscillations in AEC have been reported in the mycelia of *N. crassa* (see

Section 2.3.1) in rhythmically conidiating cultures grown on agar (Delmer and Brody, 1975; see Figs. 5.3 and 5.4), as well as in liquid cultures not exhibiting rhythmic conidiation (Schulz et al., 1985). A circadian variation in ATP content in the chloroplasts of *A. mediterranea* also has been reported (Vanden Driessche, 1970b). In contrast, no rhythmicity in AEC was found in autotrophic cultures of *E. gracilis* (Section 2.2.3; see Fig. 2.13c) during a DD free-run of the dark motility rhythm. It is possible, of course, that rhythmicity might be detectable only in different cellular compartments and not in total cell extracts. This point of caution is particularly germane given the lack of correlation found by Nakashima (1984b) between ATP content and the magnitude of the phase shifts generated by pulses of cyanide and antimycin A, drugs that presumably affected the AEC (see preceding subsection).

This question of the central (causal) role of energy charge and energy metabolism in circadian clock function, as posited earlier by such workers as Sel'kov (1975) and Wagner (1976b) and more recently by Lloyd et al. (1982a), is addressed further in Section 5.2.2).

4.3.2 Inhibitors of Macromolecular Synthesis

Given the numerous sites of action of the various chemical agents and metabolic inhibitors discussed in the preceding subsection and the resultant ambiguity in interpretation of their effects on clock functioning and mechanism, it is not surprising that attention quickly turned to the use of more specific inhibitors of macromolecular synthesis in addition (step) or pulse experiments designed to test their efficacy in affecting τ or ϕ of circadian oscillators. We now review briefly the evidence derived from experimental perturbation of the clock for the relative importance of transcription, translation, and posttranslational events for circadian rhythmicity in several model organisms *(Acetabularia, Euglena, Gonyaulax, Neurospora, Aplysia)* whose temporal organization was surveyed in Chapter 2 and Sections 4.1 and 4.2 (Edmunds, 1983, 1984c; Ehrhardt et al., 1980; Hastings and Schweiger, 1976; Jacklet, 1984; Jerebzoff, 1986; Schweiger and Schweiger, 1977; Schweiger et al., 1986; Sweeney, 1983; Wille, 1979).

Transcription of DNA

The results from the experiments involving the transplantation and implantation of the nucleus of *Acetabularia* and its role in phase determination of the photosynthesis rhythms in this alga (see Section 2.2.1, Fig. 2.3) raised the question of whether gene expression was necessary for the generation of circadian rhythms (e.g., as called for by the chronon model; see Section 5.3.1). This led naturally to the testing of the effects of several inhibitors of transcription and translation, which, on the hypothesis of the requirement of daily copying of the genome, should affect circadian timing. It is only relatively recently that a coherent pattern has begun to emerge.

Initial observations (Vanden Driessche, 1966b) showed that in *A. mediterranea* the rhythm of PA, as well as that of chloroplast shape, was abolished in nucleate cells, but not in anucleate ones, treated with dactinomycin, an irreversible inhibitor of DNA-dependent RNA synthesis. This result was confirmed for anucleate cells of *A. crenulata*. Furthermore, inhibitors of protein synthesis (puromycin and chloramphenicol) at the time also appeared to be ineffective in altering rhythms in anucleate cells (but see Mergenhagen and Schweiger, 1975b), despite the fact that they did inhibit incorporation of labeled precursors into the macromolecules (Sweeney et al., 1967). One was thus confronted with a set of most unusual paradoxes (Sweeney, 1974b) (see Section 5.4.1): (1) the rhythm continued for months in the absence of a nucleus, yet phase was determined by the transplanted nucleus, (2) dactinomycin inhibited rhythmicity in intact cells but not in anucleate ones, and (3) inhibition of RNA synthesis in nucleate cells appeared to affect the rhythm without mediation of protein synthesis (but see next subsection).

In order to substantiate these findings, a reinvestigation of the effects of inhibitors was begun using the continuous flow-through system for monitoring the PA rhythm in individual *Acetabularia* cells (see Section 2.2.1). Similar results were obtained with dactinomycin (Mergenhagen and Schweiger, 1975b); in the presence of this inhibitor, the rhythm disappeared after about 14 days in nucleate cells (Fig. 4.16/1) but continued without interruption in anucleate algae (Fig. 4.16/2). Thus, continuous transcription of the nuclear genome was not necessary for persistence of the PA rhythm. This would seem to contradict the nuclear exchange experiments (see Section 2.2.1), but the apparent paradox can be resolved in view of the fact that in *Acetabularia* long-lived mRNA can be stored in the cytoplasm. The inhibition of rhythmicity by dactinomycin in nucleate cells only and the lack of effect of enucleation can be explained in at least two ways (Mergenhagen and Schweiger, 1975b). (1) The inhibition of mRNA production (in intact cells, where the nuclear genome is the site of attack) would be complete enough to block expression of rhythmicity but would not be sufficient to prevent degradation or labilization of the mRNA in the cytoplasm and subsequent disappearance of periodicity. (2) The nucleus in some other way degrades cytoplasmic mRNA, a process that would not be inhibited by dactinomycin. In either case, differences in the control system for mRNA degradation would exist, with short-lived, cytoplasmic mRNA species decaying in nucleate algae but not in anucleate cells.

Perhaps, however, the clock is coupled to transcription of the organellar genome. In an important experiment, Vanden Driessche et al. (1970) demonstrated that rifampicin, an inhibitor of organellar RNA synthesis in *Acetabularia* (preventing transcription of chloroplastic DNA by compet-

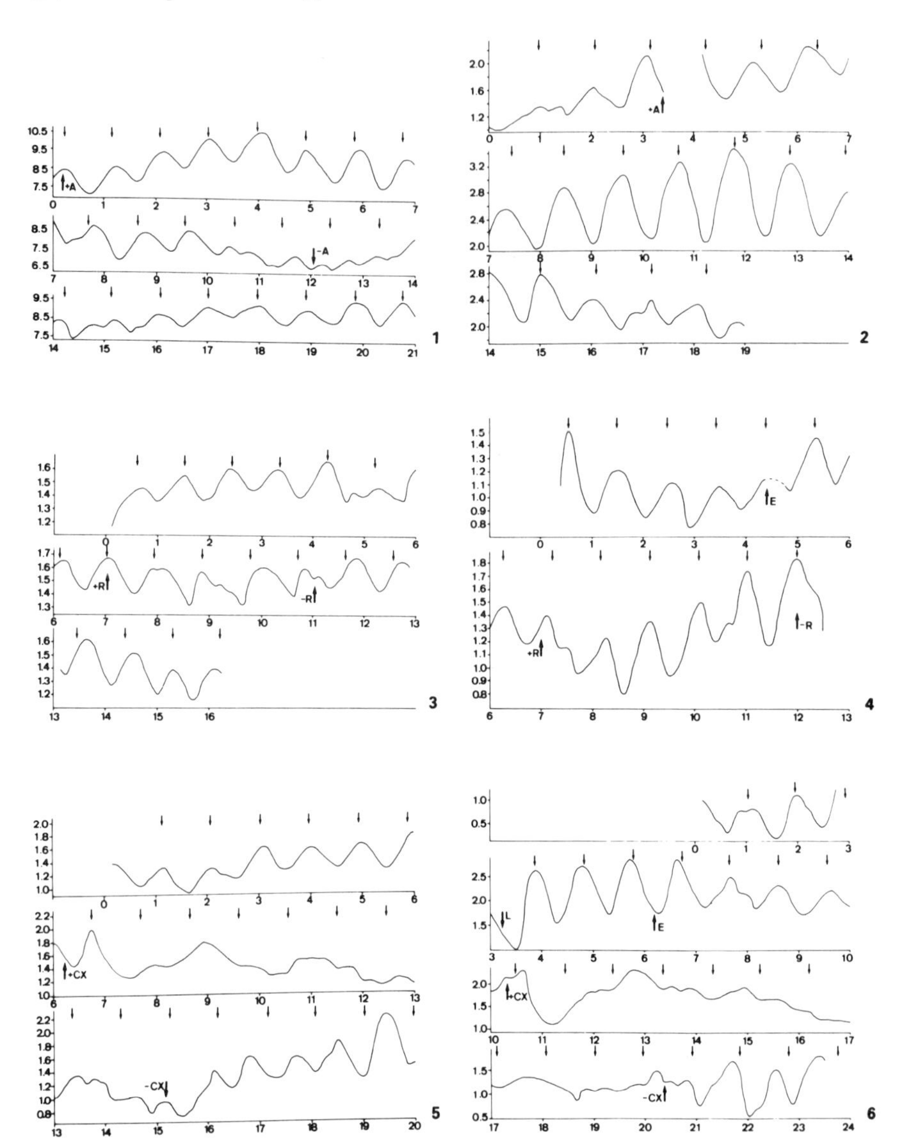

FIGURE 4.16. Effects of inhibitors of macromolecular synthesis on circadian rhythmicity in *Acetabularia*. *Abscissa:* Time (days). *Ordinate:* O_2 evolution in relative units. Small arrows give the circadian time (noon). Large arrows indicate when the inhibitor has been added to and removed from the medium. (1) Effect of dactinomycin (5 mg/ml) on the photosynthesis rhythm of a single nucleate cell of *A. mediterranea*. (2) Effect of dactinomycin (5 mg/ml) on the photosynthesis rhythm of a single anucleate cell of *A. mediterranea*. The cell has been enucleated at day zero by removing the rhizoid. (3) Effect of rifampicin (25 mg/ml) on the photosynthesis rhythm of a nucleate *Acetabularia* cell. (4) Effect of rifampicin (25 mg/ml) on the photosynthesis rhythm of an anucleate *Acetabularia* cell. E, enucleation by removing the rhizoid. (5) Effect of cycloheximide (0.1 mg/ml) on

itively inhibiting the binding of RNA polymerase to RNA), was also ineffective in altering rhythmicity in either nucleate or anucleate cells even though RNA synthesis was inhibited by at least 90% at the inhibitor concentrations used. Comparable results (Fig. 4.16/3,4) were obtained by Mergenhagen and Schweiger (1975b). Thus, continuous daily transcription either from nuclear DNA or from organellar DNA was not essential for the functioning of the circadian oscillator in *Acetabularia*.

This conclusion is consistent with results obtained on the *G. polyedra* system (see Section 2.2.4), although their interpretation is complicated by the fact that the parameter serving as a measure of rhythmicity itself was inhibited. Early work (Karakashian and Hastings, 1963) indicated that the inhibition of DNA synthesis by amethopterin, novobiocin, or mitomycin C did not appear to affect directly the rhythm of glow luminescence, either when continually present (see Fig. 4.20) or when given in pulses, although the amplitude of the rhythm was often reduced or the rhythm damped. Similarly, dactinomycin (which cannot be removed from the cultures, thereby precluding short treatments) perhaps only slightly delayed the glow rhythm before the rhythmicity was completely suppressed (see Fig. 4.20); the rhythms in photosynthesis and cell division were also strongly inhibited at a 10-fold higher concentration of the drug. Dactinomycin must be present during the subjective night (at about CT 18) in order to show inhibition of the next cycle of the glow rhythm, implying a circadian variation in drug sensitivity. If the inhibitor were added later, the next cycle of the glow rhythm was expressed (Karakashian and Hastings, 1963). This observation might suggest that determination of a subsequent cycle occurs at about CT 18, perhaps by the synthesis of particular mRNAs.

In this regard, it is interesting to note that Walz et al. (1983) have reported that the amount of total RNA in stationary-phase cultures of *G. polyedra* maintained in LL exhibits circadian rhythmicity (maximum at CT 18), as measured by the fluorescence of cells stained with acridine orange. The rhythm could be phase-shifted by a 6-h light signal in parallel with that of stimulated bioluminescence. Ribosomal RNA synthesis, as measured by the incorporation of ^{32}P into extracted RNA, was also rhythmic. New RNAs appeared at CT 18, as shown by acrylamide/agarose gel electrophoresis. These new species then disappeared 3 to 4 h later. Although these findings do suggest a role of transcription in the expression of circadian rhythmicity in *Gonyaulax,* they do not necessarily implicate transcription as a part of the clock, since the changes in RNA could be merely clock-controlled (Sweeney, 1983). Nevertheless, there are two indirect lines of evidence that support the role of these oscillations in RNA

◁

the photosynthesis rhythm of an nucleate *Acetabularia* cell. (6) Effect of cycloheximide (0.1 mg/ml) on the photosynthesis rhythm of an anucleate *Acetabularia* cell. L, ligating; E, enucleation by removing the rhizoid. (Reprinted from Mergenhagen and Schweiger, 1975b, with permision of Academic Press.)

in the clock mechanism itself (Walz et al., 1983): (1) Cycloheximide causes the largest $+\Delta\phi$s around CT 18 (Walz and Sweeney, 1979; Dunlap et al., 1980), and (2) the glow rhythm is damped by the presence of dactinomycin at about CT 18 of the preceding cycle (Karakashian and Hastings, 1963). If the transitory RNA-species discovered by Walz et al. (1983) were truly clock elements, however, dactinomycin pulses given at the time of their synthesis should phase shift the clock. This is not so: Schröder-Lorenz and Rensing (1986) recently found that although there is a circadian rhythm of RNA synthesis in *G. polyedra* (with two maxima that precede by 2 to 4 h the rhythm of RNA content reported earlier), 2-h pulses of this inhibitor (80 $\mu g\ l^{-1}$ or 800 $\mu g\ l^{-1}$) did not phase-shift the glow rhythm in a phase-dependent manner. Therefore, there was no evidence for an essential role of RNA synthesis.

Jacklet (1984) has found that continuous (step) exposure of the isolated *Aplysia* eye to camptothecin, a reversible inhibitor of RNA synthesis, lengthened τ of the CAP rhythm in a dose-dependent way, although pulses of this drug did not cause phase-dependent $\Delta\phi$s. Perhaps RNA synthesis influences a parameter of the clock, such as availability of RNA for protein synthesis, but does not directly affect the state variables characterizing the pacemaker.

Translation and Protein Synthesis

If the results from perturbations by inhibitors do not provide much evidence implicating transcription in circadian clock mechanisms—indeed, the persistence of the rhythm of photosynthesis for 7 weeks or more in enucleated *Acetabularia* cells would seem to preclude this possibility (Sweeney and Haxo, 1961; Schweiger et al., 1964a)—perhaps translation may be an important process. Feldman (1967) provided some of the earliest convincing evidence that protein synthesis is required for clock function. The period of the phototaxis rhythm of *E. gracilis* (see Section 2.2.3) lengthened (reversibly) by the addition of CHX, an inhibitor of protein synthesis on 80S eukaryotic ribosomes (see Fig. 2.7). The increase in τ was proportional to the concentration of inhibitor and was positively correlated with protein synthesis inhibition (as measured by [^{14}C]phenylalanine incorporation). The effects of CHX appeared to be on the clock itself rather than on some variable controlled by the clock, as confirmed (see Fig. 2.8) by assaying the position of the light-sensitive oscillation with 4-h resetting light signals after CHX addition in DD (as predicted from the PRC for light pulses already derived for this unicell; see Fig. 2.6). In the same organism, CHX is also capable of suppressing phase-shifting by temperature steps (Brinkmann, 1971). Since this early work with *Euglena*, evidence has accumulated from work with inhibitors in several systems (notably, *Acetabularia, Gonyaulax, Neurospora* and *Aplysia*) that continuous translation is required for the maintenance of rhythmicity and that there may be protein(s) essential to the clock mechanism itself (Edmunds,

1984c; Ehrhardt et al., 1980; Hastings and Schweiger, 1976; Jacklet, 1984; Schweiger and Schweiger, 1977; see Section 5.2.4).

Acetabularia

Having found that transcription of either the nuclear or the organellar genome does not seem to play an essential role in circadian timekeeping, Mergenhagen and Schweiger (1975b), turning then to translational controls, found that both CHX, which inhibits protein synthesis on cytosolic 80S ribosomes, and puromycin (PUR), which attacks 80S ribosomes as well as the 70S ribosomes in the cell organelles by mimicking the 3′-terminal of aminoacyl tRNA and interrupting chain elongation as it becomes incorporated into nascent peptides, reversibly inhibited the expression of the rhythm of PA in both nucleate and anucleate cells (Fig. 4.16/5,6). These experiments thus indicated that the observed inhibition was associated with the 80S ribosomes but did not exclude the involvement of the 70S ribosomes. [Note the discrepancy with respect to the much earlier results obtained with PUR on groups of cells, which did not show sensitivity to this inhibitor (Sweeney et al., 1967)]. This question was resolved by using chloramphenicol, which specifically inhibits protein synthesis only on 70S ribosomes. The inhibitor was ineffective in both nucleate and anucleate cells (Mergenhagen and Schweiger, 1975a).

The simplest conclusion that could be reached from these series of inhibitor experiments is that gene expression at the translational level—specifically on 80S ribosomes—is required for the operation of the circadian oscillator in *Acetabularia* (Fig. 4.17). Either protein synthesis would be part of the clock mechanism itself, or a vital, short-lived component would require continual resynthesis. [For generalization to other systems, see Vanden Driessche (1975), Mergenhagen (1976) and Sargent et al. (1976).] These results support the coupled translation-membrane model for the genesis of circadian rhythmicity (Schweiger and Schweiger, 1977; see Section 5.4.2), but it must be noted that a requirement for protein synthesis in itself does not necessarily exclude a transcriptional component (as in the chronon scheme) because there may be feedback regulation of transcription by translational products, whose inhibition would alter clock phase (Ehret and Trucco, 1967; see Section 5.3.1)

This hypothesis of translational control was further substantiated by a systematic study of the phase-shifting effects of CHX pulses on the circadian rhythm of PA in *Acetabularia* (Karakashian and Schweiger, 1976b), an obvious extension of Mergenhagen and Schweiger's earlier work (1975b) showing that the damped rhythm reappeared with shifted phase after removal of CHX (or PUR) from the medium. At 20°C, 8-h pulses of the drug shifted the rhythm when applied between CT 11 and 23 (subjective night) but were ineffective if applied during the subjective day (Fig. 4.18A). The magnitude of the engendered phase delay was significantly correlated with the duration of the pulse. The CHX-sensitive phase exhibited tem-

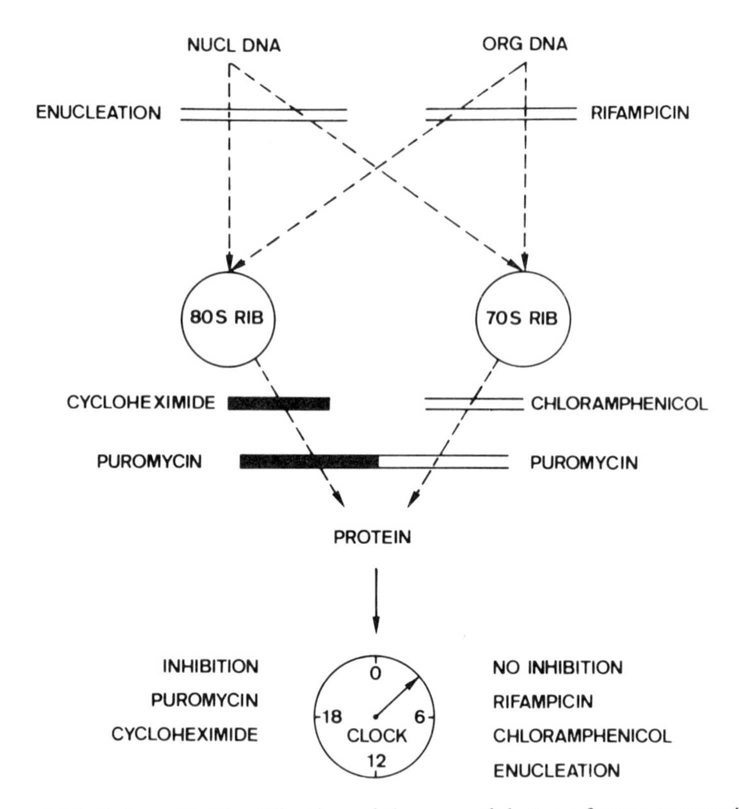

FIGURE 4.17. Schematic identification of the essential step of gene expression in the circadian photosynthesis rhythm in *Acetabularia mediterranea*. The open bars represent the absence of effects; the black bars indicate that the clock is affected. (Reprinted from Schweiger and Schweiger, 1977, with permision of Academic Press; see Schweiger, 1977.)

FIGURE 4.18. Temperature-dependence of the cycloheximide-sensitive phase of the circadian cycle in *Acetabularia mediterranea*. (A) Phase-response profiles of cells maintained at 20°C and 25°C and exposed to single treatments with cycloheximide (0.1 mg/ml). The effect on phase of each treatment is plotted according to the midpoint of the cycloheximide pulse given. All phase changes were arbitrarily classified as delays and are expressed in circadian time. ●——●, phase response to cycloheximide treatments given at various times in the circadian cycle at 25°C; ○— —○, phase response to cycloheximide treatments given at various times in the circadian cycle at 20°C. For orientation in the circadian cycle, the time of the

A

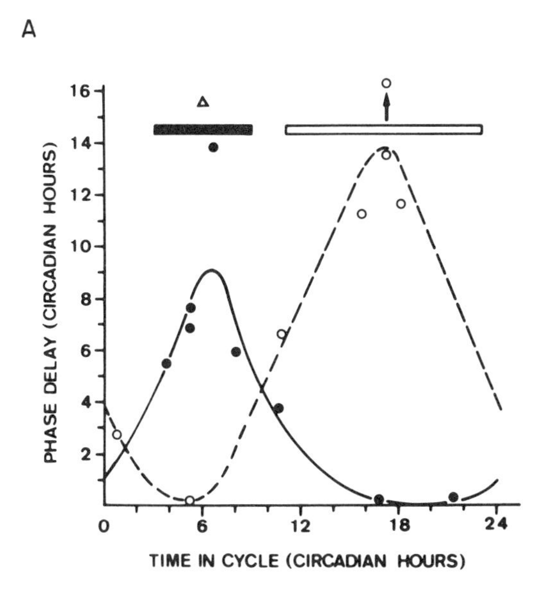

B

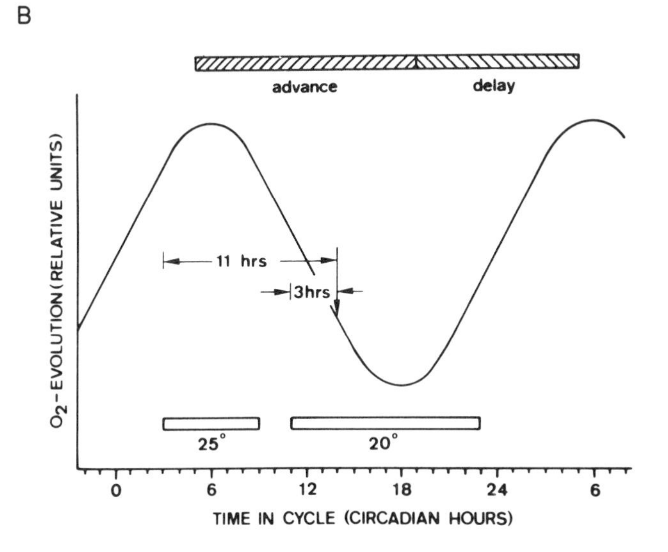

maximum of O_2 production (▲) is shown at hour 6. The putative sensitive phases identified by topological analysis are shown by the bars at the top of the figure. Arrow indicates a cycloheximide-induced phase delay in an enucleated cell. (Reprinted from Karakashian and Schweiger, 1976c, with permission of the authors.) (B) Temporal relationship between circadian oxygen evolution (curve), cycloheximide sensitivity at 25°C and 20°C (open bars), and phase response to dark pulses (hatched bars). The time between the early cycloheximide-sensitivity phase and the reversal of the oxygen evolution curve is 3 and 11 h at 20°C and 25°C, respectively. (Reprinted from Schweiger and Schweiger, 1977, with permission of Academic Press.)

perature dependence (Karakashian and Schweiger, 1976c): at 25°C the phase of CHX sensitivity for phase-shifting was 4 to 6 h shorter and occurred about 8 h earlier in the cycle than for cells kept at 20°C (Fig. 4.18A). In contrast, Mayer and Knoll (1981) have reported that the CHX-sensitive phases of the circadian clock in the pulvinus of *Phaseolus* are temperature-compensated. The period of the PA rhythm, however, was typically temperature-(over)compensated ($Q_{10} \cong 0.8$), and at both 20°C and 25°C the maximum and minimum of O_2 evolution occur, respectively, at CT 6 and CT 18. Because CHX shifts the phase of the rhythm in a temperature-dependent manner, however, there would be a much longer interval between the hypothesized production of an essential protein (presumably on 80S ribosomes) and the manifestation of their effect on the PA rhythm at 25°C than at 20°C (Fig. 4.18B).

That the observed action of CHX on the *Acetabularia* clock was the result of specific inhibition of translation has been further buttressed by the observation that G418, a translation inhibitor related to kanamycin and neomycin, inhibited the oscillation in O_2 evolution, and pulses of G418 phase-shifted the rhythm just as CHX did (Schweiger et al., 1986). Cells transformed by a gene construction coding for a neomycin-phosphorylating phosphotransferase, microinjected into an isolated nucleus that was then reimplanted into an anucleate fragment, served a controls. The rhythm was not affected by G418 in these transformed, antibiotic-resistant cells, whereas nontransformed *Acetabularia* were, indicating that enzymatic phosphorylation of the inhibitor inactivated the action of G418 on translation as well as on the rhythm.

These results, therefore, collectively suggest a dependence of the rhythm of PA on the synthesis of a protein—perhaps a membrane protein—that would provide an essential component of the circadian timing mechanism and that would appear at different times of the circadian cycle, depending on the temperature. The large effects of temperature observed indicated that protein synthesis does not per se generate the τ of the photosynthesis rhythm (Karakashian and Schweiger, 1976c). Alternatively, it is possible that the translational capacity of the 80S ribosomal complex might vary on a circadian time scale (making 80S ribosomal function a key component of the clock) and that proteins synthesized at different times of day would not differ simply because of the varying availability of message (Taylor et al., 1982b).

Indeed, Leong and Schweiger (1978) reported a periodically appearing membrane protein (polypetide P39, or polypeptide X) that was synthesized in the presence of CHX and was retained in the chloroplast membrane fraction. Moreover, the timing and amount of synthesis of this polypeptide coincided, respectively, with the timing and the duration of the phase of the PA rhythm that was sensitive to phase-shifting by pulses of CHX and depended on temperature in a similar manner. This protein was purified and found to have a molecular weight of about 39,000 on the basis of analytical gel electrophoresis, amino acid composition, and peptide map-

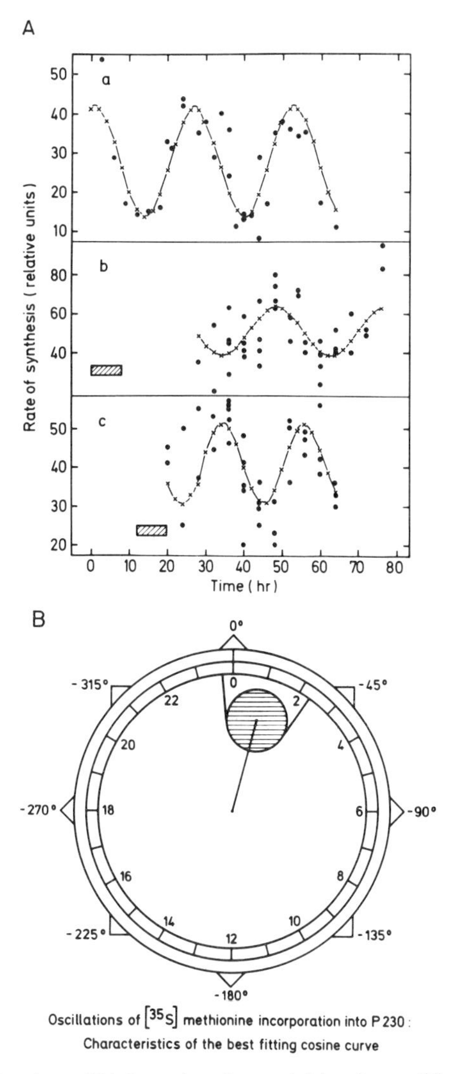

FIGURE 4.19. Identification of high-molecular-weight polypeptide that may be part of the circadian clockwork in *Acetabularia mediterranea*. (A) Effect of cycloheximide pulses on the incorporation of [^{35}S]methionine into p230. Every 4 h, 20 cells were incubated with [^{35}S]methionine for 1 h. After gel electrophoresis and fluorography of the chloroplast-fraction proteins, the fluorographs were scanned, and the areas under the peaks were calculated and plotted against time (●). The best-fitting cosine curve (x——x) was calculated. *(Top)* Control cells. *(Middle)* Cells incubated with 7 m*M* cycloheximide for 8 h after the last dark period and then washed several times. *(Bottom)* Cells treated with cycloheximide during the period 12 to 20 h after the last dark period. (B) Oscillation of [^{35}S]methionine incorporation into p230. Graphic presentation of the characteristics of the best-fitting cosine curve. The inner cycle represents the cycle of 360° subdivided into circadian hours (hrc; 24 hrc = 26.0 h). The hatched area corresponds to the 95% confidence area for acrophase and mesor. The 95% confidence limits for the acrophase are given in hrc. (Reprinted from Hartwig et al., 1985, with permission of the authors.)

ping (Leong and Schweiger, 1979). Furthermore, the proportion of intermediate amino acid groups was high, whereas the proportion of hydrophilic amino acid groups was well-balanced by that of hydrophobic amino acid groups, a property characteristic of membrane proteins. The question, of course is whether polypeptide P39 is vital part of the clock machinery, as opposed to merely an associated one, particularly given the fact that CHX did not inhibit its synthesis and CHX pulses did not phase-shift the oscillations in synthesis of P39.

More recently, Hartwig et al. (1985, 1986) have detected another protein (p230) with $M_r \cong 230{,}000$ in the chloroplast fraction of both nucleate and anucleate *Acetabularia* cells that fulfills three requirements of an essential clock protein. Its rate of synthesis in LL at 20°C (measured by the incorporation of [^{35}S]methionine) exhibited highly significant ($p < 0.01$) circadian oscillations, its synthesis was inhibited by CHX, and 8-h pulses of CHX phase-shifted the rhythm of p230 synthesis (Fig. 4.19). As judged by scanning densitometry of fluorographs of NaDodSO_4-polyacrylamide gels, the synthesis of practically all other proteins in the same fraction was independent of circadian time. Chloramphenicol had no inhibitory effect on [^{35}S]methionine incorporation; this strongly suggests that p230 is translated on 80S ribosomes. There was a clear phase relationship between the overt rhythm of O_2 evolution and the synthesis of p230 in nucleate cells. As anticipated, the maximum of the latter preceded by 6 h that of the former. Interestingly, the synthesis of p230 reached its maximum approximately 4 h earlier in enucleated *Acetabularia* cells than in intact cells.

It is provocative that DNA sequences homologous to the nuclear *per* locus of *Drosophila* (see Section 4.4.1) occur in the *chloroplast* genome of *Acetabularia*, which presumably codes for the p230 protein. If chloroplast DNA was digested with the restriction enzymes *EcoRI* or *HindIII*, the homologous sequences were found in a band respectively corresponding to 2.0×10^3 and 2.5×10^3 base pairs (Li-Weber and Schweiger, unpublished data, cited in Schweiger et al., 1986). The significance of these results is discussed further in Section 4.4.2).

Gonyaulax

Protein synthesis on 80S ribosomes seems to be essential for proper functioning of the circadian clock in *Gonyaulax* also. Early work (Hastings, 1960; Karakashian and Hastings, 1963) demonstrated that continuous exposure of cultures to PUR blocked rhythmicity of glow luminescence, whereas treatment with chloramphenicol did not (Fig. 4.20). Indeed, the amplitude of the persisting rhythm increased after exposure of cells to the latter drug. Both inhibitors blocked cell division. Similarly, 8-h pulses of PUR generated small but significant -$\Delta\phi$s, but those of chloramphenicol had no effect.

Quite large (9 to 12 h) advance and delay $\Delta\phi$s were obtained for the rhythm of stimulated luminescence (Walz and Sweeney, 1979) and for the

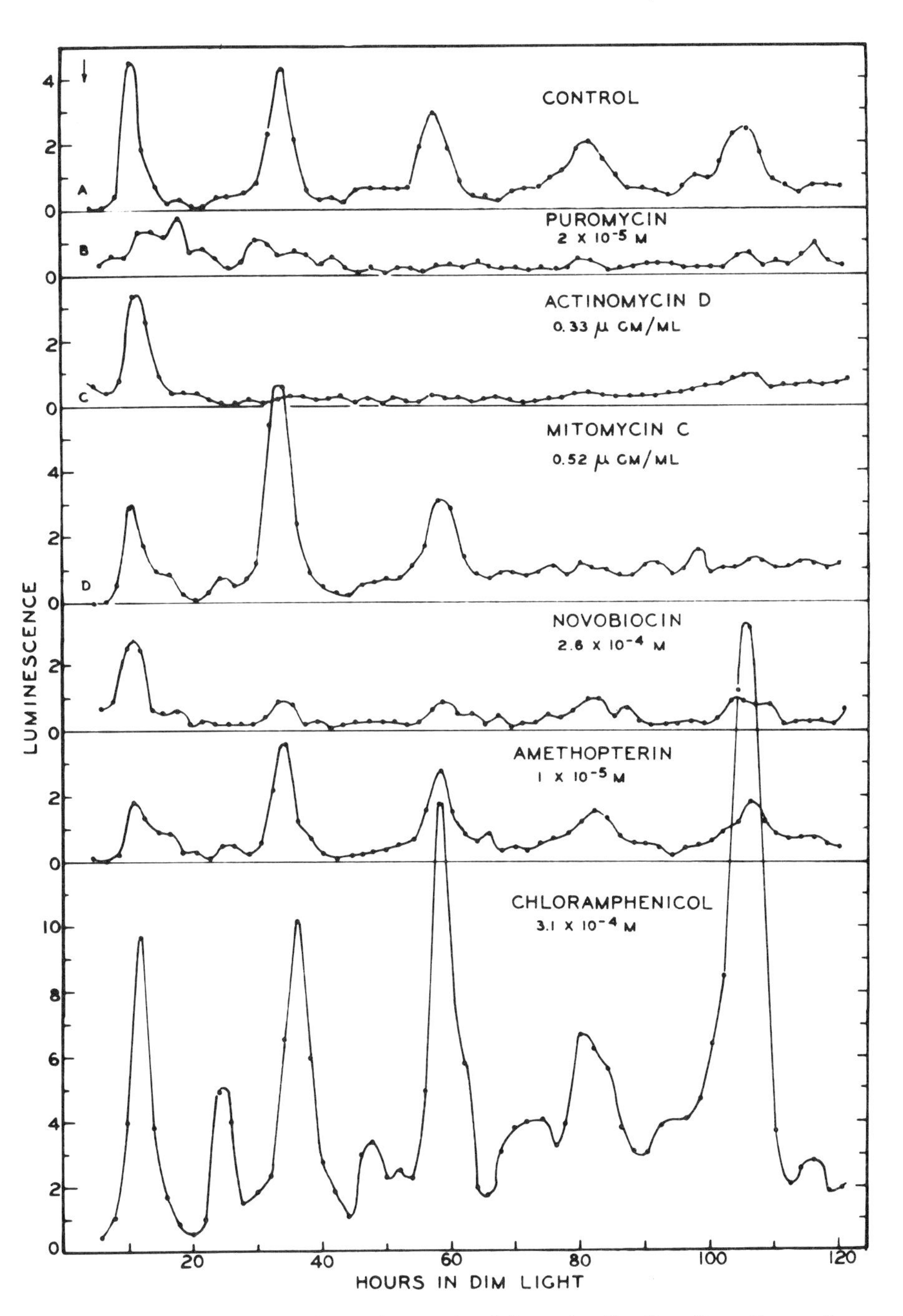

FIGURE 4.20. Recordings of the glow emitted from the dinoflagellate *Gonyaulax polyedra* during exposure to substances that affect macromolecular synthesis. The cell suspensions were kept in continuous illumination except while recording was in progress. The inhibitors were added at the time indicated by the arrow. Luminescence in arbitrary units. (Reprinted from Karakashian and Hastings, 1963, with permision of Rockefeller University Press.)

rhythm of glow luminescence (Dunlap et al., 1980) when *Gonyaulax* was pulsed with CHX, comparable to those found in *Acetabularia* (Karakashian and Schweiger, 1976c). Indeed, in *Gonyaulax*, the PRC derived from the action of 3-h CHX pulses (0.36 or 3.6 μM) was almost identical to that for 3-h light pulses; both showed greatest sensitivity to the perturbing stimulus in the middle of the subjective night, with the breakpoint at CT 18 (Walz and Sweeney, 1979).

Taylor et al. (1982a) have found that 1-h pulses of CHX, ANISO, and streptimidone (STREPT)—all inhibitors of 80S ribosomal synthesis [ANISO binds to the ribosomal subunit at the peptidyltransferase complex and noncompetitively blocks peptide bond formation (Vasquez, 1979)]—at appropriate concentrations caused strong advance or delay $\Delta\phi$s of up to 12 h in the bioluminescence glow rhythm in *Gonyaulax*. In these studies, the breakpoint occurred at CT 12-13. Even minute-long pulses of ANISO were sufficient to yield large phase shifts (Taylor and Hastings, 1982). Several derivatives of ANISO that are chemically slightly different from the parent molecule and that are inactive as inhibitors of protein synthesis were without phase-shifting effect (Taylor et al., 1982a). Dose-response curves (DRCs) and PRCs for phase-shifting the circadian oscillator in *Gonyaulax* were derived for 1-min ANISO pulses (Taylor et al., 1982b). With increasing drug concentration, PRCs appeared to be displaced to the left to earlier phases (representing an additional delay superimposed on the phase-dependent shift), and no saturation was evident in the phase-shifting effect. These results suggested that sufficiently high doses of ANISO caused an immediate strong (type 0) phase shift (Winfree, 1980) and then held the clock stationary for a time interval that increased with drug concentration.

Finally, the three-dimensional, phase-resetting surface (initial phase at time of pulse, final phase after $\Delta\phi$ has occurred, concentration of inhibitor) was found to be a right-hand helix, with the axis at a critical initial phase near CT 12, and a critical concentration near 0.2 μM (Fig. 4.21). Critical pulses of ANISO (CT, dose) significantly disrupted the glow rhythm in cell populations, and in many instances the cultures appeared nearly arrhythmic (Fig. 4.22) (see also Walz and Sweeney, 1979). Although it was not determined whether the arrhythmicity was due to a scattering of the cells comprising the population around the circadian cycle (population desynchronization) or to individual cell arrhythmia, the fact that the residual glow in a critically pulsed culture was substantially less than in one noncritically pulsed favors the latter interpretation (Taylor et al., 1982b). In this case, the data would support the notion that the circadian oscillator is governed by limit cycle dynamics, and the critical pulse then would correspond to placing the system at a singularity point, S* (Winfree, 1970, 1980). Similar findings have been reported for light perturbations of the cell division rhythms of *Euglena* (Malinowski et al., 1985), insects, and higher plants (see Section 3.2.1).

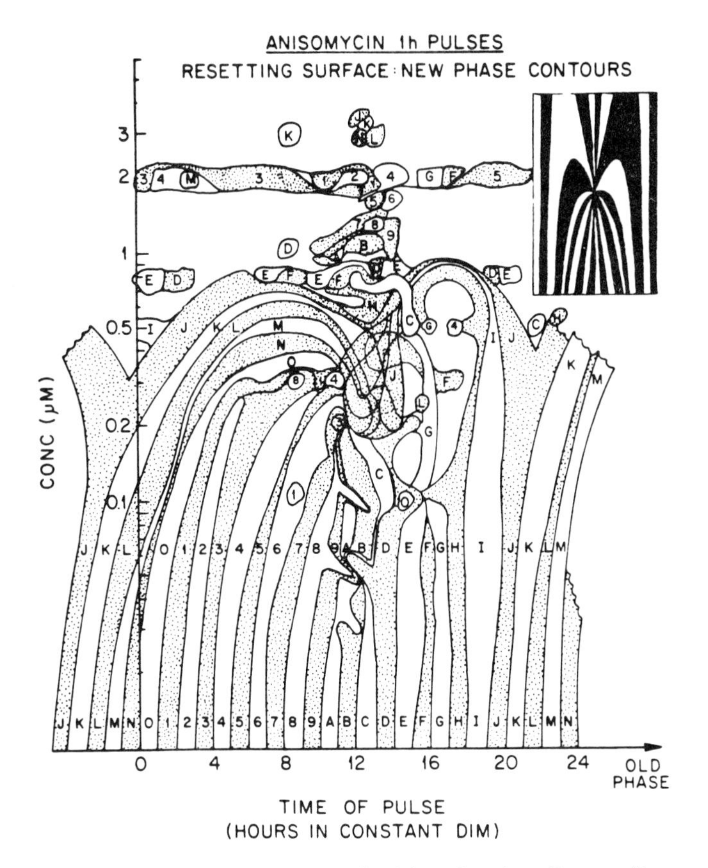

FIGURE 4.21. Critical pulses of anisomycin drive the circadian oscillator toward its singularity in *Gonyaulax polyedra*. Phase-resetting surface for 1-h pulses of anisomycin, shown as contour plots of equal new phase (phase to which the clock is shifted by the drug pulse). Numbers and letters (0 to 9, A to N, where A = 10, B = 11, and so forth) within the iso-new phase contours indicate the value of the new phase. The three-dimensional resetting surface is a plot of new phase plotted above the (drug concentration, old phase) plane. Old phase is the time in the cycle at which the pulse was given. The surface is a right-handed helix rising above the plane. The helix axis is placed at a critical (concentration, old phase) locus (0.28 ± 0.05 micromolar delivered 12 ± 1 h, or *n* periods later), such that the organism presumably becomes arrhythmic when a pulse with these values is delivered. The contours were hand drawn through pooled data from 17 experiments (DRCs and PRCs), representing some 500 data points. There are no data in the surrounding white regions. *Inset:* A model calculation to indicate how the contour plot might appear if a large number of measurements were performed. Overlapping contours in the center of the main diagram represent the experimental uncertainty in defining new phase because of the steepness of the helical surface near the critical point. (Reprinted from Taylor et al., 1982b, with permission of Springer-Verlag.)

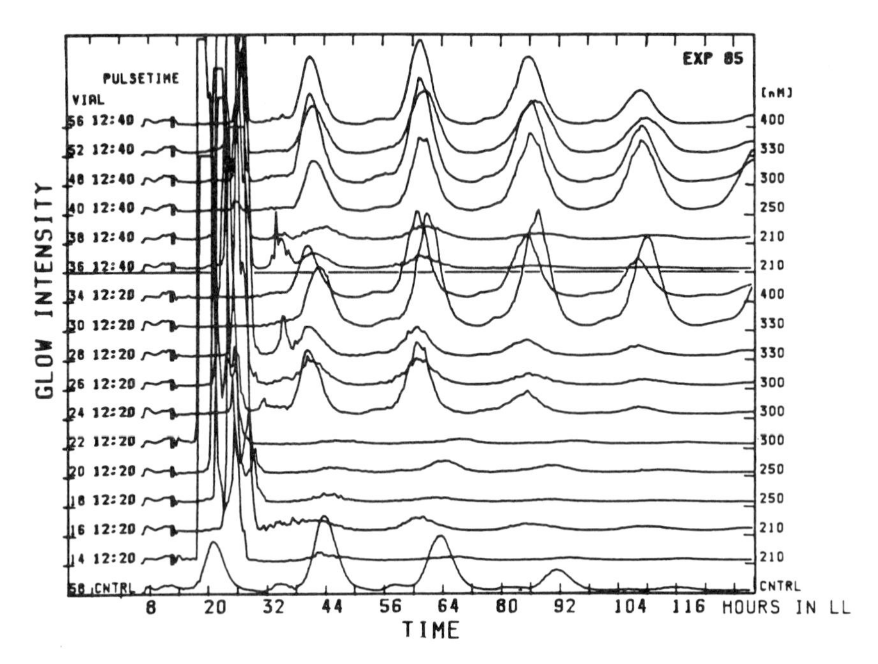

FIGURE 4.22. Dose-response experiments. Effect of anisomycin pulsed for 1 h (black bars) on the subsequent circadian rhythm of bioluminescent glow in *Gonyaulax polyedra*. Pulses applied at times 12:20 and 12:40 after entry into constant dim light, following the dark interval of a LD: 12,12 cycle. Numbers at right are concentrations in n*M*. Pulses in the critical region (CT, dose) appear to induce nearly complete arrhythmicity. (Reprinted from Taylor et al., 1982b, with permission of Springer-Verlag.)

The question of whether the circadian clock of *Gonyaulax* truly is held stationary after a strong pulse of ANISO (perhaps due to a lingering, residual inhibition of protein synthesis for some hours after washout), as suggested by Taylor et al. (1982b), has been examined by Hobohm et al. (1984) using a double-pulse technique. In this methodology, used in the past to investigate the effects of light pulses on the clocks of various circadian systems [see, e.g., Pittendrigh (1965)], a series of cultures were given two successive pulses of ANISO, the first (2 to 3 h, 2 to 2.5 μmol/l ANISO) being given to all at the same circadian time of the rhythm of glow bioluminescence and the second (2 h, 0.3 μmol/l ANISO) being given to each culture at a different CT (Fig. 4.23a; see Fig. 4.25a). Control cultures received only the initial pulse. The composite results of this second pulse yielded a PRC, taken as an indication of the progress of the oscillator underlying the overt glow rhythm. Because putative ANISO pulse-induced clock stationarity would depend strongly on the possible residual effects of the drug after the pulse had been terminated by centrifugation and resuspension in fresh medium of cell aliquots, rates of protein synthesis also

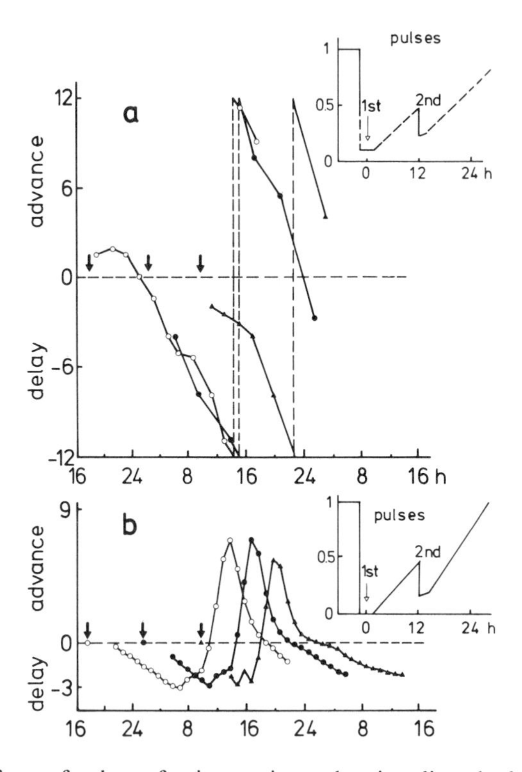

FIGURE 4.23. Effects of pulses of anisomycin on the circadian clock of *Gonyaulax polyedra*. (a) Phase-response of *Gonyaulax* circadian glow rhythm to two subsequent pulses of anisomycin. Cells were treated with with 2 m*M* anisomycin (3 h) as initial pulse (arrow) at different phases of the oscillation, given in hours after onset of LL (abscissa). They were then centrifuged and resuspended in fresh medium. At different times thereafter, a second pulse (0.2 m*M* anisomycin, 2 h) was applied in the same way. The subsequent phase-shift due to the second pulse (ordinate) was measured in comparison to cultures that had received the initial pulse only (control, baseline). *Inset:* Form of a particular pulse sequence as estimated from biochemical determinations of protein synthesis inhibition. *Ordinate:* Inhibition. *Abscissa:* Time after pulse in hours. (b) Phase-response of the model oscillator to two subsequent pulses (resetting of the variable X) at different phases of the oscillation. Phase shifts to the second pulse (ordinate) with respect to a control (first pulse only, baseline) depending on the phase of the first pulse (arrow, abscissa). *Inset:* Form of a particular pulse sequence, chosen similar to the experimental pulse forms. (Reprinted from Rensing and Schill, 1985, with permission of Springer-Verlag; see Hobohm et al., 1984.)

were determined in culture samples by assaying the incorporation of a tritium-labeled mixture of amino acids for several hours after the pulse.

These second pulses of ANISO induced $\Delta\phi$s of the glow rhythm analogous to those evoked in single-pulse control cultures but whose sign and magnitude were dependent on the CT at which the second pulse was given

(Fig. 4.23a). Rhythmicity appeared to be absent for about two cycles after the ANISO pulses, and the $-\Delta\phi$s in the different cultures receiving a second pulse corresponded roughly to the interval between the first and second pulses (Hobohm et al., 1984; Rensing and Schill, 1985). The PRCs for the second pulses differed in their phase, depending on the CT at which the initial pulse was given in different series of experiments (Fig. 4.23a). Plotting the time intervals between the first pulse and the delay advance transition (breakpoint) of the PRC derived from the phase-shifting effects of the second pulses as a function of the phase (CT) of the first pulse revealed a highly significant rhythmic variation (Fig. 4.23a).

In order to interpret the results of the double-perturbation experiments, a model circadian oscillator of the limit cycle type (Drescher et al., 1982; Goodwin, 1965; see Section 5.2.4), based on translation control and having three components (an enzyme or other protein species, its product, and a translation inhibitor whose activity depends on the product and which, in turn, determines the rate of enzyme synthesis), was used to simulate the data under conditions approximating the experimental design (Hobohm et al., 1984; Rensing and Schill, 1985). In this simulation, the actual value of the variable (x) representing the concentration of a protein was set at 1, 10, 40, or 90% of the actual value. The model oscillator showed responses to perturbations (Figs. 4.23b, 4.24b) quite similar to those obtained experimentally (Figs. 4.23a, 4.24a): the phasing of the PRCs to second pulses of ANISO (expressed by the interval between the beginning of the PRC and the crossing of the zero line between the delay and advance portion of the curve) depended strongly on the phase of the first pulse.

In the model, the first pulse of ANISO does not discontinue the oscillations of the other components of the model oscillator. Rather, the perturbed variable x representing protein concentration responds differently, depending on the phase of the initial pulse and that of the other continuously oscillating variables. In turn, this results in differences in the interval between the first pulse and the next maximum of x, reflected in the breakpoint of the PRC derived from the action of the second pulse. Hypothetical responses of the model oscillator (Fig. 4.25) to these double-perturbation treatments and their associated PRCs (in which $\Delta\phi$s are all plotted as delays to avoid the artificial discontinuity of the breakpoint) include (1) oscillation set to zero and restarted immediately after initial pulse (in which case the response of the oscillator to the second pulse would be identical to that to the first), (2) oscillation set to zero and held stationary for a time interval and then restarted (in which case PRCs for the second pulse would give identical curves except for their displacement in time), and (3) oscillation variably phase-shifted by initial pulse but then continued with no interruption. As we have seen, the experimental results yielded PRCs in disagreement with cases (1) and (2), but in accord with the predictions of case (3), namely, a continuously but immediately phase-shifted oscillation (Fig. 4.24). As Hobohm et al. (1984) note, the temporary

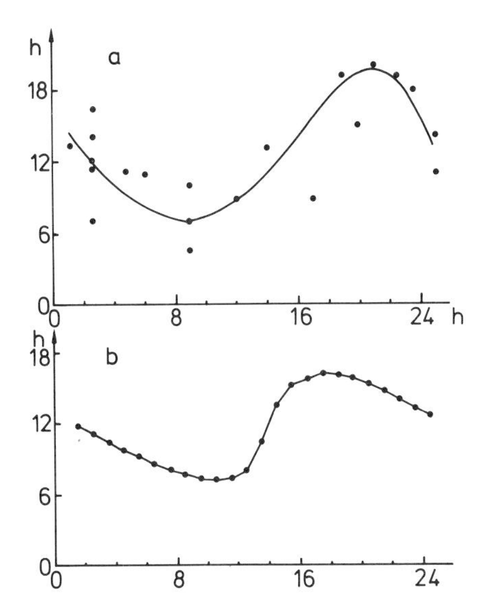

FIGURE 4.24. Phasing of the PRC to second pulses of anisomycin as a function of the phase of the circadian rhythm of glow bioluminescence at which the first pulse is applied *(abscissa)* in *Gonyaulax polyedra*. Phasing of PRCs is expressed as the interval between first pulse and the delay-advance (breakpoint) transition of the PRC. *Ordinate:* Interval duration in hours. (a) *Gonyaulax* circadian oscillator. (b) Model oscillator. (Reprinted from Rensing and Schill, 1985, with permission of Springer-Verlag; see Hobohm et al., 1984.)

stop or holding of the clock proposed by Taylor et al. (1982b), perhaps caused by the residual effects of ANISO, would not result necessarily in an absence of an oscillation but rather in a decrease in its amplitude.

From the similarity of the responses to perturbing pulses of ANISO shown by both the *Gonyaulax* clock and the model oscillator, the functional structure of the model based on translational control (Drescher et al., 1982) may represent a possible structure of the circadian oscillator (see Section 5.2.4). If the model is adequate, one might expect the concentration of a protein X, essential for the oscillation, to vary rhythmically and to have a maximum concentration between CT 14 and CT 20, the time interval when maximum $\Delta\phi$s of the variable x can be obtained (Rensing and Schill, 1985).

Support for these predictions recently has been afforded by the demonstration of a free-running circadian rhythm in total protein synthesis (maximum between CT 6 and CT 18) as well as in synthesis on 70S and on 80S ribosomes in stationary- (or infradian-)phase *G. polyedra* maintained in dim LL, as assayed by [^{3}H]leucine incorporation and confirmed by light and electron microscopic autoradiography (Volknandt and Hardeland, 1984a). In the presence of chloramphenicol, the maximal value

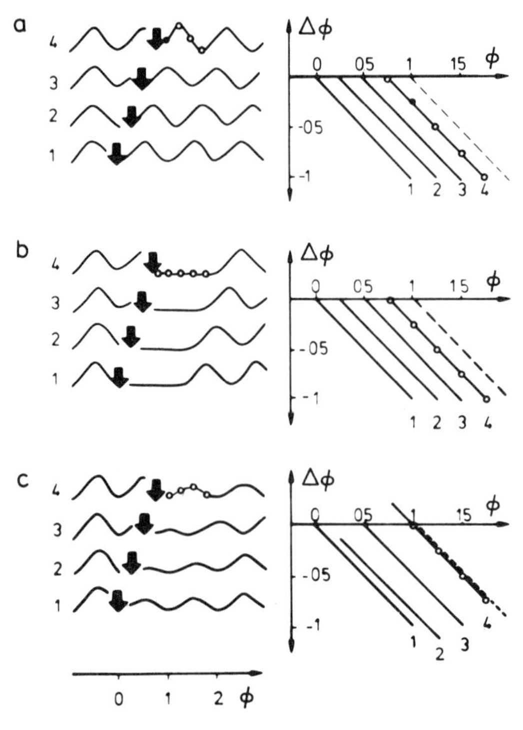

FIGURE 4.25. Hypothetical responses of the circadian oscillator in *Gonyaulax polyedra* to an initial perturbing pulse of anisomycin (arrow) at different phases (left) and phase-response curves to a series of second pulses (circles) applied at different phases during a period (0-1) after the initial pulse (right). The PRC for the assumed first period after the perturbing pulse is estimated after applying a simple procedure. A second pulse (for example ¼ period after the initial pulse of oscillation, see filled circle in curve a_4) sets the oscillation to zero and lets the new oscillation start from there. This can cause a phase shift, in this case a shift of ¼ period compared to the reference oscillation that received only the first pulse. The PRCs are given on the right (Δϕs are all plotted as delays). *Left abscissa:* Phase (ϕ) at which initial and second pulses are applied. *Right ordinates:* Phase shifts (Δϕ) as fractions of the period length. *Right abscissae:* Phase at which initial pulse (starting point of response curves) and second pulses are applied. *Dashed lines:* Repeats of a PRC beyond one period. (a) Oscillation set to zero and restarted after the initial pulse. (b) Oscillation set to zero and held stationary for a time interval and then restarted. (c) Oscillation phase-shifted by initial pulse but continued. (Reprinted from Hobohm et al., 1984, with permission of Pergamon Press.)

for the uninhibited 80S protein synthesis occurred at CT 18, whereas if 80S translation was inhibited by either CHX or ANISO, the highest value for the remaining synthesis of 70S proteins was found at CT 6–9. Several individual bands of protein, labeled and separated by SDS-polyacrylamide gel electrophoresis, showed rhythms of synthesis or of concentration or both. Similar biochemical and autoradiographic studies also have demonstrated circadian-clock control of the synthesis and degradation of total cellular proteins (Cornelius et al., 1985) and of total protein synthesis in the cytoplasm and chloroplasts (Donner et al., 1985) of growing cultures of *Gonyaulax*. The maxima of these rhythms also occurred during the subjective night, although in very slowly dividing cultures, the amplitude of the rhythms was markedly reduced (suggesting that the majority of the proteins were cell cycle-dependent), and τ was slightly shortened (leading to earlier phasing after several days in LL). Confirmatory results have been obtained by Schröder-Lorenz and Rensing (1986), who also found that altered ambient temperatures (17 to 25°C) changed the phase of the rhythm of protein synthesis as well as of the PRC for ANISO pulses. In sum, the maxima reported for the rhythms in protein synthesis occurred at approximately the same time in the circadian cycle at which maximal $\Delta\phi$s of the glow rhythm had been obtained with pulses of either CHX (Dunlap et al., 1980) or ANISO (Taylor et al., 1982b; Rensing and Schill, 1985).

Neurospora

The generalization that protein synthesis on 80S ribosomes is important for circadian clock function—already discussed for *Acetabularia, Euglena,* and *Gonyaulax* (see preceding subsections)—also applies to the bread mold. Early work (Sargent, 1969) indicated that continuous exposure of *N. crassa* to CHX did not alter the τ of the conidiation rhythm, but this may have been due to the severe inhibition of growth and conidiation by the drug, making the rhythm difficult to assay. The subsequent development of liquid culture techniques (Nakashima, 1981; Perlman et al., 1981; see Section 2.3.1) facilitated inhibitor-pulse experiments. Thus, Nakashima et al. (1981a) demonstrated that short, 4-h perturbations of CHX (1.6 μg/ml), which inhibited protein synthesis completely (assayed by [^{3}H]leucine incorporation), generated large $\Delta\phi$s (up to a 12 to 15 h advance) of the conidiation rhythm. The PRC derived from the data (Fig. 4.26; see Fig. 4.15d) showed that the $\Delta\phi$s were primarily phase advances, with the maximum sensitivity's occurring at approximately the middle of the subjective day (CT 5). Little or no $\Delta\phi$ was observed at CT 19 (midsubjective night). The phasing of the breakpoint, or crossover point between maximal delays and advances, of the PRC for CHX pulses (at about CT 4) was not the same as that (at about CT 18) of the PRC for light signals (Nakashima and Feldman, 1980), although it was almost identical to those found for a variety of respiratory inhibitors (Fig. 4.15; see Fig. 4.33). Inhibition of protein synthesis was the same at both day and night phases of the circadian cycle, indicating that the phase-dependence of the CHX-induced $\Delta\phi$s could not be attributed to variations in inhibitor effectiveness. Dose-response curves for the effects of CHX (at CT 5 and CT 19) on phase shifting and on inhibition of protein synthesis showed a striking parallel for these two phenomena; for example, the concentration of drug that

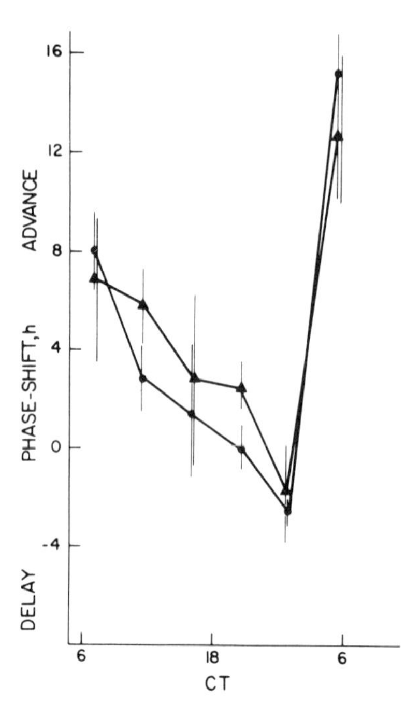

FIGURE 4.26. Phase-response curves for pulses of cycloheximide (CHX) and the circadian rhythm of conidiation in *Neurospora crassa*. Sets of six mycelial discs were incubated with CHX (1.6 μg/ml) for 4 h in liquid medium starting at times shown. Each disc was then transferred to a race tube, and the average phase of first conidial band was determined. *Error bars*, ±SD. Circadian time (CT) is based on a 22-h free-running period, with CT 12 defined as the time of light–dark transition. ▲, 20°C; •, 25°C. (Reprinted from Nakashima et al., 1981a, with permission of American Physiological Society.)

approximately halved the inhibition of protein synthesis found for 1.6 μg/ml pulses yielded only 5 to 6 h $+\Delta\phi$s. The PRC was the same at 20 and 25°C, in contrast to the temperature-dependence of the PRC for CHX pulses on the rhythm of O_2 evolution in *Acetabularia* (Fig. 4.18A).

These results support the view that synthesis of one or more proteins at specific phases of the circadian cycle is necessary for the normal operation of the *Neurospora* circadian clock (Nakashima et al., 1981a). That CHX alters the oscillator through its inhibition of 80S protein synthesis, and not through some different, unknown target, has been indicated strongly by the parallel resistance of the clock to phase-shifting by CHX pulses and of 80S protein synthesis to inhibition by CHX in mutants (*cyh-1* and *cyh-2*) of *Neurospora* with alterations in their 80S ribosomes but an otherwise normal clock (Nakashima et al., 1981b; see Section 4.4.3). Furthermore, the direct proportionality observed between inhibition of protein synthesis and phase-shifting of the clock, across a 100-fold range of CHX concentrations, makes it unlikely, though not impossible, that the effects on the clock are caused by a secondary change in the pool sizes of amino acids or energy or other parameters because such indirect effects would probably not exhibit such a close correspondence with the degree of inhibition of protein synthesis (Nakashima et al., 1981a). Jacklet (1980a,b) has reached similar conclusions for the circadian clock in the isolated eye of *Aplysia* (see next subsection).

Rather unexpected results concerning the manner in which protein synthesis-mediated phase-shifting occurs have been found in a comparison of CHX-induced $\Delta\phi$s in clock mutants (at the *frq* locus) (see Table 4-1) and nonclock mutant *(bd)* strains (Dunlap and Feldman, 1982; Feldman and Dunlap, 1983). The long-period mutant *frq-7* (in contrast to the short-τ and intermediate long-τ *frq-1*, *-3* and *-4* mutants) appears to be relatively insensitive to phase-shifting by CHX at both 25 and 30°C, even at 10-fold higher drug concentrations than those used in the *bd* strain to cause large $\Delta\phi$s and inhibit protein synthesis. The CHX-sensitive phase was not altered, nor was the inhibition of protein synthesis itself (as shown by DRCs done at three different CTs). Yet the clock in *frq-7* is functional. It is light-entrainable and free-runs, although its temperature compensation of τ is somewhat impaired (Gardner and Feldman, 1981). Thus, in this mutant the normal clock requirement for the synthesis of circadian phase-specific clock protein(s) appears to have been bypassed or eliminated, although it is possible that there is a connection between temperature compensation and protein synthesis (Feldman and Dunlap, 1983).

The relative rates of synthesis of various cellular proteins, pulse-labeled with [^{35}S]methionine at two different circadian times, have been determined by two-dimensional gel electrophoresis and autoradiography. No differences were found among samples taken from different times of the cycle, despite the resolution of more than 100 distinct polypeptides (Perlman, 1981). Thus, the search for an essential clock protein in *Neurospora* has not been successful thus far, in contrast to the discovery by Hartwig et al. (1985) of the cyclically synthesized, CHX-inhibited, and CHX-phase-shiftable protein p230 in *Acetabularia*.

We note that Cornelius and Rensing (1986) found that *N. crassa* shows a maximal capacity to synthesize three major HSP (99, 81, and 69 kDa) at about CT 12 (the phase at which the temperature decreases under normal environmental conditions). Mycelia in the stationary phase of growth were subjected to 3-h heat-shocks (42°C) at various CTs. Samples were then incubated with [^{35}S]methionine and fractionated, and the labeled proteins were separated and identified by polyacrylamide gel electrophoresis and fluorography. HSP have been associated with acquisition of thermotolerance in many organisms (review by Nover, 1984), and many participate more generally in the control of cellular states, which the CO may endogenously control by anticipating periodic environmental changes. It is interesting to note that many of the treatments that induce HSP synthesis (as, for example, pulses of DCCD, or arsenite or anaerobiosis) also perturb and phase shift CRs, such as that of conidiation (Rensing, unpublished). Since both systems—heat-shock response and the circadian clock—rely on protein synthesis, Rensing et al. (1986) suggest that translational control is the most likely link between them. In support of this hypothesis, they have found a linear dose dependence of the phase shifts of the conidiation rhythm induced by 3-h pulses of elevated temperature (30°C, 35°C, 40°C) that was matched by that for the inhibition of both total protein synthesis and HSP-induction at elevated temperatures. Thus, increases in environmental temperature may act on circadian oscillators by a process related to the heat-shock response.

Aplysia

The circadian pacemaker generating rhythmicity in the frequency of CAPs in optic nerve impulses from the isolated eye of the sea hare (see Section 4.1.1) also seems to require protein syntheis on 80S ribosomes for its proper operation. This system, like that of *Neurospora* (see preceding subsection), has proved particularly interesting not only because it is open to biochemical analysis but also because the extent of inhibition of protein synthesis by drug pulses has been measured directly and correlated with the phase-shifting effects of such pulses at the same concentration (Jacklet, 1984).

Earlier work (Jacklet, 1980a) demonstrated that ANISO (10^{-8} *M*) applied continuously inhibited protein synthesis by about 10% and caused a slight lengthening of τ and a reduction in amplitude of the CAP rhythm. If the concentration of the drug were increased (10^{-7} *M*), protein synthesis was inhibited by 50%, τ was increased to 30 h, and the rhythm was subsequently abolished. At 10^{-6} *M*, ANISO inhibited synthesis by 90%, and the rhythm was lost immediately (although CAP activity continued at a moderate rate). This loss of rhythmicity was irreversible after several days of exposure of the eye to the inhibitor. These results resemble those of Feldman (1967) for the effects of CHX on the rhythm of phototaxis in *Euglena* (see Fig. 2.7).

Phase-dependent shifts of the CAP rhythm were also induced by single pulses of PUR (Rothman and Strumwasser, 1976) and ANISO (Jacklet, 1977). The PRCs derived from the data (Fig. 4.27) were similar; the sensitive phase began in the early subjective night and extended into the subjective day. The magnitude of the phase delays produced by ANISO, however, was much greater than those caused by PUR, whereas the opposite was true for advances, and the transition between phase delays and advances (breakpoint, or crossover point) for PUR-induced $\Delta\phi$s phase-led that for ANISO by several hours.

In an attempt to explain these seeming discrepancies, which raised questions not only about the methodology used but also about the specific modes of action of the two drugs, Lotshaw and Jacklet (1981, 1986) un-

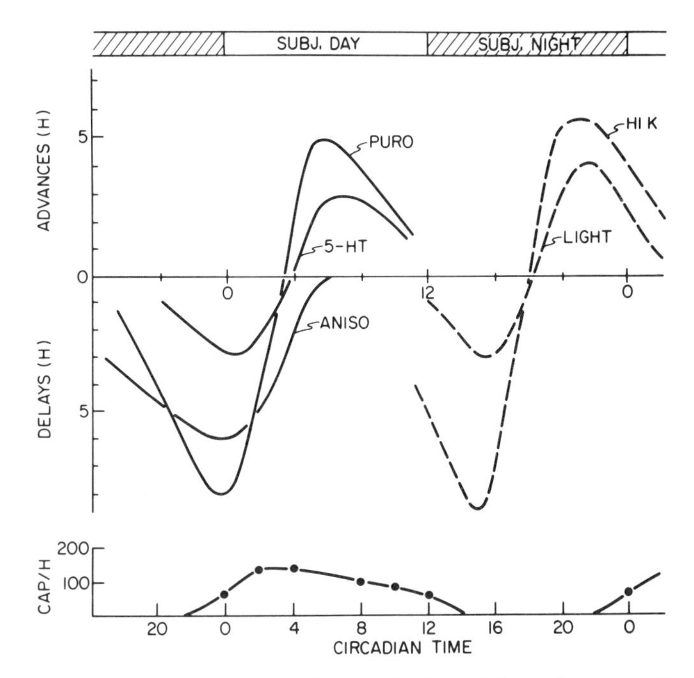

FIGURE 4.27. Phase-response curves for the rhythm of the *Aplysia* eye. Responses are plotted at midpoint of pulses for puromycin, PURO, $2 \times 10^{-4} M$; anisomycin, ANISO, $5 \times 10^{-7} M$ (see Lotshaw and Jacklet, 1981); 5-HT, $10^{-5} M$ (see Eskin, 1982b); light, 10,000 lx, and high K^+, 90 m*M* (see Jacklet and Lotshaw, 1981). Bottom graph is CAP frequency rhythm for phase reference. CT 0–12 is subjective day, CT 12–24 is subjective night. Breakpoints from delays to advances occur at CT 4 for PURO and 5-HT, at CT 18 for light, high K^+, and strophanthidin (not shown). PRCs for low temperature, metabolic inhibitors, and calcium ionophore A23187 resemble the ANISO curve. (Reprinted from Jacklet, 1984, with permission of Academic Press.)

dertook a more detailed study of the effects of these inhibitors on both protein synthesis and phase-shifting capacity in the *Aplysia* eye. ANISO pulses (6 h, initiated at CT 21) produced phase delays of the CAP rhythm that were proportional in magnitude to the duration and percentage of protein synthesis inhibition (measured as [^{3}H]leucine incorporation into trichloroacetic acid-insoluble material). The PRC for 6-h ANISO pulses at a concentration (0.5 μM) just sufficient for maximum (90%) inhibition of protein synthesis—higher concentrations did not further increase inhibition but did produce greater $\Delta\phi$s, perhaps due to slower recovery of the system—consisted solely of delays throughout the subjective night (maximum $-\Delta\phi$ between CT 18 and CT 2). Pulses initiated between CT 2 and CT 12 did not phase shift the rhythm (Fig. 4.27). In contrast, 6-h PUR pulses imposed at different CTs, at a drug dosage (230 μM) producing maximal $\Delta\phi$s and inhibition of protein synthesis, generated both phase delays and advances (Fig. 4.27). The former occurred with increasing magnitude throughout the subjective night (CT 12–2), and the latter occurred between CT 2 and CT 8. The magnitude of both PUR pulse-induced advances (initiated at CT 2) and delays (initiated at CT 21) was directly proportional to the degree of inhibition of protein synthesis. Phase advances, however, required greater drug dosages than did delays for shifts of the same magnitude (100 μM pulses of PUR, for example, generated only delays).

These results, therefore, confirmed and extended the earlier studies (Rothman and Strumwasser, 1976; Jacklet, 1977) and clearly demonstrated that both PUR and ANISO can induce nearly identical, all-delay PRCs if the doses of the drugs are appropriately adjusted and the PRCs are determined under the same conditions. The effectiveness of ANISO in inducing delays earlier in the subjective night than PUR might be attributed to its greater inhibition of protein synthesis, although maximal delays produced by PUR were larger than those for ANISO and perhaps were augmented by the direct phase-shifting effects of peptidyl-puromycin (nonsense peptides) produced by the inhibitory action of PUR (Vasquez, 1979). This hypothesis was supported by the observation that ANISO could block PUR-induced effects in the isolated *Aplysia* eye. Simultaneous, 9-h pulses of PUR (230 μM) and ANISO (0.5 μM) applied at CT 21 (where normally an equivalent pulse of PUR by itself produced a significantly greater phase delay than did ANISO) resulted in a phase delay whose magnitude was equal only to that caused by ANISO alone (Lotshaw and Jacklet, 1986). The effectiveness of the ANISO block was inferred from its inhibition of amino acid incorporation and inhibition of PUR-specific proteins synthesized independently of circadian phase during recovery from the two-drug pulse. (These proteins, labeled with [^{35}S] methionine and isolated by two-dimensional, SDS-polyacrylamide gel electrophoresis and autoradiography, included species that had molecular weights of 110, 94, 90, 70, and 50 kDa.) The PUR-induced phase advance, however, was not affected by the concomitant treatment with ANISO. This fact, in addition to the finding that an advance required a greater concentration of drug than did an equivalent PUR-induced delay, suggested that $+\Delta\phi$s and $-\Delta\phi$s are mediated by different mechanisms.

Experimental perturbation of the *Aplysia* eye in vitro with inhibitors of translation on 80S ribosomes, therefore, supports the hypothesis of a circadian phase-dependent requirement for protein synthesis during the subjective night of the pacemaker. This proposition is further supported by the finding that derivatives of ANISO ineffective in inhibiting protein synthesis were equally ineffective in phase-shifting the CAP rhythm (Jacklet, 1980b), a conclusion similar to that reached by Taylor et al. (1982a) for *Gonyaulax*.

Comparison of the PRCs for pulses of PUR and ANISO with PRCs for low temperature (Benson and Jacklet, 1977a,b,c; Jacklet, 1981), for the metabolic inhibitors NaCN and DNP, for the calcium ionophore A23187 (Eskin and Corrent, 1977), and for the putative neurotransmitter 5-HT (Corrent et al., 1978; Eskin, 1982b; see Fig. 4.12A) implicated in a nonphotic entrainment pathway coupling to the clock (see Section 4.2.2) revealed a common phase sensitivity. Crossover points between $-\Delta\phi$s and $+\Delta\phi$s centered at CT 3–4 in the case of PUR and 5-HT (Fig. 4.27). This correspondence suggests that all these treatments induce phase shifts (predominantly delays in *Aplysia*) by affecting the same clock parameter, which would be associated with protein synthesis (delays being produced by its blockage; advances, when they do occur, perhaps by its enhancement or by alterations in several proteins) or cAMP or both (Jacklet, 1984). Indeed, we have already seen that there is a requirement for protein synthesis in the regulation of the CAP rhythm by the serotonergic pathway: 6-h pulses of 5-HT significantly increased the synthesis of a 34-kDa protein species at phases where 5-HT delayed or advanced the ocular rhythm but had no effect at phases where 5-HT did not generate $\Delta\phi$s, and pulses of forskolin and 8-benzylthio-cAMP (both of which mimic the phase-shifting action of 5-HT by increasing the concentration of cAMP) likewise mimicked the effect of 5-HT on the 34-kDa protein (Eskin et al., 1984b; Yeung and Eskin, 1987) (see Section 4.2.2). In contrast, the PRCs for pulses of light, high K^+, or strophanthidin—all of which generate both advances and delays and are thought to affect membrane depolarization or ion transport (Na^+ or K^+) or both—cluster at CT 18 (Fig. 4.27), a time when the pacemaker neurons are hyperpolarized and inactive.

Similar, although not identical, groupings of PRCs for inhibitors of protein synthesis (CHX, ANISO) and for light have been found in *Gonyaulax* (CT 12/CT 18) (Taylor et al., 1982a; but see Walz and Sweeney, 1979) and *Neurospora* (CT 3–4/CT18) (Feldman and Dunlap, 1983). As a consequence, a reasonable assumption would be that the circadian oscillator in these organisms may be governed by limit cycle dynamics with only two state variables, one being associated with 80S ribosomal synthesis, the other being the variable affected by light (Cornelius and Rensing, 1982; Taylor et al., 1982b).

4.3.3 Membrane-Active Agents

Given the explosive growth of the field of membrane biology, it is not surprising that various workers have suggested that membranes and cellular compartments may be involved in the generation of circadian rhythms. Indeed, several models for circadian oscillators have been proposed (see Section 5.4) that derive from experimental results showing that a number of agents that affect ion transport also alter period and phase of overt rhythms and that incorporate the requirement for protein synthesis for clock function (see Section 4.3.2). In a sense, it is almost a truism to invoke membranes in circadian timekeeping—reflecting, perhaps, the bandwagon effect. It is difficult, although not impossible (Scholübbers et al., 1984) to imagine a biological process not involving membranes in some manner. What concerns us here is to identify roles of membranes that are essential to the clock mechanism.

Membranes could play a role in the circadian clock mechanism in two ways: (1) the membrane function, although not a part of the clock itself and thus not necessarily oscillatory, may control τ by controlling the velocities of parts of the oscillatory process, or (2) the membrane function may constitute an element, or gear, of the clock and, therefore, undergo rhythmic changes. The membrane function in either role could reside in the lipid (or other) composition of the membranes, in the activity of membrane-bound (transport) proteins or enzymes, in membrane potential, or in the configuration and conformation of membranes of any or all cell compartments. For either of the two roles, temporary changes (elicited by chemical pulses) or permanent changes (caused by step perturbations) of the membrane function should cause, respectively, either a permanent phase shift of the rhythm or a permanent change in its period (Tyson, 1976). As was noted earlier in the rationale for drug perturbation experiments, the ever present problem is that it is difficult to ascertain whether the $\Delta\phi$s engendered are the result of an effect on a postulated primary target (a gear) or on some other site (a nongear) only secondarily affected by the drug (Goto et al., 1985).

Evidence for involvement of membranes in the circadian clock mechanism (ultradian glycolytic oscillations occur in cell-free, soluble yeast extracts; see Sections 1.3.2 and 5.2.1) derives from two sources: (1) primarily descriptive observations of structural and functional changes of membranes (composition and organization, physicochemical and physiochemical properties, ion movements and transport) during the circadian cycle and their analysis (in both normal systems and in biochemical mutants) and (2) reports of the effects of continuously present (step addition) or pulsed membrane-active chemicals, drugs, and antibiotics on the period and phase of overt rhythms (reviewed by Edmunds, 1980c; Edmunds and Cirillo, 1974; Engelmann and Schrempf, 1980; Sweeney, 1976b; Sweeney and Herz, 1977). In this section about the dissection of the clock using chemical perturbations we are concerned primarily with the second line of attack. A brief review of the results obtained with some of the key experimental systems (including *Euglena, Gonyaulax, Neurospora, Phaseolus,* and *Aplysia*) follows, which will set the stage for the consideration of membrane models for circadian clocks (see Section 5.4).

Alcohols

Continuous exposure of *Gonyaulax* cultures to moderate concentrations of ethanol (0.1%), which may act at the membrane level, caused a shortening of the period of the glow rhythm (Taylor et al., 1979), contrary to the period-lengthening effects usually reported for other circadian systems (Edmunds, 1980c), including the dark motility rhythm of *Euglena* (Brinkmann, 1976a). Ethanol pulses (0.1% caused phase shifts of both the glow rhythm (Taylor et al., 1979) and of the rhythm of stimulated luminescence (Sweeney, 1974a). In the latter case, the sign and magnitude of the $\Delta\phi$ obtained were dependent on the CT at which the 4-h ethanol pulse was applied, the PRC thus derived (Fig. 4.28A) closely resembling that for light—the universal entraining agent—perturbations in similarly maintained cells (but almost 180° out-of-phase with that for the ionophore valinomycin) (Fig. 4.28B).

Furthermore, Sweeney (1976b) investigated the relationship between alcohol chain length and the corresponding ability to cause phase shifts: methanol was found to be the most effective and longer-chain alcohols the least effective (Sweeney and Herz, 1977). Thus, the notion that alcohols exert their effect on the clock through a nonspecific attack on membranes, with alcohols' that tend to partition themselves more into the lipid phase of cell membranes being more effective phase shifters, was not supported (Njus et al., 1976; see Section 5.4.1). This conclusion was reached also by Brinkmann (1976a) for the motility rhythm in *Euglena*, in which it appeared that alcohols must be metabolized in order to exert their effects (nonmetabolizable methanol and propanol, for example, did not cause $\Delta\phi$s). One would anticipate on Brinkmann's hypothesis that acetaldehyde, an immediate byproduct of ethanol metabolism, should also be able to generate $\Delta\phi$s. This was empirically confirmed by Taylor and Hastings (1979) for the glow rhythm of *Gonyaulax*; they found that aliphatic aldehydes having a chain length of one to four carbon atoms (formaldehyde, acetaldehyde, proprionaldehyde, butyraldehyde) had a significant phase-shifting effect that lessened as the carbon chain lengthened. Nevertheless, the results are difficult to interpret, since *Gonyaulax* does not use any of the many carbon sources assayed, including ethanol and acetate, as metabolic carbon but relies solely on photosynthetically fixed CO_2 (Sweeney, 1983).

In an attempt to reconcile these findings, Volknandt and Hardeland (1984b) measured the effects of aldehydes and alcohols at phase-shifting concentrations on protein synthesis ([^{3}H]leucine incorporation) in vivo and in an in vitro translating system (predominantly representing 80S protein synthesis) in the *Gonyaulax* system. Aldehydes (one to four carbon atoms) inhibited protein synthesis considerably, both in vivo and in vitro at two different circadian times (CT 3–4, CT 7–8); their efficiency was negatively correlated with their chain length, a finding that corresponded well with their phase-shifting potency. Aldehydes inhibited both 70S and 80S in vivo translation, as differentiated by incubation of the cells with either CHX or chloramphenicol. In contrast, alcohols (methanol, ethanol), although given in much higher concentrations, had only weak effects on

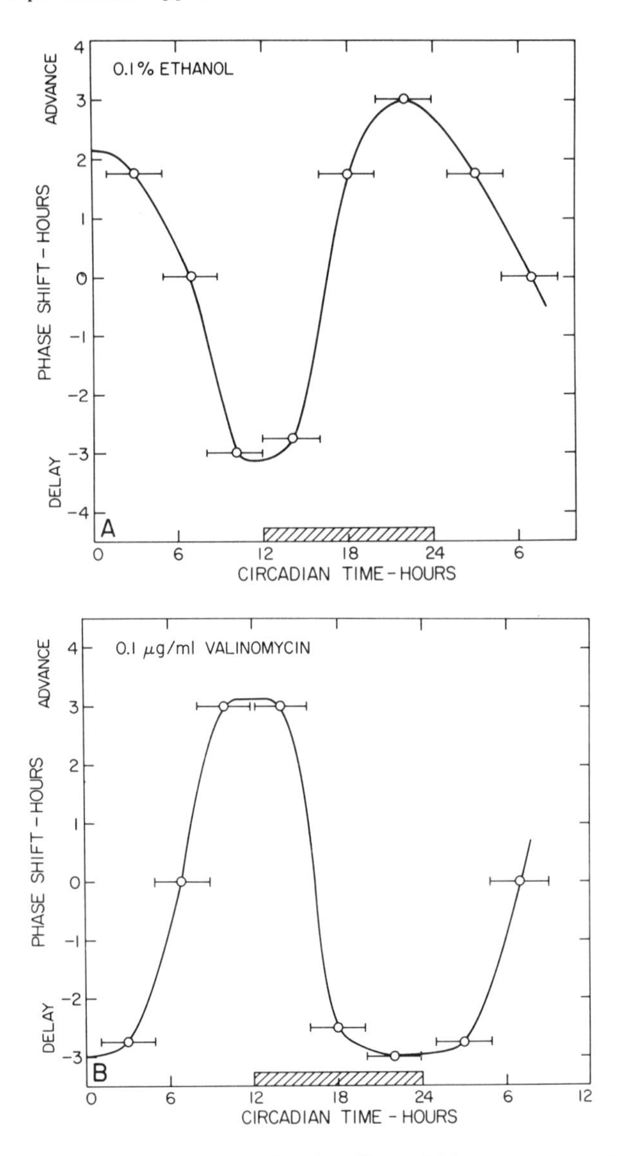

FIGURE 4.28. Phase-response curves for the effect of 4-h exposures to 0.1% ethanol (A) and to 0.1 μg/ml valinomycin in 0.1% ethanol (calculated with reference to ethanol only) (B) on the phase of the circadian rhythms in stimulated bioluminescence in *Gonyaulax polyedra*. Note that phase advances ($+\Delta\phi$) are plotted upward. Subjective dawn corresponds to circadian time 0. The subjective dark period is represented by the hatched bar on the abscissa. (Reprinted from Sweeney, 1974a, with permission of American Society of Plant Physiologists.)

protein synthesis. Ethanol 0.1% did not inhibit in vivo translation at all, despite the fact that pulses at this dosage (or even less) reset the circadian clock by hours (Fig. 4.28A). The phase-shifting action of aldehydes (but not of alcohols), therefore, can be explained on a common basis with that of CHX and ANISO (see Section 4.3.2), namely, through the inhibition of 80S protein synthesis.

Fatty Acids

If membranes are essential to the clock, a change in membrane composition should influence properties of the oscillator. One of the key features of a specific membrane model for circadian rhythmicity (Njus et al., 1974; see Section 5.4.1) is that ion fluxes are held constant at different temperatures by constant membrane fluidity (thus providing a basis for the temperature-compensation of τ characteristic of circadian rhythmicity), whose constancy, in turn, would arise from changes in the fatty acid composition of membrane lipids at different temperatures (Thompson, 1980).

Roeder et al. (1982) have demonstrated circadian oscillations in the fatty acid profile of phospholipid (membrane lipid) fractions in the growing mycelial front of both the bd^+ and *bd* strains of *Neurospora* (see Section 2.3.1). The mole percentages of linoleic (18:2) and linolenic (18:3) acids [but not those of palmitic (16:0), stearic (18:0) or oleic (18:1) acids, or of total lipid content] were found to vary with amplitudes of 5 to 8% in a mutually out-of-phase relationship. Similar oscillations were found in the *csp-1* strain, which did not express the conidiation rhythm under the growth conditions employed. [The content of 18:3 in the *bd* strain of *Neurospora* has been found to be inversely correlated with temperature, whereas the 18:1 content was roughly correlated with temperature. The complete readjustment of the fatty acids required about 6 to 8 h in temperature-shift experiments (Vokt and Brody, 1985)]. Preliminary indications of circadian changes in fatty acid compositions also were obtained for *Gonyaulax* (Adamich et al., 1976) and *Phaseolus* (Gardner and Galston, 1977).

The question, of course, is whether these oscillations in fatty acids play any role in the clock mechanism. One approach to this problem is to change or otherwise perturb fatty acid metabolism and to correlate the resulting changes in lipid composition with alterations, if any, in clock properties. For example, fusaric acid, which interferes with the lipid moiety of membranes, shifts the phase of the circadian leaf movement rhythm in the cotton plant, *Gossypium hirsutum* (Sundararajan et al., 1978), and filipin, which interferes with sterol stabilization of phospholipid membrane layers, inhibits photoperiodic induction of flowering in *Xanthium* (Mitra and Sen, 1976).

More specifically, Brody and Martins (1979) showed that addition of unsaturated fatty acids to the medium of the cel^- mutant *(bd csp cel)* of *Neurospora* (see Section 4.4.3 and Table 4.2), which has a defect in the fatty acid synthetase complex and thus a partial requirement for exogenously supplied fatty acids, caused a striking increase in the period (τ = 21.5 h) of the sporeforming rhythm (Fig. 4.29). Linoleic acid (18:2) was most effective (yielding values of τ as long as 40 h), followed bylinolenic acid (18:3) and oleic acid (18:1), which lengthened τ to 33 and 26 h, respectively. The period-lengthening effect of 18:2 was proportional to its concentration up to 0.13 m*M* and also was reversed by the addition of palmitic acid (16:0) to the medium (Fig. 4.29), although neither this saturated acid nor stearic acid (18:0) altered τ when added to the medium alone. The 40-h hyphal-branching rhythm in cultures fully supplemented with 18:2 was entrainable by LD: 12,12 to a 24-h period. Finally, supplementation had no effect on prototrophic control strains *(bd csp cel$^+$)*.

Similarly, shorter-chain (8 to 13 carbons), saturated fatty-acid supplements also could be used to manipulate, or titrate, the period of the conidiation rhythm in the *cel* mutant (but not in the wild type) in much the same manner that unsaturated fatty acids were used to lengthen τ (Mattern and Brody, 1979). Nonanoic acid (9:0) gave the maximum effect at a concentration of 5×10^{-4} *M*, increasing τ to 33 h. As was found for unsaturated fatty acids, the effect was fully reversible by the addition of excess 16:0 and the lengthened τ could be entrained to a 24-h *T* by LD:16,8. The different responses of *cel$^+$* and *cel$^-$* to the supplements could not be attributed to differential use or incorporation by the two strains. Their incorporation of [^{14}C]lauric acid (12:0) was approximately the same. Indeed, only 2 to 5% of the label appeared in the membrane lipid fraction, most

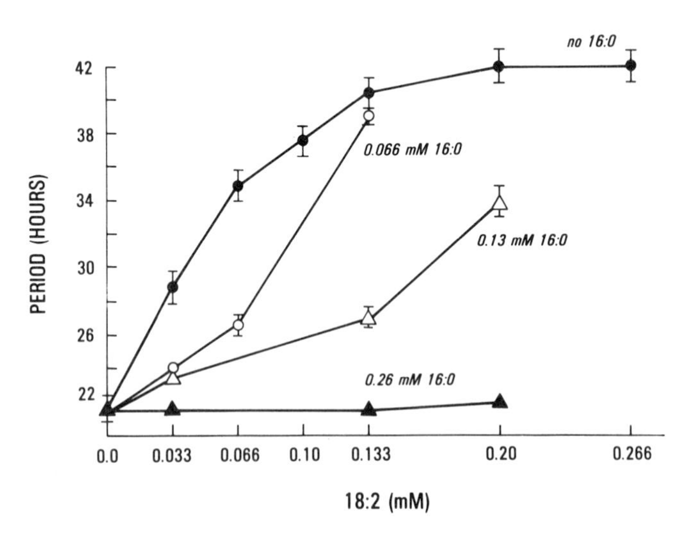

FIGURE 4.29. Periodicity of *bd csp cel* cultures of *Neurospora crassa* as a function of 18:2 and 16:0 concentrations. (Reprinted from Brody and Martins, 1979, with permission of American Society for Microbiology.)

of the rest being β-oxidized in the mitochondria and liberated as CO_2. Normal-length saturated fatty acids (>13 carbons) did not perturb τ and afforded protection against the effects of unsaturated and short-chain supplements. From these results, therefore, it would appear that the *cel*⁻ mutation renders the *Neurospora* clock sensitive to the accumulation or metabolism of exogenous, short-chain fatty acids, which in the wild type are incorporated, activated to fatty acyl coenzyme As, and transferred to the mitochondria to the same extent without affecting its period (Mattern and Brody, 1979). An explanation of the results based on a simple fatty-acid deficiency in the mutant is not sufficient, even though this hypothesis was attractive in view of the fact that the fatty acid synthetase complex of *cel* is known to be deficient in 4′-phosphopantetheine prosthetic groups (see Mattern et al., 1982).

The *cel*⁻ mutation also resulted in the loss of temperature compensation of the conidiation rhythm in cultures grown below 22°C (see Section 4.4.3 and Table 4.2); over the range 14 to 22°C, a Q_{10} of 2.2 and periods of about 40 h were observed (Mattern et al., 1982). Above 22°C, however, *cel*⁻ exhibited temperature compensation (Q_{10} = 1.1), with τ values (18 h) approximating those of the wild type. Thus, the mutant displayed a threshold temperature, or breakpoint, for the temperature compensation of its rhythm. Supplementation of *cel*⁻ strains with unsaturated (18:1, 18:2, 18:3) or short-chain (12:0) fatty acids raised this threshold to 26°C, causing the rhythm to have an even greater dependence on temperature, whereas supplementation with long-chain (16:0, 18:0) unsaturated fatty acids lowered it to 18°C, restoring compensation to that characteristic of the wild type.

All of the foregoing data, then, suggest a role for fatty acids, as lipid components or as cellular metabolites, in determining both period length and the degree of temperature compensation of the overt rhythm of conidiation ion *Neurospora*. But are they part of the clock mechanism itself? If they are not incorporated significantly into *cel*⁻ membranes, how would they impose an altered fluidity upon them, as called for by Njus et al.'s membrane model (1974) for circadian rhythms (see Section 5.4.1)? Perhaps mitochondrial metabolism of the supplements, a localized membrane change that might affect the activity of the fatty acid ATP synthetase complex, might account for the clock alterations (Dieckmann and Brody, 1980; Feldman and Dunlap, 1983; Section 4.4.3).

To clarify the mechanism by which fatty acids exert their effect on the *Neurospora* oscillator, Mattern (1985a) examined the effects of unsaturated fatty acid isomers on clock properties of the *cel*⁻ mutant. If 18:1, 18:2 and 18:3—each of which has a *cis* double bond at the Δ^9 position—lengthen τ of *bd csp cel* by direct perturbation of its membrane fluidity, other fatty acids having a similar structure should exert a similar effect. This was not the case. Positional isomers of 18:1 (having exactly the same number of carbons and *cis* double bonds), such as petroselinic (Δ^6) and vaccenic

(Δ^{11}) acids, and longer-chain isomers, such as eicosenoic (Δ^{11}-20:1) and erucic (Δ^{13}-22:1) acids, did not lengthen τ of the conidiation rhythm. In contrast, τ was lengthened dramatically with γ-18:3, which has a double bond at the Δ^6 position, just like the inactive Δ^6-18:1. These data, therefore, do not support a direct physical action (such as an increase in membrane disorder by incorporation of fluidizing fatty acids supplements) as the source of the τ-lengthening effect, although indirect changes in membrane properties could not be excluded. Rather, the presence of unsaturation at the Δ^9 position may be a critical structural feature for chronobiotic activity (Mattern, 1985a).

In a further attempt to distinguish between enzymatic and membranar modes of action of fatty acids on the *Neurospora* clock, Mattern (1985b) has begun an investigation of the effects of functional group fatty-acid analogs of 18:1, 18:2, and 18:3 that mimic some, but not all, of the properties of unsaturated and short-chain saturated fatty acids used in the earlier work to lengthen τ and to alter the degree of temperature compensation. These knobbed analogs, prepared by solvomercuration of oleic acid esters, have an alkyl group introduced in the center of the fatty acyl chain [corresponding to that of the alcohol solvent: for example, methoxy-18:0 (methoxystearic acid)], which should fluidize membranes upon their incorporation much as do the kinks of *cis*-unsaturations or the misalignments of occasional short chains. On the other hand, their enzymatic metabolism would be expected to be greatly diminished. Thus, if incorporation of these knobbed analogs into the *cel*$^-$ mutant is accompanied by an alteration of clock properties, membrane involvement in the clock mechanism would be favored. In fact, of the analogs tested, only the monoepoxy, monomethoxy, dibromo, and hexabromo stearic acids significantly lengthened τ. Other analogs, which should have had comparable abilities to disrupt lipid bilayer packing, were without effect, even though some were incorporated to a greater extent (for example, the dimethoxystearic and trimethoxystearic analogs) than the active ones (such as monomethoxystearic acid). Once again, the results did not support a direct alteration in membrane fluidity as the mode of action of the τ-lengthening fatty acids.

Recently, however, Mattern (personal communication, 1985) has found that alkyne bonds are equally potent as alkene bonds in perturbing rhythmicity. Thus, 9-octadecynoic acid, the triple-bond analog of 18:1, caused lengthening of τ of the conidiation rhythm in a manner similar to that demonstrated (Brody and Martins, 1979) for normally double-bonded, unsaturated fatty acids. This alkyne fatty acid accumulated preferentially in *cel*$^-$ but not in the wild type and is one of the best candidates thus far for exerting a direct, fluidizing effect on membranes of *cel*$^-$.

In addition to citing evidence that supports membrane involvement in the generation of circadian rhythmicity (Engelmann and Schrempf, 1980), it is important to detail results of clock perturbation studies that are not consistent with an essential role of membranes in the oscillator mechanism.

Nakashima (1983) has further examined the question of whether the fatty acid composition of phospholipids is important for the function of the *Neurospora* clock, as suggested by the earlier findings of Brody and Martins (1979) and Roeder et al. (1982). The effects of 2-phenylethanol (PE), which has been reported to inhibit the synthesis of phospholipids that contain saturated fatty acids in *E. coli* and *Tetrahymena* and to thereby alter the lipid composition of their membranes, and of the related compounds phenylmethanol (PM) and 3-phenyl-1-propanol (PP) on the conidiation rhythm and on the fatty acid profile of the phospholipids in the mycelial mat were determined. The period of the rhythm was shortened in proportion to the concentration of PE (> 1 m*M*) used. PM (10 m*M*) and PP (3 m*M*), however, did not alter τ despite the fact that all three compounds greatly reduced the mycelial growth rate as a function of their increasing concentration. Furthermore, PE (6 m*M*) did not affect temperature compensation of τ or the amplitude of the PRC derived for 2-min light pulses, although the crossover point occurred slightly earlier. In contrast, the ratio of linolenic (18:3) to linoleic (18:2) acid decreased in proportion to the concentrations of PE, PM, and PP in all of the phospholipids examined, as did the proportion of phosphatidic acid + diphosphatidylglycerol to total phospholipids. Period-shortening by PE, therefore, cannot be explained by the change in phospholipid fatty acid composition. Other species that were not perturbed by the compounds, such as the ratio of unsaturated to saturated fatty acids or the proportion of other phospholipids, could play some role in clock function (Nakashima, 1983).

Feldman et al. (1986) have examined the fatty acid composition of the clock mutant *frq-9,* which exhibits a complete loss of temperature compensation—τ of the conidiation rhythm ranges from 34 h at 18°C to 16 h at 30°C (see Section 4.4.1 and Table 4.1). The fatty acid profiles of this mutant, assayed at 18°C, 25°C, and 30°C, did not differ significantly from those of the wild-type strain at any of these temperatures. These results, taken together with those of Nakashima (1985) just described, do not support the direct involvement of fatty acid composition in the clock (Mattern et al., 1982) (see Section 4.4.3 for *cel*$^-$ and *oli*^{r} mutants).

Scholübbers et al. (1984) have examined membrane properties of whole cells of *G. polyedra* by measuring fluorescence polarization (P) with the probe dye molecule 1,6-diphenyl-1,3,5-hexatriene. The P-values obtained by this technique are assumed to be roughly proportional to the microviscosity of the membrane, which, in turn, is a function of the protein-to-lipid ratio, the degree of fatty acid saturation and chain length, and the ratios of cholesterol to phospholipid and lecithin to sphingomyelin—that is, to the gross state of all membranes and lipids present. Circadian changes of P were found in both exponentially increasing and stationary (infradian) cultures of *Gonyaulax*. The fact that the amplitude was larger in dividing cells indicated an interaction with the CDC itself, although this coupling was not obligatory, since rhythmicity also occurred with an unaltered τ and amplitude (in comparison to stationary cells) even in cultures that had been treated with colchicine (380 or 700 μM) to block possible division in out-of-phase subpopulations of cycling and dying cells. Considerable differences (Q_{10} = 2.5 to 3.0) in P were found in cultures kept at different temperatures ranging from 15 to 27.5°C, although τ of the glow rhythm was temperature compensated (Q_{10} = 0.9 to 1.1) over this temperature range. Similarly, values of P depended on the density and growth rate, whereas τ of the glow rhythm did not vary. Temperature pulses (4 h) of different magnitude (−9, −5, +7°C) given at different CTs yielded maximal advances of 6, 4, and 2 h, respectively. The PRCs derived from the data were identical in phase, a finding not anticipated on the hypothesis that pulses perturb the clock by perturbing membrane properties. Finally, 4-h pulses of membrane-active anaesthetics (lidocaine, procaine, halothane, chlorpromazine) that changed P had little or no phase-shifting effect on the glow rhythm (in contrast to the $+\Delta\phi$s reported earlier by Balzer and Hardeland, 1981). The fact that no correlation was found in these experiments between either permanent or transitory changes of membrane properties and τ or the temperature-induced phase-shifting behavior, respectively, of the glow rhythm, therefore, does not support a role for membrane changes in the *Gonyaulax* clock, although the results do not exclude involvement of other membrane functions, such as the activity of membrane-bound enzymes and membrane potentials (Scholübbers et al., 1984).

Ions

The fact that single cells can simultaneously exhibit circadian rhythmicity in quite different processes, such as those of photosynthesis and bioluminescence in *Gonyaulax* (see Section 2.2.4), suggests that membrane-bound compartmentalization is important for temporal organization in this representative cell type. Since both of these processes, as well as others, are known to be affected by changes in the ionic environment and are probably membrane-bound systems, it is not at all surprising that ion transport across membranes has been proposed to be one component of a limit cycle, circadian oscillator (Njus et al., 1974; Sweeney, 1974b) (see Section 5.4).

Potassium and Other Monovalent Ions

To test the general assumption of a central role for ion transport in the clock mechanism, a number of membrane-active probes have been assayed for their effect on period or phase in the *Gonyaulax* system (Sweeney and Herz, 1977; Sweeney, 1981), although not without seemingly inconsistent results. [Circumstantial evidence, comprising documented circadian rhythms in intracellular K^+ content and membrane ultrastructure (see Table 2.5), had already favored such an involvement of membranes in this microorganism.] Thus, 4-h to 6-h pulses of the ionophore valinomycin (which permits passage of K^+), dissolved in low concentrations (0.1 μg/l) in ethanol and given at different CTs, caused small but reproducible phase shifts in the rhythm of stimulated luminescence (Sweeney, 1974a). The PRC thus derived (Fig. 4.28B) was 180° out-of-phase with that for 4-h pulses of 0.1% ethanol alone (Fig. 4.28A) and with that for light signals. The valinomycin-induced $\Delta\phi$s could be increased at some CTs by an increase in the concentration of K^+ in the medium. It is interesting to note that this PRC was not in phase with that (Fig. 4.30) derived for the phase-shifting effects of 5-h pulses of valinomycin (in ethanol) administered via the transpiration stream on the leaf movement rhythm of *Phaseolus multiflorus* (Bünning and Moser, 1972).

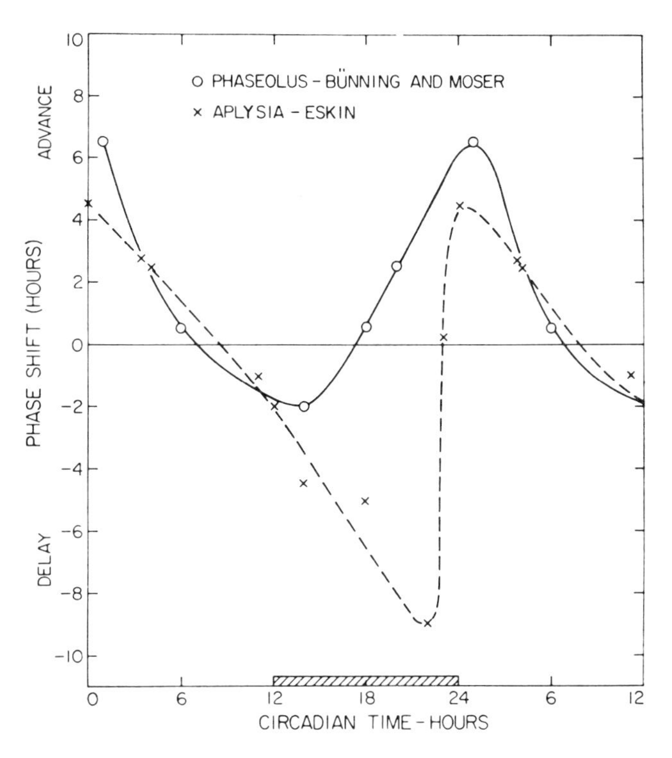

FIGURE 4.30. Phase-response curve for the effect of valinomycin (in 0.1% ethanol) on the phase of the leaf movement rhythm in *Phaseolus multiflorus* calculated with reference to the ethanol control (open circles) (see Bünning and Moser, 1972). Phase-response curve for the effects of high potassium-ion pulses on the phase of the rhythm in firing rate of the optic nerve of isolated *Aplysia californica* eyes (crosses) (see Eskin, 1972). The data have been replotted so that circadian time 0 corresponds to dawn in each case and so that phase advances are plotted upward. Subjective night (the time of the dark period in the previous entraining light-dark cycle) is shown as a shaded bar on the abscissa. (Reprinted from Sweeney, 1974a, with permission of American Society of Plant Physiologists.)

Contrary to what one might have anticipated on the membrane hypothesis, however, other ionophores, such as gramicidin and A23187 (an ionophore for divalent cations), had no effect on the phase of the rhythm of stimulated bioluminescence in *Gonyaulax* (Sweeney and Herz, 1977), but note below the positive results obtained in *Neurospora* and *Aplysia*. The potent respiratory inhibitor and uncoupler of membrane ion gradients, CCCP, produced rather small phase shifts and then only if applied during the early subjective day (Sweeney, 1976b; Taylor and Hastings, 1979), although it did produce large phase shifts during the subjective day in *Neurospora* (Fig. 4.15a). Solvents, such as acetone and dimethyl sulfoxide (DMSO), which might be expected to attack membranes, failed to shift the phase of the stimulated bioluminescence rhythm (Sweeney and Herz, 1977; Sweeney, 1981). Finally, the fact that alcohols that penetrate membranes more easily are not as efficient in phase-shifting as those that do not (Sweeney, 1976b) seemed inconsistent with a membrane theory for the generation of circadian oscillations.

In *Neurospora*, Sato et al. (1985) have reported circadian rhythms of both K^+ and Na^+ content in the growing front of the mycelium. In this mold, the polyene antibiotic nystatin interacts with membrane steroids, particularly ergosterol of the plasma membrane, causing loss of K^+ and inhibition of growth. Pulses (3 h) of nystatin yielded large phase delays (16 h) in the late subjective night and smaller advances (3 h) in the early subjective day. These $\Delta\phi$s were attenuated in the ergosterol-deficient mutants *erg-1* and *erg-3* (see Table 4.2), a result suggesting that nystatin's phase-shifting action is mediated through plasma membrane steroids and, perhaps, by K^+ ion movements (N. Koyama, unpublished results; cited in Feldman and Dunlap, 1983). Pulses of the K^+ ionophore valinomycin, which also might be hypothesized to cause K^+ leakage and resultant depolarization of the plasma membrane, yielded a PRC similar to that for nystatin (N. Koyama). As Feldman and Dunlap (1983) caution, however, valinomycin is known to affect mitochondrial function and protein synthesis also, so the interpretation that ion movements are an essential part of the clock is ambiguous (note similarity in phasing of the PRCs for several respiration inhibitors and CHX to that for nystatin, although magnitudes vary; see Figs. 4.15 and 4.33). Other ionophores, such as gramicidin, nonactin, nigericin, and monesin, influence circadian rhythms in fungi also, inducing a growth rhythm in the mycelium of *Podospora anserina* that becomes entrainable by LD cycles (Lysek and von Witsch, 1974a,b).

We have already noted that potassium flux and other ionic movements are associated with the circadian rhythm of leaf movement in many leguminous plants, such as *Phaseolus, Samanea, Albizza, Mimosa,* and *Trifolium* (see Section 4.1.1) and that their isolated pulvini are self-contained, light-entrainable systems for the generation of these turgor-based rhythmicities (reviewed by Satter and Galston, 1981; Gorton and Satter, 1983). Since total K^+ in a pulvinus of *Albizza* or *Samanea* remains constant during leaflet movement, different regions of their pulvinar microstructure have been examined using electron microprobe analysis of cryostat-sectioned, lyophilized tissue sections and other techniques. These earlier studies indicated that leaf movements are a consequence of a massive, diurnal shuttling of K^+ and Cl^- ions and other solutes into and out of extensor and flexor cells on opposing sides of the pulvinus and that this rhythmic redistribution of ionic species continues during long dark intervals (Satter et al., 1970, 1974a, 1977). More recently, using a scanning electron microscope equipped with an energy-dispersive x-ray analyzer, Campbell et al. (1981) have demonstrated that the apoplast (cell walls and intercellular spaces) is an important pathway for pulvinar ion redistribution wherein solutes must cross the plasmalemma of two or more cells. It was proposed that K^+ in the extensor would cross the plasma membrane into the apoplast and subsequently diffuse to the flexor and be taken up across the plasma membrane into flexor cells. The plasma membrane, therefore, would play a crucial role in controlling K^+ flux. Satter et al. (1982) have found that hydrophobic barriers to apoplastic ion diffusion exist between extensor and flexor, implying that symplastic transport through the cytoplasmic continuum (cytoplasm and intercellular cytoplasmic bridges, or plasmodesmata) also is necessary for ion redistribution.

Active extrusion of protons by an electrogenic proton pump has been postulated to energize K^+ accumulation in *Samanea* (Satter and Galston, 1981). Indeed, the activity of a proton pump has been correlated with K^+ flux and turgor changes. Uptake of K^+ into flexor cells of intact pulvini was associated with net H^+ release from excised flexor tissue into the medium, whereas the loss of K^+ from the extensor side was accompanied by net H^+ uptake into excised extensor tissue (Iglesias and Satter, 1983a). Likewise, both the sensitivity of *Samanea* leaflets to dark-promoted closure as well as the magnitude and direction of dark-promoted H^+ fluxes varied rhythmically in LL (Iglesias and Satter, 1983b). The mechanism whereby these rhythmic changes in dark-promoted proton extrusion are generated is not known, nor is it clear whether cytoplasmic pH varies as a consequence. If so, a plethora of pH-sensitive cellular processes might be influenced or even driven, including many overt circadian rhythmicities [although this does not seem to be the case for *Neurospora*, where no correlation was found between intracellular pH and circadian periodicity (Johnson, 1983)]. The H^+ extrusion data for *Samanea* is particularly interesting given the feedback model of Chay (1981) for biological oscillations, in which changes in H^+ transport, cytoplasmic pH, and pH-sensitive H^+-producing or H^+-transporting enzymes play an essential role (see Table 5.1). Ions other than protons, such as Ca^{2+}, which plays a key role in cell function and affects transport-related processes like membrane phosphorylation, Ca^{2+} -ATPase activity, and the opening of K^+ diffusion pathways (Cheung, 1980, 1981) also may be involved in the pulvinar pacemaker and be environmentally regulated by phytochrome.

The ion fluxes underlying circadian leaf movements in leguminous plants display rhythmic and light-mediated sensitivity to metabolic inhibitors, low temperature, and chemical agents that interfere with ion transport (Satter and Galston, 1981). For example, the leaf movement rhythm of *P. coccineus* can be phase-shifted by 5-h pulses of valinomycin (in ethanol) administered via the transpiration stream (Bünning and Moser, 1972). The PRC for such perturbations (Fig. 4.30) was similar to those light signals and for 2-h pulses of ethanol and 4-h to 5-h pulses of high concentrations of KCl in this plant, although the K^+ pulses produced only advances (Bünning and Moser, 1973) but was not in phase with the *Gonyaulax* PRC for valinomycin (Fig. 4.28B). Similarly, the continuous or pulsed application of several drugs affecting cAMP metabolism, perhaps by perturbing the concentration of intracellular Ca^{2+}, also altered τ or generated phase shifts of the leaf movement rhythm in both *P. coccineus* (Mayer and Scherer, 1975; Mayer et al., 1975) and *T. repens* (Bollig et al., 1978).

There is also considerable evidence that the circadian periodicity of stomatal opening and closure is attributable to clock-controlled ion movements within the stomatal apparatus that generate rhythmic changes in osmotic potential and turgor pressure in the constituent epidermal guard cells (Raschke, 1975; Snaith and Mansfield, 1985, 1986). The degree of stomatal opening in epidermal strips taken from *Commelina communis* L. was reduced in the presence of potassium iminodiacetate (a nonabsorbable anion) instead of potassium chloride, a finding that suggested that chloride uptake was important for full expression of the rhythm (Snaith and Mansfield, 1986). This conclusion was supported by the fact that the anion channel inhibitor 4,4′-diisothiocyano-2,2′-stilbenedisulfonic acid (DIDS) caused a similar reduction in stomatal aperture during the day phase of the rhythm. Guard cells treated with sodium chloride instead of potassium chloride also displayed reduced stomatal opening. This result suggested that potassium movements might contribute to the rhythm, and studies of ionic fluxes using ^{86}Rb as a tracer showed greater effluxes during the night phase. Thus, the stomatal rhythm may be the result of a varying ability of the guard cells to accumulate or retain both chloride and potassium ions (Snaith and Mansfield, 1986).

Potassium itself produced circadian-phase dependent $\Delta\phi$s of the circadian rhythm in compound action potential (see Section 4.1.1) recorded from the isolated eye of *Aplysia* (Eskin, 1972; Jacklet and Lotshaw, 1981). The PRC (Fig. 4.30) for pulses of high external K^+ (90 m*M*) was similar to that for light signals, which induce sodium-dependent membrane depolarization (Eskin, 1977) and for perturbations with strophanthidin, a membrane depolarizer and an inhibitor of the sodium–potassium pump (Eskin, 1979b), with the breakpoint for each centering at about CT 18 (Fig. 4.27). These common effects suggest that membrane depolarization or associated ion fluxes directly or indirectly perturb the underlying pacemaker variables (Jacklet, 1985). In contrast, pulses of the ionophore

A23187, inducing mitochondrial Ca^{2+} transport, yielded a PRC comprising only delays (at CT 04, in the early subjective day) (Eskin and Corrent, 1977). It is interesting to note that brief treatments with manganese, A23187, dinitrophenol, and sodium cyanide—all inhibitors of oxidative metabolism and mitochondrial function—as well as low temperature, ANISO, PUR, and serotonin, all produced phase delays during the late subjective night and early subjective day, with the breakpoints of their respective PRCs clustering at CT 4 (Jacklet, 1981, 1985; see Fig. 4.27). This would suggest that the agents affect a common clock parameter, perhaps associated with protein synthesis (see Section 4.3.2).

In addition to K^+, other monovalent ions also influence circadian rhythms. Li^+, for example, which is known to compete with Ca^{2+} and Mg^{2+} and to affect membrane ATPases, permeability, receptors, and neurotransmitters, as well as other cellular structures and metabolic events, lengthens τ in a number of circadian systems, ranging from microorganisms, insects, rodents, and higher plants to humans, for whom it is used also as an antidepressant (Engelmann and Schrempf, 1980; Rinnan and Johnsson, 1986). If this ion (7 m*M*) is supplied to the root system of *P. vulgaris* L., τ of the ultradian rhythm (100 min at 25°C) of circumnutation is increased; the effect of this treatment is reversed by K^+ ions (Millet et al., 1984). The presence of Li^{2+} increased τ of the circadian rhythm of K^+ uptake in *Lemna gibba* G3 (Kondo, 1984b), an effect that was negated by the presence of Na^+ (but not by Ca^{2+}, Mg^{2+}, or Rb^+). The period of the K^+ uptake rhythm was lengthened also by the substitution of Rb^+ for K^+ in the medium, although this effect was dependent on temperature and light—in one case (20°C, bright light) τ was shorter than that of the controls in K^+ (Kondo, 1984a). The degree of temperature compensation was affected also by this exchange of Rb^+ for K^+; the Q_{10} increased to 1.13 for the temperature range 20 to 30°C, a value that is fairly high for a circadian rhythm (Kondo, 1984a). Since the total ionic concentration was kept constant in these experiments with *Lemna*, it might be argued that the observed Rb^+ effect was due to a lowered concentration of K^+.

Finally, Rinnan and Johnsson (1986) have compared the effects of different univalent alkali ions (Li^+, Rb^+, Cs^+, Na^+, and K^+) on the circadian leaf movement rhythm in isolated stalks and leaflets of *Oxalis regnellii*. When continually added to the nutrient medium, Li^+ (as expected) and, less effectively, Cs^+ (up to 40 m*M*) always lengthened τ, Rb^+ either shortened τ at lower concentrations (2.5 or 5.0 m*M*) or lengthened it at higher concentrations, and Na^+ and K^+ (up to 40 m*M*) had no effect. Pulses of Rb^+ (4-h, 50 m*M*) given at different CTs generated mostly advances (maximum about 1 h). Thus, Rb^+ and K^+ behave differently and cannot be assumed to be interchangeable in their action on circadian systems. The effects of Rb^+ were interpreted on Burgoyne's model (1978) for circadian rhythmicity, in which the rate of synthesis of a key membrane protein is feedback-controlled by the intracellular concentration of monovalent ions (see Section 5.4.3): Rb^+ ions would stimulate a K^+-ATPase membrane protein (or a modulator of its activity), which in turn would caused increased K^+ uptake and changes in turgor pressure coupled to leaf movements. The increased K^+ concentration then would switch off synthesis of the transport protein more rapidly than in the control, leading to a shorter period of the rhythmic pumping at low concentrations of Rb^+ ions, whereas higher concentrations would delay the time of resynthesis of transport protein and lengthen the period (Rinnan and Johnsson, 1986).

Calcium

Since calcium acts as a second messenger, coordinating many kinds of intracellular reactions (Cheung, 1981), it is possible that it might also be involved in the mechanism of circadian clocks. We have already noted that although the calcium ionophore A23187 had no effect on the phase of the stimulated luminescence rhythm in *Gonyaulax* (Sweeney, 1976b; Sweeney and Herz, 1977), it was quite effective in phase-shifting the nerve cell activity rhythm in *Aplysia* (Eskin and Corrent, 1977), yielding an all-delay PRC in the early subjective day (Fig. 4.27). Woolum and Strumwasser (1983) have demonstrated that an increase in the intracellular level of Ca^{2+} caused by the presence of different agents (such as lithium, manganese, and lanthanium) increased the period in a dose-dependent manner not only of the circadian pacemaker in the *Aplysia* eye but also of the high-frequency neural pacemaker ($\tau \leqslant 1$ min) of this gastropod. They interpreted these results as meaning that the frequency of the circadian oscillator (normal conserved) is not directly homeostatically regulated against changes in Ca^{2+} but rather depends on the constancy of the intracellular environment.

Light signals impinging on the *Bulla* eye preparation (see Section 4.2.2) cause membrane depolarization, which can be mimicked by direct electrical stimulation of a single basal retinal neuron. Depolarization may be causing changes in the ionic flux of Ca^{2+}, allowing for its entry, perhaps by opening and closing Ca^{2+} channels. If the extracellular Ca^{2+} concentration was reduced from 10 m*M* to 10^{-4} m*M* with the chelator EGTA, light-induced phase shifts of the *Bulla* ocular rhythm were blocked (McMahon and Block, 1986). In contrast, a 2-h pulse (CT 14–16) of the convulsant agent pentylenetetrazole (50 m*M*), which is known to directly modulate Ca^{2+} levels by releasing Ca^{2+} from intracellular stores, generated phase delays (-1.0 h) in the *Bulla* eye comparable to those produced by light pulses (Khalsa and Block, 1986). These results, therefore, indicate that $\Delta\phi$s can be generated despite reduced extracellular Ca^{2+}. Mediation by a calcium-dependent mechanism has been implicated (Earnest and Sladek, 1986) in the circadian rhythm of vasopressin release in response to membrane polarization that has been reported for perfused rat SCN explants in vitro (Earnest and Sladek, 1985; Reppert and Gillette, 1985), and the concentration of calcium ion may be important (Shibata et al., 1986) for maintaining and regulating the rhythm of metabolic activity (labeled 2-deoxyglucose uptake) that has been reported for rat hypothalamic-slice preparations (Newman and Hospod, 1986).

Phase-shifting by transitory perturbations (increases) of intracellular Ca^{2+} has been reported also in *Trifolium* (Bollig et al., 1978) and in *Chlamydomonas* (Goodenough and Bruce, 1980). Likewise, Nakashima (1984a) has found that 3-h pulses of A23187 (1 μM) generate $\Delta\phi$s of the conidiation rhythm of *Neurospora* in a Ca^{2+}-free medium. The crossover point between delays and advances of the PRC for this agent was at CT 12 (Fig. 4.31a). A $-$gDϕ of 10 h was obtained at CT 10 and a $+\Delta\phi$ at CT 14. Phase-shifting by A23187 was inhibited completely by the addition of $CaCl_2$ (0.1mM) to the medium but not by $MgCl_2$ at any concentration examined. The directionality of the phase-shifting could be regulated by the concentration of external calcium near CT 10 (but not at other CTs) in the presence of the respiratory inhibitor antimycin at concentrations that inhibited respiration by 90% and lowered ATP content by 85% (see Section 4.3.1). Thus, calcium at concentrations lower than 0.3 mM resulted in $-\Delta\phi$s, but resulted in $+\Delta\phi$s at concentrations higher than 10 mM. These results indicated, therefore, that Ca^{2+} is important in clock function at CT 12, perhaps involving mitochondrial Ca^{2+} transport.

Calcium can bind with calcium-binding proteins, such as calmodulin, which in conjunction with mitochondrial calcium transport, plays a pivotal

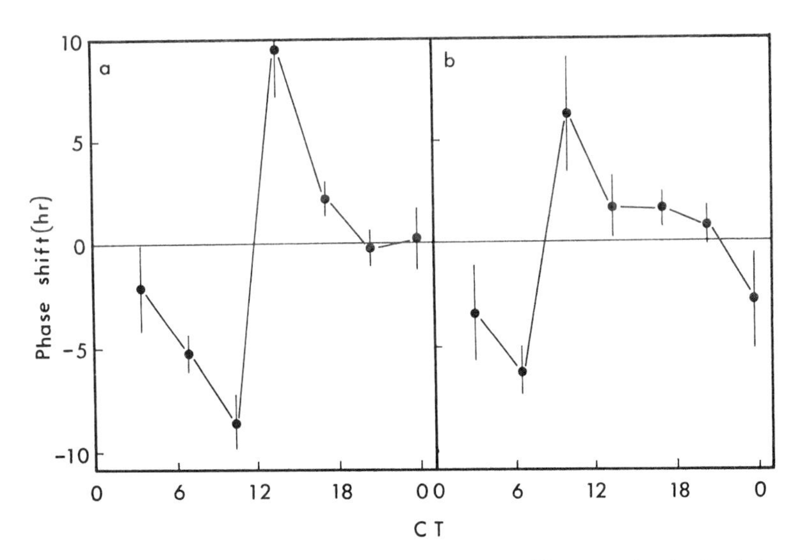

FIGURE 4.31. Phase-response curve for A23187 and trifluoperazine in the *bd* strain of *Neurospora crassa*. Mycelial discs were treated with A23187 (1 μM) in the absence of calcium (a) or with trifluoperazine (30 μM) (b) for 3 h and transferred in race tubes. (Reprinted from from Nakashima, 1985, with permission of Hokkaido University Press.)

role in cellular regulation (Cheung, 1980). Nakashima (1985, 1986) has examined the effects of four calmodulin antagonists, TFP, CPZ, imipramine (IMP), and alprenolol (ALP). As shown in Figure 4.31b, 3-h pulses of TFP (30 μM) produced CT-dependent $\Delta\phi$s. Maximum $-\Delta\phi$ occurred at CT 6, maximum $+\Delta\phi$ at CT 9, and the crossover point at about CT 8. CPZ was less effective, generating $\Delta\phi$s of the same magnitude as TFP at a concentration of 300 μM, and IMP and ALP had little effect. This order of effectiveness in phase-shifting activity was qualitatively paralleled by the efficacy of the drugs in inhibiting calmodulin-induced activation of phosphodiesterase, a finding that suggests that calmodulin antagonists cause $\Delta\phi$s in *Neurospora* by affecting calmodulin-dependent reactions (Nakashima, 1985). This conclusion was buttressed by the observation that 20-mM pulses of W7 or W13, both specific calmodulin antagonists, resulted in large $\Delta\phi$s, whereas their ineffective, respective chlorinated analogs, W5 and W12, did not. However, the amount of calmodulin itself extracted at different phases in DD and assayed by its stimulation of phosphodiesterase activity, showed no circadian variation, a fact that suggested that the cyclic changes in sensitivity to calmodulin antagonists might be due to variation in the synthesis and levels of calmodulin-binding proteins. This hypothesis is consistent with the earlier finding of Nakashima et al. (1981a) that in *Neurospora* the most sensitive phase for CHX-inhibition of protein synthesis occurs at about CT 4 (Fig. 4.26), a time that just precedes the phase of maximum sensitivity to TFP (Fig. 4.33, and with the observation that the crossover point of the PRC for the latter is closely followed by the calcium-sensitive phase at CT 12 (Fig. 4.31). Thus, calcium metabolism seems to be important for clock function during the late subjective day, and its accumulation may be catalyzed by calmodulin-binding proteins (Nakashima, 1985).

Nakashima (1985, 1987) has compared the PRCs for low- and high- temperature pulses and for pulses of TFP in both wild type *(bd) Neurospora* and in the clock mutant *frq-7*, whose τ is longer (28 h at 26°C) and which shows incomplete temperature compensation (Table 4.1). On a real-time basis, the phase affected (i.e., extended) by the *frq-7* mutation was found to lie between the TFP-sensitive and low temperature-sensitive phases, a finding that suggests that each phase of the *Neurospora* clock may be controlled by a different genetic locus (see Section 4.4) and that the product of the *frq* gene may be related to calcium metabolism (Nakashima, 1985). Further support for this idea comes from the recent findings (Nakashima, 1985, personal communication) that in mutant strains of *Neuropsora* that have a Ca^{2+} dependency for growth (Table 4.2), τ is affected also by the concentration of external Ca^{2+} *(Ca4)* and even entrainability and rhythmicity itself *(Ca23)*.

That Ca^{2+}, mitochondrial Ca^{2+} transport, and calmodulin may play a key role in the clock mechanism also has been postulated by Goto et al. (1985), based on the results of a biochemical analysis of the circadian oscillator underlying the overt rhythm of cell division in *Euglena* (see

Section 3.2.1) and the circadian variations in the intracellular contents of NAD^+ and NADP(H) that have been discovered in this unicell (see Fig. 5.14). Inasmuch as Ca^{2+}-calmodulin activates NAD^+ kinase in many plants, including green algae, and in the sea urchin, and because the circadian rhythms in the activities of NAD^+ kinase and $NADP^+$ phosphatase appear to be generated by a rhythm in the in vivo level of this complex in *Lemna* (Goto, 1984), it would seem to be a likely candidate for a regulatory clock element (gear) in *Euglena* also.

To test this possibility, attempts were made to perturb the clock either by directly causing a transitory decrease in $[Ca^{2+}]$ with 2-h to 3-h pulses of chlortetracycline (200 μM), a membrane-permeable chelator of Ca^{2+}, or by transitorily inhibiting Ca^{2+}-calmodulin by similar short pulses of W7 (20 μM) and chlorpromazine (50 μM), both calmodulin inhibitors (Goto et al., 1985; Edmunds and Laval-Martin, 1986). All three drugs yielded pronounced $\Delta\phi$s of the cell division rhythm. The PRCs obtained (see Fig. 5.15B) suggested that both cytosolic Ca^{2+} and calmodulin constitute clock gears according to the rationale given earlier (see Section 4.3). If so, there should be another gear directly regulating the level of Ca^{2+} in *Euglena*. The main regulatory sites for many noncircadian systems are known to be the plasmalemma, the endoplasmic reticulum, and the mitochondria. To test the possibility that the mitochondrial Ca^{2+}-transport system might be a gear (Goto et al., 1985), cultures were pulsed (2 to 3 h) with either nitrogen (600 ml min^{-1}), dinitrophenol (100 μM), or sodium acetate (10 mM), all of which affect electron transport, ATP hydrolysis, and mitochondrial Ca^{2+}-efflux. Each of these agents phase-shifted the division rhythm (see Fig. 5.15C). The mitochondrial Ca^{2+}-transport system, therefore, would appear to be another gear of the oscillator in *Euglena*.

These findings suggest that NAD^+, the mitochondrial Ca^{2+}-transport system, Ca^{2+}, calmodulin, NAD^+ kinase, and $NADP^+$ phosphatase represent clock gears that might constitute a self-sustained, circadian oscillating loop in *Euglena* and other eukaryotic microorganisms, as well as in higher plants and animals. A model has been proposed (Goto et al., 1985; see Section 5.2.5)—not for the clock but for one oscillator in what is most probably a cellular clock-shop. This regulatory scheme (see Fig. 5.16) includes the following three steps: (1) NAD^+ (or stimulated photoreceptor) would enhance the rate of net Ca^{2+} efflux from the mitochondria or net Ca^{2+} influx across the plasmalemma into the cytoplasm, resulting 6 h (90°) later in a maximal concentration of cytosolic Ca^{2+}, (2) Ca^{2+} would immediately form an activated Ca^{2+}-calmodulin complex in the cytoplasm (maximal level at 90°), (3) this active form of Ca^{2+}-calmodulin would decrease the rate of net production of NAD^+ by both activating NAD^+ kinase and inhibiting $NADP^+$ phosphatase in the cytoplasm so that the rate would become maximal 12 h later (at 270°) when Ca^{2+}-calmodulin reaches its minimum level. After 6 h more when the in vivo level of NAD^+ becomes lowest (at 0°), the regulatory sequence would be closed, and the cycle would repeat. This model shares several aspects in common with the calcium cycle model of Kippert (1987), in which the Ca^{2+} concentration gradient between cytoplasm and mitochondrial matrix alters cyclically, and with that of Lakin-Thomas (1985) based on oscillations in intracellular Ca^{2+} compartmentation (see Section 5.2.5).

Calcium and calmodulin may also play an important role in the regulation of the circadian rhythm of cell shape in *Euglena* (see Section 2.2.3). When cultures were grown in Ca^{2+}-free medium, cells assumed round shapes within 10 min, this effect was reversible by the restoration of Ca^{2+} to the medium (Lonergan, 1984). Cultures grown in low concentrations of Ca^{2+} (10 μM) did not display the typical circadian rhythm in cell shape even though those of photosynthesis and cell division were unaffected. Elevating the intracellular levels of Ca^{2+} by addition of A23187 prevented the cells from undergoing changes in shape, although the shape found at the time of the addition of the ionophore was maintained. In contrast, addition of the calmodulin antagonists TFP and CPZ always caused the cells to round, thereby also blocking the rhythm of daily shape change. Interestingly, pulses of TFP apparently do not elicit $\Delta\phi$s of the cell shape rhythm (Lonergan, personal communication, 1985), nor do 2-h pulses of TFP, CPZ, or W7 shift the phase of the rhythm of O_2 evolution (Lonergan, 1986a). Instead, photosynthesis is uncoupled from the clock, suggesting that calmodulin itself is not part of the clock controlling these rhythms (see previous contrary findings for the effects of CPZ and W7 on division rhythmicity) but must have some other clock-related (coupling?) function. In fact, Lachney and Lonergan (1985) have implicated stable, pellicle-associated microtubules in the modulation of cell shape in *Euglena*, and Lonergan (1985) has localized by immunofluorescence techniques actin, myosin, calmodulin, and tubulin beneath the pellicle (see Section 2.2.3). The data are consistent with an actomyosin contractile system controlled by calmodulin (Lonergan, 1985).

Calcium and calmodulin are both accumulated at the apex of growing *Acetabularia*; their gradients disappear at the time of cap formation (Cotton and Vanden Driessche, 1987). Both morphogenetic processes are dramatically affected by TFP (10^{-7} M to 10^{-5} M), and 6-h pulses (2.5×10^{-6} M) of this calmodulin antagonist either accelerated or delayed growth and cap formation, depending on the CT at which they were applied. Thus, both the polarity and circadian rhythmicity are regulatory components in the morphogenesis of *Acetabularia* (Vanden Driessche and Cotton, 1986; see Section 6.5.2). Similarly, pulses of W7, CPZ, and TFP given at CT 3 and CT 9 generated $-\Delta\phi$ and $+\Delta\phi$, respectively, in rat hypothalamic brain slices (Section 4.1.1; R. Y. Moore, unpublished results, 1987). We note, however, that Khalsa and Block (1987) found that W7 (100 μM), TFP (10 μM), and a third calmodulin antagonist, calmidazolium (10 μM), did not block light-induced phase shifts of the *Bulla* ocular rhythm—results that indicate that it is unlikely that calmodulin itself mediates these $\Delta\phi$s in this system.

In addition to the massive, cyclic redistributions of K^+ and Cl^- and the rhythmic H^+ extrusion discussed previously, Ca^{2+} and calmodulin also may be involved in the pulvinar pacemaker underlying circadian leaf movement in *Samanea* and other leguminous plants. Thus, Roux et al. (1981) have proposed a model for phytochrome action based on Ca^+ fluxes: conversion to P_{fr} would promote the diffusion of Ca^{2+} from the cellular exterior and from the mitochondria into the cytoplasm. This would increase free cytoplasmic Ca^{2+}, in turn activating calmodulin, whose activity could be coupled to several transport-related processes, such as membrane phosphorylation, Ca^{2+}-ATPase activity, and the opening of K^+ diffusion pathways (Cheung, 1980, 1981). Phytochrome, K^+, and the clock interact to regulate the circadian rhythm of leaf movement in *Albizzia* (Satter and Galston, 1971a,b) and *Samanea* (Satter et al., 1974b), and there is some evidence that the perturbation of Ca^{2+} flux can phase-shift the leaf movement rhythm in *P. coccineus* (Mayer and Scherer, 1975; Mayer et al., 1975) and *T. repens* (Bollig et al., 1978). It is also interesting to note that the phosphatidylinositol cycle has been hypothesized to mediate the effects of light on leaflet movement in *Samanea saman* pulvini. Hydrolysis of membrane-localized phosphoinositides, accompanied by an increase in cytosolic free Ca^{2+}, would provide a mechanism for phototransduction in the motor cells (Morse et al., 1987b,c).

There is considerable evidence that oscillations in Ca^{2+} levels may play an important role in the generation of ultradian rhythms. For example, Yada et al. (1986) have investigated the higher-frequency oscillations in membrane potential with repeated hyperpolarizations that occur in cultured, secretory epithelial Intestine 407 (I-407) cells (as well as in fibroblastic L cells, macrophages, sympathetic neurons, and hamster eggs). Periodic activation of Ca^{2+}-dependent K^+ conductance has been shown to be a common mechanism. The hyperpolarizations were inhibited in I-407 cells not only by K^+-channel blockers (tetra- and nonethylammonium) but also by inhibitors of the Ca^{2+}-activated K^+ channel (quinine, quinidine). Cyclic increases in cytosolic free Ca^{2+} concentrations ($> 1 \times 10^{-6}$ *M*), measured with Ca^{2+}-selective microelectrodes, were found to coincide with the cyclic membrane hyperpolarizations. The oscillation in potential, therefore, appeared to be brought about by the oscillation of intracellular free Ca^{2+} level, which induced periodic activation of the Ca^{2+}-dependent K^+ channels. Neither the deprivation of extracellular Ca^{2+} nor the application of Ca^{2+}-channel blockers (Co^{2+}, Ni^{2+}) abolished the oscillation in potential. Mitochondrial inhibitors (KCN, NaN_3, antimycin A, *p*-trifluoromethoxyphenylhydrazone [FCCP], DNP) inhibited the potential oscillation, whereas glycolytic inhibitors (NaF, iodoacetic acid) were without effect, as were caffeine and oxalate, which affect microsomal Ca^{2+} transport. From these findings, Yada et al. (1986) concluded that the cytosolic Ca^{2+} oscillation results from cyclic releases of Ca^{2+} from the intracellular storage site, which, in turn, depends on mitochondrial activity.

Membrane ATPases

If active transport of ions, particularly by the plasma membrane, is an essential part of the clock mechanism in *Neurospora*, as well as in other organisms, inhibitors of plasma membrane ATPase would be expected to shift the phase of the oscillator because ATPases are usually involved in ion transport. Indeed, the plasma membrane ATPase functions as a proton pump in this microorganism, and many substrates are transported into the cell via the electrochemical proton gradient thus established. On this rationale, Nakashima (1982b) examined the phase-shifting effect of 3-h pulses (50 μM) of DES, an inhibitor of plasma membrane ATPase in several plants, including *Neurospora*, which also completely inhibits light-induced phase-shifting of the conidiation rhythm (Nakashima, 1982a) and of the structurally similar compounds dienestrol (DIE), hexestrol (HEX), diethylstilbestroldipropionate (DESP), and dienestroldiacetate (DIEA). DES was the most effective compound, giving a maximal phase advance of about 10 h, followed closely by DIE and HEX, and finally by DESP and DIEA, which generated only a few hours of phase advance. Despite these differences in their DRCs the PRCs derived for DES, HEX, and DIE were almost identical in their phasing (Fig. 4.32), having crossover points between $-\Delta\phi$ and $+\Delta\phi$ centering at about CT 4, which corresponds to those of the PRCs for CHX and respiratory inhibitors (Figs. 4.15, 4.33). The activity of isolated plasma membrane ATPase was inhibited by DES and partially by HEX but not by DIE, DESP, or DIEA (even though these three drugs, especially DIE, generated $\Delta\phi$s). Mycelial O_2 consumption (a measure of mitochondrial function and ATP synthetase activity) was in-

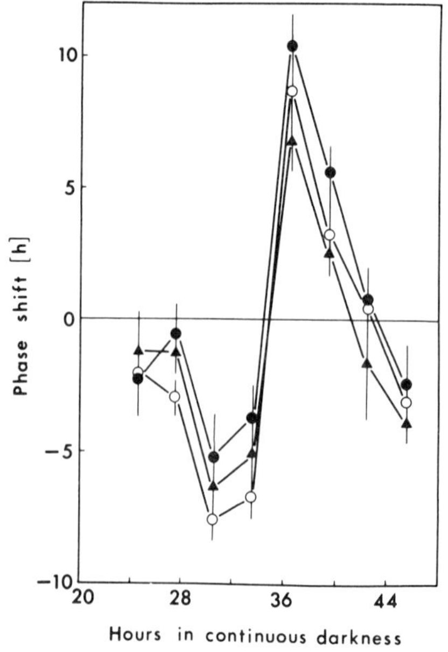

FIGURE 4.32. Phase-response curves for diethylstilbestrol (DES), hexestrol (HEX), and dienestrol (DIE) and the conidiation rhythm in *Neurospora crassa*. Mycelial discs were treated with 50 μM of DES (●), HEX (○), and DIE (▲) for 3 h and then transferred to race tubes after being washed in fresh medium. Control series were treated with the same concentration of ethanol and transferred to race tubes at the same time. Each point is the average phase of six discs ±SD. (Reprinted from Nakashima, 1982b, with permission of American Society of Plant Physiologists.)

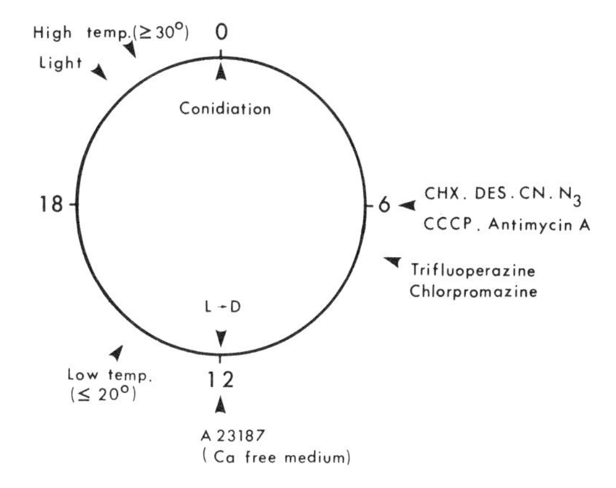

FIGURE 4.33. Time-map of crossover points of phase-response curves for pulses of metabolic inhibitors (cycloheximide, diethylstilbestrol, and antimycin A), light, high temperature, A23187, and trifluoperazine in the *bd* strain of *Neurospora crassa*. Individual crossover points were plotted on the map. CT 12 was the time of transition from light to dark. (Reprinted from Nakashima, 1985, with permission of Hokkaido University Press.)

hibited equally by DES, DIE, and HEX (all of which had different DRCs for phase-shifting activity), whereas DIEA and DESP had little effect.

Thus, because of the lack of correspondence between the the DRCs of DES and the related compounds for plasma membrane ATPase inhibition and their PRCs for phase-shifting the conidiation rhythm, the $\Delta\phi$s generated by the drugs could not have been directly due to the inhibition of plasma membrane ATPase (and, by implication, to the ion fluxes it perturbed). By similar reasoning, because of the dissimilarity between the DRCs for O_2 consumption and the phase-shifting PRCs for the compounds, it seemed unlikely that the DES-induced $\Delta\phi$s were caused indirectly by a lowering of plasma membrane ATPase activity via inhibition of respiration or of O_2 consumption. The rate of O_2 consumption before addition of DIE, in fact, was almost the same as that after its addition at a concentration that resulted in almost maximum phase-shifting of the rhythm of conidiation. Consequently, the phase-shifting action of DES cannot be taken as evidence that plasma membrane ATPase (or mitochondrial ATP synthetase) is a component of the *Neurospora* clock (Nakashima, 1982b).

Mitochondrial membrane ATPase activity apparently is not involved in the mechanism underlying light-induced phase shifts of the conidiation rhythm of *Neurospora* (Nakashima, 1982a). Neither oligomycin nor venturicidin—both mitochondrial ATPase inhibitors—affected phase-shifting by light despite the fact that they were incorporated into the cells (as indicated by growth inhibition). On the other hand, 3-h pulses of azide and oligomycin, as well as of DCCD (which inhibits both mitochondrial and plasma membrane ATPases), themselves generated $\Delta\phi$s and yielded similar PRCs (Nakashima, 1982a; Schulz et al., 1985; see Section 4.3.1).

The observation that oligomycin-resistant *oli^r* mutants of *Neurospora* (Table 4.2) have shortened τs that are correlated with the degree of their resistance suggests some role for mitochondrial function in circadian timekeeping (Dieckmann and Brody, 1980; Brody, 1981; Brody et al., 1985). This role, rather than being a direct one of general energy metabolism, may involve structural changes in the mitochondria and its membrane-bound proteins, ionic gradients across the mitochondrial membrane, HSP synthesis, or phosphate transfer reactions (see Section 4.3.1). In this regard, Brody et al. (1985) have reported provocatively that point mutations in several *oli^r* strains affect the primary structure of the proteolipid, or DCCD-binding protein (yeast subunit 9), found in the F_0 membrane portion of the mitochondrial ATP synthetase complex and thereby lead in some manner to a shortened period.

Cyclic AMP

Adenosine 3′,5′ cyclic monophosphate (cyclic AMP, cAMP) has been implicated in a great variety of regulatory schemes in different organisms, and oscillations in internal Ca^{2+} and cAMP concentration (arising from an instability in their common control loops) have been proposed to form the basis of pacemaker activity and other high-frequency biological rhythms (Rapp and Berridge, 1977). It has been invoked as an essential element in a biochemical model for a circadian clock (Cummings, 1975; see Table 5.1). The enzymes that are responsible for cAMP metabolism, phosphodiesterase (PDE) and adenylate cyclase (AC), are often associated with cellular membranes, and hence a brief discussion of cAMP metabolism and oscillators is germane to the dissection of circadian clocks using membrane-active agents.

Early work with *Neurospora* (Feldman, 1975) indicated that the PDE inhibitors aminophylline, theophylline and caffeine all caused a dose-dependent increase in τ of the conidiatiton rhythm of the wild type *(bd)* and the clock period mutants *frq-1*, *-2* and *-3* (Table 4.1). The drug concentrations used did not inhibit growth and raised the level of intracellular cAMP. Aminophylline was most effective, lengthening τ by up to 5 h in continuous application. Isobutylmethylxanthine (IBMX), a synthetic PDE inhibitor, was somewhat less effective (increasing τ by about 2 h), whereas quinidine, an inhibitor of AC that would be expected to decrease the level of endogenous cAMP, had virtually no effect or only a slight shortening effect on the period (Perlman, 1981; Feldman and Dunlap, 1983). With the liquid culture system developed for treating mycelial discs (Perlman et al., 1981), 3-h pulses of quinidine (10 m*M*), caffeine (10 m*M*), aminophylline (10 m*M*), and IBMX (5 m*M*), in order of decreasing phase-shifting effectiveness, yielded equivalent PRCs, producing delays in the late subjective night, followed by advances of approximately equal magnitude in the early subjective morning. The high efficacy of quinidine perturbations (which produced $+\Delta\phi$ and $-\Delta\phi$ of 8 to 10 h) was somewhat paradoxical given its lack of tonic effect on τ when it was continually administered.

The simplest explanation of these results is that the drugs exerted their effects via their activity on the PDE, AC, and cAMP system: resultant tonic increases in the intracellular level of cAMP (resulting from the inhibition of PDE by the methylxanthine inhibitors) would cause the circadian clock to run more slowly. Tonic decreases (as with quinidine or in *crisp-1*) would be without significant effect, and transient decreases produced by drug pulses would generate advance and delay $\Delta\phi$s of similar magnitude, although the possibility of a common mode of action via the inhibition of protein synthesis or some other process was not excluded (Feldman and Dunlap, 1983). The question always arises, of course, as to whether the perturbations of τ and ϕ were the direct, primary result of the drugs on their presumed targets (in this case cAMP, presumed to be a clock gear) or merely the indirect, secondary (or even tertiary) consequence of the inhibition (Goto et al., 1985; see Section 4.3).

In a genetic approach to this problem (see Section 4.4.3), the role of cAMP levels in the normal operation of the *Neurospora* clock was examined in partial revertants of the morphological mutant *crisp-1* (Table 4.2) that did not exhibit the tight conidial banding pattern (so that the rhythm could be measured). Although these mutants had been shown to have greatly reduced levels of AC and cAMP, they entrained in the usual fashion to LD: 12,12 and free-ran with an unaltered period (Feldman et al., 1979). These results, therefore, appear to demonstrate clearly that the maintenance of significant pools of these two biochemical species at wild-type levels is unessential for clock operation. Nevertheless, recent results obtained with two groups of rhythmically conidiating, orthophosphate-regulated, repressible cyclic-PDE mutants of *Neurospora, cpd-1, cpd-2* (Table 4.2), in which there is also a reduced level of cAMP (brought about by a reduction in AC activity, or an increase in the activity of Mg^{2+}-stimulated cyclic PDE or both) but in which τ is unaffected, have demonstrated a circadian oscillation in cAMP whose peak corresponds to the time of the conidiation band (Hasunuma, 1984a,b; Hasunuma and Shinohara, 1985). These findings, coupled with those showing a rhythmic oscillation of energy charge (Delmer and Brody, 1975; see Fig. 5.4), a shortened period in oligomycin-resistant *(bd oli^r)* strains (Dieckmann and Brody, 1980; Table 4.2), and a lengthened period caused by inhibitors of PDE (Feldman, 1975), all suggest a rhythmic flow of ATP from mitochondria to cAMP in *Neurospora* (Hasunuma, 1984a,b).

Results from higher plant and animal systems are also consistent with the hypothesis that regulation of cAMP levels might be part of the oscillator mechanism (Edmunds, 1980c; Engelmann and Schrempf, 1980). For example, continuous application of the methylxanthine PDE inhibitors theophylline (0.1%) or theobromine (0.1%) increased the period of the rhythm of leaf movement by several hours in *P. coccineus* (Keller, 1960), and pulses (4 h, 10 m*M*) of theophylline (Mayer et al., 1975) or caffeine (Mayer and Scherer, 1975) given at different CTs generated advance and delay $\Delta\phi$s. Similar studies (Bollig et al., 1978) with the leaf movement rhythm of *T. repens*, however, did not support this model: 4-h perturbations with cAMP (1 m*M*) and imidazole (10 m*M*) each generated -$\Delta\phi$s only, even though imidazole activates PDEs in animals and microorganisms and thus would be expected to decrease endogenous cAMP levels. Theophylline produced both advances and delays, whereas pulses of cAMP itself produced only delays, although both presumably elevated the intracellular level of cAMP.

Theophylline can serve as a circadian *Zeitgeber* for the rhythm of locomotor activity in the rat (Ehret et al., 1975). Cyclic AMP appears to be involved in mediating the output of the circadian pacemaker in the avian pineal gland (Takahashi and Zatz, 1982), and the evidence that cAMP mediates the phase-shifting effects of serotonin on the circadian rhythm of compound action potential in the isolated *Aplysia* eye already has been detailed (see Section 4.2.2). In the latter example, treatments that should increase the level of endogenous cAMP mimicked the action of serotonin: pulses of the cAMP analog benzylthio-cAMP (Fig. 4.12B), of the PDE inhibitors papaverine and Ro-20-1724 (Fig. 4.12C,D), and of forskolin (Fig. 4.13), a highly specific activator of AC, all advanced or delayed the rhythm at phases where perturbations by serotinin had similar effects (Eskin et al., 1982; Eskin and Takahashi, 1983). Thus, either cAMP is part of an input pathway by which serotonin entrains the *Aplysia* oscillator, or it (along with the cAMP generating and degrading system) is part of the clock mechanism itself (Lotshaw and Jacklet, 1983). Circadian pacemaker control of membrane potential could occur via regulation of ion transport or of changes in the conductance of specific ions, perhaps by cAMP- or Ca^{2+}-dependent protein phosphorylation (Jacklet, 1984; Lotshaw and Jacklet, 1987). Finally, 1-h pulses of the cAMP analogs 8-benzylamino-cAMP and 8-*p*-chlorophenylthio-cAMP, which are potent stimulators of cAMP-dependent protein kinases (see Section 4.2.2), can reset the phase of the circadian rhythm of electrical activity in rat hypothalamic brain slices. The PRC (comprising both $+\Delta\phi$ and $-\Delta\phi$) for 1-h pulses given at eight different CTs was 12 h out-of-phase with that for light pulses administered in vivo (Prosser and Gillette, 1986).

4.4 Dissection of the Clock: Molecular Genetics

Genetic analysis has proved to be a powerful approach toward the elucidation of biochemical pathways, gene organization, and gene regulation and, more recently, for the dissection of more complex systems, such as bacteriophage assembly, cell division cycles, and behavorial responses. There is no reason a priori, why a similar approach should not be useful in discovering the mechanism(s) that underlie circadian rhythmicity [reviewed for invertebrates by Konopka (1981) and for microorganisms by Feldman (1982, 1983); see Hall and Rosbash (1987)].

Such a genetic approach has comprised two major avenues of attack. (1) Clock mutants can be isolated that manifest alterations in the length of the free-running period, in phase, or in sensitivity to various environmental *Zeitgeber*, such as light and temperature. These mutants then can be analyzed genetically, physiologically, or biochemically and can be used to localize the clock to specific organs or tissues. (2) Biochemical mutants with known metabolic lesions can be isolated, and the effect of such mutations on the functioning of the clock can be determined (Feldman, 1982). We now consider three systems—*Chlamydomonas*, *Neurospora*, and *Drosophila*—in which both clock and biochemical mutants have been isolated and whose analysis has resulted in major progress in the genetic dissection of circadian and ultradian clock mechanisms.

4.4.1 Isolation of Clock Mutants

Chlamydomonas

After to the discovery of the rhythm of phototaxis (see Section 2.2.2), clock mutants with both shorter and longer periods than the wild type have been isolated and characterized genetically (Bruce, 1972). For example, two short-period mutants, designated $w^{-}c$ and 90^{-}, displayed periods of about 21 to 22 h and temperature compensation comparable to the wild type. Similarly, a long-period mutant, $Lo^{-}104$, had a period of approximately 27 h. More recently, four long-period mutants, designated *per 1,-2,-3,-4,* were isolated after nitrosoguanidine mutagenesis and were analyzed in greater detail (Bruce, 1974). These mutants had τs ranging from about 26 to 28 h and unimpaired temperature compensation. Unlike the clock mutants of species of *Neurospora* and *Drosophila* in which the mutations are allelic (Bruce, 1976; Feldman, 1982), the long period characteristic of the *per* mutants seems to be unlinked and to be controlled by several single genes at separate loci. Thus, crosses between single mutants, as well as crosses involving three or four mutant genes, yielded progeny with both parental and recombinant period lengths, including not only normal (wild type) periods, but also extra-long periods (30 to 33, 33 to 36, and 37 to 40 h in the double, triple, and quadruple mutants, respectively).

One of the most provocative findings was that the period-lengthening effect seemed to be additive [although Lakin-Thomas and Brody (1985) have reinterpreted the data as indicative of a multiplicative effect]. If one gene lengthened the period by m h and a second by n h, the period of the double mutant was lengthened by $m + n$ h. Bruce (1974) has interpreted these findings as suggesting that the mutations all affect the same rhythmic system rather than independent, autonomously oscillatory systems. It is interesting to note that this additive effect would be a logical consequence of tape-reading-type models (Ehret and Trucco, 1967; Edmunds and Adams, 1981) for circadian clocks (see Section 5.3), in which mutations affecting τ would involve the addition or deletion of tape segments, although the observed lack of clustering of clock mutations at a single locus would not.

Mergenhagen and Hastings (1977) have selected for a number of strains of *Chlamydomonas* having metabolic deficiencies (e.g., niacin-requiring, thiamine-requiring, or erythromycin-resistant) that show an elongation in period (usually by several hours) or an abnormal expression (e.g., arrhythmic behavior) of the rhythm of phototaxis. Crosses of these mutants with short-period mutants should help clarify whether the observed period deviation is caused by the change in metabolism for which the strain was selected (Mergenhagen, 1980).

Neurospora

A considerable amount of genetic data has been accumulated on the filamentous bread mold, *N. crassa*, which has proved to be an excellent system for studying circadian rhythms (reviewed by Feldman et al., 1979; Feldman, 1982; Feldman and Dunlap, 1983). Its most intensively examined rhythm is that of conidiation (see Section 2.3.1), which can be assayed on agar medium in either race tubes or Petri dishes, in which the culture produces alternating areas of conidia (or bands) and mycelia (see Fig. 2.20). At least 20 clock mutants for the conidiation rhythm have been isolated from *Neurospora* mutagenized with either *N*-methyl-*N'*-nitro-*N*-nitrosoguanidine or ultraviolet irradiation (Table 4.1). These mutants, all bearing (not indicated) the mutation *bd* (band), which expresses a clear pattern of rhythmicity but does not affect the clock itself (Sargent et al., 1966), have either shorter or longer periods than the wild type (where τ = 21.6 h at 25°C), and more than 13 of these have been well characterized both genetically and physiologically (Feldman et al., 1979; Feldman, 1982). Seven mapped to a single locus *(frq)* on the right arm of linkage group VII and exhibited periods ranging from 16.5 to 29.0 h. All of these *frq* mutants showed incomplete dominance when tested against the wild-type *frq*$^+$ allele and displayed otherwise normal growth and development. Provocatively, their periods were not randomly arrayed but appeared to differ (at least at 25°C) from that of the wild type by some integer multiple of 2.5 h, suggesting the existence of a quantal element in the clock mechanism (perhaps the number of copies of the *frq* locus) that can be repeated a variable number of times each cycle (see Section 3.1.2). An eighth mutant *(frq-9)* was recessive to the *frq*$^+$ allele and to the other *frq* mutants (Loros et al., 1986). Five additional mutants, each at a different locus unlinked to *frq* or to each other, have been isolated (Feldman and Atkinson, 1978; Feldman et al., 1979). All showed significantly shorter or longer τs than the wild type (Table 4.1).

In order to determine whether the clock genes act independently or interact with each other, a number of clock double mutants (Table 4.1)

TABLE 4.1. Circadian clock mutants (conidiation rhythm) in the bread mold, *Neurospora crassa*.

Strain	Period (h)			Mutagen[a]	Linkage group	Dominance
Mutants at the *frq* (frequency) locus						
frq-1	16.5			NG	VII R	Incomplete[e]
frq-2	19.3			NG	VII R	Incomplete[e]
frq-3	24.0			NG	VII R	Incomplete[e]
frq-4	19.3			NG	VII R	Incomplete[f]
frq-6	19.2			NG	VII R	Incomplete[g]
frq-7	29.0			NG	VII R	Incomplete[g]
frq-8	29.0			NG	VII R	Incomplete[g]
frq-9	16–34			NG	VII R	Recessive[h]
Mutants at other loci[b]						
chr (chrono)	23.5			NG	VI L	Incomplete[i]
prd-1 (period)	25.8			NG	III C	Recessive[j]
prd-2	25.5			UV	V R	Recessive[i]
prd-3	25.1			UV	I C	Recessive[i]
prd-4	18.0			UV	I R	Incomplete[i]
Strains carrying multiple clock mutations[i,g,h]						
frq-3, prd-1	28.5	(28.2)[c]	[28.7][d]			
frq-3, prd-2	28.1	(27.9)	[28.3]			
frq-3, prd-3	30.6	(27.5)	[27.9]			
frq-1, prd-4	13.8	(13.5)	[13.8]			
frq-2, prd-4	16.1	(16.0)	[16.1]			
frq-1, prd-1	19.3	(20.7)	[19.7]			
frq-2, prd-1	22.8	(23.5)	[23.1]			
frq-7,chr,prd-1	38.5	(35.1)	[37.7]			

For further details, see Feldman (1982) and Feldman and Dunlap (1983).
[a]NG, *N*-methyl-*N*′-nitro-*N*-nitrosoguanidine; UV, ultraviolet irradiation.
[b]*prd-1,-2,-3*, and *-4* were formerly *frq-5*, UV IV-2, UV IV-4, and UV V-7, respectively.
[c]Values in parentheses: theoretical additive period lengths (h).
[d]Bracketed values: multiplicative τ lengths predicted by Lakin-Thomas and Brody (1985).
[e]Feldman and Hoyle (1973)
[f]Feldman and Hoyle (1976)
[g]Gardner and Feldman (1980)
[h]Loros et al. (1986)
[i]Feldman et al. (1979)
[j]Feldman and Atkinson (1978)
[k]Lakin-Thomas and Brody (1985)

were constructed (Feldman et al., 1979; Gardner and Feldman, 1980). In nearly every case, the mutations appeared to be not only cumulative (e.g., two long-period mutants produced a still longer τ than either exhibited individually) but additive as well. If correct, this finding would suggest little or no interaction between the paired genes. More recently, Lakin-Thomas and Brody (1985) have reanalyzed these and other data on the interaction between clock mutations (including those crossed into the biochemical mutants *cel* and *oli*^{r}, both of which showed altered clock properties; see Section 4.4.3), and have considered them to be indicative of a multiplicative effect, in which each mutation would alter the period by a fixed ratio or percentage of the period (rather than by a fixed number of hours), and the result would be the product of the individual effects. In every case (Table 4.1), the multiplicative prediction was at least as close to the observed value as was the additive prediction, and in many cases it was closer (as with *frq-7 chr prd-1* and *frq-1 prd-4*, whose periods were farthest from those of the respective control strains). Lakin-Thomas and Brody (1985) speculate that this nonadditivity of effects might implicate biochemical pathways and reaction rates in clock processes; τ might be viewed as the output (flux) of the oscillator, and the series of alleles could be assumed to represent modifications in one step that alter the magnitude of the output.

An analysis of the PRCs of the *frq* mutants has revealed that their periods have not been altered uniformly thoughout the circadian cycle, but rather the changes were confined to a 7-h portion of the wild-type oscillation corresponding to the early subjective night (Feldman, 1982). This phase was compressed to as little as 2.5 h in *frq-1* and expanded to as long as 14.5 h in *frq-7* and *frq-8*. Thus, it may be possible to construct a temporal map describing the timing and sequence of the expression of various genes during the circadian cycle. Furthermore, in *prd-1*, the amplitude of the PRC was greatly reduced at 25°C but was almost normal at 20°C, whereas wild-type phase-shifting was severely inhibited at 34°C, suggesting that light-induced phase-shifting in *Neurospora* may be itself a temperature-dependent phenomenon (Nakashima and Feldman, 1980; Feldman, 1982).

Other altered temperature responses as well have been noted in clock mutants. In wild-type *N. crassa*, the conidiation rhythm is temperature-compensated below 30°C ($Q_{10} \cong 1$) but above 30°C is less well so ($Q_{10} =$ 1.3 to 1.7) (Sargent et al., 1966; Nakashima and Feldman, 1980). In contrast, whereas the short-period *frq* mutants also are characterized by this pattern of temperature compensation, the longer-period mutants *(frq-3,-7,-8)* are not (Gardner and Feldman, 1981), with the breakpoint temperature's being lowered from 30°C to 25°C *(frq-3)* or even to below 18°C *(frq-7,-8)*, where a Q_{10} of 1.3 was found over the entire 18 to 34°C temperature range. The mutant *frq-9* shows a complete loss of the temperature-compensation feature, with its τ ranging from 34 h at 18°C to 16 h at 30°C (Loros and Feldman, 1986). These differences in the capacity for temperature compensation displayed by the clock mutants could be interpreted as reflecting changes in the quantity of some gene product that would alter the rate of a metabolic reaction vital to clock timing (Gardner and Feldman, 1981).

Drosophila

Circadian Clock Mutants

One of the most intensively investigated and best understood circadian rhythms is that of eclosion (adult emergence from the pupal case) in the fruitflies *D. pseudoobscura* and *D. melanogaster* (reviewed by Pittendrigh, 1960, 1974). This model system was used to develop the formal properties of circadian oscillators (see Fig. 1.2), the mechanism of their entrainment by light cycles (Pittendrigh, 1965), and their role in photoperiodic time measurement (Pittendrigh, 1981; Pittendrigh et al., 1984)—principles that have since been applied to a great variety of other circadian systems.

At least 9 ethylmethane sulfonate-induced or nitrosoguanidine-induced clock mutants on the X-chromosome of *D. melanogaster* have been isolated that map to a single locus designated *per (period)* in the 3B1-2 banding region, one of the most extensively investigated regions of the fruit fly genome (J.C. Hall, personal communication, 1986; Konopka, 1984). Five of these alleles are particularly germane (Fig. 4.34): per^{s} shortens the normal 24-h period of the circadian rhythms of locomotor activity and eclosion free-running in constant infrared light and constant temperature to about 19 h, per^{L1} and per^{L2} lengthen τ to about 29 h, and per^{o1} and per^{o2} completely abolish rhythmicity (Konopka, 1979; Konopka and Benzer, 1971;

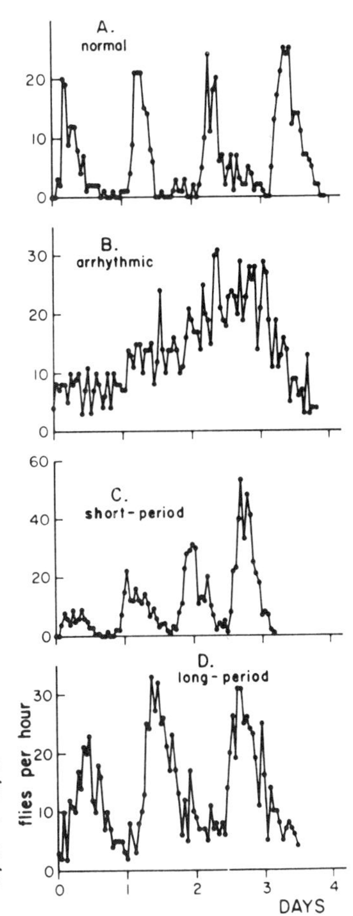

FIGURE 4.34. Eclosion rhythms of *Drosophila melanogaster* in DD (20°C) for populations of rhythmically normal and mutant *(per)* flies, previously exposed to LD: 12,12. (Reprinted from Konopka and Benzer, 1971, with permission of the authors.)

Smith and Konopka, 1981, 1982; but see Dowse et al., 1987). In addition, two other clock mutants at other loci on the X-chromosome have been identified (Konopka, 1984). One of these mutations, *and (andante)* is located at 10E3 and lengthens τ of the eclosion and locomotor activity rhythms by about 2 h, whereas the other, designated *clk (clock)*, lies distal to the centromere and shortens τ of the activity rhythm by about 1.5 h.

Smith and Konopka (1981) have examined the circadian rhythm phenotypes of eight available chromosome aberrations having a breakpoint in the 3B1-2 region. Two duplications and five deficiencies produced either a wild-type or an arrhythmic clock phenotype. The only available translocation with a 3B1-2 breakpoint, *T(1;4)JC43*, resulted in either a very long-period rhythm of locomotor activity (τ = 31 to 39 h) or partial or complete arrhythmicity. These results suggested that the deficiencies either had breakpoints within the *per* locus itself that caused phenotypic arrhythmicity or outside the *per* locus, thus leaving it intact and resulting in a wild-type rhythm, and that they shortened, lengthened, or eliminated periodicity by respectively increasing, decreasing, or eliminating *per* activity. This hypothesis was supported by an analysis of the effects of dosage alterations at the *per* locus on the circadian period of both circadian rhythms (Smith and Konopka, 1982). This locus, like other X-linked loci in *Drosophila*, is dosage-compensated in males, that is, one dose of per^+ in a male yielded the same τ as two doses in a female. An extra dose in a male, however, shortened τ of each rhythm by about an hour, and a reduction of one dose in a female lengthened τ by about an hour. Since τ of rhythms in per^s is significantly shorter than that of males carrying as many as five doses of per^+, the nature of the per^s mutation appears to result in increased activity of the *per* gene product and not merely in its overproduction (Smith and Konopka, 1982).

A more recent, quantitative analysis of the data by Coté and Brody (1986) indicates that τ is an inverse logarithmic function of the level of the *per* gene product but that it is relatively insensitive to this level (a presumed 300% increase in level produces only a 4.6% decrease in τ). This analysis may provide an explanation for the partial dominance of per^s over wild type and the dominance of the wild type over per^L (Konopka and Benzer, 1971; Smith and Konopka, 1981), since it suggests that the per^L gene product is nearly inactive whereas that of per^s is more than 34 times as active as the wild-type product. These authors note that the *per* gene product could play three roles in the regulation of circadian rhythmicity, although the available evidence is insufficient to reject any of them: (1) as an unrelated substance only secondarily affecting rhythmicity, (2) as a component of the intracellular clock (a part of the biochemical pathway that generates periodicity within each cell), or (3) as a component of the intercellular clock (forming or maintaining communication among rhythmic cells that would lead to mutual synchronization). All three roles would be consistent with the implication that the circadian τ is relatively insensitive to the level of the *per* gene product (but see Baylies et al., 1987; Yu et al., 1987b; Section 4.4.2). This insensitivity, in turn, might be attributed to the fact that flux through a metabolic pathway is dependent on the levels of all the enzymes in the pathway (and even in other, coupled pathways), and, therefore, any given enzyme would contribute only a relatively small share toward the control of flux through the pathway. On this hypothesis, τ would be analogous to metabolic flux, and the *per* gene product alone would have comparatively little effect on τ in relation to the concerted effect of all other elements of the oscillator (see Section 5.2.1).

Dowse et al. (1987), using digital techniques for signal analysis (correlogram, maximum entropy spectral analysis [MESA]), subsequently detected weak, shorter circadian rhythms and ultradian periodicities in the locomotor activity of free-running per^{o} males and of females lacking the *per* locus (per^{-}; heterozygous for two deficiencies, each of which deletes the gene). The dominant periods ranged from 4 to 22 h, and most of the significantly rhythmic flies exhibited multiple periodicities, unlike the wild type. The authors hypothesized that the per^{+} gene product mediates the coupling of multiple ultradian oscillators to produce wild-type circadian rhythms. Dowse and Ringo (1987) have obtained further data in support of this notion. Signal-to-noise ratios (SNR) were computed for the *per* mutants (in order to characterize the precision of their rhythms of locomotor activity), and the following progression was obtained: per^{o} (noisiest) $> per^{L} > per^{+} > per^{s}$. The SNR was found to decrease as τ increased in per^{s}, per^{+} and per^{L}; per^{o} typically had multiple ultradian periodicities and the lowest SNR. At least 70% of per^{L} individuals also had ultradian rhythms. In this scheme, τ would be a function of the coupling strength, or tightness, among ultradian oscillators, increasing as coupling loosens. Thus, ultradian rhythms would become apparent under weak coupling or in the absence of coupling (see Section 4.4.2; Bargiello et al., 1987).

The response of *per* mutants to light and temperature is altered also (Konopka, 1979). Thus, in per^{s} the 12-h subjective day (insensitive to light signals in the wild type) has been shortened by about 5 h, which reflects the decrease in τ of this mutation. The duration of the light-sensitive portion of the PRC for light perturbations is unaffected. Since locomotor activity normally occurs during the light-insensitive phase of the cycle, one might expect a decrease in the duration of activity in per^{s}, which is the case. Remarkably, the mutant exhibits a resetting curve (PRC) for light signals (40 min, 300 ft-cd, cool white fluorescent) of the type 0 (strong) variety (Winfree, 1980), yielding large $\Delta\phi$s of up to 12 h [analogous to that for *D. pseudoobscura* (Pittendrigh, 1965)], whereas the wild type has a type 1 (weak) PRC with maximal $\Delta\phi$s of only 2 to 3 h (Konopka, 1979; Winfree and Gordon, 1977). In wild-type *D. melanogaster*, as well as in per^{L1}, τ increases slightly with increasing light intensity (between 0.001 and 0.1 lx), but an inverse relationship was found in per^{s} (Konopka, Pittendrigh, and Orr, cited in Feldman, 1982). For increasing intensity above 0.1 lx, both mutants and the wild type exhibited increases in τ, although this period-lengthening effect was greater in per^{L1} than in per^{s}. The photosensitivity of both mutants with regard to maximal period lengthening or for the production of arrhythmia was greater than that of the wild type. These alterations in light sensitivity in the mutants have been incorporated into a membrane model for the circadian clock (similar to that of Njus et al., 1974; Sweeney, 1974b; see Section 5.4.1), in which the per^{s} mutation would increase the activity of an ion pump so that a critical ion gradient would be attained more quickly (Konopka and Orr, 1980). Temperature compensation of the activity rhythm in both per^{s} and per^{L1} is affected reciprocally: τ shortened with increasing temperature in the former but lengthened in the latter, a finding that perhaps reflects a differentially sensitive, multioscillator system (Konopka, Pittendrigh, and Orr, cited in Feldman, 1982).

The oscillator controlling locomotor activity in the fruit fly (produced by the output of motorneurons in the thoracic ganglia) could be located in the thoracic nervous system or in the brain. To distinguish between these possibilities, Konopka et al. (1983) used gynandromorphs that showed genetic mosaicism for an X-chromosome carrying the per^{s} allele in order to construct a two-dimensional, blastoderm fate map of external cuticle structures upon which a focus for the site of action of the mutation could be located. When the head of the fly was mutant genotype, the shortened τ of the activity rhythm characterizing per^{s} was exhibited, regardless of the genotype of the rest of the body. These results were consistent with a brain location for the driving oscillator. Indeed, some flies with mosaic heads appeared to have two components in their rhythmic activity output (as determined by periodogram analysis), suggesting that the oscillators in the two halves of the brain were of different genotype and capable of functioning independently. The coupling between brain and thoracic ganglia was found to be humoral rather than electrical. Implantation of a per^{s} brain into the abdomen of an arrhythmic per^{o1} individual conferred a short-period rhythm on the host (Handler and Konopka, 1979), reminescent of Truman's loose-brain experiments (1972) with the silk moth (see Section 4.1.1). Konopka and Wells (1980) have discovered that the morphology and location of a particular cluster of four neurosecretory cells, normally located at the posterior of each side of the brain of the wild-type fly, were significantly and similarly affected (scattered in ectopically dorsal locations) in both the arrhythmic per^{o1} mutant of *D. melanogaster* and several chemically mutagenized, arrhythmic strains of *D. pseudoobscura*.

Ultradian Clock Mutants

The circadian rhythm mutations in *D. melanogaster* discussed in the preceding subsection also affect short-term fluctuations in the male fly's courtship song (Kyriacou and Hall, 1980). The lovesong is produced by vibration of the extended wings during display and consists of pulses of tone produced generated at intervals of approximately 30 to 40 msec (Fig. 4.39A). This interpulse interval (IPI), however, is not constant but was found to fluctuate rhythmically with a free-running period of approximately 50 to 60 sec (Fig. 4.35A). Unlike the arrhythmia observed for the eclosion rhythm, the song rhythm in stocks raised for five generations in LL was maintained and was no different from the pattern observed in LD: 16,8. Finally, the song rhythm, unlike many ultradian rhythms (see Section 5.2.1), was temperature-compensated: τ was 53 sec at 16°C and 51 sec at 35°C.

In *Drosophila simulans*, a sibling species, the IPI is typically 45 to 55 msec, and τ of its IPI cycle is approximately 30 to 40 sec (Kyriacou and Hall, 1980). The difference in τ between the two species probably serves as a sexual isolating mechanism. Kyriacou and Hall (1986), using two types of hybrid males produced from reciprocal crosses (both of which had an autosome from each parent, but whose X-chromosome was derived from one or the other of the parental species), found that the species-

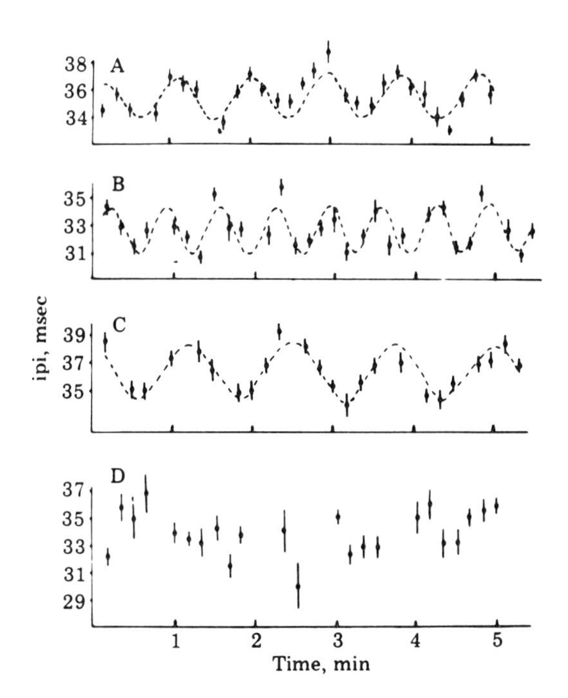

FIGURE 4.35. Courtship song profiles of the *per* mutants and a per^+ male (*yellow* on a Canton-S genetic background) of *Drosophila melanogaster*. (A) per^+, τ = 56 sec. (B) per^s, τ = 40 sec. (C) per^L, τ = 76 sec. (D) per^o, τ = ?. (Reprinted from Kyriacou and Hall, 1980, with permission of the authors.)

specific differences in the lovesong rhythm appeared to be due to sex-linked genes, whereas the basic IPI difference was autosomally inherited. Hybrid females showed selective preference for artificially generated songs (by an electronic song simulator) carrying intermediate, hybrid characteristics. It would be interesting to know whether a *D. melanogaster* male deleted of its per^+ gene but carrying the *D. simulans* per^+ allele would sing with a *D. melanogaster* or *D. simulans* period.

The three major allelic *per* mutations that affect the circadian rhythms of eclosion and locomotor activity strikingly affect the 60-sec song rhythm in a parallel fashion (Fig. 4.35B,C,D). Thus, per^s shortened τ to about 40 sec; per^{L1} lengthened τ to about 76 sec, and per^{o1} abolished rhythmicity. Preliminary results from genetic mosaic analysis indicate that the focus for the courtship song oscillator fate-maps to the thoracic nervous system (Konopka and Hall, personal communication, 1985), in contrast to the circadian clock located in the head (Konopka et al., 1983). These results suggest that both circadian rhythms and a very short, ultradian rhythm are influenced by the same gene, with the period's being regulated by the amount of *per* product in the cell. [The effects of *per* on the circadian and the male courtship song rhythms can be dissociated, however, as recently demonstrated by Yu et al. (1987b); see Section 4.4.2.] Other rhythms also appear to be affected by *per* mutations: circadian fluctuations of membrane potential in the isolated, fluorescent dye-labeled larval salivary glands of *D. melanogaster* were seemingly absent or decreased in amplitude in some per^{o1} individuals, an observation that might explain behaviorial arrhythmicity (Weitzel and Rensing, 1981), and larval heartbeat is erratic in this mutant (Livingstone, 1981).

Provocatively, mutations affecting action potential seem to stop the ultradian clock underlying the *D. melanogaster* song rhythm (Kyriacou and Hall, 1985). Flies carrying the temperature-sensitive mutations nap^{ts} (no-action-potential, temperature sensitive) or $para^{ts1}$ (paralytic, temperature sensitive) are immobilized rapidly at restrictive temperatures $> 35°C$ and $> 29°C$, respectively, but this paralysis is reversible, both mutants recovering in seconds when returned to the permissive temperature of 25°C. Heat pulses (30 sec) were applied to mutant males that had begun their courtship songs in order to switch off the nervous system. On recovery at 25°C, each male was permitted to continue its lovesong, which was analyzed to determine if any phase shift in the song cycle had occurred (relative to controls that were incapacitated for 30 sec with short pulses of low temperature or CO_2). The high temperature-pulsed mutants all exhibited phase delays of ~30 sec. These results and others suggested that shutting down the nervous system led to a stoppage of the song clock proportional to the duration of the heat stimulus, thereby indicating that we are dealing with a neural ultradian oscillator. It would be interesting to attempt to localize presumptive neural foci using genetic mosaics in a manner analogous to that used by Konopka et al. (1983) for the *per* gene.

Likewise, it would be useful to detemine whether the *nap*^{ts} and *para*^{ts} mutations lead to a temperature-sensitive arrest of the circadian clock in fruit flies.

4.4.2 Recombinant DNA Studies: Cloning Clock Genes

In the preceding section we saw that the *per* locus of *D. melanogaster* has a fundamental role in the construction or maintenance of both a circadian and an ultradian clock. But how is it possible that such a gene could mutate so that it alters or abolishes the oscillator(s) underlying these quite disparate types of rhythmicity? And why does an excess of the *per* gene product speed up the clock and a deficit slow it down or obliterate it? Partial answers to these key questions have been derived from the isolation and molecular analysis of the *per* locus itself (Rosbash and Hall, 1985; Young et al., 1985; Hall, 1986; Hamblen et al., 1986).

Isolation of a Clock Gene

The *per* region DNA was isolated almost simulataneously by two different groups, one (at The Rockefeller University) using chromosome walking/jumping (Bargiello and Young, 1984), the other (at Brandeis University) using microexcision techniques (Reddy et al., 1984). Each method depended in part on knowing with considerable precision the location of the gene on the X-chromosome (afforded, for example, by the cytogenetic studies of Smith and Konopka, 1981). The mapping of cloned DNA segments with respect to X-chromosomal breakpoints of mutant strains aided in localizing the molecular boundaries of the *per* locus.

Thus, Bargiello and Young (1984) isolated 90 kilobase pairs (kb) of DNA from the 3B1-2 X-chromosomal region containing the *per* gene of *D. melanogaster*. Southern-blotting experiments were carried out, in which DNA from wild-type flies was hybridized with DNA from flies carrying the chromosomal aberrations, probed with the same types of DNA sequences from the *per* region. The results from this physical characterization indicated that DNA affecting circadian behavorial rhythms was found in a 7.1-kb *Hin*dIII fragment that mapped just to the left of the 3B1-2 breakpoint *T(1;4)JC43*, which Smith and Konopka (1981) had shown appears to cause lower than normal expression of *per*. Northern blots, probed with the same types of *per*-region sequence, revealed that a single 4.5-kb poly(A)$^+$ RNA was transcribed from this DNA in wild-type pupae and adult flies. This transcript was eliminated by a segmental aneuploid *per* mutant that retained some rhythmic activity. This mutant, however, substituted two novel transcripts (11.5 and 0.9 kb). The simplest interpretation of these results is that the 4.5-kb transcript is required for the expression of wild-type rhythmic behavior and that the novel RNAs provided the residual *per* locus activity found in the weakly rhythmic, *T(1;4)JC43* flies (Bargiello and Young, 1984).

Reddy et al. (1984), applying microexcision techniques to obtain clones from the 3B1-2 region of the X-chromosome of *D. melanogaster,* delimited the clock gene by mapping chromosomal breakpoints in or near the *per* locus. As Bargiello and Young (1984) also had found, one of the most interesting variants was part of the *T(1;4)JC43* aberration, which mapped to the middle of the cloned DNA, and attention consequently was concentrated on an analysis of transcripts from this region. Five transcripts, of sizes 4.5 kb, 2.7 kb, 1.7 kb, 1.0 kb, and 0.9 kb, were detected, although the last two were virtually indistinguishable under normal conditions of electrophoresis. The 0.9-kb transcript, located to the right of the *JC43* breakpoint, appeared at the time to be strongly implicated in *per*'s control of biological rhythms. Two independently isolated arrhythmic mutations at the *per* locus (per^{o1}, per^{o2}) dramatically reduced the level of this transcript. Furthermore, the level of the 0.9-kb transcript was strongly modulated by an entraining LD: 12,12 cycle, in contrast to the levels of the other transcripts, which were relatively constant when compared to each other or to the level of mRNA transcribed from the locus for alcohol dehydrogenase. This diurnal cycling of the 0.9-kb species free-ran with a circadian period for at least 3 days in DD.

There seemed to be, therefore, an unexpectedly large number of mRNA sources derived from a relatively small genetic interval comprising, perhaps, only two loci. Paradoxically, the level of the 4.5-kb transcript itself did not oscillate, whereas that of the 0.9-kb species did, although the findings of Bargiello and Young (1984) suggested that the former played a key, if not primary, role in the generation of rhythmicity. Was this simply an unusual coincidence, or do other trancripts also cycle? Is the *per* locus complex, with multiple transcripts from both the right and left of the *JC43* breakpoint contributing to the control of behavorial rhythmicity, or, alternatively, does *per* lie entirely to the left of the breakpoint and ultimately drive the 0.9-kb transcript, causing it to cycle (Rosbash and Hall, 1985)? Once again, we face the problem of distinguishing between putative clock gears and the hands coupled to them (Goto et al., 1985; see Section 4.3).

Restoration of Biological Rhythms in Transgenic Drosophila

The clock gene obviously could not be precisely identified and localized along the 15-kb region of chromosomal band 3B1-1 by determining the portion of the *per* locus that coded for the particular transcript giving rise to the relevant clock protein because such a hypothetical polypeptide was completely unknown at the time. Rather, the problem was approached (but not completely resolved) by mutant rescue experiments in which various DNA fragments from the *per* region were transduced into mutant host flies, taking advantage of powerful germ-line transformation techniques using P-element strategies (Rosbash and Hall, 1985; Young et al.,

1985; Hall, 1986; Hamblen et al., 1986). Thus, when the 7.1-kb *Hind*III DNA fragment of a *per*$^+$ fly, which Bargiello and Young (1984) had shown encoded the 4.5-kb RNA transcript, was introduced into the genome (Fig. 4.36) of an arrhythmic *per*^{o} fly (Fig. 4.37b), circadian rhythmicity of both eclosion (Fig. 4.37c) and locomotor activity was restored (Bargiello et al., 1984; see Fig. 4.38). The transformants could now be entrained by LD: 12,12 and were able to free-run in DD, although the amplitude of the rhythm in the rescued fly (Fig. 4.37c) was lower than that of the wild-type (Fig. 4.37a) and τ was 1 to 2 h longer. Perhaps the transformed P1.48C strain produced less than wild-type levels of *per* gene product. This transforming DNA complemented *per* locus deletions and was transcribed, forming a single 4.5-kb poly(A)$^+$ RNA comparable to that produced by wild-type flies (Bargiello et al., 1984).

Another set of transforming DNA fragments homologous to all or part of the important 4.5-kb RNA species has been generated and tested for their ability to rescue not only the circadian phenotype of *per*^{o} hosts but also the ultradian song rhythm (Zehring et al., 1984; Hall, 1986; Hamblen et al., 1986). In addition to the 7.1-kb fragment, four other overlapping DNA fragments (8.0, 10.2, 12.8, 14.6 kb) from the *per* region (Fig. 4.38) were found to effect a partial rescue of both circadian locomotor activity and ultradian courtship behavior in normally arrhythmic *per*^{o1} hosts (Fig. 4.39b). Four further subsegments of the *per* region (5.8, 7.3, 9.0, 9.8 kb) did not rescue mutant behavorial phenotypes. Once more, correction of the mutant phenotype was not complete. Although this observation could mean that the *per* locus is complex, comprising several components, it is more likely, perhaps, that none of these fragments, including the segment of the *per* region left of the *JC43* break, contained all of the information necessary to produce a 4.5-kb transcript (which afforded almost complete rescue; Fig. 4.37c) or, alternatively, completely encoded a rather long, necessary transcript. Interestingly, two fragments (14.6 and 8.0 kb) did not cover a significant portion of the 7.1-kb piece from which the 4.5-kb RNA transcript was transcribed (Fig. 4.38) despite the fact that both were partially successful in rescuing the mutant (Fig. 4.39a,b). Thus, the entirety of the 4.5-kb transcript and, by inference, its protein product, does not appear to be necessary for quasinormal functioning of the *per* clock gene (Hall, 1986; Hamblen et al., 1986). Coverage of the transcript source for the 0.9-kb RNA species (shown to oscillate in its abundance over the course of a day in the wild type but not in *per*^{o}), immediately adjacent to the source for the 4.5-kb species, in several transformants carrying a *per*^{o} mutation did not restore rhythmicity (Hamblen et al., 1986).

In summary, the germ-line transformation experiments indicated that the genomic region encoding the 4.5-kb transcript is the core of the *per* locus, even though (1) the entirety of the transcript is not necessary for rather strong rhythmicity, its either terminus being dispensable for quasinormal *per* functioning, and (2) it does not seem to be completely sufficient, in transformants, to produce wild-type behavorial phenotypes. In contrast,

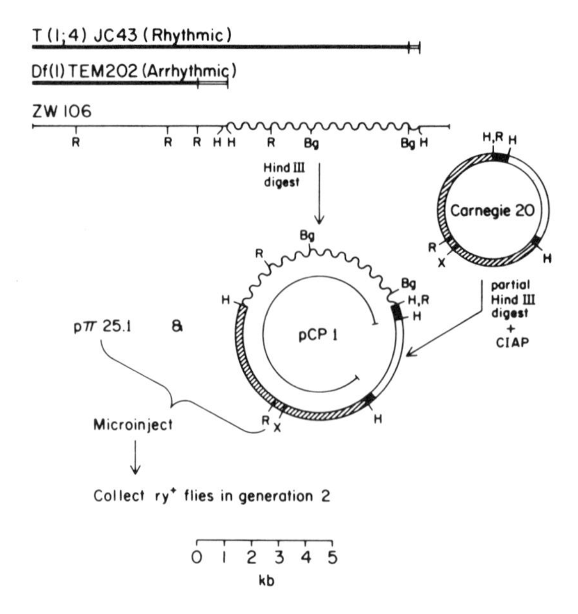

FIGURE 4.36. Physical map of *per* region DNA of *Drosophila melanogaster* and the construction of transforming DNA, pCPl. The positions of two chromosomal rearrangements, *T(1;4)JC43* and *Df(1)TME202*, are shown relative to phage ZW106. Solid bars represent DNA still present and in the wild-type configuration, and the location of each chromosomal rearrangement breakpoint is shown as an open bar. *Methods:* Transforming DNA was constructed as follows: ZW106 was digested to completion with the restriction endonuclease *Hin*dIII (H, *Hin*dIII; R, *Eco*RI; Bg, *Bgl*II; X, *Xho*I). The 7.1-kb fragment (shown as a wavy line in ZW106) was purified from a low-melting point agarose gel. Carnegie 20 was partially digested with *Hin*dIII and treated with calf intestinal alkaline phosphatase (CIAP) to prevent self-religation. The 7.1-kb fragment was ligated to the partially digested vector and introduced into *Escherichia coli* (HB101). Transforming DNA sequences were mapped with restriction endonucleases, and one, pCPl, was chosen for microinjection into *Drosophila*. pCPl contains the recombinogenic ends of a truncated P-factor (■), *per* locus DNA (~), a wild-type *rosy* gene (▨), and a combination of DNA sequences from *Escherichia coli* and the *Drosophila melanogaster white* locus (□). The part of pCPl that is inserted into the *Drosophila* genome by P-factor-mediated transposition includes both per^+ and $rosy^+$ DNA, and is represented by |—| . pCPl was mixed with pπ25.1, which mobilizes the P-factor associated with pCPl in *trans*, at a 5:1 concentration ratio (350 μg ml^{-1} final concentration), and the mixture was injected into *Drosophila* embryos of the genotype *y* per^0; ry^{42}. Germ-line integration of pCPl DNA sequences results in a heritable change in eye color (from *rosy* to wild type). Because both per^+ and $rosy^+$ DNA sequences are integrated as a unit, flies with ry^+ eye color were subsequently tested for expression of behavioral rhythms. (Reprinted from Bargiello et al., 1984, with permission of Macmillan Journals Limited.)

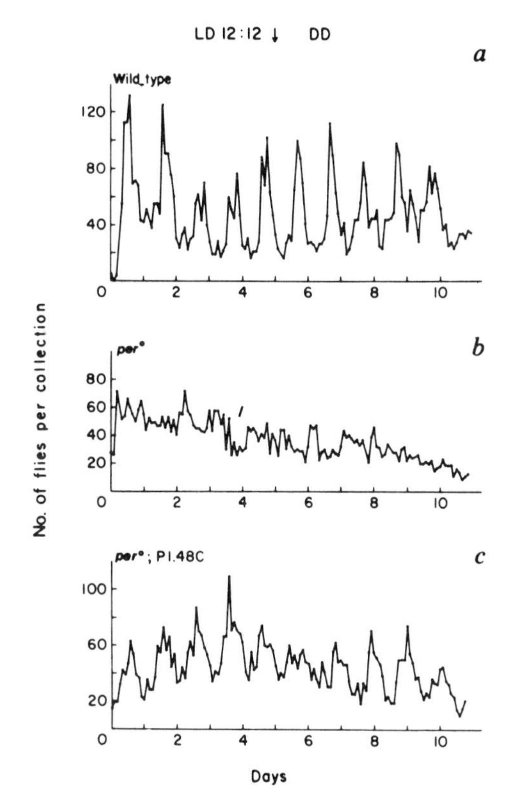

FIGURE 4.37. Temporal profiles of eclosion in *Drosophila melanogaster* for populations of: *(a)* wild-type females (*y* $per^0/+$; $ry^{42}/+$); *(b)* per^0 males and females (*y* per^0;ry^{42}/ry^{42}); and *(c)* per^0; P1.48C males and females (*y* per^0; P1.48C/+ ;ry^{42}/ry^{42}). Eclosion was monitored simultaneously in the three different populations by manually collecting (at 2-h intervals) the adults that had emerged in culture bottles maintained at 25°C. Developing cultures were entrained to LD: 12,12 for 3 days before the start of collections. This LD cycle was continued for the first 5 days of collections, then cultures were maintained in DD for the remainder of the experiment. The LD to DD transition is indicated by the arrow above record *(a)*. Collections during DD were made under a low-intensity safelight (a 15 W incandescent light with a Kodak GBX-2 filter), which does not affect *Drosophila* circadian rhythms. During entrainment to LD: 12,12, 24-h period lengths were estimated for the rhythms of wild-type *(a)* and per^0; P1.48C *(c)* populations by measuring the time elapsed between medians of successive peaks of eclosion. By the same procedure, wild-type and per^0; P1.48C populations have τs (in DD) of 25 ± 1.1 h and 26.5 ± 1.9 h, respectively. Similar estimates (24 to 25 h for wild-type and 27 to 28 h for per^0; P1.48C) were obtained from a spectral analysis of the data. Spectral analysis revealed no periodicity in the eclosion profile for per^0 flies. (Reprinted from Bargiello et al., 1984, with permission of Macmillan Journals Limited.)

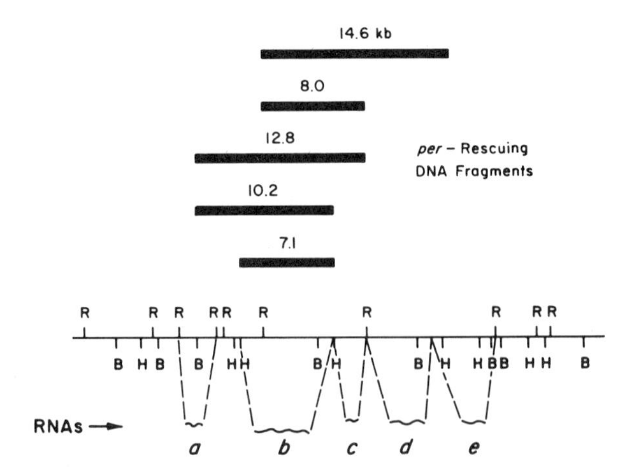

FIGURE 4.38. Restoration of rhythmicity to *per* mutants of *Drosophila melanogaster* carrying DNA fragments from this clock locus. These pieces of X-chromosomal DNA—each designated by its length in kilobases—were subcloned eventually into vectors that allow insertion of such fragments into sites within the genome of the fly. The host flies (for the appropriate microinjections of the DNA) carried an arrhythmic per^0 allele. The transformed flies that eventually resulted had the effects of this mutation (or, in subsequent tests, homozygosity for a completion deletion of the gene; see Smith and Konopka, 1981; Bargiello and Young, 1984; Reddy et al., 1984) rescued by any of these five types of fragment that were originally cloned from normal per^+ flies. The 7.1-kb piece has been manipulated by one group of investigators (Bargiello et al., 1984), who tested circadian rhythmicity of eclosion (see Fig. 4.37), and the other four pieces by another group (Zehring et al., 1984), who have analyzed the ultradian courtship song rhythm as well (see Figs. 4.35 and 4.39). Appearing beneath the map of the *per* region shown here—which includes the relevant restriction sites cut by the endonucleases Eco RI (R), Bam HI (B), or Hind III (H)—are polyadenylated RNA transcripts that are complementary to the various subsegments of the locus (Bargiello et al., 1984; Reddy et al., 1984). The approximate lengths of these transcripts are *(a)* (which is male-specific), 1 kb, *(b)* 4.5 kb, *(c)* 1 kb, *(d)* and *(e)* 2 to 3 kb. Another RNA species (size, about 1 kb) that was thought to be complementary to the same part of the locus that gives rise to *(b)* (Reddy et al., 1984) has proved to be an artifact (J. C. Hall, personal communication, 1985:). (Courtesy of J. C. Hall, 1985; see Zehring et al., 1984; Hall, 1986.)

the normally oscillating, 0.9-kb RNA species is neither necessary nor sufficient for rhythmicity, despite the fact that arrhythmic mutations abolish the cycling in its abundance. Therefore, the effects of the per^o mutation on the molecular oscillation must be indirect (Hamblen et al., 1986).

Recently, Yu et al. (1987a) mapped point mutations in the per^{ol} and per^s loci to single nucleotides. Chimeric DNA fragments consisting of well-defined wild-type and mutant DNA subsegments were constructed, in-

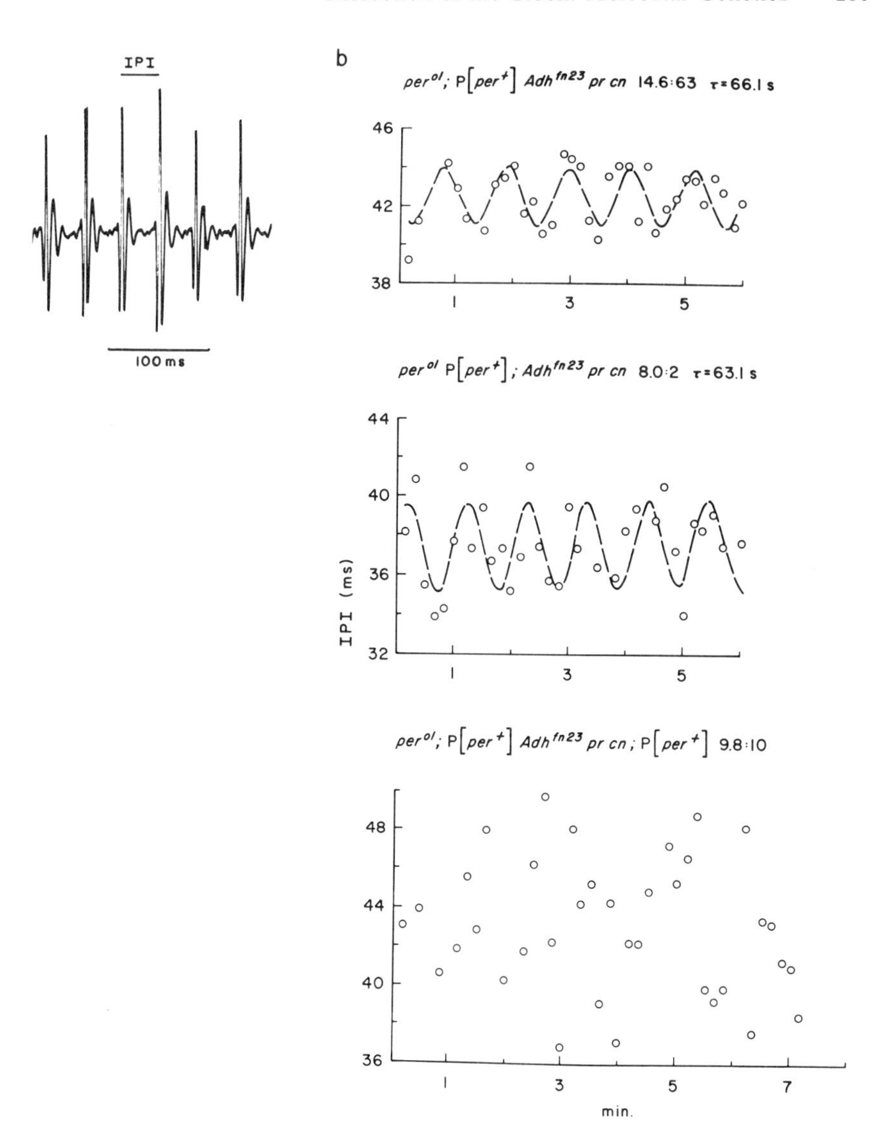

FIGURE 4.39. Oscillations of interpulse intervals in the male's courtship song in *Drosophila melanogaster*. Males of the transformed genotypes had their songs recorded and analyzed. (a) An example of the pulse train, with eight tone pulses, hence seven interpulse intervals (IPIs), is depicted. In this species of *Drosophila*, pulses are typically generated at the rate of ~ 20 to 30/sec (see Kyriacou and Hall, 1980). (b) The two rhythmic songs (for a male from a 14.6 transformed line, and one from an 8.0 line) are indicated by the sinusoidal (dashed) lines, accompanied by the best estimates of the periods (τ). No line was drawn with respect to the IPI fluctuations for the male from a 9.8 transformed line because there was no discernible pattern to these changing IPIs. (Reprinted from Zehring et al., 1984, with permission of M.I.T. Press.)

troduced into flies by germ-line transformation, and assayed for their biological activity (circadian eclosion rhythm). These experiments further localized both *per*^{o1} and *per*^{s} to a 1.7-kb DNA fragment that was mostly coding DNA. Sequencing of this subsegment from each mutant showed that each reflected a single basepair substitution: *per*^{o1} was completely accounted for by a nonsense mutation (a stop codon) in the third coding exon of the 4.5-kb RNA transcript, whereas *per*^{s} was a missense mutation in the fourth coding exon. The predicted gene protein sequence (see next subsection) of the *per* product of the two mutations, respectively, would be affected at amino acids 464 and 589, the latter corresponding to a serine-to-asparagine substitution. Similar results also have been obtained by Baylies et al. (1987).

Other interesting discoveries have been made during the course of these transformation-rescue experiments. From a preliminary behavorial study of 30 transformed lines, Baylies et al. (1987) observed that τ ranged from about 25 to 40 h. To determine the molecular basis of this variation, seven such lines were further analyzed. Although circadian rhythms were restored when arrhythmic *(per*$^{-}$*) Drosophila* were transformed with *per* locus DNA, flies that had received identical transforming DNA fragments yielded rhythms whose value for τ varied by more than 12 h. Transcription studies revealed a 10-fold variation in the level of *per* RNA among transformed flies; these levels were inversely correlated with period length, so that flies with lowest levels of *per* product had the mostly slowly running clocks. Thus, τ would appear to be set by the level of gene product (cf. Coté and Brody, 1986; Section 4.4.1). But because *per*^{s} and *per*^{L} both produce wild-type levels of *per* mRNA (Bargiello and Young, 1984), Baylies et al. (1987) suggested that the amino acid substitutions in these two mutants respectively increase or decrease the stability of the *per* product; or, alternatively, *per*^{s} protein would be hyperactive and *per*^{L} protein hypoactive.

In the same vein, Yu et al. (1987b) found that the length of the Gly-Thr run of the *per* gene (see next section) varies between 17 and 23 repeats in *per*$^{+}$ alleles. To ascertain if this variable region of the *per* locusis functionally significant, these workers contructed an in-frame deletion of this region that removed the entire Gly-Thr repeat and used it to transform *per*^{o1} flies. Surprisingly, this mutant construct rescued the circadian rhythm phenotype, but yielded a short love-song rhythm of only 40 s; this restored ultradian τ is more characteristic of *per*^{s}, contrasting with that (60 s) of both *per*$^{+}$ and of *per*^{o1} flies transformed with a normal *per* gene having an intact Gly-Thr repeat region. Thus, the effects of *per* on the circadian and the male courtship-song rhythms could be dissociated. Perhaps this polymorphism of the *per* gene within the Gly-Thr tract could explain how one gene product can differentially affect two different classes of rhythm. The species-constant circadian period could be conserved if the oscillation depended on the absolute amount of *per* product (which might oscillate if the protein were unstable and blocked translation of its own mRNA), whereas the highly species-specific, ultradian song period might derive from variations in the concentration of *per* product (generated by changes in its stability caused, for example, by different lengths of the Gly-Thr repeat region). If this reasoning is valid, a molecular basis for a temporal sexual isolating mechanism in courtship behavior would have been provided within the fine structure of the *per* gene (Yu et al., 1987b).

The Product of the *per* Locus

Recent work based on transcript mapping and DNA sequencing of the 7217 bases in the biologically active segment of *per*-locus DNA that encodes for the 4.5-kb RNA species supports the notion of a long transcript required for clock function. Using a complementary DNA clone that hybridized to the 1.5-kb *Bam*H/*Hin*dIII interval of the *per* gene lying to the left of the *JC43* break, Jackson et al. (1986) were able to place the transcript map onto the DNA restriction map. The transcript visualized included eight exons originating over a slightly longer region than the 7.1-kb *Hin*dIII fragment that was used in the earlier transformation studies and could be interepreted to have a single open reading frame with an AUG start codon in exon 2 and a UGA stop codon in exon 8. None of the fragments used for transformation could have been expected to encode this full-length transcript, all (or a large part) of which probably alone constitutes the *per* gene.

The conceptual translation of the open reading frame determined from the *per* DNA sequence yielded a polypeptide of 1127 amino acids (Jackson et al., 1986; Yu et al., 1987a; Citri et al., 1987). Several abnormal phenotypes displayed by some transformed flies, characterized by long-period rhythms, could be associated with changes in the sequence of untranslated portions of the transcription unit. In each case, the change in DNA sequence failed to affect the structure of the *per* protein; therefore, the long τ's were probably generated by altered regulation of *per* protein synthesis. Surprisingly, nearly half of the predicted gene product (47%) comprised only five amino acids (serine, threonine, glycine, alanine, proline), and these often formed simple repeats, such as polyalanine and polyglycine tracts up to 17 residues in length. The hypothesized *per* protein also included a potential site for phosphorylation by cAMP-dependent protein kinase. Inasmuch as various chemical agents that affect cyclic nucleotide levels can phase-shift circadian rhythms (see Section 4.3.3), the same agents perhaps alter the phosphorylation of a *per* protein (Jackson et al. 1986).

The putative *per* protein sequence was compared to all available sequences in the Dayhoff and Doolittle database library (Jackson et al., 1986). Extensive regions of homology to the core protein of a rat chondroitin-sulfate proteoglycan were detected; this core (104 amino acids in length) is known to contain 49 alternating Ser-Gly residues and to play a role in the glycosylation of the core protein. Thus, this unusual repeat sequence, together with the Thr-Gly repeat tracts that were also found in the *per* protein, might form a site serving for glycosylation, a process with an unknown role in clock function.

A similar DNA sequencing of a portion of the 4.5-kb transcript's source (Reddy et al., 1986) also implicated an unusual, multiresidue, Gly-Thr repeat in the protein that was inferred to be encoded within the *per* locus. By recombinant DNA techniques, a small subregion of the coding sequence, corresponding to 154 amino acids of the putative product, was cloned and expressed in bacteria as part of a fusion protein, which was then used to immunize rabbits. When the resultant antisera were characterized and used to probe protein preparations from *D. melanogaster*, an antigen was detected in wild-type flies but not in a per^- mutant. Biochemical characterization of this antigen indicated that it indeed was a proteoglycan, as Shin et al. (1985) and Jackson et al. (1986) also independently had concluded.

Although not much is known about the function of proteoglycans, their distribution has been studied intensively. They typically are found in extracellular locations or in association with cell surfaces. James et al. (1986) have examined the temporal and spatial expression of the 4.5-kb mRNA

FIGURE 4.40. DNA homologous to the *Drosophila per* locus is present in vertebrates. *(a)* Total genomic DNA from yeast (0.5 μg), *Drosophila* (0.5 μg), chicken (5 μg), cat (10 μg), mouse (10 μg), and human (10 μg) was digested with *Bam*HI, subjected to electrophoresis on a 0.75% agarose gel, and transferred to a nitrocellulose filter. A 1.3-kb *Pst*I/*Bam*HI fragment fom *per* was nick-translated and used as a hybridization probe [for location of probe, see *(d)* below]. Hybridization was at 40°C in 5 × SSC, 50 formamide, 7 m*M* Tris pH 7.5, 1 × Denhardt's solution, 25 μg ml^{-1} salmon sperm DNA, and 10% dextran sulfate. Washes were in 0.2 × SSC, 0.1 SDS at 45°C. Numbers indicate sizes in kb. *(b, c)* Mouse cp2.2 and cp35.3 DNAs were digested with *Bam*HI, subjected to electrophoresis, and transferred to nitrocellulose as in *(a)*. Blots were hybridized with nick-translated *per* 7.1-kb *Hin*dIII fragment [see *(d)* below], and washed as in *(a)*. *(d)* Map of *per* 7.1-kb *Hin*dIII fragment (left) and a 6-kb *Bam*HI fragment from cp2.2 bearing homology to *per* (right). Hatched regions indicate cross-hybridizing restriction fragments. The *per* 1.3-kb *Pst*I/*Bam*HI fragment used as a probe in *(a)* includes the 1.0-kb *Eco*RV/*Bam*HI fragment shown in *(d)*. H, *Hin*dIII; B, *Bam*HI; R, *Eco*RI; E5, *Eco*RV; P, *Pst*I. *(e)* Total genomic DNA from the indicated species (10 μg DNA from rat) was treated as in *(a)* except that the nitrocellulose blot was hybridized with a 0.5-kb *Stu*I/*Eco*RI fragment from mouse clone cp2.2. The DNA sequence of this probe has been determined (Shin et al., 1985). Hybridizations to mouse and rat DNAs are shown twice so that a long exposure of the human DNA hybridization can be presented. Numbers in *(a, b, c)* and *(e)* indicate sizes, in kb, of some hybridizing DNA segments as measured against restriction fragments of λ *c*I857 DNA. (Reprinted from Shin et al., 1985, with permission of Macmillan Journals Limited.)

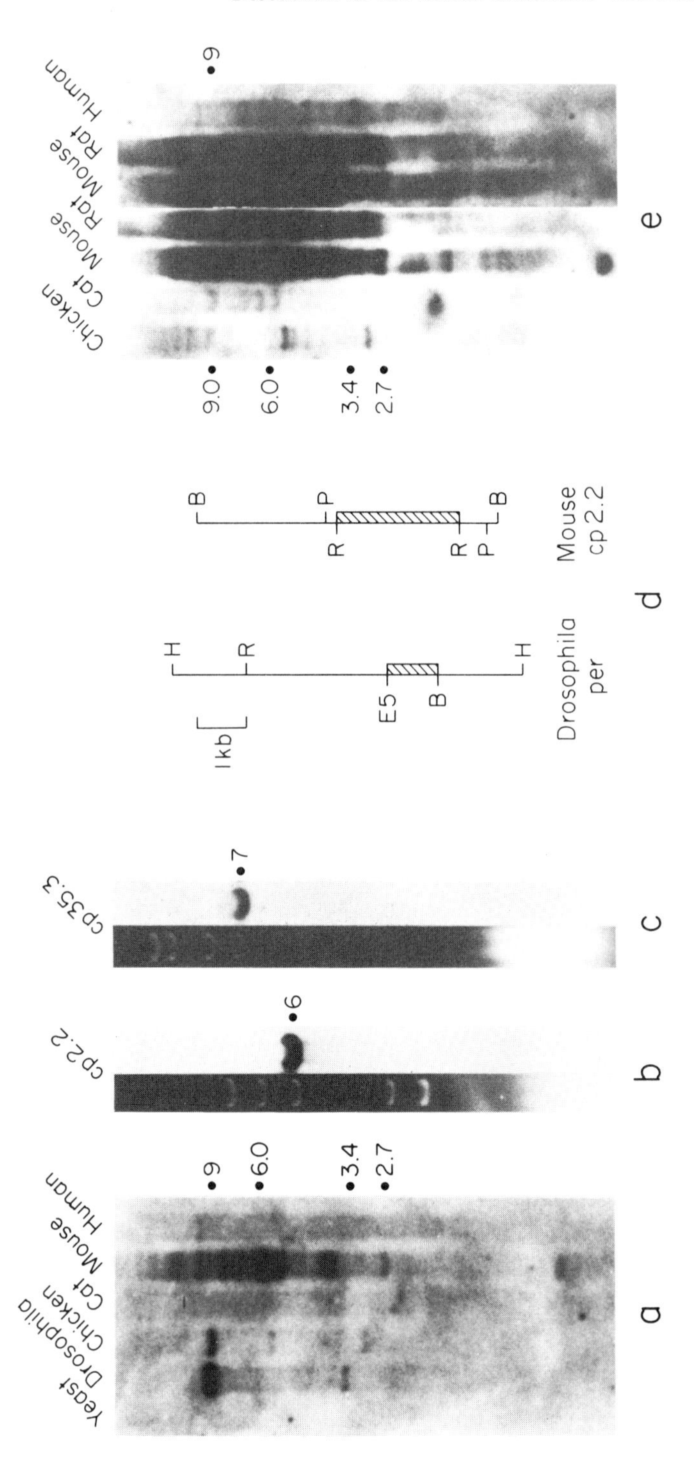
a
Yeast
Drosophila
Chicken
Cat
Mouse
Human
•9
•6.0
•3.4
•2.7
b
cp2.2
•6
c
cp35.3
•7
d
1kb
H
R
E5
B
H
Drosophila
per
R
B
P
R
P
B
Mouse
cp2.2
e
Chicken
Cat
Mouse
Rat
Mouse
Rat
Human
9.0•
6.0•
3.4•
2.7•
•9

that is transcribed from the *per* locus and that presumably codes for proteoglycan-like clock protein(s). Both Northern blot analyses and in situ hybridizations to tissue sections revealed significant expression of this transcript in *D. melanogaster* embryos of different developmental stages. Although none was detectable in embryos from 0 to 6 h after egg laying, significant amounts were present thereafter, from gastrulation until the embryos hatched some 22 to 24 h later. Expression of the 4.5-kb mRNA then was undetectable in all larval instars and through the first half of the pupal stage, whereupon *per* locus activity recommenced (Bargiello and Young, 1984; Reddy et al., 1984; Young et al., 1985). This expression of the *per* clock gene was limited to the central nervous system of the developing embryo and was localized within the brain and ventral ganglia. It is significant that these sites correspond, respectively, to the foci for the eclosion and courtship song oscillators localized by fate-map, genetic mosaic analysis (Konopka et al., 1983; Konopka and Hall, personal communication, 1985; see Section 4.4.1). In addition to the physiological role that the 4.5-kb mRNA species and the nervous system-specific proteoglycan for which it codes might have in maintaining biological rhythms, James et al. (1986) suggest that they participate in development by establishing mechanisms necessary for eventual expression of clock functions.

The *per* gene product also appears to regulate the degree of intercellular communication by altering the number and organization of functional gap junctions in *Drosophila* (Bargiello et al., 1987). Direct measurement of the coupling coefficient, apparent space constant, and junctional conductance in the larval salivary glands of per^o, per^+ and per^s demonstrated that the extent of coupling was correlated inversely with τ, being weakest in per^o. These provocative results are consistent with the hypothesis (Dowse et al., 1987; Dowse and Ringo, 1987) that the per^+ gene product mediates the coupling of multiple ultradian oscillators to produce wild-type circadian rhythms—a suggestion that might explain weak or erratic rhythms of membrane potential in isolated salivary glands (Weitzel and Rensing, 1981) and larval heartbeat (Livingstone, 1981) in the per^o mutant (see Section 4.4.1).

DNA homologous to the *per* locus has been found in a number of vertebrates, and the unusual coding sequence from this *Drosophila* clock gene appears to have been conserved (Shin et al., 1985). Thus, the peculiar, tandemly repeated sequence forming a portion of the 4.5-kb *per* transcript was found to be strongly homologous to mouse DNA and, to a lesser extent, to DNA in chicken and man (Fig. 4.40). Cloned DNAs from mouse and *Drosophila* were related by long, uninterrupted, tandem repetitions of the sequence ACNGGN, which at the *per* locus were predicted to code for poly(Thr-Gly) tracts as long as 48 amino acids in length. These repeated sequences were also transcribed in the mouse, and several long poly(Thr-Gly) tracts appeared to be encoded by cloned mouse DNA.

Thus, these findings raise the attractive possibility that homologous DNA sequences play a role in the generation of both ultradian and circadian biological rhythms in many species (Shin et al., 1985; Young et al., 1985). Sequences homologous to those in *per* probes from *Drosophila* recently have been reported in spinach, rape, and *Acetabularia* (Li-Weber et al., 1987). In *Acetabularia*, however, they were found not in nuclear DNA but rather in the chloroplast genome. A 1.175 kb *Eco*RI-*SAl*I chloroplast DNA fragment containing the homologous sequence was subcloned into pUC12 and sequenced; its core consisted of a repetitive, tandemly arranged sequence of 43 units of the hexamer GGA-ACT coding for glycine and threonine. These results imply that were these homologous sequences to be expressed in *Acetabularia*, they would be translated on the 70S ribosomes of the chloroplasts rather than on the 80S ribosomes of the cytosol, which is in apparent contradiction to the results obtained with inhibitors in this algal cell (see Section 4.3.2) and to the coupled translation–membrane model (see Section 5.4.2) of Schweiger and Schweiger (1977). Schweiger et al. (1986) suggest a possible resolution of this paradox: the activity of the high-molecular-weight, chloroplastic polypeptide (p230) identified by Hartwig et al. (1985) and thought to be an essential clock protein with a highly specialized membrane function (Fig. 4.19) would be affected by the product of the *per* locus, which, in the case of *Acetabularia*, would be a near-neighbor protein (but not an essential clock element) also encoded in the chloroplast genome. This situation perhaps reflects an ancient phase of symbiosis between a eukaryotic host cell and a prokaryotic, photosynthetic endosymbiont in which the relevant genes have not yet been translocated to the nuclear genome, as they have been in the fully integrated chloroplasts of higher plants. Finally, DNA sequences that were isolated from *Neurospora* were found to hybridize with the 8.0-kb "rescuing fragment" from the *per* locus of *Drosophila* under conditions of moderate stringency, and preliminary data indicate that these sequences are homologous (Feldman, personal communication, 1987; see next subsection).

Molecular Analysis of *Neurospora* Clock Genes

The techniques of molecular genetics also have been used to determine how the circadian clock controls the expression of the *frq* gene in *Neurospora* (see Section 4.4.1). The first step was the characterization of the *frq* locus itself. A set of contiguous, overlapping cosmid and phage clones that generated a physical map extending 130 kb in either direction from the *oli* marker (a total of more than 8 map units) on linkage group VIIR was identified by using chromosome-walk methodology (Dunlap et al., 1987a). Transformation and phenotypic rescue of recessive mutants further identified the *frq* and *for* (formate) genes; the physical map for 200 kbp in the *oli* region that included these loci agreed with the previously determined genetic map. Transformation experiments using successively smaller subclones allowed localization of *frq* to an 8-kbp region of DNA. Phenotypic rescue of *frq-9* (see Table 4.1) restored not only rhythmic conidial banding with a wild-type τ, but also temperature compensation of τ over a 20 to 30°C temperature range and wild-type carotenogenesis.

The next step was to isolate "timed target" (clock-controlled) genes whose level of expression is regulated by the circadian oscillator. The *frq* gene was cloned (Dunlap et al., 1987b) by using complementary DNAs corresponding to mRNAs present at different times (CT 1, CT 13) to isolate cDNAs by subtractive and differential hybridization. These time-specific cDNA populations were then used to probe cDNA and genomic libraries. A putative clock-controlled gene identified by this protocol was verified by using the cloned DNA to probe Northern blots of RNA that had been isolated every 4 h from frq^+ (τ = 21.5 h) and *frq-7* (τ = 29 h) strains maintained for 12 to 56 h in DD. In these tests, the mRNAs derived from wild-type clock-controlled genes was found to cycle 2.5 times from maximum to minimum concentration, whereas mRNAs from the same genes in *frq-7* cycled only twice; consequently, the two mRNA species were out of phase over a 40-h time span. Thus, these cDNA clones clearly represented clock-controlled genes, rather than merely developmentally controlled genes, that were responding to culture conditions.

Are *Neurospora* clock-gene components similar to those of *Drosophila*? Feldman (personal communication, 1987) found that DNA sequences of *Neurospora* hybridize with the 8.0-kb rescuing fragment from the *per* locus of *Drosophila* (see preceding section) under conditions of moderate stringency. Preliminary sequencing data indicate that these sequences are homologous. One must now identify the *frq* gene product and determine if the biochemical functions encoded in these *Neurospora* genomic fragments are similar to those encoded in those of *Drosophila*.

4.4.3 Alteration of Clock Properties in Biochemical Mutants

In addition to the isolation and characterization of clock mutants discussed earlier, another related approach in the dissection of clock mechanisms is the isolation of biochemical mutants with known metabolic lesions and the determination of the effects of such mutations on the functioning of the clock (reviewed by Feldman, 1982, 1983). One of the best systems for illustrating this line of attack is afforded by *N. crassa* (Feldman and Dunlap, 1983).

Although the great majority of auxotrophic and morphological mutations that have been characterized in *Neurospora* do not affect the circadian clock, several mutants with various biochemical lesions have been discovered that exhibit significantly altered clock properties (Table 4.2). Conversely, certain biochemical mutants have been found in which clock

TABLE 4.2. Some biochemical mutants in *Neurospora crassa* in which clock properties (rhythm of conidiation) have been examined.

Strain	Mutation	Reference
Mutations affecting respiratory and photosensitive pigments		
al-1; al-2 (albino)	No detectable carotenoids; damping of rhythm in constant light unaltered	Sargent and Briggs (1967)
poky	Respiratory mutant; reduction in nonmitochondrial cytochrome; higher threshold intensity for inhibition of banding in LL	Brain et al. (1977a,b); Britz et al. (1977)
rib-1; rib-2	Riboflavin auxotrophs; reduction in levels of FAD and FMN in mycelia; reduction in light sensitivity of clock for phase-shifting and damping	Paietta and Sargent (1981)
nit-1; nit-2; nit-3	Reduction in activity of nitrate reductase; no effect on photosuppression or phase-shifting of conidiation rhythm	Paietta and Sargent (1982)
Mutations affecting cyclic 3′,5′-AMP		
NG 6-3; NG 6-11 revertants of *crisp-1*)	No colonial morphology; reduced levels of adenylate cyclase and cAMP; entrainment and free-running period unaffected	Feldman et al. (1979); (see Feldman, 1975)
cpd-1; cpd-2	Reduced levels of cAMP caused by reduction in activity of Mg^{2+}-stimulated cyclic PDE or of AC; period unaffected; rhythm in cAMP	Hasunuma (1984a,b); Hasunuma and Shinohara (1985)
Mutations affecting Ca^{2+} metabolism		
Ca4; Ca23	Ca^{2+}-dependency for growth; growth rate slower than wild type; τ in *Ca4* affected by changes in Ca^{2+} concentration; *Ca23* arrhythmic, does not entrain to LD or temperature cycles	Nakashima (personal communication, 1985)
Mutations affecting cysteine biosynthesis		
cys-X; cys-4; cys-12	Cysteine auxotrophs; shortened period	Feldman and Widelitz (1977); Feldman et al. (1979)
Mutations affecting fatty acids synthesis		
cel	Defective fatty acid synthetase complex; deficient in synthesis of palmitic acid (16:0); addition of unsaturated or short-chain saturated fatty acids lengthened period; loss of temperature compensation below 22°C, exacerbated by supplementation with linoleic acid (18:2), but restored in part by addition of 16:0	Brody and Martins (1979); Mattern and Brody (1979); Brody and Forman (1980): Mattern et al. (1982); Roeder et al. (1982); Lakin-Thomas and Brody (1985); Mattern (1985a,b); Coté and Brody (1987)

TABLE 4.2. Continued.

Strain	Mutation	Reference
Ergosterol-deficient mutants		
erg-1; erg-3	Steroid deficiency in plasma membrane; normal clock; growth rate resistant to nystatin, but phase-shifting by pulses of nystatin attenuated	Koyama, cited in Feldman and Dunlap, 1983
Oligomycin-resistant mutants		
oli^r	Resistance to drug oligomycin due to mutations in the DCCD-binding protein of mitochondrial ATPase; τ shortened in proportion to degree of resistance; introduction into *cel* restores temperature compensation below 22°C and negates effect of linoleic acid supplementation	Dieckmann and Brody (1980); Brody and Forman (1980); Brody (1981); Lakin-Thomas and Brody (1985); Brody et al. (1985)
Cycloheximide-resistant mutants		
cyh-1; cyh-2	Cycloheximide-resistant 80S ribosomes; clock unaltered, but pulses of CEX could no longer phase-shift rhythm	Nakashima et al. (1981b)

For further details, see Feldman (1982) and Feldman and Dunlap (1983).

characteristics have not been changed, thereby excluding a particular pathway or sequence as a key part of the oscillator itself.

Respiratory and Photosensitive Pigments

One approach has been to attempt to identify the photoreceptor(s) for the clock (see Section 4.2.1). Thus, Sargent and Briggs (1967) first showed that the action spectrum for the damping of the conidiation rhythm of *Neurospora* in LL resembled the absorption spectra of both flavins and carotenoids—typical of the blue-light photoreceptors that have been implicated in a multitude of responses in many organisms (Briggs, 1976; Senger, 1980). The carotenoids, however, seemed to be excluded by the fact that LL-induced damping was unaffected in an albino double mutant *(al-1, al-2)* in which these pigments were not detected (Sargent and Briggs, 1967). Additional evidence supporting the role of flavins in photoreception, for example, flavoprotein-mediated reduction of a *b*-type cytochrome (Muñoz and Butler, 1975), has been afforded by studies on several other biochemical mutants of *Neurospora* that affect light sensitivity. Thus, in the respiratory mutant *poky* in which the concentration of nonmitochondrial cytochrome is decreased, a 50-fold higher threshold intensity for the inhibition of banding in LL was found (Brain et al., 1977a,b; Britz et al.,

1977). Similarly, Paietta and Sargent (1981) have demonstrated a lowered light sensitivity of the clock for both the damping and phase-shifting responses in two riboflavin auxotrophs, *rib-1* and *rib-2*, grown under conditions where riboflavin was growth-limiting and the concentrations of mycelial flavin-adenine dinucleotide and flavin mononucleotide were decreased. On the other hand, three other light-insensitive mutants *(lis-1, lis-2, lis-3)* that did not exhibit photosuppression of the conidiation rhythm were found to have normal complements of flavin and cytochrome *b*; their defects, therefore, are not similar to those for *rib* and *poky* (Paietta and Sargent, 1983). Nitrate reductase, a molybdo-flavoprotein containing a *b*-type cytochrome has been hypothesized to serve as the photoreceptor for blue light-stimulated conidiation (Klemm and Ninnemann, 1979). This enzyme, however, does not appear to be important for the LL-suppression and the light-induced phase-shifting effects, since three mutants *(nit)* deficient in nitrate reductase activity (as well as cultures grown under conditions in which enzyme activity was repressed) continued to exhibit normal photoresponses (Paietta and Sargent, 1982).

Cyclic AMP

Because a number of studies had suggested the involvement of cAMP in circadian clock function [see Section 4.3.3 and Cummings' model (1975), Section 5.2.5], a genetic approach to testing this hypothesis promised to be fruitful in *Neurospora* (Feldman, 1975). The role of cAMP in the normal operation of the clock underlying conidiation was examined in partial revertants of the morphological mutant *crisp-1* (Table 4.2) that did not exhibit the tight conidial banding pattern (so that the rhythm could be measured). Although these mutants had been shown to have greatly reduced levels of AC and cAMP, they were entrained in the usual fashion by LD: 12,12 and free-ran with an unaltered period (Feldman et al., 1979). These results, therefore, appear to demonstrate clearly that the maintenance of significant pools of these two biochemical species at wild-type levels is unessential for clock operation. Nevertheless, recent results obtained with two groups of rhythmically conidiating, orthophosphate-regulated, repressible cyclic-phosphodiesterase mutants of *Neurospora, cpd-1, cpd-2* (Table 4.2), in which there is also a decreased level of cAMP (brought about by a reduction in AC activity or an increase in the activity of Mg^{2+}-stimulated cyclic PDE or both) but in which τ is unaffected, have demonstrated a circadian oscillation in cAMP whose peak corresponds to the time of the conidiation band (Hasunuma, 1984a,b). These findings, coupled with those showing a rhythmic oscillation of energy charge (Delmer and Brody, 1975; see Fig. 5.4), a shortened period in oligomycin-resistant *(bd oli^r)* strains (Dieckmann and Brody, 1980; Table 4.2), and the lengthened period caused by inhibitors of PDE (Feldman, 1975; see Section 4.3.3) suggest a rhythmic flow of ATP from mitochondria to cAMP in *Neurospora* (Hasunuma, 1984a,b).

Fatty Acid Metabolism and Other Biosynthetic Pathways

In contrast to the preceding mutant strains in which clock properties were not affected, mutations affecting calcium metabolism, cysteine biosynthesis, and other biochemical pathways have been shown to affect the period, entrainability, and even rhythmicity itself in *Neurospora* (Table 4.2). For example, the cysteine auxotroph *cys-X* grew at a slower rate on limiting amounts of sulfur (furnished as methionine), but its clock ran faster, that is, was significantly shortened (Feldman and Widelitz, 1977). Similarly, phase-shifting (but not the duration of τ itself) of the conidiation by pulses of nystatin also was attenuated in the ergosterol-deficient mutants *erg-1* and *erg-3* (Feldman and Dunlap, 1983).

Perhaps even more dramatic was the demonstration that the period could be effectively titrated by the addition of unsaturated fatty acids to the medium of the *cel*$^-$ mutant of *N. crassa,* which has a defect in the fatty acid synthetase complex, and thus a partial requirement for exogenous fatty acids, as well as a loss of temperature compensation of the conidiation rhythm in cultures grown below 22°C (see Section 4.3.3). Brody and Martins (1979) showed that the addition of unsaturated fatty acids caused a striking increase in the period of the conidiation rhythm. Linoleic acid (18:2) was most effective, yielding values of τ as long as 40 h, followed by linolenic acid (18:3) and oleic acid (18:1). Although saturated fatty acids, such as palmitic acid (16:0) and stearic acid (18:0), did not alter the period when added to the medium alone, palmitic acid could reverse the period-lengthening effects of linoleic acid. In contrast, shorter-chain saturated fatty acids (C_8 to C_{13}) did lengthen the period of the *cel*$^-$ mutant, but not of the wild type, from about 21 h to as long as 40 h, depending on the supplement (Mattern and Brody, 1979; Mattern et al., 1982). Interestingly, addition of unsaturated (18:1, 18:2, 18:3) or short-chain (12:0) fatty acids to the medium caused a further loss of temperature compensation in *cel*$^-$, whereas supplementation with long-chain (16:0, 18:0) unsaturated fatty acids restored compensation to that characteristic of the wild type (Mattern et al., 1982; see Section 4.3.3).

The different responses of the wild type and of *cel*$^-$ to exogenous fatty acids could not be attributed to their differential use or incorporation by the two strains. Their incorporation of [^{14}C]lauric acid (12:0) was approximately the same. More recent results on the effects of unsaturated fatty acid isomers (Mattern, 1985a) and functional group fatty acid analogs (Mattern, 1985b) support a more indirect—perhaps enzymatic—mode of action of fatty acids on the clock as opposed to a direct (membranar), fluidizing effect (but see Section 4.3.3). Perhaps mitochondrial metabolism of the fatty-acid supplements, a localized membrane change that might affect the activity of the fatty acid ATP synthetase complex, could account for the clock alterations in *cel*$^-$, as discussed below. It is interesting to note that the phospholipid fatty acid composition of *cel*$^-$ in liquid shaker culture, as well as the mitochondrial phospholipids, has been found to be distinctly different from that of a *cel*$^+$ control strain (Coté and Brody, 1987).

Oligomycin and Cycloheximide Resistance

Several drug-resistant mutants of *N. crassa* have been successfully examined for altered clock properties (Table 4.2). Thus, *oli*^{r} nuclear mutants, resistant to oligomycin (which inhibits mitochondrial ATPase), have been found to have periods several hours shorter than the wild type at 22°C, and the extent of this shortening is correlated with the degree of oligomycin resistance (Dieckmann and Brody, 1980; Brody, 1981). These results suggest that ATP production, and perhaps mitochondrial function in general, may be important for the function of circadian clocks (see Sections 4.3.1, 4.3.3). In this regard, Brody et al. (1985) provocatively have reported that point mutations in several *oli*^{r}strains affect the primary structure of the proteolipid, or DCCD-binding protein (yeast subunit 9), found in the F_0 membrane portion of the mitochondrial ATP synthetase complex (which transmits energy from the electrochemical proton gradient to the F_1 moiety of the enzyme) and thereby lead in some manner to a shortened period. Furthermore, Brody and Forman (1980) discovered that if the *oli*^{r} mutation was introduced into the *cel*$^-$ strain, the sensitivity of the latter to the addition of linoleic acid (increase in τ, loss of temperature compensation) was eliminated (Brody and Martins, 1979; Mattern et al., 1982; Lakin-Thomas and Brody, 1985). They speculate that linoleic acid alters the clock via its effects on mitochondrial ATPase, perhaps by binding to the ATPase F_0 portion. The *oli*^{r} mutation might negate these effect by causing a conformation change in the protein, thereby blocking the binding site.

Nakashima et al. (1981b) have shown that although the clock is not altered in the mutants *cyh-1* and *cyh-2*, which are resistant to CHX inhibition of protein synthesis on 80S ribosomes, neither of these two mutants could be phase-shifted by pulses of this inhibitor. In contrast, the wild type is well known to be sensitive to CHX, and a PRC has been derived for its action (Nakashima et al., 1981a) that is similar to those found in *G. polyedra* (Walz and Sweeney, 1979; Dunlap et al., 1980), *A. mediterranea* (Karakashian and Schweiger, 1976b), and other organisms (see Section 4.3.2). This finding demonstrated that the target of CHX is indeed the 80S ribosomes and not some other unknown target. A cautionary note is in order: *frq-7* (in contast to the other *frq* mutants) appears to be CHX-insensitive to CHX pulses for phase-shifting yet is CHX sensitive for protein synthesis and growth and manifests a light-entrainable, free-running rhythm of conidiation (although temperature compensation is somewhat impaired). Thus, in this mutant, the normal clock requirement for 80S protein synthesis (see Section 4.3.2) seems to have been bypassed or eliminated (Feldman and Dunlap, 1983).

4.5 Characterizing the Coupling Pathway: Transducing Mechanisms Between Clocks and Their Hands

Another time-honored approach toward elucidating the mechanism underlying an overt, physiological rhythm, or expression (hand) of the clock, is to attempt to thread one's way back through the biochemical pathways mediating the rhythm—the so-called transducing mechanisms (Sweeney, 1969b)—until one arrives at their point of coupling to the oscillator (Edmunds, 1976; Johnson and Hastings, 1986). The circadian rhythms of bioluminescence and of photosynthesis of *G. polyedra* (whose physiological characteristics have been discussed in Section 2.2.4 and which were dissected by chemical perturbations in Section 4.3) will serve here as excellent minicase histories.Other illustrative systems include the circadian rhythms of photosynthesis in *E. gracilis* (see Section 2.2.3), of the intracellular contents of NAD^+, NADP(H), NAD^+ kinase, and $NADP^+$ phosphatase and of Ca^{2+} and mitochondrial Ca^{2+} transport in *Lemna* and *Euglena* (see Sections 4.3.3,5.2.5), the rat-liver tyrosine aminotransferase and β-hydroxy-β-methylglutaryl (HMG) coenzyme A (CoA) reductase pathways, the compound action potentials produced by the eye in the molluscs *Aplysia* and *Bulla* (see Section 4.1.1), and the pineal gland (see Section 4.1.1) and the role it plays in photoperiodism via the clock-controlled enzyme N-acetyltransferase (see Section 5.2.4), which limits the overall rate in the synthetic pathway for melatonin. The ultradian glycolytic oscillator (see Section 5.2.1) is also a well-understood biochemical system.

4.5.1 Rhythms of Bioluminescence in *Gonyaulax*

The reaction responsible for light production involves the oxidation of dinoflagellate luciferin (substituted, open-chain tetrapyrrolic structure) by molecular oxygen, catalyzed by a specific luciferase enzyme:

$$\text{Luciferin} + O_2 \xrightarrow{\text{luciferase}} \text{light } (\lambda = 475 \text{ nm}) + \text{products}$$

The absolute level of luciferin (Bode et al., 1963), the binding capacity of its specific binding protein (Sulzman et al., 1978), and the activity of luciferase (Hastings and Bode, 1962) have all been shown to be under circadian clock control. Their activities in extracts made in the middle of the night phase are typically 5 to 10 times greater than those in similar extracts made from day-phase cells, and these cyclic changes continue in cells under free-running conditions of continuous dim illumination.

Earlier work (McMurry and Hastings, 1972b) on the cause of rhythmic luciferase activity ruled out simple explanations involving differences in enzyme extractibility or extractable inhibitors or activators, leaving open the alternatives of cyclic synthesis and degradation of enzyme (constant specific activity) or cyclic covalent modification of the polypeptide, thus altering its activity (cyclic specific activity). Dunlap and Hastings (1981) have purified unproteolyzed, higher-molecular-weight luciferase from both day-phase and night-phase cells and compared the two preparations with respect to several physicochemical, enzymatic, and immunological criteria. A given amount of antiluciferase inactivated the same amount of luciferase activity in both extracts, indicating that their specific activities were the same and suggesting that the luciferase was the same polypeptide in day and night preparations but that there were different amounts of the enzyme in each. These findings were confirmed by Johnson et al. (1984), who directly measured luciferase protein with antibody to luciferase and found that the cyclic activity of the enzyme corresponded to a rhythm in the concentration of immunologically reactive luciferase protein. Thus, the circadian rhythm of luciferase activity could be attributed to circadian clock-modulated synthesis or degradation or both of the luciferase polypeptide.

Clock control of enzyme turnover may constitute a general adaptive strategy enabling *Gonyaulax* to conserve nitrogen (normally limiting in the natural environment, in contrast to energy supply) by degrading enzymes whose function is temporarily unnecessary and by recycling their amino acids while expending abundant photosynthetic energy (Dunlap et al., 1981). Further, control of luciferase through its turnover may represent an efficient means of modulating a metabolic pathway by regulation of the activity of a rate-limiting enzyme, particularly if it is rapidly degraded and thus has a half-life that is only a small fraction of the circadian period (Johnson et al., 1984).

Of course, the question now arises (as it always does) as to the next step in the quest for the elusive clock. In particular, is the control of the synthesis of luciferase and luciferin binding protein (LBP) transcriptional (in which case their respective concentrations should be rhythmic) or translational (where constant mRNA concentrations would be translated rhythmically)? A molecular genetic attack on this problem involves the cloning of luciferase and LBP cDNAs and using them as probes to measure their luciferase mRNAs as a function of circadian time. Just such an approach has been taken in the *Gonyaulax* system (Morse et al., 1987a; Milos et al., 1987). The LBP cDNA was isolated by immunologic screening of a cDNA library that had been subcloned into an expression vector. The identity of the cDNA was confirmed by in vitro translation of a mRNA hybrid selected from total RNA by the LBP clone. Northern hybridization of the cDNA to mRNA, isolated at seven different CTs, showed that the amounts of the LBP mRNA were invariant. A putative luciferase cDNA also was isolated, and similar experiments demonstrated constant levels of the corresponding luciferase mRNA over a 24-h time span. In a complementary approach, in vitro translation of the mRNA that was used in the Northern blots showed that the synthesis of LBP at all time points was identical. Thus, the regulation of the circadian rhythms in the amounts of LBP and luciferase is exerted at the translational level.

4.5.2 Rhythms of Photosynthesis in *Gonyaulax*

In the study of transducing mechanisms from the putative circadian oscillator to observed rhythmicities, photosynthesis is particularly attractive because it comprises a number of relatively well understood processes that can be measured individually. Its periodicity, together with that of bioluminescence, was one of the first rhythms to be examined in detail in *Gonyaulax* (see Section 2.2.4).

Since no rhythmicity was observed in total cellular chlorophyll content, attention turned to other transducing mechanisms that might be involved in the expression of the rhythm of PC. Earlier efforts concentrated on the dark reactions comprising the Calvin scheme (Sweeney, 1969b, 1972; Bush and Sweeney, 1972), but results were not promising. For example, although ribulose-l,5-biphosphate carboxylase sometimes showed fluctuations in activity, there was not sufficient correspondence to satisfy the rates of CO_2 fixation at all circadian stages investigated in *Acetabularia* sp. none of the Calvin cycle enzymes were found to have rhythmic activities (Hellebust et al., 1967; see Section 2.2.1).

Inasmuch as the enzymes responsible for carbon fixation did not vary over the circadian cycle, experimentation focused on the light reactions of photosynthesis and the activities of PSI and PSII, and results have made it clear that they are regulated by the clock (Sweeney, 1981). Thus, from analysis of the rates of photosynthesis as a function of irradiance, a temporal change in relative quantum yield in dim LL was found, although total chlorophyll, half-saturation constants and the size and number of particles on the thylakoid freeze-fracture faces were constant (Prézelin and Sweeney, 1977; Prézelin et al., 1977).

Govindjee et al. (1979) then examined fluorescence transients at different times in *Gonyaulax* cells maintained in either LD or LL, and no rhythmic changes were found. On the other hand, the intensity of Chl *a* fluorescence (both initial and peak values) was about twice as high during the day phase of the circadian cycle in LL than during subjective night, and these changes were positively correlated with the rhythm of O_2 evolution (Sweeney et al., 1979). This periodicity in fluorescence persisted in the presence of 10 μM DCMU, which blocks electron flow from PSII during photosynthesis, indicating that the cause was not due to a change in net electron transport between PSII and PSI. Although the rhythm of fluorescence could arise from differences in the efficiency of spillover energy from strongly fluorescent PSI, such spillover should occur unimpaired at 77°K, but this was not the case. The rhythmicity in PC was abolished at this temperature (Sweeney et al., 1979). The results could indicate that the nonradiative decay of Chl excitation is less during the day than at night, although the reason for such a change remains obscure. Perhaps circadian ion fluxes across the thylakoid membrane generate reversible conformational changes that would couple and uncouple entire photosynthetic units in the light-harvesting pigment–protein complex and thus induce a circadian rhythmicity in PC (Prézelin and Sweeney, 1977; Sweeney, 1981).

In *E. gracilis* (see Section 2.2.3), although the individual activities of PSI and PSII do not appear to change significantly with the time of day (Walther and Edmunds, 1973; Lonergan and Sargent, 1979), the rate of light-induced electron flow through the entire electron chain (water to methyl viologen) was found to be rhythmic in both whole cells and isolated chloroplasts (see Fig. 2.17B), the highest rate of flow coinciding with the highest rate of O_2 evolution (Lonergan and Sargent, 1979). Evidence consistent with the notion that the coordination of the two photosystems may be the site of circadian control of rhythms in PC was obtained with studies of low-temperature fluorescence emission from PSI andII following preillumination, respectively, with light wavelengths of 710 or 650 nm, whereas there was no indication that changes in total Chl, the ratio of Chl *a* to Chl *b*, or the size of the photosynthetic units were responsible (Lonergan and Sargent, 1979).

5
Biochemical and Molecular Models for Circadian Clocks

We now have completed our survey of circadian organization in eukaryotic microorganisms (see Chapters 2 and 3)—with appropriate reference to prokaryotic and higher systems and to ultradian rhythms—and discussion of some the most important experimental findings from attempts to elucidate clock mechanisms (see Chapter 4). It is instructive to consider briefly some of the models for the biochemical and molecular bases of these oscillators that have been posited over the years (reviewed by Edmunds, 1976, 1983; Engelmann and Schrempf (1980); Sweeney, 1983; Jacklet, 1984; Jerebzoff, 1986; Schweiger et al., 1986; Vanden Driessche, 1987; see the still useful proceedings of the Dahlem Workshop on this subject, edited by Hastings and Schweiger, 1976). Despite the fact that models in this fields are plentiful, they do serve as useful foci, often leading to important experimental advances. If these results then falsify the construct, so much the better: a good model sows the seeds of its own destruction! Furthermore, it is perhaps useful to introduce the potential initiate to this field to some of the extant notions if for no other reason than to provide reassurance that it is still virgin territory. We have a long way to go to placate the editorial writer who plaintively queried, "Why is so little known about the biological clock?" and expressed hope that it soon would be "wound up" (*Nature* 231: 97–98, 1971). Indeed, the very fact that so many models have been put foward should serve as a warning that the biological clock possibly is much more complex than we earlier had envisioned.

There are several different classes of model for endogenous, self-sustaining circadian clocks (Table 5.1), which are neither mutually exclusive nor jointly exhaustive. They can be grouped into several main categories (Edmunds, 1976, 1983): (1) strictly molecular models, which rely on the properties of molecules themselves for generating persisting 24-h rhythms, (2) feedback loop, network models for oscillations in energy metabolism and in other biosynthetic pathways, in which longer periods would be generated (frequency demultiplication) by energy reservoirs or depots, appropriate allosteric constants, and turnover numbers of key enzymes, or by negative cross-coupling among individual oscillators (or even among

TABLE 5.1. Some molecular models for circadian clocks.

Model	Key elements	References
Molecular (in vitro)	Periodicity in the structure and the properties of molecules themselves (e.g., alternation between two conformational subunit states resulting from posttranscriptional phosphorylation and dephosphorylation)	Queiroz-Claret and Queiroz (1981); Brulfert et al. (1986)
Network (biochemical feedback loops)	Glycolytic oscillator: By suitable selection of allosteric constants and turnover numbers of key enzymes, frequency can be controlled over a large range (in principle, even 24 h)	Hess and Boiteux (1971); Goldbeter and Caplan (1976): Goldbeter and Nicolis (1976); Hess (1976); Das et al. (1982)
	Cell energy metabolism = the clock: Appropriate choice of depot (deposition effect) would allow self-oscillatory reactions on a circadian time scale	Sel'kov (1975, 1979); Reich and Sel'kov (1981)
	Coupled oscillators: Cross-coupling among high-frequency oscillations in energy metabolism could generate circadian rhythmicity in energy transduction	Winfree (1967); Goodwin and Cohen (1969); Pavlidis and Kauzmann (1969); Wagner and Cumming (1970); Pavlidis (1971); Wagner (1976a,b); Vanden Driessche (1973); see Lloyd et al. (1982a)
	Cyclic AMP model: cAMP, ATP, AC and PDE are oscillating variables that would exhibit limit cycle behavior by allosteric feedback of AMP on AC and PDE	Cummings (1975)
	Mitochondrial Ca^{2+} cycle: NAD^+, NAD^+ kinase and $NADP^+$ phosphatase, Ca^{2+}, the mitochondrial Ca^{2+} transport system, and calmodulin represent clock gears, which, in ensemble, would constitute a self-sustained, negatively cross-coupled, circadian oscillator	Goto et al. (1985); Edmunds and Laval-Martin (1986); Edmunds et al. (1987); see Kippert (1987), Lakin-Thomas (1985)
	Heterodyne endosymbiont hypothesis: Two prokaryotic colonists of a putative ancestral eukaryotic cell emitted chemical pulses with slightly different short periods; their coincidence would have yielded a longer, circadian period	Levandowsky (1981); see Kippert (1985a, 1987)
Transcriptional (tape-reading)	Chronon model: Sequential transcription of long, polycistronic, DNA complexes in eukaryotic chromosomes, coupled to rate-limiting, time-consuming, temperature-independent diffusion steps by mRNA to the ribosomes for translation would yield circadian periods	Ehret and Trucco (1967); Barnett et al. (1971a,b): Wille et al. (1972)
	Chronogene–cytochron model for cell cycle clocks: Programmable, sequential transcription of segment of chromosomal DNA without requirement for translation	Edmunds and Adams (1981)

TABLE 5.1. Continued.

Model	Key elements	References
Membrane	Molecule X actively transported into organelles, changing configuration and transport capacity of membranes; passive diffusion then occurs until X is evenly redistributed	Sweeney (1974b)
	Limited cycle behavior in which an ion concentration gradient and membrane transport activity are the oscillating variables; slow translational diffusion of membrane proteins; cross-coupling; temperature compensation by changes in membrane lipid saturation	Njus et al. (1974, 1976); Njus (1976); see Konopka and Orr (1980)
	Membrane oscillator hypothesis: Membrane-bound photoreceptors modulate membrane-bound energy transduction; energetic state of membranes, in turn, determines photoreceptor sensitivity	Wagner and Cumming (1970); Wagner et al. (1974a,b)
	Membrane transport: Rhythmic interplay of membrane pumps, leaks, and porters; role of cell division cycle	Satter and Galston (1971a,b; 1973; 1981); Edmunds and Cirillo (1974)
	Proton gradient model: Membrane-bound carrier actively transports H^+ to the outside; key enzyme having pH-dependent activity profile translocates H^+ from exterior to interior (or produces H^+ as a product); sustained oscillations occur when the two competing processes balance	Chay (1981)
	Coupled translation–membrane model: assembly, transport, and insertion (loading) of essential proteins into membranes	Schweiger and Schweiger (1977); Schweiger et al. (1986)
	Monovalent ion-mediated translational control model: Intracellular monovalent ion concentration feedback-regulates the synthesis and insertion of membrane proteins, resulting in changes in ion concentration	Burgoyne (1978)

Updated from Edmunds, 1984c, with permission of Academic Press.

prokaryotic endosymbionts) within a cell, (3) transcriptional (tape-reading) models, wherein the transcription of a long, polycistronic piece(s) of DNA with associated rate-limiting diffusion steps leading to translation would meter circadian time, and (4) membrane models, in which the transport activity and other properties of various membranes in the cell (intimately related to state transitions in the fluid mosaic membrane itelf, which, in turn, affect membrane structure) would comprise, in ensemble, a stable limit cycle oscillator that ultimately would account for the properties of circadian rhythms.

These various models (Table 5.1) sometimes are overlapping, for they often incorporate several different notions from each other. This is not surprising, since each has strengths and shortcomings for which it attempts to compensate by hybridization.

5.1 In Vitro Molecular Models

Slow, self-sustaining oscillations in the catalytic capacity of the enzymes phosphoenolpyruvate carboxylase (E.C. 4.1.1.31, PEPC) and malate dehydrogenase (E.C., 1.1.1.37, MDH) have been reported to arise spontaneously in crude or desalted extracts of leaves of the succulent *Kalanchoe blossfeldiana* (Queiroz-Claret and Queiroz, 1981). Statistical treatment of the data (Queiroz-Claret et al., 1985) with a curve-fitting model (sum of a polynomial trend and a Fourier series) indicated periods ranging from 16 h to 27 h, depending on the experiment, for enzyme solutions maintained in DD or LL for more than 100 h and sampled at intervals of 1 to 1.5 h (Fig. 5.1). The period was temperature-independent over a 5 to 17°C range of constant temperatures. These periodicities are in contrast to those oscillations produced by feedback or feedforward regulatory mechanisms occurring during the interval in which an enzyme is active (see Section 5.2). The authors speculate that slow, synchronous transitions among several conformational states of enzymes in solution might occur, although the possibility that regulatory proteins present in these crude extracts may play a role still exists.

Rapid autonomous changes of proteins in solution in the order of seconds or minutes have been reported. For example, Shnoll and Chetverikova (1975) found synchronous, reversible alterations in the in vitro activity of actomyosin and creatine kinase preparations, and Mandell and Russo (1981) have given evidence for multiple, conformational kinetic oscillators underlying the quick periodic changes observed in the activity of striatal tyrosine hydroxylase in complex biological extracts. Spectral analysis of the slower oscillations in PEPC and MDH (Queiroz-Claret et al., 1985) indicated, however, that these long periods did not derive from combinations of the more rapid fluctuations also seen (Fig. 5.1). Shnoll and Chetverikova (1975) also observed that the rapid fluctuations in enzyme activity disappeared when enzyme extracts were incubated in the presence of substrates. Thus, the in vivo role of such in vitro phenomena is not at all clear.

It is interesting to note that the regulatory properties of PEPC—a key enzyme in the control of the diurnal cycle of crassulacean acid metabolism (CAM)(Ting and Gibbs, 1982)—themselves change during the diurnal LD cycle (as well as in DD). Thus, PEPC extracted from CAM plants during

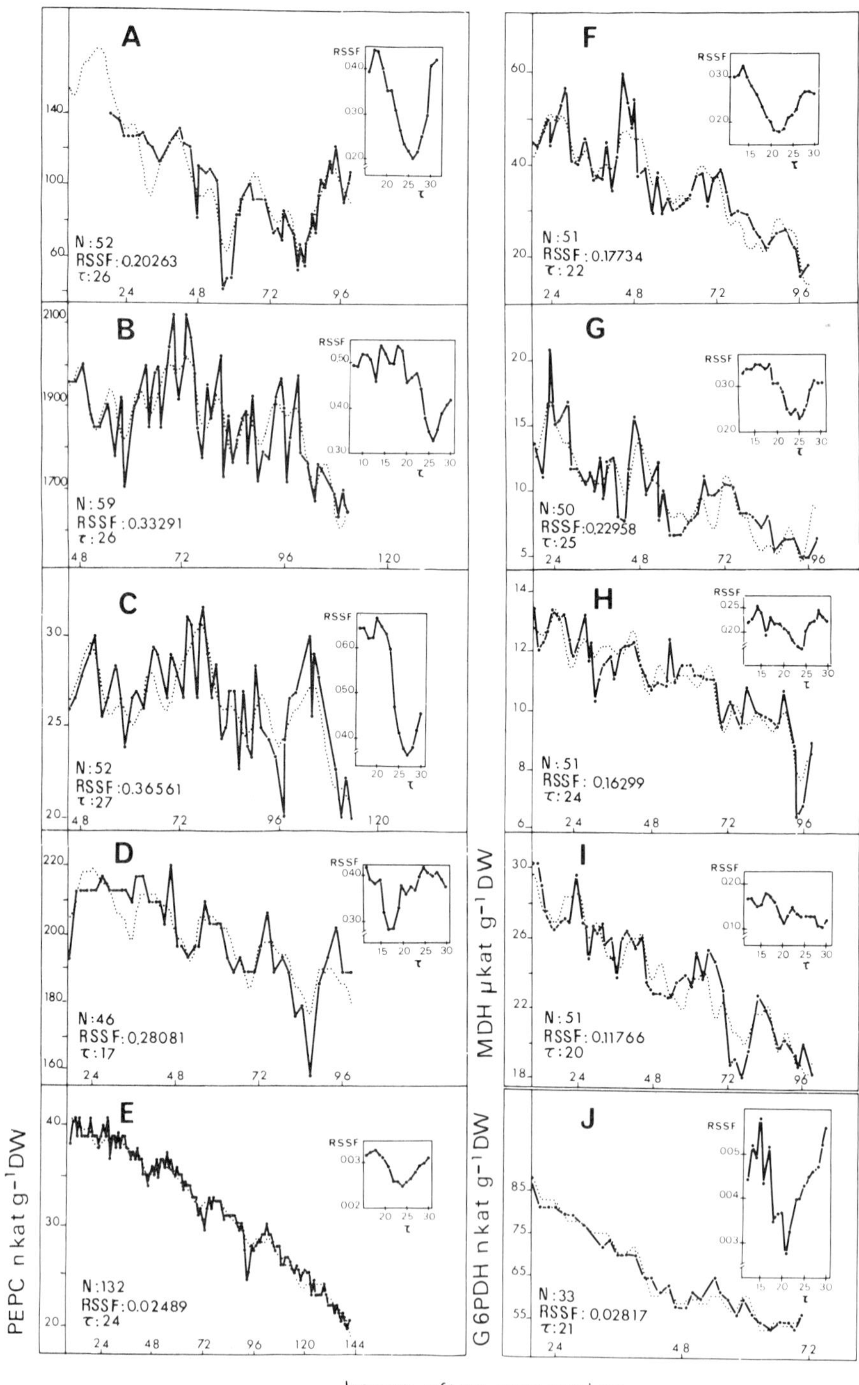
A
N:52
RSSF:0.20263
τ:26
B
N:59
RSSF:0.33291
τ:26
C
N:52
RSSF:0.36561
τ:27
D
N:46
RSSF:0.28081
τ:17
E
N:132
RSSF:0.02489
τ:24
PEPC nkat g⁻¹DW
F
N:51
RSSF:0.17734
τ:22
G
N:50
RSSF:0.22958
τ:25
H
N:51
RSSF:0.16299
τ:24
I
N:51
RSSF:0.11766
τ:20
MDH μkat g⁻¹DW
J
N:33
RSSF:0.02817
τ:21
G6PDH nkat g⁻¹DW
hours after extraction

the night has a high affinity toward the substrate PEP and is only weakly sensitive to inhibition by malate and to activation by G6P.The reverse situation holds true for PEPC extracted during the day. The night form is supposedly the active one, permitting high rates of nocturnal CO_2 fixation, whereas the day species is inactive. Thus, the futile cycling of PEP by CO_2 that would otherwise arise from malate decarboxylation during the day is prevented.

Although the mechanism(s) underlying these changes in PEPC activity is unknown, titration of crude and purified PEPC extracted at different times of day from the CAM plants *K. blossfeldiana* (Brulfert et al., 1982) and *Bryophyllum fedtschenkoi* (Nimmo et al., 1984, 1986) with antibody raised against it indicated that the daily rhythm did not derive from a rhythm in protein synthesis or degradation or both (as was found for *Gonyaulax polyedra*; see Section 4.5.1), but rather from changes in the specific activity of the enzyme. Nimmo et al. (1984) proposed that these changes in the regulatory properties of PEPC at the posttranslational level were the consequence of a reversible phosphorylation–dephosphorylation process. This hypothesis has received further support in *Kalenchoe* spp., in which only the night form of partially purified PEPC, extracted from the leaves of plants that had been previously supplied with exogenous [^{32}P]orthophosphate, was phosphorylated (Brulfert et al., 1986). Moreover, in vitro addition of exogenous acid phosphatase (E.C. 3.1.3.2) to desalted extracts containing the night form resulted within 30 min in a shift in PEPC enzymatic properties similar to those observed in vivo, and indeed, Nimmo et al. (1987) recently demonstrated persistent circadian rhythms in the phosphorylation state of PEPC from *B. fedtschenkoi* leaves as well as in its sensitivity to inhibition by malate. This cyclic phosphorylation leading to covalent modifications of PEPC would be caused, in turn, by the antagonistic, rhythmic activity of the appropriate kinases and phosphatases. However, the changes in regulatory properties of the enzyme could be the result of noncovalent, spontaneous changes in the protein molecules themselves, as discussed previously, which in some manner would generate a rhythmic, differential sensitivity to phosphorylation and dephosphorylation.

◁

FIGURE 5.1. Oscillations of phosphoenolpyruvate carboxylase (A, B, C, D, E), malate dehydrogenase (F, G, H, I) and glucose-6-phosphate dehydrogenase (J) activity in aliquots sampled successively from extracts maintained in constant conditions from the leaves of *Kalanchoe blossfeldiana*. N, number of samples. Dotted lines, fitting model; RSSF, residual sum of squares with a fitting by the model corresponding to the optimum period τ (see inset plots). Enzyme activity expressed in nanokatals or microkatals per g dry weight (Kat = mol·s^{-1}). (Reprinted from Queiroz-Claret et al., 1985, with permission of Swets and Zeitlinger B.V.)

5.2 Biochemical Feedback Loop and Network Models

In the early part of the twentieth century, it was commonly believed that the evolution of any physiochemical system invariably leads to a steady state of maximum disorder as a result of the application of the second law of thermodynamics. As a consequence, any coherent behavior, such as self-sustained oscillations, was thought to be ruled out. The development in the 1940s of the concept of open systems (as opposed to closed), typified by living organsisms, enabled one to predict in principle the spontaneous appearance (self-organization) of temporal dissipative structures, including autonomously oscillatory systems (Prigogine, 1961). Stable, sustained oscillations of the limit cycle type can appear in certain nonlinear systems, usually beyond the domain of stability of a steady state. Positive or negative feedback, sometimes combined with cross-catalysis, is a necessary physical prerequisite (Nicolis and Portnow, 1973; Nicolis and Prigogine, 1977).

These theoretical notions were buttressed in the 1950s and 1960s by the discovery of a number of chemical and biochemical oscillations (reviewed by Hess and Boiteux, 1971; Nicolis and Portnow, 1973; see Chance et al., 1973; Goldbeter and Caplan, 1976; Hastings and Schweiger, 1976; Estabrook and Srere, 1981; Gurel and Gurel, 1983). For example, one of the most striking, purely chemical, oscillating systems is the Belousov-Zhabotinskii reaction (Winfree, 1974; Tyson, 1976), in which analogs of malonic acid are oxidized by bromate in the presence of Ce (or Fe or Mn) ions. Sustained oscillations in the concentrations of the reactants appear spontaneously if the reaction is carried out in a well-stirred, homogeneous medium, with sharp, reproducible periods and amplitudes. Other chemical clock reactions, in which a sudden marked change occurs after a precise waiting period, also have been described (Bose et al., 1986). At the biochemical level, three of the earliest oscillators all involved energy supply: photosynthesis in chloroplasts, respiration in mitochondria, and glycolysis in the cytoplasm of yeast (and even in cell-free preparations). In glycolysis, the inner mitochondrial membrane was found to possess distinct control properties, a finding that demonstrated that oscillations in complex biochemical systems can be reduced to the fundamental properties of enzymes or membranes and can finally be recovered in structural changes at the molecular level (Hess and Boiteux, 1971). The cyclic-AMP oscillator in the cellular slime mold *Dictyostelium* also is a well-defined, oscillating metabolic system.

Of course, these metabolic oscillations typically have periods measuring in minutes (see Fig. 1.1) and are highly dependent on temperature. In subsequent sections, we briefly examine mechanisms whereby such short-period, ultradian, biochemical oscillations might be frequency-demultiplied to generate epigenetic (τ = 1 h or more), circadian, and infradian oscillations, as well as means by which temperature compensation of the period could be incorporated into an evolving system (but see Jerebzoff, 1987).

5.2.1 Ultradian Metabolic Oscillators

The glycolytic oscillations observed in the yeast *Saccharomyces cerevisiae* represent one of the best understood soluble biochemical oscillators and will serve as a model for the escapement of one type of biological clock (reviewed by Hess and Boiteux, 1971; Goldbeter and Caplan, 1976; Goldbeter and Nicolis, 1976; Hess, 1976). Similarly, the cAMP oscillations ($\bar{\tau} \cong 7$ to 10 min) found in aggregating cells of *Dictyostelium discoideum* reflect a well-characterized, cellular metabolic oscillator that probably involves the cell membrane (Gerisch, 1976, 1986; Goldbeter and Caplan, 1976).

The Glycolytic Oscillator in Yeast

If yeast cells or yeast extracts are given a continuous addition of glucose as substrate, sustained oscillations occur in NADH fluorescence (which can be continuously monitored) and in other elements of the glycolytic system (Fig. 5.2; see Fig. 1.3). For a chemical system to exhibit oscillatory behavior, certain general types of reaction pathways must exist (Nicolis and Portnow, 1973). If X and Y represent any two substances whose net rates of production are V_x and V_y, respectively, oscillations can occur if the following three conditions are met: (1) one of the chemical (X) must activate its own production (assuming that X remains fixed), (2) the other substance (Y) must tend to inactivate its own net production, normally true of most chemical reactions, since increasing concentration increases the rate of removal of that chemical; and (3) there must be a cross-coupling of opposite character so that increasing the concentration of X activates the net production of Y, and increasing Y inhibits the net production of X (or vice versa). The period of the resulting oscillation will be dependent at the very least on the kinetic rate constants associated with these reactions. Let us consider a more specialized case of this general mechanism to account for the oscillatory behavior of the glycolytic system in yeast.

Elements of the Oscillator

The enzyme PFK (E.C. 2.7.1.11) has been found to periodically generate its products, ADP and fructose-1,6-bisphosphate (F-1,6-P). The cyclic changes in PFK activity, in turn, are propagated along the entire enzymatic reaction sequence of glycolysis through the adenylate system. The allosteric properties of PFK and the positive feedback exerted on it by a reaction product appear to be responsible for the resulting temporal dissipative structure (Hess and Boiteux, 1971; Goldbeter and Caplan, 1976;

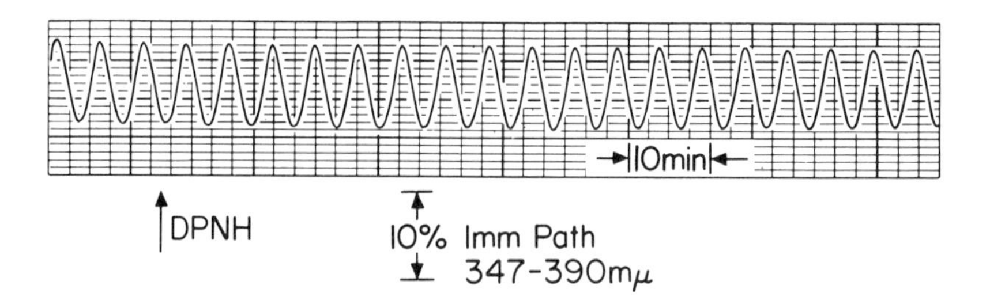

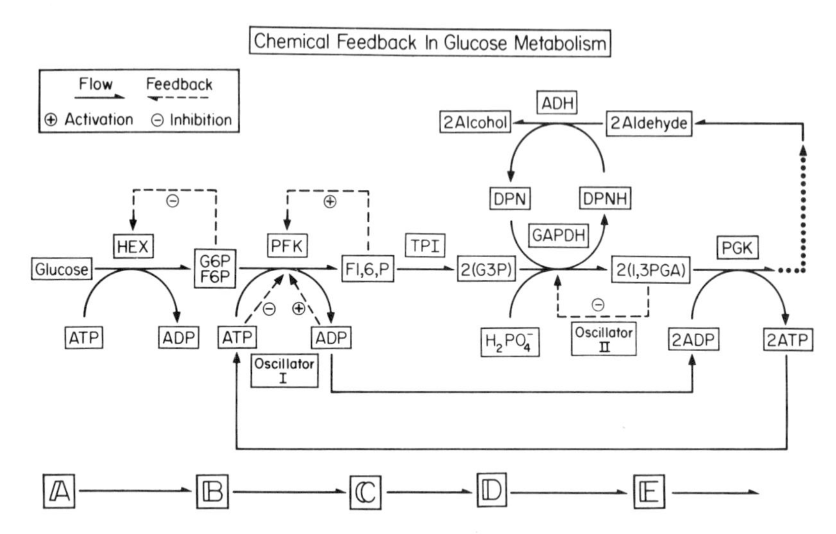

FIGURE 5.2. Feedback model for the sinusoidal oscillations of NADPH (DPNH) in a cell-free extract of the yeast enzyme system. The period is about 5 min. The amplitude of the curve represents about 10% change. A simplified diagram of the pathway from glucose to ethyl alcohol with the consequent reduction and oxidation of $NADP^+$ (DPN) is given in the lower panel. Feedback activation and inhibition are designated by the broken arrows (+ or −). Hypothesized oscillator sites are represented by boxes. Letters A, B, C, D, and E at the bottom of the lower graph refer to intermediates in the pathway. (Reprinted from Chance et al., 1967, with permission of IEEE.)

Goldbeter and Nicolis, 1976). Figure 5.2 indicates some of the key enzymes that operate in the series of reactions that convert glucose in alcohol in this microorganism. Only a few of the 30-odd enzymes and five intermediates (A, B, C, D,E) in this pathway are included. The basic reactions of concern here can be tabulated sequentially in a simplified version of the actual events, in both a generalized (left column) and a specific (right column) form (Higgins, 1964; see Pye, 1969):

(a) $A \rightarrow B$ (a′) $GLU \rightarrow F6P$
(b) $B + E_1^* \rightarrow E_1^* \cdot B$ (b′) $F6P + PFK^* \rightarrow PFK^* \cdot F6P$

(c) $E_1^*{\cdot}B \rightarrow E_1 + C$ (c′) PFK*·F6P → PFK + F-1,6-P
(d) $C + E_1 \leftrightarrow E_1^*$ (d′) F-1,6-P + PFK ↔ PFK*
(e) $C + E_2 \rightarrow E_2{\cdot}C$ (e′) F-1,6-P + TPI → TPI·F-1,6-P
(f) $E_2{\cdot}C \rightarrow E_2 + D$ (f′) TPI·F-1,6-P → TPI + 2(G3P)

Reaction (a) represents the conversion of glucose (GLU; continuously provided to the system directly or via hydrolysis of trehalose and designated as intermediate A in Fig. 5-2) to fructose 6-phosphate (F6P) (intermediate B). In the second step [Reaction (b)], E_1^* represents an activated form of the enzyme PFK (PFK*), which combines with its substrate F6P to form an enzyme-substrate complex; this complex then breaks down by Reaction (c) to release the free activated enzyme and the product of the reaction (F-1,6-P; intermediate C), with the concomitant production of ADP, as shown in Figure 5-2. Finally, in Reactions (d) and (e), F-1,6-P serves as the substrate for a second enzyme E_2, triosephosphate isomerase (TPI), to eventually produce glyceraldehyde-3-phosphate (G3P) (designated as intermediate D); this product then continues through the glycolytic pathway leading to the formation of alcohol.

Thus far, the enzymatic reactions constitute simply a linear sequence and contain no step that would cause oscillation. The crucial step is given by Reaction (d), in which F-1,6-P reacts with an inactive form of PFK (E_1) to produce the active form, E_1^* (PFK*). [In point of fact, it appears that ADP, the other byproduct of Reaction (c) in addition to F-1,6-P, is actually the critical activator; see Sel'kov, 1968.] This ability of F-1,6-P (or ADP) to activate PFK results not only in a net increase in its own production by Reactions (b) and (c), but also in a decrease (inhibition) in the net rate of production of of F6P by Reaction (a). Thus, the coupled reaction sequences now satisfy the general requirements for oscillatory behavior, with F-1,6-P (or ADP) corresponding to hypothetical substance X, and F6P to substance Y.

Operationally, if the concentration of ADP in the cell extract or suspension happened to be low (under constant glucose infusion), the activity of PFK would also be low due to the lack of ADP-activation [by Reaction (d)] and F6P would start to pile up [via Reaction (a)]. As a result, the rate of Reactions (b) and (c) would increase, since F6P is the substrate for the catalyzing enzyme PFK, and, in turn, the concentrations of the products F-1,6-P and ADP would also rise. The increase in ADP concentration would then activate PFK [Reaction (d)] and cause the former to increase even more and thus provide further activation of PFK. Eventually, the velocity of Reactions (b) and (c) will exceed the rate of F6P production [Reaction (a)] and cause the concentration of F6P to fall and then limit the PFK-mediated reaction. Finally, the pool of ADP would be depleted and the cycle then would be complete.

Experimental studies on the yeast system, as well as simulation studies on an analog computer (in which one ascribes appropriate rate constants and glucose concentrations to the foregoing set of reactions and represents the basic feedback property inherent in the system by two or more differential equations), reveal that it has a number of properties (see Fig. 1.2) that are formally similar to those characterizing circadian systems (Boiteux et al, 1975; Hess, 1976). The self-sustained oscillations in NADH absorbance typically have a free-running period of 2 to 70 min. Both τ and amplitude are flux-rate dependent, and oscillations are observed only at critical flux range (outside of this limited range of chosen constants, the system simply proceeds to a nonoscillatory state, i.e., the rhythm damps out). The rhythm can be entrained by a periodic source of substrate that is sinusoidally applied. Entrainment by the fundamental frequency of substrate input occurs only within limits corresponding to the domain $0.75 < T/\tau < 1.25$. This range of entrainment for glycolysis is comparable to that observed for light cycles in circadian systems, in which T of the driving *Zeitgeber* typically must fall between 18 and 30 h (Bruce, 1960; Pittendrigh, 1965). The glycolytic oscillations can be phase shifted by pulses of metabolites; the addition of 0.7 m*M* ADP to a yeast extract generates a delay ($\Delta\phi = -1.5$ min). On the other hand, the temperature sensitivity of the glycolytic oscillator is high, unlike the temperature compensation of τ observed for circadian rhythms.

Although one can demonstrate that periods of up to 1 h or so might be obtained by decreasing the enzyme activity, the possibility of achieving circadian periods in this manner is ruled out by the PFK model (Goldbeter and Nicolis, 1976). By dilution of yeast extracts with buffer while maintaining metabolite concentrations at their initial levels, Das and Busse (1985) obtained a maximum period of about 6 h. Coupling enzyme oscillations by diffusion also fails to elongate the period of the glycolytic oscillator. Boiteux et al. (1975) note, however, that at least in principle the frequency can be shifted to any biologically relevant value by the appropriate fitting of source rate, sink, and enzyme activity (see Section 5.2.2). Likewise, temperature effects on enzymatic processes might be minimized if the enzymes operated in the first-order range of their rate law that contained the temperature-sensitive rate constants only in the form of ratios (Pavlidis and Kauzmann, 1969; see next subsection). Despite these speculations, it seems more likely that the circadian period is generated by an entirely different set of rate constants than those found for the high-frequency glycolytic oscillations or by a frequency reduction of an entirely (and unknown) character (but see next subsection and Section 5.2.3). The important point that emerges from these elegant studies of the glycolytic oscillator is that the mechanism of an endogenous, self-sustaining, fully autonomous (ultradian), biological clock can be accounted for quite satisfactorily, in concrete terms, and without having to resort in any way to alternative, extrinsic timing hypotheses.

Modeling Circadian Clocks with a Glycolytic-Type Oscillator

As alluded to earlier, one encounters difficulties in trying to generate circadian periods and to account for temperature compensation with ultradian metabolic oscillators. Nevertheless, a number of theoretical frequency-reduction mechanisms, mostly borrowed from physics and engineering principles, have been proposed (Tyson et al., 1976). These include (1) subharmonic resonance, or frequency demultiplication, by which an endogenous oscillator will synchronize to a subharmonic of a driving frequency if the latter is close to an integral multiple of the free-running frequency, (2) beats, by which the summed output of two uncoupled sinusoidal oscillators of frequencies ω_1 and ω_2 exhibits amplitude modulation of frequency $\frac{1}{2}(\omega_1 - \omega_2)$, and the amplitude goes through a maximum periodically at twice this frequency $(\omega_1 - \omega_2)$, (3) a frequency filter, or counter, of some sort (although this option obviously generates another black box), (4) long feedback loops, producing low-frequency oscillations from high-frequency subunits (which may include enzymes, mRNAs, membranes, and the like), but with the resultant low frequencies depending on the special properties of the feedback loop and not particularly on the mechanisms of the oscillations in the subunits themselves, which in fact, do not even have to be oscillatory (see Sections 5.2.4, 5.3, and 5.4), (5) feedback quenching, wherein a cellular oscillator, which can be expected to modify the levels of other constituents, would have its frequency reduced (by a factor of 10 or more) by the feedback of some of these constituents and, under certain conditions, their temporary quenching effect, on the oscillating system (Gilbert, 1978c), (6) coupled conservative oscillators (whose ampitude and period do not depend on initial conditions), which might exhibit a low-frequency collective mode of oscillation (Goodwin, 1963, p. 119), (7) mutual coupling of two oscillators having free-running periods τ_1 and τ_2, which, depending on the strength of the coupling, will adopt a common period (and in some, will display multiply periodic oscillations), and (8) elastic coupling, whereby a large frequency reduction can be obtained by the coupling of n oscillators of endogenous frequency ω_0. The sum of the outputs of the same subunits oscillates at a frequency given by the product of ω_0 and the square root of $[1-(n-1)r]$, where r is the coupling constant (Pavlidis, 1969; see Section 5.2.3).

There are several general types of objection, however, to each of these hypotheses as a possible explanation of circadian rhythms (Tyson et al., 1976). First, the mechanism underlying the high-frequency component(s) bears little relevance to the properties of the low-frequency output. Second, the degree of reliable reduction in the frequency is often too small. Third, the predicted PRCs for light (and other) perturbations do not have correct shape.

In spite of these difficulties, Pavlidis and Kauzmann (1969) were able to demonstrate that a rather simple biochemical oscillator, given certain assumptions (as always!), could simulate quantitatively the formal behavior of circadian oscillators. Their system involved two enzymes whose concentrations were assumed to be decreasing functions of temperature. This resulted in a temperature-independent period of the oscillations. Necessary feedback was achieved through an allosteric effect, and light was assumed to contribute to the activation of one of the enzymes, leading to a system obeying Aschoff's circadian rules (Hoffmann, 1965), entraining to LD cycles, and yielding PRCs for light and temperature pulses similar to those observed in various organisms (Pittendrigh, 1965). The set of detailed reactions comprising the model are as follows (Pavlidis and Kauzmann, 1969):

$$G \xrightarrow{1} X \qquad \text{(a)}$$

$$X + E_1^{+} \xrightarrow{2} E_1X \xrightarrow{3} E_1 + Y \qquad \text{(b)}$$

$$Y + E_1 \underset{5}{\overset{4}{\rightleftarrows}} E_1^{+} \qquad \text{(c)}$$

$$Y + E_2 \xrightarrow{6} E_2Y \xrightarrow{7} E_2 + Z \qquad \text{(d)}$$

X is a substrate, Y is an activator, E_1 and E_2 are enzymes (E_1^{+} being an active form), G is a substance in constant supply, which is transformed into X according to Reaction (a), and Z is the final product. X is transformed into Y in an enzyme-catalyzed reaction (b) obeying Michaelis-Menten kinetics, and Y is then transformed into Z by a similar Reaction (d). Reaction (c) describes the activation of E_1 by Y, perhaps by a ligand-induced conformational change or allosteric effect. Reaction (c) provides the feedback required for self-sustained oscillations. Each of the reactions is characterized by its own rate constant.

The chemical kinetics of the set of reactions can be described by the pair of differential equations developed by Higgins (1967):

$$\frac{dx}{dt} = k_1 g - v \qquad (1)$$

$$\frac{dy}{dt} = v - \frac{k_7 e_2 y}{y + (k_7/k_6)} \qquad (2)$$

where:

$$v = \frac{k_3 e_1 x y}{y[x + (k_3/k_2)] + (k_3/k_2) \cdot (k_4/k_5)} \qquad (3)$$

In these equations, the lower-case letters denote the concentrations of the corresponding substances (given in Reactions a, b, c, d); e_1, is the total concentration of both the active and inactive forms of E_1, whereas e_2 relates only to the active form of E_2. The values for k denote the rate constants for the various reactions; Reaction (c) was assumed to reach equilibrium much faster than the others.

It should not be at all surprising to discover that G corresponds to glucose, X to F6P, Y to F-1,6-P (or to ADP), E_1 to PFK, and E_2 to aldolase—the elements of the glycolytic oscillator discussed in Section 5.2.1 (Fig. 5.2), for this is just what was intended. As we have noted, choice of appropriate values of the parameters involved yields self-sustained oscillations (Higgins, 1964) with a limit cycle of the same form that has been used successfully to model the overt behavior of circadian rhythms (Pavlidis, 1968, 1973). Because almost all of the rate constants involved enter the differential equations (1, 2) through their ratios only, the corresponding coefficients could depend much less on temperature than the rate constants themselves; indeed, the Q_{10} of the coefficients will equal the ratio of the Q_{10}s of the rate constants, which in itself could lead to partial temperature independence. For example, if Q_{10} (k_7 = 2.4 and that of k_6 = 2.1, then Q_{10} (k_7/k_6) = 1.14. The period of the oscillations could still be somewhat temperature dependent, however, because some of the rate constants (k_1, k_3, k_7) do not enter as ratios. Nevertheless, temperature compensation could be achieved if it were assumed that (1) Reaction (a) is a diffusion process whereby X (F6P) would be produced from some external source of G (glucose) entering the cell at a constant rate, with the result that k_1 would be temperature independent, and (2) the enzyme concentrations E_3 and E_7 are decreasing functions of temperature, since they enter the equations as products with their corresponding rate constants (for formal details on the manner in which this might occur, see Pavlidis and Kauzmann, 1969). Light was assumed to cause a decrease in the concentration of Y, possibly by contributing to the activation of E_2.

This leaves us with the problem of the generation of the long circadian period. Pavlidis and Kauzmann (1969) ruled out a very slow diffusion process as a mechanism for frequency reduction. They also considered the possibility that very low concentrations of E_1 and E_2 (i.e., a biochemical oscillator starved of enzymes) together with a low diffusion rate k_1 might suffice. Despite the fact that some lengthening of the period can be obtained experimentally (Das and Busse, 1985; Hess, 1976), the possibility of achieving a 24-h τ in this manner is ruled out by the PFK model (Boiteux et al., 1975; Goldbeter and Nicolis, 1976). Pavlidis and Kauzmann (1969) suggested that appropriate frequency reduction perhaps could be obtained through the mutual, inhibitory coupling of a large number of individual oscillators, resulting in the appearance macroscopically of a single oscillator. In a review of the earlier clock literature, Vanden Driessche (1973) also finds this notion of a population of interacting oscillators compatible with the experimental biological evidence. The coupling itself could be among several different reactions catalyzed by one enzyme within a cell, but it could involve equally well different intracellular structures (e.g., membranes or organelles) or even cell populations. We consider coupled oscillators in more detail in Section 5.2.3. Since the enzymes of the set of reactions are proteins, we also will examine periodic enzyme synthesis, which often represents a cyclic epigenetic event that is characterized by longer periods (see Section 5.2.4).

The Cyclic AMP Oscillator of the Cellular Slime Mold

The cellular slime mold *D. discoideum* exists in two states; independently growing amebae and a multicellular fruiting body. The transition between these two morphological stages occurs after starvation. Aggregating cells communicate by periodic cAMP pulses released from the cells at intervals of a few minutes (see reviews by Gerisch, 1976, 1986; Goldbeter and Caplan, 1976). The amebae move toward center-founding cells, chemotactically oriented by cAMP-binding to receptors on the cell surface. Within 30 sec after the activation of cell surface receptors by cAMP, increases in the concentrations of at least three diffusible factors that are capable of processing signals from the plasma membrane to intracellular targets can be detected: cGMP by activation of guanylate cyclase, cAMP by activation of adenylate cyclase, and intracellular Ca^{2+} by stimulation of Ca^{2+} influx into the cells. Spirals or concentric waves of chemotactic activity are propagated toward the periphery of a field. These dynamic, spatiotemporal patterns, examples of dissipative structures (Prigogine, 1961), are thought to arise through (1) the ability of each cell to release cAMP rhythmically, (2) the stimulation of cAMP release by cAMP pulses, and (3) the coupling of the individual oscillators (i.e., each cell) by the diffusion of cAMP among the cells. Interference between a cell's own production of cAMP and its response to cAMP produced by other cells is avoided by the temporal separation of the two functions. As noted previously, production and response change rhythmically with periods of several minutes. If a clock is a timekeeping device, the *Dictyostelium* oscillator can be used as a clock, since under defined conditions τ is relatively constant during a section of the developmental sequence (although it is not temperature compensated). In a mutant in which development is arrested during the oscillating stage, the oscillations continue for hours with a τ of about 2.5 h (Gerisch, 1971).

The molecular basis of this oscillatory generation and propagation of cAMP signals has been investigated intensively in an actively stirred, homogeneous cell suspension, in which the spatial pattern is eliminated but the temporal aspect is preserved (Gerisch, 1976, 1986). Under these conditions, the cells produce cAMP pulses in phase at intervals of about 7 min without any external stimulation. The elements of the biochemical network responsible for the periodic pulsing of cAMP include AC, cAMP PDE, cell-surface receptors, and possibly specific sites for the transport of cAMP through the plasma membrane (thus, it is a cellular system, unlike the soluble glycolytic oscillator). Early development is characterized by increases in the number of cAMP receptors, the basal activity of AC, and the activity of cAMP PDE on the cell surface. All of these developmental changes are controlled, in turn, by cAMP. PDE is an important component of the signal system because it is responsible for the destruction of cAMP after a pulse of synthesis and, thereby, permits the cells to recover from a refractory state induced by cAMP.

Goldbeter (1975), simulating cAMP oscillations in a homogeneous medium, has proposed a model for their generation that may be valid for cAMP rhythmicity within the cellular compartment (Goldbeter and Caplan, 1976). It is based on the cross-activation of two membrane-bound enzymes,

AC and ATP pyrophosphohydrolase (PPH), which participate in the synthesis of cAMP, transforming ATP into cAMP and 5'AMP (via the PDE reaction), respectively. Strongly cooperative activation (a type of autocatalytic control) of AC by 5'AMP and of ATP-PPH by cAMP has been observed in *Dictyostelium*, although only the first type of interaction would be required for sustained oscillations in the synthesis of cAMP to occur around a nonequilibrium, unstable, stationary state, a requisite in accord with the experimental results. The model nicely simulates the waveform of the nonsinusoidal, spike-shaped cAMP oscillations observed in cells. The fact that the period of the oscillations varied by less than a factor of 2 for various parameter values agrees with the observation that aggregation centers release cAMP pulses with a period of 3 to 5 min.

As noted previously, the activation of ATP-PPH by cAMP is not a necessary requirement for oscillatory behavior in Goldbeter's model (1975), although the positive feedback exerted by cAMP on AC is indispensable for the periodic activity of the enzyme. The mechanism underlying the cAMP oscillations, therefore, formally resembles that of the glycolytic oscillator in that both invoke positive feedback exerted on an allosteric enzyme by a reaction product (Goldbeter and Nicolis, 1976).

5.2.2 Cell Energy Metabolism

In the preceding discussion (see Section 5.2.1) of soluble, biochemical, ultradian oscillators, attention was focused on energy supply and the glycolytic metabolic network (not forgetting those of photosynthesis and respiration). The role of cellular energy metabolism (CEM) in the generation of circadian rhythms has been examined also: the activity of metabolic pathways transducing information between clock and hands could be controlled by feedback regulation and by energy supply depending on the metabolic state of the cell (see Section 4.3.1). Sel'kov (1975, 1979) believes that cell energy metabolism *is* the cellular clock.

Energy Charge

An integrated view of the coupling of metabolism and energy supply has been put foward in the adenylate control hypothesis of Atkinson (1968, 1969), who developed the concept of energy charge (EC) to indicate quantitiatively the energy state of the cell. Overall cellular EC was defined as (ATP + 0.5 ADP)/(ATP + ADP + AMP), if one assumes the absence of compartmentalization of AMP. Subsequently, it has been demonstrated that the rates of energy-expending reactions increase as EC increases, whereas the inverse holds true for energy-producing reactions. There is increasing evidence both in vivo and in vitro that key enzymes of energy-transducing pathways are allosterically controlled by EC.

On the hypothesis that the circadian rhythm in CO_2 production (see

Section 2.3.1) might reflect changes in respiratory metabolism and mitochondrial function, Delmer and Brody (1975) sampled the edges of the growing front of a mycelial mat at different ages in *Neurospora crassa* cultured in Petri plates and enzymatically assayed the levels of adenine nucleotides (see Section 4.3.1). Although the concentrations of ADP and ATP showed no obvious oscillation, that of AMP oscillated between 0.5 and 6.0 μmol/g (residual dry weight), with a peak at the midpoint of the conidiating phase (Fig. 5.3). This oscillation had many of the properties of a true circadian rhythm: τ was 22 h, it could be phase-shifted by light, and it damped out in LL. In turn, the rhythm in the level of AMP gave rise to an oscillation between 0.65 and 0.93 (Fig. 5.4) in overall cellular EC. The authors suggested that the underlying cause of the oscillation in AMP level could be a rhythmic, partial uncoupling of mitochondrial oxidative phosphorylation, although it might also be considered to be merely a reflection of the morphological conidiation rhythm and the accompanying cycle in the synthesis and processing of adenine nucleotides (Feldman and Dunlap, 1983). However, Schulz et al. (1985) have excluded conidiation as a causal factor; the concentrations of ATP and ADP (and thus total EC) in mycelia kept in liquid culture (where rhythmic conidiation

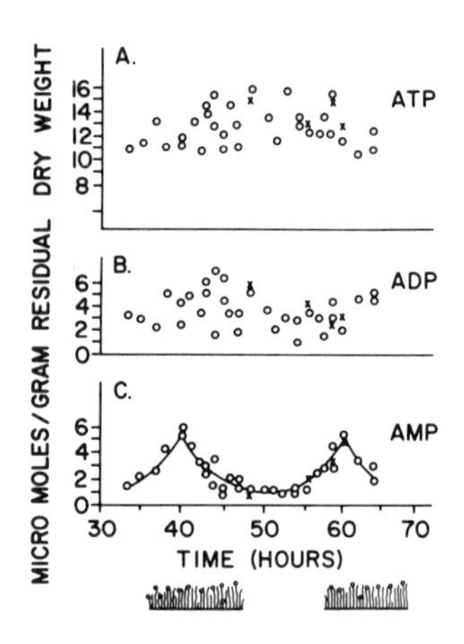

FIGURE 5.3. Adenine nucleotide levels in the growing front of the *bd* mutant of *Neurospora crassa* as a function of time. Open circles indicate values obtained via enzymatic assay whereas Xs represent the values obtained by liquid chromatography. The shape of the AMP curve between 40 and 60 h was determined by a computer, using a standard deviation of ±0.50 for each value. Wavy line areas at the bottom of the figure indicate the periods of conidiation. The edge of the mycelial mat at any given time is indicated by the points on the graph; for example, at 42 h, the growing front of the mycelia was at the midpoint of the conidiating region. (Reprinted from Delmer and Brody, 1975, with permission of American Society for Microbiology.)

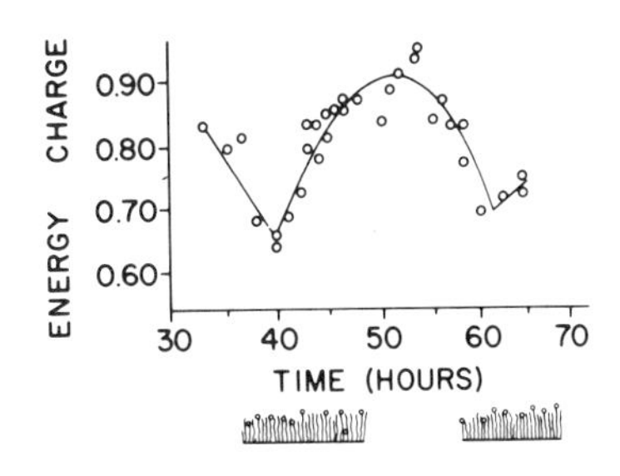

FIGURE 5.4. Energy charge in the growing front of the *bd* mutant of *Neurospora crassa* as a function of time. The shape of the curve between 40 and 60 h was determined by a computer using a standard deviation of ± 0.03 for each value. (Reprinted from Delmer and Brody, 1975, with permission of American Society for Microbiology.)

does not occur) underwent circadian changes in DD, exhibiting a maximum at CT 0 to CT 6—the time of maximal activity of Krebs cycle enzymes and CO_2 production (see Section 4.3.2).

Brody and Harris (1973) have also observed spatial differences in the levels of oxidized and reduced pyridine nucleotide levels (NAD^+, $NADP^+$, NADH, and NADPH) in the two morphologically distinct areas of the growing front of *N. crassa* formed as a result of the rhythmic conidiation process. Although the total pyridine nucleotide content of these two areas was the same, the level of NAD^+ was higher in the conidiating area, and the levels of the other three nucleotides were lower. More recent data (Dieckmann, 1980) indicated that there was a significant overall decrease in NAD^+ in aging cultures without regard to circadian time and that changes in nucleotide levels at the growing front were small. Also, the redox ratio appeared to be constant across the circadian cycle, indicating that the spatial differences observed in pyridine nucleotide levels probably were not due to a clock-controlled rhythmicity in their levels.

Brinkmann and coworkers measured the EC and the concentration of G6P under various sinusoidal temperature cycles within and beyond the limits of entrainment for the motility rhythm of *Euglena gracilis* (see Section 2.2.3). In contrast to G6P, EC never synchronized (see Fig. 2.13a), and neither showed rhythmicity in autotrophic cultures during a free-run at a constant 27.5°C (see Fig. 2.13c), indicating that these variables could not be responsible for generating the motility rhythm. Similarly, although the redox state of $NADP^+$ (but not that of NAD^+) was synchronized by a temperature cycle (see Fig. 2.14a), there was no significant oscillation in the redox state of pyridine nucleotides during a free-run. Of course, rhythmicity might be detectable only in separate compartments and not in total cell extracts.

The role of energy metabolism in circadian clock mechanisms has been intensively investigated in *Chenopodium rubrum* (Wagner, 1976b, 1977, 1985). In seedlings kept in LL and constant temperature, net photosynthesis, chlorophyll and betacyanin accumulation (Wagner and Cumming, 1970), adenylate EC (Wagner et al., 1974c), redox ratio (Wagner and Frosch, 1974), and the activities of adenylate kinase isozymes and enzymes of the tricarboxylic acid cycle and the oxidative pentose phosphate pathway (glyceraldehyde-3-phosphate, malate, glutamate, G6P, and gluconate-6-phosphate dehydrogenases) (Deitzer et al., 1974) oscillated with period lengths of 12 to 30 h. The rhythms (fluctuations?) of EC and redox ratio were able to synchronize the other rhythms only in diurnal LD cycles but not during the LL free-run. Wagner and Cumming (1970) suggested that the basis of circadian rhythmicity could be energy transduction based on conformational changes of the flexible, lipoprotein ion-exchange membranes, with phytochrome's acting as a membrane operator, capable of phase-shifting rhythmic energy transduction. Thus, the interaction of environmental signals with the clock would occur at membrane-organized receptors that modulate membrane-bound energy transduction. In turn, the energy-dependent state of membranes would determine the sensitivity of these receptors (Wagner et al., 1974a,b; Wagner, 1985; see Section 5.4).

The Deposition Effect and the Generation of Circadian Periods

Sel'kov and coworkers have provided a different emphasis over the years (Sel'kov, 1975, 1979; Reich and Sel'kov, 1981). The main function of CEM would be to stabilize ATP concentration within a wide range of metabolic loads. Such a function could not be achieved if there were uncontrolled intermediate cycling in the key futile cycles of CEM, such as between F6P and F-1,6-P between acetyl-CoA and fatty acid, or between glutamate and glutamine. The only way to suppress the recycling would be by the temporal organization of all futile cycles of CEM. This organization could be achieved easily by allosteric regulations, such as product activation and substrate inhibition of antagonist enzymes of the key futile CEM cycles (see Section 5.2.1). These control mechanisms switch the antagonist enzymes on and off reciprocally to produce self-oscillatory changes of the enzyme activities.

Now the concentration of cellular F6P in the F6P–F-1,6-P cycle (in which F-1,6-P activates PFK and inhibits F-1,6-biphosphatase) is strongly buffered by large amounts of reserve carbohydrates, such as glycogen, starch, and trehalose. Similarly, the fatty acid concentration in the acetyl-CoA–fatty acid cycle is buffered by reserve lipids, and glutamine content of the glutamate–glutamine cycle is buffered by the total amino acid and reserve protein content within the cell. Such a buffering action can be represented schematically by an open, self-oscillatory futile cycle between two intermediates (I_1 and I_2) coupled to a reversible deposition mechanism (between D and I_1) that equilibrates intermediate I_1 with its depot form D. Theoretical analysis of this deposition effect and choice of appropriate rate constants of the partial reactions (entry of I_1 into and exit from the depot) that comprise it indicate that the period of the oscillating system can be increased significantly with respect to that of the initial system involving cycling between the intermediates without deposition (Sel'kov, 1975). This is especially true for nonlinear cases, when the deposition reactions are enzymatic and are controlled allosterically by D and I_1. Indeed, mathematical analysis of a kinetic model representing the F6P–F-1,6-P futile cycle linked with the glycogen deposition mechanism showed that values of τ of the order of 22 to 26 h and of mean glycogen content (expressed as glycosyl units) of the order of 5 to 10 m*M* could be obtained. Thus, the deposition effect could form a theoretical basis for the generation of very slow—even circadian—rhythms within cellular energy metabolism, which the ultradian glycolytic oscillator in itself could not (see Section 5.2.1).

The autonomous futile cycles and their corresponding deposition mechanisms all have an antisymmetric, stoichiometric structure that allows a very fine compensation for any nonspecific perturbation that uniformly affects the opposing partial reactions of the deposition (D–I_1) or futile (I_1–I_2) cycles. For example, if an increase in temperature activates both opposing pathways, any excess of D produced by the accelerated synthetic pathway leading from I_2 and $I_1$to D would be exactly consumed by the activated catabolic pathway leading from D to I_2. The result, of course, is that τ would be completely independent of the temperature perturbation, although the amplitudes of the oscillations in the concentrations of D, I_1, and I_2 would not (increasing with increasing temperature). Even if the opposing pathways were not equally dependent on perturbing factors, a partial compensatory effect would be attained. Thus, the CEM model provides a mechanism for achieving temperature compensation of the period—once again, an attribute not evidenced by the simple, soluble glycolytic oscillator. Other regulatory mechanisms of the negative feedback type inherent in CEM can stabilize effectively the period against metabolic perturbations, such as changes in the activity of ATPase (Sel'kov, 1975, 1979).

Temperature-Compensated Ultradian Clocks in Respiration

In addition to the mathematical analysis of cycles in CEM carried out by Sel'kov and coworkers (see preceding subsection), experimental evidence has been obtained that suggests the participation of respiratory oscillations in timekeeping (Lloyd et al., 1982b). For example, synchronous cultures of the soil amoeba *Acanthamoeba castellanii*, obtained by centrifugal selection, showed significant oscillations of O_2 consumption (measured polarographically), adenine pools, and total cellular protein (Lloyd et al., 1982a; see Marques et al., 1987). Both the period (τ = 72 min) and the phase of these oscillations were approximately the same, although incubation temperatures in different experiments varied between 20 and 30°C, and cell division time (*g*) increased from 7.8 to 16 h (yielding a $Q_{10} \cong 2$). Common phase reference points between the rhythms and the CDC permitted the construction of reproducible CDC maps; at any fixed temperature there were an integer number of respiratory cycles per CDC. Thus, these CDC-dependent epigenetic oscillations (τ > 1 h) are unlike typical metabolic oscillations (see Section 5.2.1) but rather resemble circadian rhythms in that they have a temperature-compensated periodicity.

Similar observations have been made in synchronized cultures of *Tetrahymena pyriformis* (Lloyd and Kippert, personal communication, 1986), in which the 30-min respiratory rhythm of strain A2 has a Q_{10} of 1.12 and is quite insensitive to brief perturbations (such as short-term anaerobiosis). This is particularly interesting in view of the energy reserve escapement mechanism hypothesized for circadian clocks in this ciliate, as well as in higher eukaryotic organisms (Ehret and Dobra, 1977; Ehret et al., 1977). In this unifying formulation, a gene action circadian oscillator ultimately would drive overt circadian rhythms, coupling them through causally interconnected oscillations in glycogenolysis and in biogenic amine metabolism. Glycogen would provide the reliable energy source needed, which would be metered out discontinuously, on a daily basis, by an escapement component comprising either the indoleamine or the catecholamine pathway (see Section 2.1.1; Fig. 2.2).

In summary, Lloyd et al. (1982a) argue that temperature-compensated epigenetic oscillations in energy metabolism may serve a dual timekeeping role for both CDC and circadian rhythmicities, although the mechanism by which this is accomplished is not clear (see Lloyd and Edwards, 1986, 1987). An ancillary prediction is that the CDC would be quantized, as, in fact, has been observed in *T.pyriformis, Paramecium tetraurelia,* and mammalian cells (see Sections 2.1.2, 3.1.2). Nevertheless, the fundamental oscillator appears to be in a transcriptional or translational feedback loop (see Section 5.2.4). Total cellular protein and RNA contents oscillate, and these systems would enslave the system of mitochondrial energy supply, rather than vice versa.

5.2.3 Coupled Oscillators

Despite our discussion in the preceding subsections of soluble metabolic oscillators and the ultradian rhythms in glycolysis and energy metabolism, even embellished with time-consuming deposition effects, we keep returning to a seemingly eternal theme: How is the relatively long period of circadian rhythms generated?

It is interesting to note that several possible frequency-reduction mechanisms involve the notion of coupled oscillators, no matter whether they be at the intracellular, intercellular (tissue, organ), or organismal (inter-individual) levels of organization (see Section 5.2.1). A mathematical treatment of populations of interacting oscillators—in itself fraught with difficulties because of the increased degrees of freedom and the structural instabilities inherent in higher order systems—is beyond the scope of this book (but see Goodwin, 1963; Chapters 7 and 8 in Pavlidis, 1973; Chapters 8 and 11 in Winfree, 1980).

Winfree (1967) has shown that a population of weakly coupled oscillators with slightly different frequencies tends to synchronize, under certain conditions, to a common frequency that is not very different from that observed in individual, noninteracting oscillators. In an attempt to simulate such a population, he used a group of 71 electrically coupled, flickering neon tubes. The neon oscillator was chosen for its analogy to a rhythmically flashing firefly (see Section 1.3.3), a heart cell or other pacemaker neuron (see Section 6.3.1), the cAMP oscillator in a slime mold cell (see Section 5.2.1), or any other biological relaxation-type oscillator (Winfree, 1980). When the capacitors of the neon lamps were coupled, intrinsically slower lamps (i.e., having longer τs) sped up somewhat, while faster ones slowed down slightly, with the net result that all stayed in step at one common frequency. Threshold conditions were discovered for mutual synchronization in any of a variety of modes. With stronger coupling, the population abruptly split into two (formally, a bifurcation), with a slight change in period, as a result of which the frequency of the aggregate rhythm was doubled.

Another type of more complex interactive behavior within a population of oscillators has been developed by Pavlidis (1969). In this case, the individual frequencies were taken to be either identical or different, and the oscillators were coupled to each other in an inhibitory manner. For weaker coupling, it was found in certain cases that even if each constituent oscillator continued to oscillate at a high frequency, the summated output was a low-frequency oscillation. At higher coupling strengths, the oscillators mutually entrained to a new frequency, much lower than the one at which they would have oscillated individually. In this case, a large number (n) of oscillators of endogenous frequency ω_0 are cross-coupled. The sum of the outputs of the same subunits oscillates at a frequency given by the product of ω_0 and the square root of $[1-(n-1)r]$, where r is the coupling constant (Pavlidis, 1969). Indeed, numerical simulations showed that, when the value of $(n-1)r$ was approximately 0.9, the subunits oscillated in phase at the low frequency of about $\omega_0/4$, and since there was no high-frequency component, the correct shape of the PRC for light pulses was obtained.

But what does all this mean in concrete terms? We have discussed the possibility of this type of inhibitory cross-coupling in Pavlidis and Kauzmann's attempt (1969) to model a circadian clock and its general properties (light sensitivity, temperature compensation) using a relatively simple, biochemical oscillatory system, the ultradian glycolytic oscillator (see Section 5.2.2). In a review of the earlier clock literature, Vanden Driessche (1973) also found the notion of a population of interacting oscillators compatible with the experimental biological evidence. It is not clear, however, at what level of organization the coupling occurs: among several different reactions catalyzed by one enzyme within a cell, among different intracellular structures (e.g., membranes or organelles), or among the cells within cell populations (see Section 6.2.2)? Since the enzymes of the set of reactions of the PFK model are proteins, there is the possibility that periodic enzyme synthesis, which often represents a cyclic epigenetic event that is characterized by a longer period, may be involved (see Section 5.2.4).

In his masterful theoretical consideration of temporal organization in cells, Goodwin (1963), followed by many others, modified the relatively

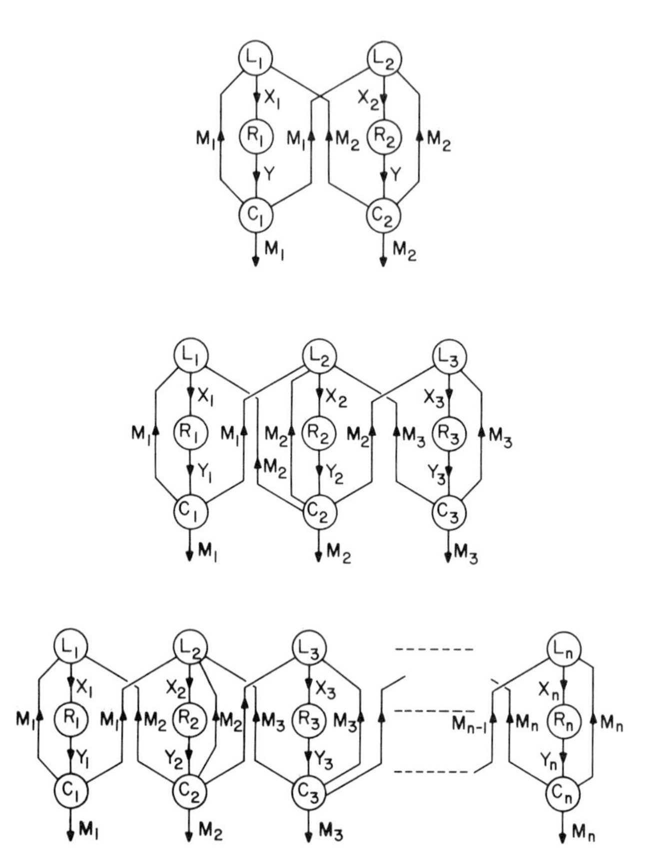

FIGURE 5.5. Metabolic feedback control circuitry for the generation of biochemical oscillations. L_1 represents a genetic locus that synthesizes mRNA in quantities represented by the variable X_1. This specific signal encounters a cellular structure R (a ribosome), where its activity results in the synthesis of a particular species of protein in quantities denoted by the variable Y_1. The protein then travels to some cellular locus, C, where it exerts an influence on the metabolic state either by enzyme action or by some other means. This activity, in turn, generates a metabolic species in quantity M_1, a fraction of which closes the control loop by returning to the genetic locus L_1, where it may act as a repressor. The three panels depict increasingly more complex cross-coupled control loops (which may have either an intracellular or an intercellular basis). (Reprinted from Goodwin, 1963, with permission of Academic Press.)

simple, general scheme for glycolytic and other oscillatory chemical reactions (Higgins, 1964, 1967; see Section 5.2.1) by building in additional transcriptional and translational regulatory events. Repression is permitted to occur between two different components as well as within single components (see Fig. 5.12), so that strong interactions are obtained, as shown in Figure 5.5 (top panel). Here a metabolite controlled by Y_1 (denoting ribosomal protein synthesis) interacts by repression with another genetic locus L_2, and a reciprocal interaction occurs from L_2 to L_1. Even more complex parallel repression networks involving many interacting, coupled components can easily be generated (Fig. 5.5, lower panels). Significantly, these components could be within one cell or could involve cross-talk (i.e., intercellular chemical communication; see Section 6.2.2) among individual cells comprising a tissue (such as in the isolated eye of *Aplysia* or *Bulla*; see Fig. 4.3). In such situations, inhibitory coupling between a large number of oscillators could result in a much lower frequency of the overall oscillation observed in the population than that of the constituent oscillators (Pavlidis, 1969). The addition of time-consuming diffusion steps for key substrates could further lengthen the period of the overt rhythm and perhaps impart a measure of temperature compensation. If one postulates that one or more key reaction sequences are photoactivable, the fundamental light-sensitive properties of circadian rhythms can be simulated (Pavlidis and Kauzmann, 1969).

It is noteworthy that the product of the *per* clock gene in *D. melanogaster* appears to regulate the degree of intercellular communication by altering the number and organization of functional gap junctions (Bargiello et al., 1987; Section 4.4.2). Direct measurement of the coupling strength in larval salivary glands of various *per* mutants that affect τ (Section 4.4.1) demonstrated that the extent of coupling was correlated inversely with τ. This provocative finding supports the hypothesis (Dowse et al., 1987; Dowse and Ringo, 1987) that the per^+ gene product mediates the coupling of multiple ultradian oscillators (which predominate in the "arrhythmic" mutant per^o) to produce wild-type circadian rhythms. In this scheme, τ would be a function of the coupling tightness among ultradian oscillators, increasing as coupling loosens. Ultradian rhythms would become apparent under weak coupling or in its absence.

5.2.4 Periodic Enzyme Synthesis and Other Cyclic Epigenetic Events

Thus far, we have considered simple feedback loops in energy supply and intermediary metabolism as mechanisms for generating periodicity. Although they are both elegant and convincing at the ultradian level within the metabolic domain (periods on the order of minutes), they become a bit strained when one endeavors to force them into a procrustian fit of the circadian bed. But somewhere between these two time domains (see

Fig. 1.1), we encounter a halfway house reflected in rhythmicities whose period, although still ultradian, is in the order of several hours. These epigenetic oscillations often comprise regulatory circuits involving transcription and translation and, therefore, extend into both the cell cycle (genetic) and circadian domains. Perhaps the mechanism(s) by which they are generated can be instructive.

Periodic Enzyme Synthesis

Implicit in the models for the soluble glycolytic oscillator (see Section 5.2.1) is the highly ultradian, periodic activation of the enzyme PFK, probably by a ligand-induced conformational change or allosteric effect (Goldbeter and Nicolis, 1976). Once we turn to dividing prokaryotic and eukaryotic cellular systems, however, we encounter more complex patterns of enzyme accumulation during the CDC (see general treatments in Mitchison, 1971; Lloyd et al., 1982b). Mitchison (1969) distinguishes four basic types: continuous exponential, continuous linear, step, and peak. The last three patterns mark specific events in the CDC: the time of doubling in the rate of synthesis of a linear enzyme, the time of doubling in the amount of a step enzyme, and the time of maximum activity of a peak enzyme.

Self-Sustained Enzyme Oscillations

During the 1960s, many examples of periodic enzyme accumulation were reported in synchronously dividing cultures of bacteria, yeast and mammalian cells (see reviews by Donachie and Masters, 1969; Mitchison, 1977; Halvorson, 1977). Some of these earlier studies must be reexamined critically, however, before one can conclude that periodic enzyme synthesis is a common feature of the CDC (Tyson, 1982), for a number of reasons. First, the inductive (forcing) methods often used to synchronize the cells (as opposed to selection methods) may have generated artifacts. Second, a true distinction between de novo synthesis and activation often was not made. Third, enzyme accumulation was often equated with enzyme synthesis, with little attention being given to enzyme turnover. Nevertheless, accepting the reports at face value, it is obvious that these enzyme periodicities would reflect the period of the CDC (i.e., the values of g for individual cells), which typically ranges from 30 min for bacteria, several hours for yeast and *Tetrahymena*, 10 to 12 h for *Euglena* and many algal cells, to 16 to 17 h for mammalian cells, all cultured under optimal conditions. We have discussed at length the interaction of circadian clocks with the CDC in many cell types, with the resultant gating of division and other CDC events—including enzyme patterns—to intervals of approximately 24 h (see Section 3.2).

It would be erroneous to conclude, however, that oscillations in enzyme activity necessarily are dependent on the driving force of DNA replication and cell division. For example, Masters and Donachie (1966) demonstrated periodic changes in the activity of aspartate transcarbamylase and ornithine transcarbamylase in cultures of *Bacillus subtilis* W23 that had been synchronized by dilution with fresh medium of cells in the stationary phase of growth. These oscillations occurred for at least two cycles ($\tau \cong 70$ min) in the presence of FUdR, which almost completely inhibited DNA synthesis. Obviously, discontinuous, autogenous enzyme synthesis did not require periodic gene replication in this microorganism. In *Schizosaccharomyces pombe*, step changes in the activity of nucleoside diphosphokinase observed during the CDC also continue after a block to the DNA division cycle (Creanor and Mitchison, 1986). Similarly, sustained oscillations in the activity of TAT have been reported in cultures of *T. pyriformis* that had been synchronized in the ultradian growth mode by a 5-h temperature cycle and that then were transferred to a constant temperature of 20°C (Kämmerer and Hardeland, 1982; see Section 2.1.1). Not only was the free-running period ($\tau \cong$ 4-5 h) of this rhythm temperature compensated (Fig. 5.6), characterized by a Q_{10} closer to 1.0 than 2.0 over steady-state temperatures of 10 to 30°C (Michel and Hardeland, 1985), but also the ultradian enzyme oscillations persisted when protein synthesis, cell growth and division were inhibited by cycloheximide (20 μg/ml) or by 1 m*M* emetine or 250 m*M* hydroxyurea (Fig. 5.7; Thiel et al., 1985). The activities of a number of enzymes (e.g., lactate dehydrogenase) that have no obligatory relation with other periodic events, such as DNA synthesis, oscillated with periods of 3 to 4 h (the period of the G_q quantal cycle; see Section 3.1.2 in cultured mammalian cells having a *g* of 11.5 to 12 h, even if DNA and RNA synthesis were inhibited (Klevecz, 1969a,b). We have noted in Section 5.2.2 that the activities of the adenylate kinase enzymes and enzymes of the tricarboxylic acid cycle and the oxidative pentose phosphate pathway oscillated with periods of 12 to 30 h in seedlings of *Chenopodium rubrum* maintained under constant conditions (Deitzer et al., 1974), although here tissue growth was occurring.

In addition to ultradian enzyme rhythms, self-sustained circadian oscillations in enzymatic activity also have been documented in many organisms, even in the absence of DNA replication and cell division. For example, a 24-h rhythm in the activity of alanine dehydrogenase has been found in nondividing populations of *E. gracilis* (Z) batch-cultured organotrophically in LD: 10,14 at 19°C (Fig. 5.8B); this rhythm persists in DD (Fig. 5.8C) but not in bright LL (Fig. 5.8A) (Sulzman and Edmunds, 1972, 1973). Analysis of the partially purified enzyme suggested that cycles of de novo synthesis and degradation may generate the observed variations in its activity (Sulzman and Edmunds, 1973; see Section 2.2.3). The activities of a number of other enzymes (see Table 2.4) exhibited similar diurnal oscillations in nondividing cultures (Fig. 5.9). Quentin and Hardeland (1986a,b) have described circadian rhythms of protein synthesis (measured by [^{3}H]leucine incorporation) in stationary cultures of photosynthetic (Z, T) and heat-bleached (WTHL) strains of *E. gracilis* in LL. Preincubation of intact cells with heat-stable cytosolic extracts resulted

A

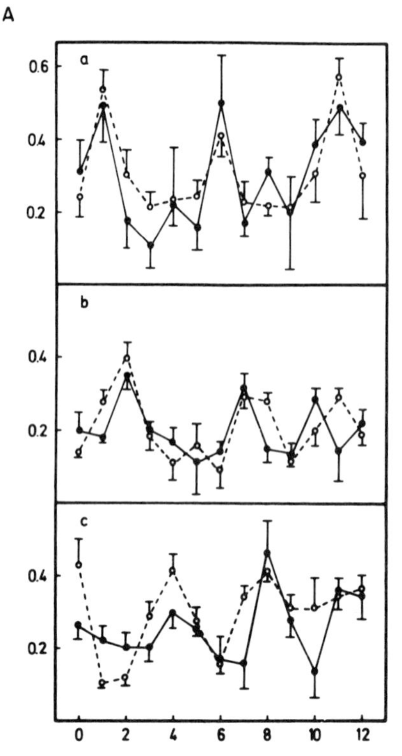

FIGURE 5.6. (A) Free-running ultradian rhythmicities of tyrosine aminotransferase activity in *Tetrahymena thermophila* at 20°C (open circles) and at 10°C (full circles). a, b, c, three independent experiments. *Abscissa:* Time (hours) after last cold pulse of previous synchronization. *Ordinates:* Enzyme activity (E). *Vertical lines:* Three-fold S.E.M.

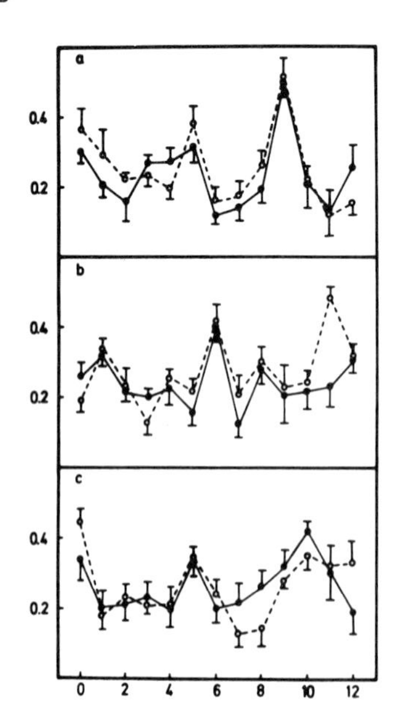

FIGURE 5.6 (B) Free-running rhythmicities of tyrosine aminotransferase activity at 20°C (open circles) and at 30°C (full circles). (Reprinted from Michel and Hardeland, 1985, with permission of Swets and Zeitlinger B.V.)

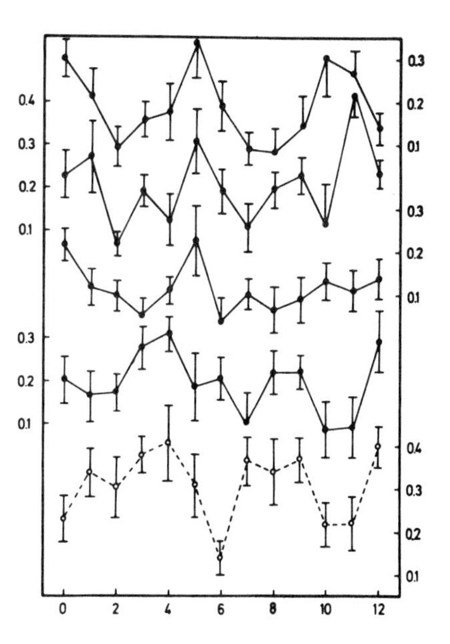

FIGURE 5.7. Persistence of ultradian rhythmicity of tyrosine aminotransferase in *Tetrahymena thermophila* in the presence of 1 m*M* emetine (full circles; four independent runs) and 250 m*M* hydroxyurea (open circles). Measurements were carried out after synchronization by temperatures cycles, at a constant temperature of 20°C. Drugs were administered at −1 h. *Abscissa:* Hours after the last cold pulse. *Ordinate:* Enzyme activity (ΔE). *Vertical lines:* Three-fold S.E.M. (Reprinted from Thiel et al., 1985, with permission of Swets and Zeitlinger B.V.)

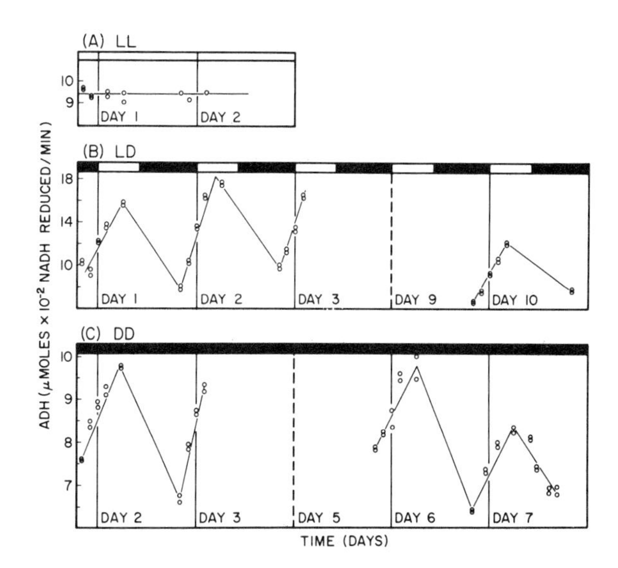

FIGURE 5.8. Circadian oscillations in the activity of alanine dehydrogenase (ADH) in nondividing, organotrophically batch-cultured *Euglena gracilis*. (A) ADH activity in culture maintained under constant bright illumination (LL). (B) ADH activity in culture in LD: 10,14 light cycle. Data from day 1, 2, 3, and 10 are shown. (C) ADH activity in constant darkness (DD) after many cycles of LD: 10,14. Data from days 2, 3, 6, and 7 following the transition from LD to DD are given. Vertical lines are spaced 24 h apart. Double points represent duplicate determinations. (Reprinted from Sulzman and Edmunds, 1972, with permission of Academic Press.)

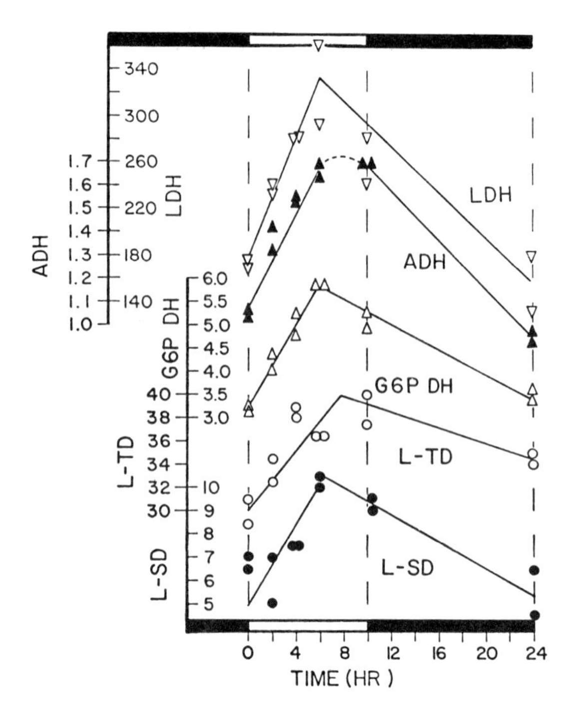

FIGURE 5.9. Oscillations (24 h) in the activities of several enzymes in nondividing, stationary cultures of *Euglena gracilis* in LD: 10,14 at 19°C on low pH glutamate-malate medium. Enzyme activities are given on the ordinates. (●) L-serine deaminase (L-SD) activity (μmol pyruvate produced · min^{-1}); (○) L-threonine deaminase (L-TD) activity (μmol α-ketobutyrate produced · min^{-1}); (△) glucose-6-phosphate-dehydrogenase (G6P DH) activity (μmol × 10^2 NADH oxidized · min^{-1}); (▲) alanine dehydrogenase (ADH) activity (μmol × 10^2 NADH oxidized · min^{-1}); (▽) lactic dehydrogenase (LDH) activity (μmol × 10^2 NADH oxidized · min^{-1}). (Reprinted from Edmunds and Halberg, 1981, with permission of Williams & Wilkins Co.)

in a stimulation (up to 40-fold) of 80S protein synthesis; the extent of this enhancement varied as a function of circadian time, indicating a rhythm of translational capacity. A similar diurnal rhythmicity in the cytosolic control of cell-free hepatic protein synthesis in vitro also has been reported (Krug and Hardeland, 1985).

The bioluminescence system of *G. polyedra* (see Section 2.2.4) affords another representative case of CDC-independent, circadian enzyme oscillations. As we noted in our discussion of the transducing mechanisms between the clock and the hands in this organism (see Section 4.5.1), the three different molecular species that participate in both the soluble and the particulate reaction systems responsible for light production are all under circadian clock control. The absolute level of the substrate luciferin, the binding capacity of its specific binding protein (LBP; Sulzman et al.,

1978), and the activity of the enzyme luciferase show circadian oscillations that persist in LL (Fig. 5.10). The rhythm of luciferase activity has been attributed to circadian clock-modulated synthesis and degradation (McMurry and Hastings, 1972b; Dunlap and Hastings, 1981; Johnson et al., 1984). The unique molecule LBP is particularly interesting in that it may provide for the rapid control of the enzymatic reaction, perhaps by providing a stable sequestering site for luciferase (Sulzman et al., 1978).

At a higher level of organization, one might consider the circadian rhythms in ocular and pineal NAT in the chick as a portrait of an enzyme clock (Binkley, 1983). We have seen that NAT, a key enzyme in the pathway for melatonin synthesis (seee Fig. 4.4) in the pineal gland circadian pacemaker (which, in turn, underlies much avian photoperiodic and circadian behavior in birds), shows persisting circadian oscillations in its activity in surgically reduced pineal glands (see Fig. 4.6) and in dispersed cells (see Fig. 4.7) cultured in vitro (see Section 4.1.1). There is a striking correlation between the cycle of NAT activity in the chick pineal gland in vivo in LD: 12,12 and the PRC for the phase-shifting effects of 4-h light pulses on the NAT rhythm (Fig. 5.11A). Based on these and other properties of pineal NAT, Binkley (1983) presents a hypothesis for an enzyme clock (Fig. 5.11B) whose salient features include enzyme synthesis and degradation, buildup of fixed amounts of an enzyme initiator; a rapid disulfide inactivation; and two enzymatic forms, one photostable, one photolabile. There are striking parallels between this scheme and that of the luciferase enzyme in the bioluminescence system of *Gonyaulax*, which suggest, perhaps, the generalizability of the NAT clock portrait.

It is worth recalling that in our previous discussion of the results obtained with inhibitors of macromolecular synthesis used to dissect circadian clocks (see Section 4.3.2), we observed that there appears to be a requirement for 80S protein synthesis in its operation. We found that a cyclically occurring p230 protein has been suggested as a candidate for a key clock protein in *Acetabularia* (Hartwig et al., 1985, 1986), and a 34-

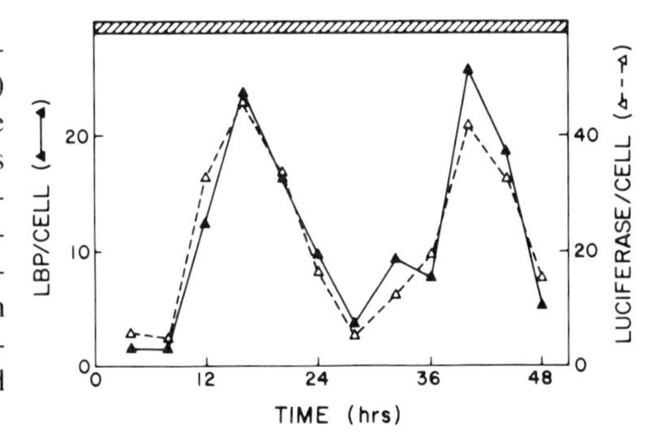

FIGURE 5.10. The extractable luciferase binding protein (LBP) (closed triangles) and luciferase (open triangles) activity rhythms of *Gonyaulax polyedra* in constant light (100 foot candles) expressed on a per cell basis in arbitrary units. Constant light began at 0 h; subjective night was approximately from 12 to 24 h and 36 to 48 h.

kDa protein may play a similar role in the CAP rhythm in *Aplysia* (Eskin et al., 1984b; Yeung and Eskin, 1987). The product of the *per* locus in *Drosophila* clock mutants (see Section 4.4.1) has been identified as a proteoglycan (Jackson et al, 1986; Reddy et al., 1986), and DNA homologous to this locus has been found in mice, chicken, and man (Shin et al., 1985), although it appears not to oscillate during the circadian cycle, as well as in *Neurospora, Acetabularia,* and higher plants (Section 4.4.2).

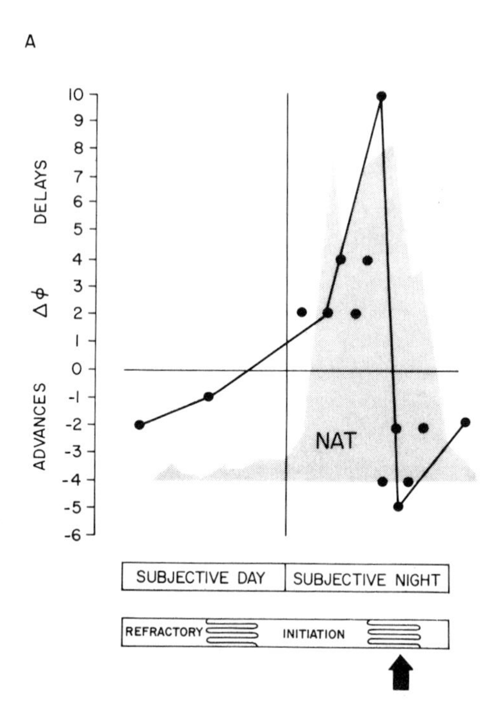

FIGURE 5.11. A portrait of a circadian enzyme clock. (A) Comparison of *N*-acetyltransferase (NAT) phase-response curve and NAT cycle. The cloud represents the NAT cycle in the chick pineal gland in vivo in LD: 12,12. The points are phase shifts of the NAT cycle following a 4-h pulse. Below the data are indicated the times when the system is refractory (dark cannot initiate or reinitiate NAT) and when initiation may occur (sensitive period, when dark can stimulate a rise in NAT). The times of the switch from delays to advances (xing) and the change from initiation to refractoriness coincide. (B) Diagram of the NAT hypothesis. The events have been listed and ordered in time (from left to right); the dashed lines below the sequence of events represent the postulated molecules in the same time frame. The change from sensitive (rapid plummet in NAT in response to light, followed by reinitiation of NAT in dark) to refractory (dark will not reinitiate NAT after its plummet) is correlated with and explains the change in circadian phase to light pulses from delay to advance. Light pulses in the late subjective night reset the NAT circadian clock to zero. (Modified from Binkley, 1983, with permission of Pergamon Press.)

B

```
CT0 = CT24, D/L transition
NAT low
Pineal refractory to dark stimulation
NE, rat nerve endings cease release
NE, rat pineal peak
Receptor site number nadir, rats
NAT, chick pineal can't initiate a rhythm in culture

 CT1-6, early subjective day
 Pineal refractory to dark stimulation
 RNA, rat peak at CT5
 Nucleus, nucleolar size peaks, rats
 Hyperpolarization
 Initiator increasing

       CT7-11, late subjective day
       Pineal no longer refractory to dark stimulation
       Protein, rat pineal peak at CT10
       Actinomycin D blocks NAT, rats
       Sufficient initiator for NAT cycle, chicks

            CT12, L/D transition
            Dark stimulates NAT release, rats
            Receptor site number peak, rats
            Depolarization
            NAT initiation begins

             CT13-17, early subjective night
             NAT high
             Sensitive to light, plummet
             Actinomycin D does not block reinitiation, rats
             De novo NAT synthesis
             NE level in pineal increasing, rats
             Photolabile NAT peak, chicks
             Repolarization
             Phase delays in response to light pulses
             NAT can be reinitiated after plummet

              CT18, xing
              Response to light pulses changes from delays to advances
              Sensitive changes to refractory
              Initiator used up

                   CT19-23, late subjective night
                   NAT cannot be initiated
                   Refractoriness begun
                   Photostable NAT peak, chicks
                   Phase advances in response to light pulses
                   Initiator used up

 -------                          INITIATOR accumulation
      ----------                  INITIATOR use
             ------               NAT photolabile, synthesis
             --------             NAT photostable, synthesis
                   ------         DISULFIDE inactivation
 ------------------------         NIS protease

 +     +     +     +     +
 0     6     12    18    24
```

Models of Microbial Enzyme Synthesis

How can these sustained oscillations in enzymatic activity be explained? The feedback control circuitry that has been proposed to underlie the soluble glycolytic oscillations in yeast (see Section 5.2.1) can be generalized to more complex systems. Monod and Jacob (1961) have described an idealized model for cyclic phenomena that uses known cellular components and genomic regulatory elements. It formally shares basic similarities with the general model for chemical oscillations described earlier. In a typical circuit (Fig. 5.12), the synthesis of an enzyme E_1, genetically determined by the structural gene SG_1, is blocked by the repressor synthesized by the regulator gene RG_1. Synthesis of another enzyme E_2, controlled by structural gene SG_2, is blocked by another repressor synthesized by regulator gene RG_2. Finally, the product P_1 of the reaction catalyzed by E_1 acts as an inducer for the synthesis of E_2, while the product P_2 of the reaction catalyzed by E_2 is a corepressor for the synthesis of E_1. Provided that adequate time constants are chosen for the decay of each enzyme and of its product, this negatively cross-coupled system will oscillate from one state to the other (Higgins, 1967).

This basic model subsequently was developed by a number of workers to explain periodic enzyme accumulation, particularly in prokaryotes, and become known as the oscillatory repression model (reviewed by Goodwin, 1965; Donachie and Masters, 1969). In brief, it proposes that the endproducts of enzyme reactions repress the synthesis of the enzyme by negative feedback, which, under constant conditions, leads to oscillations in the levels of enzyme activity during the CDC. Enzyme synthesis would be repressed when a critical level of product is reached; conversely, it would be triggered when the endproduct concentration is low. In syn-

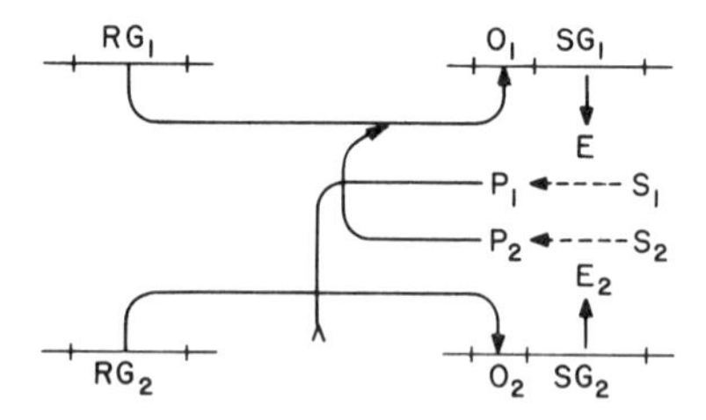

FIGURE 5.12. Model of regulatory control circuitry of cyclical biochemical phenomena. Synthesis of enzyme E_1, genetically determined by the structural gene SG_1, is blocked by the repressor synthesized by the regulator gene RG_1. Synthesis of another enzyme E_2, controlled by structural gene SG_2, is blocked by another repressor synthesized by regulator gene RG_2. The product P_1 of the reaction catalyzed by enzyme E_1 acts as an inducer for the synthesis of enzyme E_2, whereas the product P_2 of the reaction catalyzed by enzyme E_2 acts as corepressor for the synthesis of enzyme E_1. (Reprinted from Monod and Jacob, 1961, with permission of Cold Spring Harbor Laboratory.)

chronous cultures, enzyme levels would oscillate with a definite periodicity, maxima (and minima) occurring at the same stage of the CDC. To explain this, Goodwin (1966) proposed that the oscillations would be entrained and phased by a timed event, such as a pulse of mRNA synthesis at the time the corresponding gene was replicated [although Tyson (1983a) suggests that these two fundamental periods (enzyme cycle, CDC) coincidentally approximate each other simply because they are both close to the mass-doubling time of the cell culture and that the period of oscillation is largely attributable to the slow dilution of the stable enzyme during cell growth]. The oscillatory repression model requires that the system is partially derepressed (or induced) and that genes are available for transcription and translation at all times.

Most data on prokaryotic enzyme synthesis are in agreement with the oscillatory repression model. Especially noteworthy are the observations—bearing out the predictions—that (1) periodic changes in enzyme activity can occur even if DNA replication and cell division are blocked, and (2) the relative positions of the steps in the syntheses of two given enzymes during the CDC can be reversed by delaying the first enzyme with exogenous repressor (e.g., uracil) and releasing it from repression (by washing out the repressor) after the second enzyme has doubled (Masters and Donachie, 1966). This theory has been faulted by Tyson (1979), who considers that the period of an oscillation arising from a negative feedback loop is much greater than the half-life of the most stable element in the loop, an observation that would seem to constrain severely oscillatory repression as a system for generating stepwise increases in enzyme activity.

By using Monod and Jacob's basic control loop for cyclical biochemical events (Fig. 5.12), more complex models can be built. Thus, Goodwin (1963) modified the simple scheme by permitting repression to occur between two different components as well as within single components, so that strong interactions are obtained (Fig. 5.5, top panel). A metabolite controlled by Y_1 (denoting ribosomal protein synthesis) interacts by repression with another genetic locus L_2, and a reciprocal interaction occurs from L_2 to L_1. Even more complex, parallel repression networks involving many interacting, coupled components can easily be generated (Fig. 5.5, lower panels). Significantly, these components could be within one cell or could involve cross talk (i.e., intercellular chemical communication; see Section 6.2.2) among individual cells comprising a tissue. In such situations, inhibitory coupling between a large number of oscillators could result in a much lower frequency of the overall oscillation observed in the population than that of the constituent oscillators (Pavlidis, 1969; see Section 5.2.3). The addition of time-consuming diffusion steps for key substrates could further lengthen the period of the overt rhythm and perhaps impart a measure of temperature compensation. If one postulates that one or more key reaction sequences are photoactivable, the fundamental light-sensitive properties of circadian rhythms might be simulated (Pavlidis and Kauzmann, 1969).

Unfortunately, most prokaryotes do not display circadian rhythmicity (but see Section 6.1.1). Further, many results from eukaryotic systems are inconsistent with the oscillatory repression model for the regulation and timing of enzyme synthesis, perhaps due to their greater structural organization and complexity of their control systems. As a consequence, another model—the sequential transcription, or linear reading, model—has been proposed by Halvorson and coworkers based on the results of experiments primarily with the yeast *Saccharomyces* spp. (Halvorson et al., 1971; Halvorson, 1977). As its name implies, this model postulates that genes are transcribed (and the corresponding enzymes make their appearance) in exactly the same order as their linear position on the chromosome (note the formal similarity to the chronon model; see Section 5.3.1). The model makes several important predictions, some of which have been confirmed experimentally: (1) A gene is available for transcription and, therefore, for induction for only a short time during the CDC. (2) There is a unique interval in the CDC during which a particular gene is trancribed. (3) Periodic synthesis of enzymes should occur throughout the CDC. Particularly compelling evidence for the second prediction was afforded by studies of the enzyme β-galactosidase in a hybrid created from *Saccharomyces dobzhanskii* and *Saccharomyces fragilis*. Two antigenically distinct forms of the enzyme were found, each of which was synthesized periodically at a specific time during the CDC of the hybrid (Gorman et al., 1964). Had oscillatory repression been responsible, only one step increase in activity would have been expected.

Nevertheless, many data do not fit the linear reading model (especially those enzymes that are synthesized continuously and those that remain fully inducible, i.e., have unrestricted potential, throughout the yeast CDC), despite the fact that Halvorson (1977) modified the original version of the model, allowing inducer to override ordered trancription and exempting some proteins from control by this mechanism. Clearly, synthesis of enzymes during the eukaryotic CDC must be controlled by other regulatory mechanisms as well (Lloyd et al., 1982b), including translational control of long-lived mRNAs and posttranslational controls responsible for enzyme stability (e.g., stabilizing ligands, degradation) and activation (e.g., conformational changes, phosphorylation, glycosylation, and other covalent modifications).

Cornelius and Rensing (1982) have attempted to explain the common features of the PRCs for various chemical and physical perturbations on circadian rhythms in *Gonyaulax*, *Phaseolus*, *Kalenchoe*, *Trifolium*, and *Aplysia* (see Section 4.3) by effects on protein synthesis and degradation. The time of maximal phase shift in a major group of these PRCs clusters around a given time of day, suggesting a convergence on the same biochemical process essential to the clock mechanism. These agents inde-

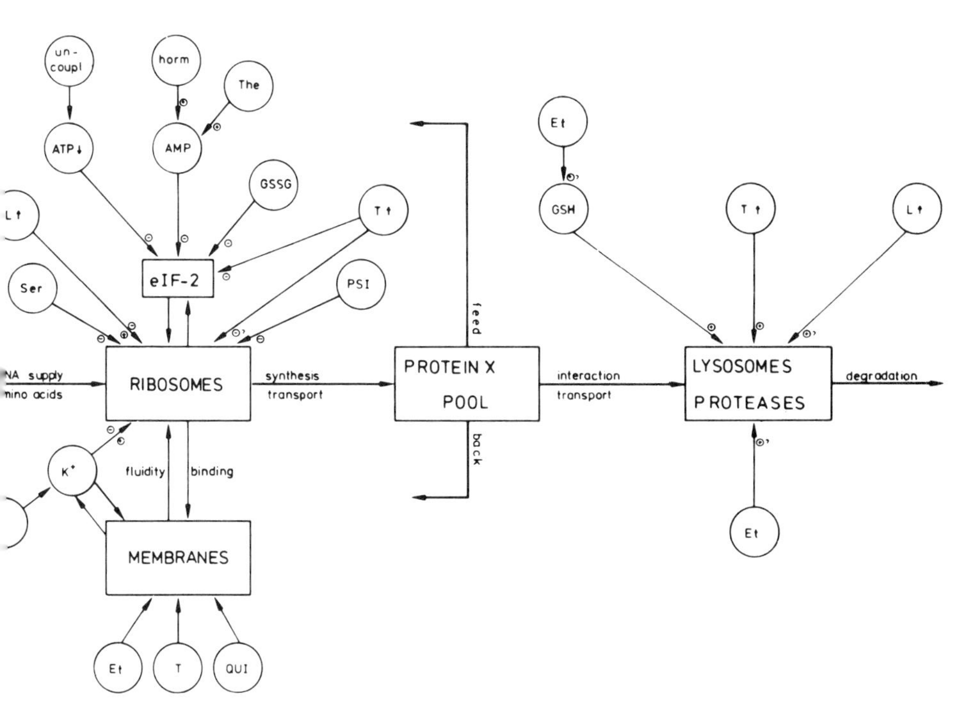

FIGURE 5.13. Influences of various treatments on protein synthesis and protein degradation. AMP, adenosine monophosphate; ATP, adenosine triphosphate; eIF-2, eucaryotic initiation factor 2; Et, ethanol; GSH, reduced glutathione; GSSG, oxidized glutathione; horm., hormones; L, light; PSI, protein synthesis inhibitors; Qui, quinidine; RNA, ribonucleic acid; T, temperature; uncoupl., uncouplers; Val, valinomycin. (Reprinted from Cornelius and Rensing, 1982, with permission of Elsevier Scientific Publishers.)

pendently have been shown to be capable of directly or indirectly affecting protein synthesis (or perhaps certain protein species or enzymes) by a variety of routes (Fig. 5.13). The hypothesis of an oscillatory change in the concentration of a protein species, here the result of alternating buildup and degradative phases Drescher et al., 1982), formally is similar to the oscillatory repression model for periodic enzyme synthesis in the field of cell cycle clocks discussed previously. Both models consist of three components: an enzyme, its product, and a translational inhibitor whose activity depends on the translational product, which in turn determines the rate of enzyme synthesis.

Other Cyclic Epigenetic Events

In addition to periodic changes in the rate of enzyme synthesis, there are a number of other biological and biochemical oscillations having periods on the order of hours, some of which are noted in passing. We have already discussed the temperature-compensated ultradian rhythms in O_2 consumption (τ = 72 min) observed during the variable-length CDC of *A. castellanii* (see Section 5.2.2), which Lloyd et al. (1982a) believe result from a transcriptional or translation feedback loop that enslaves the system of mitochondrial supply. Similar observations have been made in synchronized cultures of *Tetrahymena* (Lloyd and Kippert, personal communication, 1986), in which the 30-min respiratory rhythm has a Q_{10} of 1.12. Periodic changes in the rate of CO_2 production in synchronous cultures of the fission yeast *S. pombe* persist even if the DNA division cycle (Novák and Mitchison, 1986) or protein synthesis (Novák and Mitchison, 1987) is blocked (see Section 3.1.2). The intracellular concentrations of ATP, cAMP, and cGMP in plasmodia of *Physarum polycephalum* rhythmically sporulating at 5 to 10-h intervals in LL oscillated in phase with a period of 4 to 5 h, which was temperature compensated over a 10 to 28°C temperature range (Akitaya et al., 1985). Finally, Mano (1970, 1975) has reported cyclic protein synthesis in sea urchin embryos, correlated with cell division and accompanying changes in –SH groups (see Section 3.1.2). These oscillations ($\tau \cong$ 30 to 60 min) occur only in fertilized eggs and can be observed in reconstituted, cell-free extracts and apparently are due to translational control, since it occurs in enucleated egg fragments or in the presence of dactinomycin.

5.2.5 Circadian Biochemical Oscillators

Thus far, we have progressed from the faster metabolic frequencies ($\tau <$ 10 min), typified by the soluble glycolytic oscillator of yeast and the cAMP oscillator of the cellular slime mold, to sustained epigenetic and genetic (CDC correlated) oscillations having periods of a few hours or more. Can a biochemical model for a circadian clock, incorporating the usual feedback loops and networks encountered in the ultradian models, be far behind? With optimism, if not blind faith that such an extrapolation can be made, let us examine several different models that have been put forward.

Cyclic AMP Models

The central role of cAMP in the cellular metabolic oscillator of aggregating amoeboid cells of *Dictyostelium* was discussed in Section 5.2.1. It is not surprising, therefore, to find that the cAMP system has been incorporated into a biochemical model of the circadian clock (Cummings, 1975). This model is based on the Sutherland model of cAMP as a second messenger (Robinson et al., 1971) and includes membranes that provide compartmentalization and serve as carriers of enzyme and receptors. Just as we saw from Goldbeter's treatment (1975) of the cellular slime mold ultradian oscillator, the five principal elements would be ATP (supplied at a rather constant rate to the cell membrane), membrane-bound AC, cAMP, the cAMP degradative enzyme PDE, and its product, AMP. The mathematical modeling, however, differed somewhat: AMP was assumed to activate PDE but to inhibit AC cooperatively by an allosteric mechanism. Limit cycle oscillations would occur through these feedback effects, given the appropriate choice of parameters, and would result in the daily cycling of all five variables. The effects of light (Aschoff's rules, PRCs) could be simulated by allowing light to mediate increases or decreases in the steady-state levels of active AC. More speculatively, Cummings (1975) hypothesized that the long circadian period would result from the nonlinearity of the system, in which the cycle time would be the complex resultant of individual reaction rates, whereas the stability of the clock against random perturbations of the parameters would derive from the mutual coupling of a population of interacting cellular oscillators (as in a tissue). Temperature homeostasis of τ could result if PDE itself were a self-regulating enzyme (i.e., its endproduct would regulate its activity), as hypothesized by the model, and if the rates of the various AC reactions were temperature compensated by virtue of the properties of the membranes to which the enzyme is bound (Njus et al., 1974). It is interesting to note the blend of ideas from the glycolytic and slime mold systems (see Section 5.2.1), from the mathematical analysis of coupled oscillators (see Section 5.2.3), and from the impact of membrane models for circadian rhythms at the time (see Section 5.4.1).

This model provoked considerable general interest when it was first proposed, particularly given a rather large body of circumstantial evidence that was consistent with some of its predictions (reviewed in Section 4.3.3) (Edmunds, 1980c; Engelmann and Schrempf, 1980). For example, methylxanthines that inhibit PDE did affect circadian rhythms in some cases. On the other hand, the leaf movement rhythm of *Trifolium* was a clear exception (Bollig et al., 1978), and a mutant of *Neurospora* that had greatly reduced levels of AC and cAMP entrained normally to LD: 12,12 and free-ran with an unaltered period (Feldman et al., 1979). More recent work (see Section 4.3.3) further implicates cAMP in circadian timing in both *Neurospora* (Hasunuma, 1984a,b) and *Aplysia* (Eskin et al., 1982; Eskin and Takahashi, 1983; Lotshaw and Jacklet, 1983, 1987), although the question about whether it is actually a clock element must await clarification.

Finally, Rapp and Berridge (1977), in a theoretical study, have suggested that oscillations in calcium-cAMP control loops may form the basis of pacemaker activity and other higher-frequency biological rhythms (see Table 1.1). They hypothesized that oscillations in the intracellular concentrations of Ca^{2+} and cAMP not only are possible, arising from an instability in their common control loops, but also may be widespread. Fluctuations in membrane potential, among other cellular properties, would be a direct consequence of the rhythmic changes in second-messenger concentrations. Most metabolic processes would be influenced, given the central role of Ca^{2+} and cAMP to the regulation of metabolism. Thus, these oscillations would form the basis of several diverse ultradian biological rhythms: potential oscillations in cardiac pacemaker cells, neurones and insulin-secreting β-cells, slow waves in smooth muscle, cAMP pulses in *Dictyostelium*, rhythmic cytoplasmic streaming in *Physarum*, and transepithelial potential oscillations in *Calliphora* salivary glands. Although circadian periods were not addressed directly in this analysis, the central idea would be germane. The coupling of Ca^{2+} to cAMP is one of the most provocative features of this general model, and the role of this important ion in circadian clock models is considered further in the next subsection.

Models Based on the Mitochondrial Calcium Cycle

The experimental evidence that Ca^{2+} may be involved in circadian timekeeping already has been detailed in Section 4.3.3. That Ca^{2+}, mitochondrial Ca^{2+} transport, and calmodulin may play a key role in the clock mechanism itself has been postulated by Goto et al. (1985), based on the results of a biochemical analysis of the circadian oscillator underlying the overt rhythm of cell division in *Euglena* (see Section 3.2.1) and the circadian variations in the intracellular contents of NAD^+ and NADP(H) that have been discovered in this unicell.

But how to test this hypothesis? The difficulty in distinguishing between the hands of circadian clocks (mechanism-irrelevant events) and the gears themselves (clock-relevant processes) has impeded efforts to elucidate their mechanism(s). Phase-shift experiments have been used in attempts to ascertain whether or not a given process is an integral part of a circadian clock on the rationale that a transitory perturbation of either the state variables or the parameters that may be used to characterize an oscillation can cause a permanent $\Delta\phi$ of an overt rhythm. Unfortunately, the converse is not necessarily true; an observed $\Delta\phi$ may have occurred as a result of the effect of the drug or other perturbing agent on some site only secondarily affected by the drug rather than on its postulated primary target (see Section 4.3). Goto et al. (1985) have designated any element as a gear (G) if it can be expressed as a state variable or parameter; if not, it is a nongear (~G). Together, the set of gears would constitute a closed control loop, or oscillator. Because a ~G in its unperturbed, normal or physiological oscillatory range would not be expected to regulate the operation of the Gs themselves, its artificial perturbation within this range should not perturb circadian timekeeping, and no steady-state $\Delta\phi$ in an overt rhythm should be observed. Consequently, if an experimental alteration in the level of a target (by either the direct, pulsed addition of some component or drug known to affect the concentration of that element) does perturb the clock and generates steady-state $\Delta\phi$s, that target is most likely a G (criterion A). Whereas both the activation and the inhibition of a G should perturb timekeeping according to this criterion, this may not be the case when a ~G is similarly perturbed. To more stringently differentiate between G and ~G, therefore, a further requirement for a target to be classified as a G is that both the direct activation, and the direct inhibition of the target must perturb the clock (criterion B). In principal, these criteria are applicable to any set of processes thought to constitute an autonomously oscillatory clock (see Section 4.3).

Applying this rationale, obviously the first step was to determine whether the putative elements of a circadian clock actually oscillated with a circadian period; if not, they a priori could be excluded as Gs. Accordingly, the in vivo levels of NAD^+, $NADP^+$, and NADPH were measured in synchronously dividing and in very slowly dividing cultures of *Euglena*

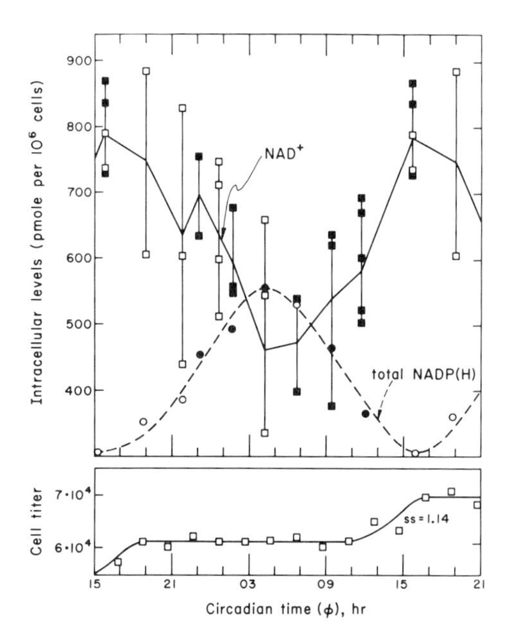

FIGURE 5.14. (Upper panel). Circadian variations in the intracellular content of NAD^+ and total NADP(H) (expressed as pmol per 10^6 cells) in very slowly dividing autotrophic cultures of *Euglena gracilis* maintained in LD: 3,3 regimens. Two out-of-phase, free-running cultures (□ and ■) were sampled at different circadian times, and their NAD^+ contents were spectrophotometrically measured using an enzymatic assay (two to five determinations for each time point). The curve connects the mean values of all data points obtained at a given CT (ϕ). Similarly, total NADP(H) was determined (○ and ●). CT 0 indicates the phase point of a free-running rhythm that has been normalized to 24 h and occurs at the onset of light in an LD: 12,12 reference cycle. (Lower panel). Representative growth curve for one of the two, very similar, analyzed cultures. Cell titer is plotted (on a log scale) as a function of CT (ϕ). *Note:* (a) The very small factorial increase (ss = 1.14) indicating that only 14% of the free-running population divided during the cycle. (b) The amount of intracellular NAD^+ increased by 70% during this fission interval. (Reprinted from Goto et al., 1985, with permission of American Association for the Advancement of Science.)

grown photoautotrophically at 25°C in LD: 3,3 (Goto et al., 1985) and were found to oscillate with a τ of 27 h and with an amplitude not directly related to the CDC (Fig. 5.14). There was also a circadian rhythm in the activity of NAD^+ kinase (peak at CT 0) in extracts of *Euglena* with a phase relationship such that it could induce the rhythm in the in vivo level of NAD^+ (Goto, 1984). No circadian oscillation of the ratios of either $NADH/(NAD^+ + NADH)$ or $NADPH/(NADP^+ + NADPH)$ occurred. A 3-h light pulse applied at CT 18 to the free-running rhythm of cell division not only generated the expected $\Delta\phi$ of the division rhythm but also shifted the phase of the oscillation in total NAD^+ content by approximately the same amount (Goto et al., 1985; Edmunds and Laval-Martin, 1986). These results suggest that the circadian oscillation in the in vivo level of NAD^+ could be ascribable, as for *Lemna* (Goto, 1984), to that of the conversion between NAD^+ and $NADP^+$, but not to that of reduction-oxidation between NAD^+ and NADH.

To determine whether NAD^+ constitutes a clock G according to the criteria given in the rationale, small (25 ml) seed cultures displaying a free-running circadian rhythm of cell division were pulsed (for 2 h) at various CTs with NAD^+ or $NADP^+$—expected to directly elevate their own in vivo levels and then to increase or decrease the rate of the reactions catalyzed by NAD^+ kinase and $NADP^+$ phosphatase—or *p*-nitrophenylphosphate, a competitive inhibitor of $NADP^+$, and then resuspended in fresh medium (effectively terminating the pulse by dilution) for subsequent automated monitoring of the cell division rhythm (Goto et al., 1985). PRCs for the division rhythm could be derived with each of these agents (Fig. 5.15A). Furthermore, a 2.5-h pulse of NAD^+ given at CT 21.7, which caused a $-\Delta\phi$ of 4 h in the division rhythm, also phase-shifted the CR in the in vivo levels of of NAD^+, NADH, $NADP^+$, and NADPH (Edmunds and Laval-Martin, 1986). These results suggest, therefore, that NAD^+ (or $NADP^+$, or NADPH), NAD^+ kinase, and $NADP^+$ phosphatase represent Gs of the underlying circadian oscillator.

What, then, is the element that regulates these enzymes, already suggested to be Gs in themselves? Inasmuch as Ca^{2+}-calmodulin activates NAD^+ kinase in many plants, including green algae, and in the sea urchin, and because the circadian rhythms in the activities of NAD^+ kinase and $NADP^+$ phosphatase appear to be generated by a rhythm in the in vivo level of this complex in *Lemna* (Goto, 1984), it would seem to be a likely candidate for a regulatory clock element (gear) in *Euglena* also. To test this possibility, attempts were made to perturb the clock either by directly

FIGURE 5.15. Phase-response curves (PRCs) for the effect of pulses of different compounds on the free-running rhythm of cell division in photoautotrophic cultures of *Euglena gracilis* in LD: 3,3. Steady-state phase shift ($\pm\ \Delta\phi$) is plotted as a function of the midpoint of the perturbation [CT (ϕ), normalized to 24 h]. The number of data points for a given PRC was varied from 5 to 10, depending on the

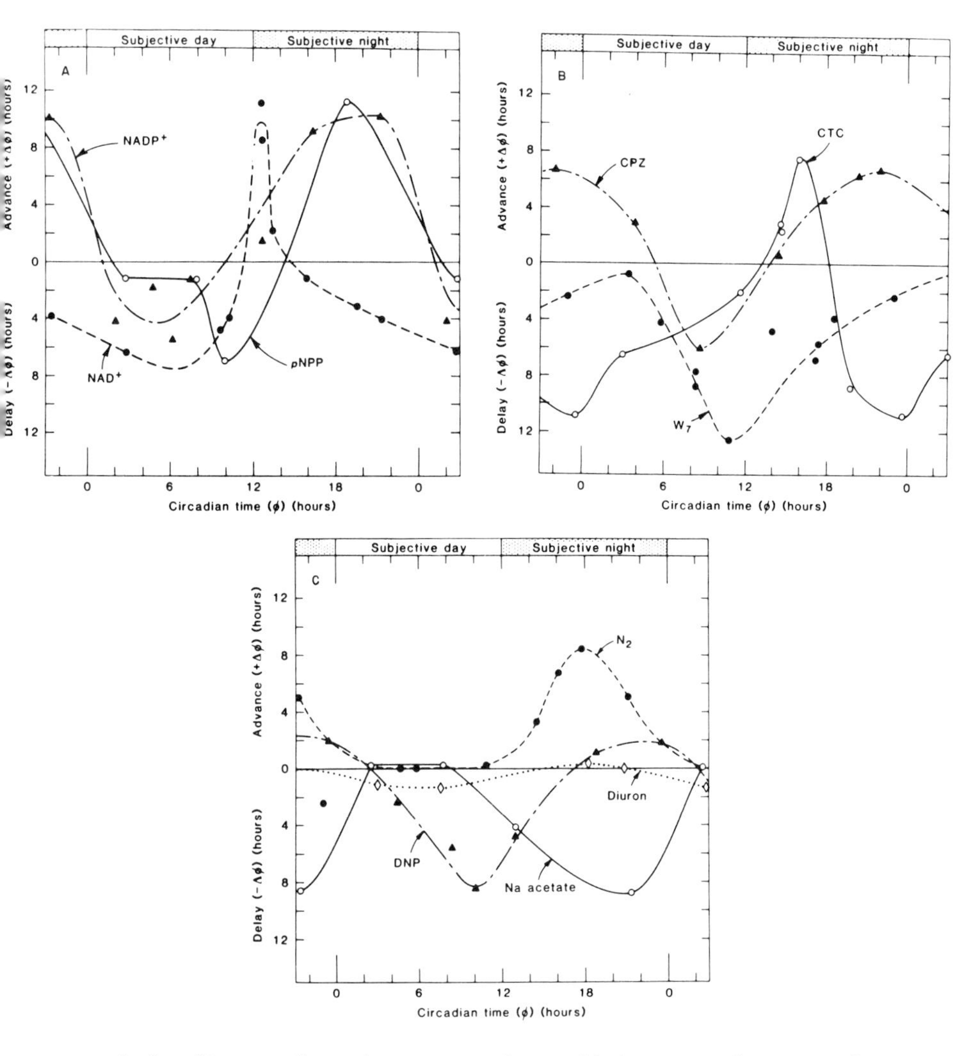

complexity of its waveform. Assays were clustered in instances of apparent sharp discontinuities (e.g., top left figure: NAD^+, CT 12) or to insure that experiment-to-experiment variation was small (e.g., top right figure: W7, CTC). A (Top left). Pulses of NAD^+ (0.5 m*M*, 2.3 h) (●----●), of $NADP^+$ (0.2 m*M*, 2 h) (▲---▲), and of *p*-nitrophenylphosphate (pNPP, 10 m*M*, 2 h) (○—○) applied during a light interval of the LD: 3,3 cycle. B (Top right). Pulses of W7 (20 μ*M*, 2.3 h) (●---●) and of chlorpromazine (CPZ, 50 μ*M*, 2.3 h) (▲---▲) applied during a light interval of the LD: 3,3 cycle, and of chlortetracycline (CTC, 200 μ*M*, 3 h) (○ --- ○) applied during a dark interval. C (Bottom). Pulses of nitrogen (N_2, 600 ml min^{-1}, 3 h) (●---●) applied during a dark interval of the LD: 3,3 cycle, and of sodium acetate (10 m*M*, 2 h) (○—○), of dinitrophenol (DNP, 100 μ*M*, 2 h) (▲---▲), and of diuron (DCMU, 10 μ*M*, 2 h) (◇ ··· ◇) applied during a light interval. (Reprinted from Goto et al., 1985, with permission of American Association for the Advancement of Science.)

causing a transitory decrease in [Ca^{2+}] with 2-h to 3-h pulses of chlortetracycline, a membrane-permeable chelator of Ca^{2+}, or by transitorily inhibiting Ca^{2+}-calmodulin by similar short pulses of W7 and chlorpromazine, both calmodulin inhibitors (Goto et al., 1985; Edmunds and Laval-Martin, 1986). All three drugs yielded pronounced $\Delta\phi$s of the cell division rhythm. The PRCs obtained (Fig. 5.15B) suggested that both cytosolic Ca^{2+} and calmodulin constitute clock gears. [Preliminary results obtained with synchronously dividing cultures of *T. pyriformis* (strain W), in which there is a circadian rhythm of cell division in LL (see Section 2.1.1), are consistent with this hypothesis. There appears to be a circadian variation in the concentration of NAD^+, and pulses of W7 (but not the inactive analog W5) and A23187, an ionophore for divalent cations, phase-shift the division rhythmicity (Goto, personal communication, 1987).]

If so, there should be another gear directly regulating the level of Ca^{2+} in *Euglena*. The main regulatory sites for many noncircadian systems are known to be the plasmalemma, the endoplasmic reticulum, and the mitochondria. To test the possibility that the mitochondrial Ca^{2+} transport system might be a gear (Goto et al., 1985), cultures were pulsed (2 to 3 h) with either nitrogen, dinitrophenol, or sodium acetate, all of which affect electron transport, ATP hydrolysis, and mitochondrial Ca^{2+} efflux. Each of these agents phase-shifted the division rhythm (Fig. 5.15C). The mitochondrial Ca^{2+} transport system, therefore, would appear to be another gear of the oscillator in *Euglena*.

These findings suggest that NAD^+, the mitochondrial Ca^{2+} transport system, Ca^{2+}, calmodulin, NAD^+ kinase, and $NADP^+$ phosphatase represent clock gears that might constitute a self-sustained, circadian oscillating loop in *Euglena* and other eukaryotic microorganisms, as well as in higher plants and animals. A model has been proposed (Goto et al., 1985:), not for the clock but for one oscillator in what is most probably a cellular clock shop. This regulatory scheme (Fig. 5.16) includes the following three steps: (1) NAD^+ would enhance the rate of net Ca^{2+} efflux from the mitochondria (or, alternatively—step 1′—, net Ca^{2+} influx across the plasmalemma) into the cytoplasm, resulting 6 h (90°) later in a maximal concentration of cytosolic Ca^{2+}. (2) Ca^{2+} would immediately form an activated Ca^{2+}-calmodulin complex in the cytoplasm (maximal level at 90°). (3) This active form of $Ca^{2''}$-calmodulin would decrease the rate of net production of NAD^+ by both activating NAD^+ kinase and inhibiting $NADP^+$ phosphatase in the cytoplasm so that the rate would become maximal 12 h later (at 270°) when Ca^{2+}-calmodulin reaches its minimum level. After six more hours when the in vivo level of NAD^+ then becomes lowest (at 0°), the regulatory sequence would be closed, and the cycle would repeat.

This proposed feedback loop can autonomously oscillate because the cross-couplings are always of opposite sign (Higgins, 1967; see Section 5.2.1). Thus, whereas the increases in NAD^+ concentration causes increases in the rate of formation of the active Ca^{2+}-calmodulin complex

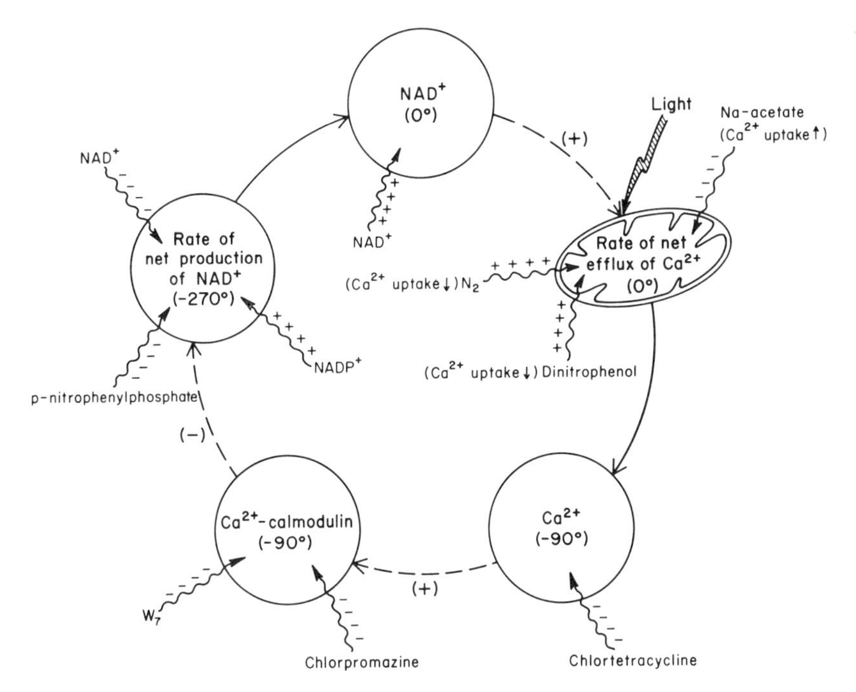

FIGURE 5.16. Proposed control loop for the circadian oscillator in *Euglena gracilis* and *Lemna gibba*. The pattern of regulation is indicated by both solid lines, which relate the reactions to their products, and dashed lines, which correspond to the sequence of steps. Each element oscillates with a circadian period (with peaks and troughs occuring 180° apart). The degrees in parentheses reflect the phases at which maximal values are attained. The maximal rate of a reaction precedes by 90° the maximal concentration of its product. The activation (+) or the inhibition (−) of each compound on the succeeding one is shown and can be considered to produce phase delays of 0° or 180°, respectively. The inhibitory effect of Ca^{2+}-calmodulin on the rate of net production of total NAD^+ derives from both its inhibitory effect on $NADP^+$ phosphatase activity and its activating effect on that of NAD^+ kinase. A number of different compounds and agents known to have either positive (activation: $^+\!\rightsquigarrow^+$ or negative (inhibition: $^-\!\rightsquigarrow^-$) effects on their targets are indicated. Their phase-shifting action on the overt, circadian rhythm of cell division is documented in Figure 5.15. (Reprinted from Goto et al., 1985, with permission of American Association for the Advancement of Science.)

by steps 1 and 2, the resulting elevation in the level of active Ca^{2+}-calmodulin complex causes decreases in the rate of formation of NAD^+ by step 3. Step 1′ would externally entrain, or synchronize, the system. In order for this oscillator to display limit cycle dynamics, additional requirements would have to be satisfied (Higgins, 1967). The increase in the concentration of NAD^+ must cause a decrease in the rate of net formation of NAD^+ (which is very likely). The elevation in the level of active Ca^{2+}-calmodulin complex must cause an increase in its own rate of formation somewhere in the oscillatory region, perhaps by the cooperative binding of Ca^{2+} to the four sites of each calmodulin molecule, which increases the binding affinity of calmodulin for Ca^{2+}. The strength of this self-coupling must be weaker than that of the cross-coupling.

The model does not purport to explain either the long period length or its steady-state temperature compensation characteristic of circadian rhythmicity, primarily because of a lack of hard data. We have seen that a limit cycle can display a circa 24-h period given a particular set of parameter values (see Section 5.2.1), some sort of deposition effect (see Section 5.2.2), or a network of interacting intracellular oscillators (see Section 5.2.3). Alternatively, other time-consuming processes involving transcription, translation or membrane-based ion transport may play a role (see Sections 5.3, 5.4) (Edmunds, 1976; Engelmann and Schrempf, 1980). Similar explanations could be invoked as compensatory mechanisms for the temperature-dependent steps (such as the NAD^+ kinase and $NADP^+$ phosphatase reactions) in the model (Goto et al., 1985). It is important to emphasize, however, that in the present scheme—in contrast to those for the ultradian oscillators discussed previously—one commences with experimentally observed, circadian rhythmicities in elements that are known to be causally and biochemically interrelated. Should additional evidence indicate that a certain putative clock element to the contrary is not a true gear (as, e.g., may be the case for calmodulin), one may merely remove it from the proposed regulatory loop without detriment to the remainder of the scheme.

This model of Goto et al. (1985) shares several aspects in common with the calcium cycle model of Kippert (1986), in which the Ca^{2+} concentration gradient between cytoplasm and mitochondrial matrix alters cyclically (attributed simply to their counteracting Ca^{2+} transport pathways), and to that of Lakin-Thomas (1985) based on oscillations in intracellular Ca^{2+} compartmentation. In his scheme, Kippert (1987) provocatively speculates that endogenous rhythms and the calcium system of intracellular signalling coevolved as a consequence of mutual interactions between partners in a cellular symbiotic consortium developing toward the eukaryotic cell, with its defined nucleoplasm and mitochondria (Levandowsky, 1981; see Section 6.1). Thus, the early, endogenous, self-sustaining clock would have provided a device for internal timekeeping and temporal coordination between still largely autonomous compartments (host and endosymbionts): A homeostatic, self-regulated, oscillatory regimen would have helped preserve this developing association from destabilizing, chaotic fluctuations that otherwise might have developed in the complex network of feedback mechanisms (Decroly and Goldbeter, 1982).

5.2.6 Infradian Metabolic Oscillators

To conclude our discussion of biochemical soup models for biological oscillators, let us briefly examine several infradian systems ($\tau > 24$ h) to see if they share common elements with circadian and ultradian clocks.

Regulation of Metabolic Pathways in Noncircadian Sporulation Rhythms

The sporulation rhythms of the fungi *Leptosphaeria michotii* and *Aspergillus niger* have variable, infradian periods that are not temperature-compensated and that can be manipulated by chemicals (reviewed by Jerebzoff, 1979). In *L. michotii,* a stable period of about 28 h can be obtained in dim LL. The action spectrum for this photoregulatory, stabilizing effect (λ = 350 to 595) has implicated a complex receptor comprising, perhaps, flavins, porphyrins, and carotenoids [which differs, therefore, from the blue-light receptor in *Neurospora* (see Section 4.2.1)]. Furthermore, the equilibrium between saturated and unsaturated carotenoids, acting perhaps as a hydrogen pool for redox reactions, and the level of cAMP seems to be implicated in the oscillator of *L. michotii*. In contrast, a stable period of about 300 h can be obtained in *A. niger* by appropriate adjustment of the ratio of glucose: K^+ in the medium. Jerebzoff (1979) has discovered a metabolic crossway controlling the sporulation rhythm in both organisms (Fig. 5.17). Connected with glycolysis, it includes the asparagine–pyruvate

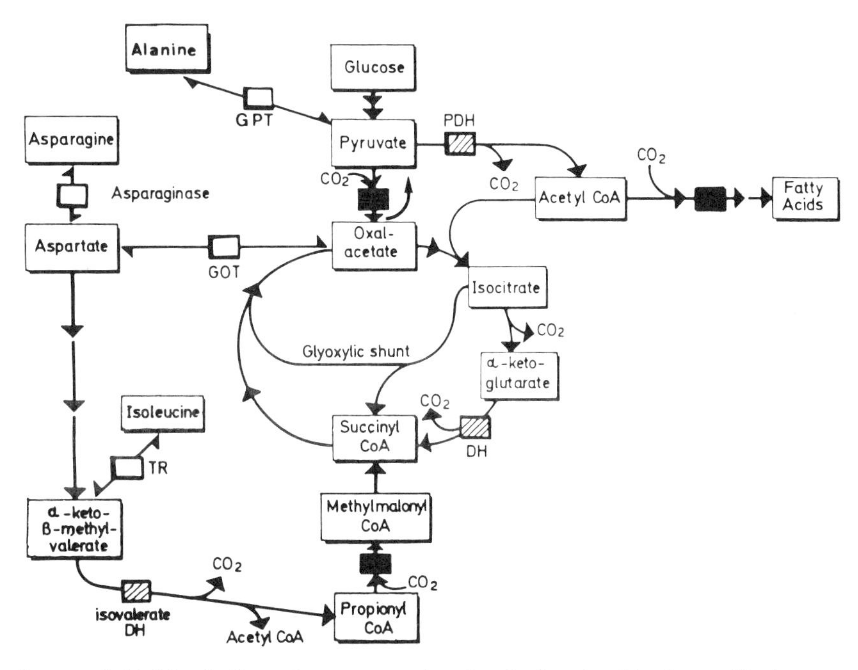

FIGURE 5.17. Metabolic pathways postulated to be involved in the oscillating systems of *Aspergillus niger* and *Leptosphaeria michotii*. Carboxylases, black rectangles; complexes responsible for oxidative decarboxylation, DH (dehydrogenases), hatched rectangles; aminotransferases, white rectangles; GOT, glutamate-oxaloacetate aminotransferase; TR, transaminase. (Adapted from Jerebzoff, 1979, with permission of La Société Botanique de France.)

pathway, the Krebs cycle, and the pathway of isoleucine synthesis and degradation (coupled with a specific K^+ requirement), several carboxylases, and multienzymatic, oxidative phosphorylation complexes. A two-variable, limit cycle model has been developed to account for rhythmic morphogenesis in these and other fungi (Hyver et al., 1979).

The most important element of this network in the two fungi appears to be the asparagine–pyruvate pathway. Indeed, in *L. michotii* the enzymatic activities of asparaginase, aspartate, and alanine aminotransferases and the intracellular aspartate pool all exhibited a 24-h periodicity (Jerebzoff-Quintin and Jerebzoff, 1980). The effects of two inhibitors of protein synthesis, CHX and PUR, given at different circadian times of the activity rhythms of the three enzymes of the pathway have been examined in *L. michotii* (Jerebzoff and Jerebzoff-Quintin, 1983). Both inhibitors generated $+\Delta\phi$s or $-\Delta\phi$s of the enzyme rhythms of varying magnitude, depending on the CT at which they were applied. CHX caused the three periodicities to desynchronize, and the sporulation rhythm was no longer normal. Nevertheless, asparaginase activity remained rhythmic (τ was shortened) even though protein synthesis was inhibited by at least 90% in the presence of CHX (measured as [U^{14}-C]asparagine incorporation into protein) without affecting asparagine uptake. Two different forms of asparaginase (having identical molecular weights but different specific activities) have been detected, their properties changing according to the phase of the in vivo activity rhythm. Both forms showed protein phosphatase and kinase activity, the level of which was dependent on a reversible phosphorylation process (Jerebzoff and Jerebzoff-Quintin, 1984; Jerebzoff-Quintin and Jerebzoff, 1985).

These results, therefore, collectively demonstrated that infradian metabolic networks—once functionally established during development—can exhibit self-sustained oscillations without the requirement for 80S protein synthesis, although the synthesis of protein regulating molecules seems to be necessary for the proper synchronization among enzyme rhythms. If this is true, the aspartate–pyruvate pathway would form an essential part of the periodic system, which might include at least two oscillating subunits, one controlling asparaginase activity, and the other controlling aspartate and alanine aminotransferase activity, whose coordination would be essential for the expression of a normal overt rhythm of sporulation (Jerebzoff and Jerebzoff-Quintin, 1983). At least two cellular compartments (cytoplasm and mitochondria) participate in the timing mechanism. Finally, posttranslational control of the oscillator is exerted through reversible phosphorylation of the enzymes regulating asparaginase activity.

Seasonal Rhythms and Photoperiodism

That organisms can measure even longer infradian periodicities is reflected in seasonal photoperiodism, wherein plants and animals perform a certain function at a quite specific time of the year (see Section 1.1). Although true circannual (circannian) clocks in certain cases might underlie these annual biological rhythms (Pengelley, 1974), it is more likely that this type of timekeeping is accomplished by what may be essentially a daily measurement of the length of the day (or night) by means of a circadian clock, since these intervals are day-specific at any given geographical location (Pittendrigh, 1981).

In a review, Queiroz (1983) noted that the restriction of the operation of particular metabolic patterns to a certain time of the year, in order to obtain optimal performance during favorable conditions or maximum resistance to adverse conditions, would confer a strong adaptive advantage on the organism. As mentioned previously, precise biological timing, afforded by an endogenous oscillator, would permit the detection of the seasonal variation in daily photoperiod and the timely triggering of adaptive responses. But how are these metabolic programs coupled to photoperiodism?

The CAM pathway (Ting and Gibbs, 1982) is the basic process for the photosynthetic incorporation of CO_2 in plants adapted to arid or semiarid, usually very hot habitats and will serve as an illustrative example. A characteristic of this process is that CO_2 is fixed during the dark span because the stomata are open only during this time, thereby preventing dehydration during the daytime. Further, CO_2 is incorporated by PEPC and stored in malate within the vacuole until the following morning, at which time the malate is decarboxylated, affording an internal source of CO_2 for photosynthesis. The CAM pathway uses the enzymes PEPC, malate (NAD) dehydrogenase (MDH), and NAD(P)H malic enzyme (ME), all three of which show diurnal variations in activity in LD that persist with a circadian period in DD (see Section 5.1). Photoperiodic control of the CAM pathway was first observed in *K. blossfeldiana*: short days, but not long ones, resulted in typical CAM-type CO_2 exchanges and the induction of the CAM enzymes accompanied by a drastically increased carbon flow through the pathway (Fig. 5.18).

This photoperiodic induction of C_4 carbon metabolism (as opposed to C_3 metabolism) by short days would be expected also to modify the glycolytic pathway because the pool of PEP required for nighttime CO_2 incorporation is furnished by glycolysis (Fig. 5.18), and indeed, this has been shown to be the case (Pierre et al., 1985). Analysis of the glycolytic pools of various intermediates during the LD cycle in the leaves of plants transferred from long days (LD: 16,8) to short days (LD: 9,15) revealed 24-h variations in their contents and that the timing of the acrophases (maximal values of the oscillations) from the onset of light changed as a function of the photoperiod (Fig. 5.19). Increases in the catalytic capacity of the enzymes and the size of the metabolic pools followed similar kinetics and were accompanied by coherent phase shifts in each. Photoperiodic control (entrainment) of both glycolysis and the CAM pathway, probably via the phytochrome system, thus assures tight coordination between the two metabolic programs, and particularly between their respective key

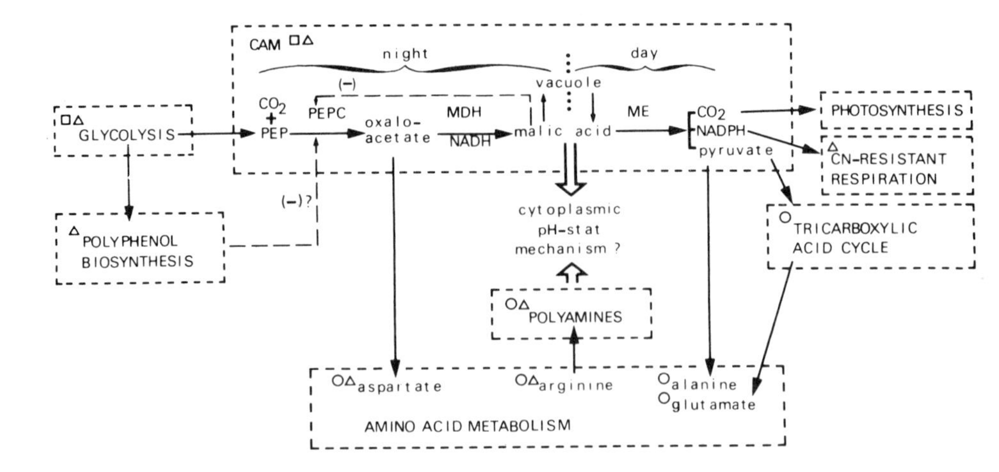

FIGURE 5.18. Photoperiodism and rhythmicity in the primary CO_2-fixation pathway (CAM pathway) and connected pathways in *Kalanchoe blossfeldiana*. (○), Metabolic pathways or components showing diurnal oscillations. (□), Systematic studies on enzymic or metabolic rhythmicity or both. (△), Direct or indirect control by photoperiodism. Enzymes of the CAM pathway; PEPC, PEP carboxylase (EC 4.1.1.31); MDH, malate dehydrogenase (EC 1.1.1.37); ME, malic enzyme (EC 1.1.1.40). (Reprinted from Queiroz, 1983, with permission of Gauthier-Villars.)

enzymes, PFK and PEPC, with regard to both diurnal and seasonal variations (Pierre et al., 1984). This photoperiodic regulation, in turn, would enable metabolic adaptation to drought (Queiroz, 1983).

In summary, we have gone full circle: from the ultradian glycolytic oscillator, through circadian and infradian metabolic rhythms, and culminating in seasonal modulation (via photoperiod) by a circadian clock of those very same glycolytic enzymes and intermediates with which we started. Can one timing mechanism embrace this entire network?

5.3 Transcriptional (Tape-Reading) Models

Another provocative approach—more stringent than the modeling of biochemical feedback-loop mechanisms discussed in Section 5.2—to the problem of transducing the higher-frequency ticks of metabolic events into the 24-h tocks of the circadian escapement has as its basis the notion that the distance between genes could be used for timing. In other words, transcription along the DNA tape could serve as a measure for biological time.

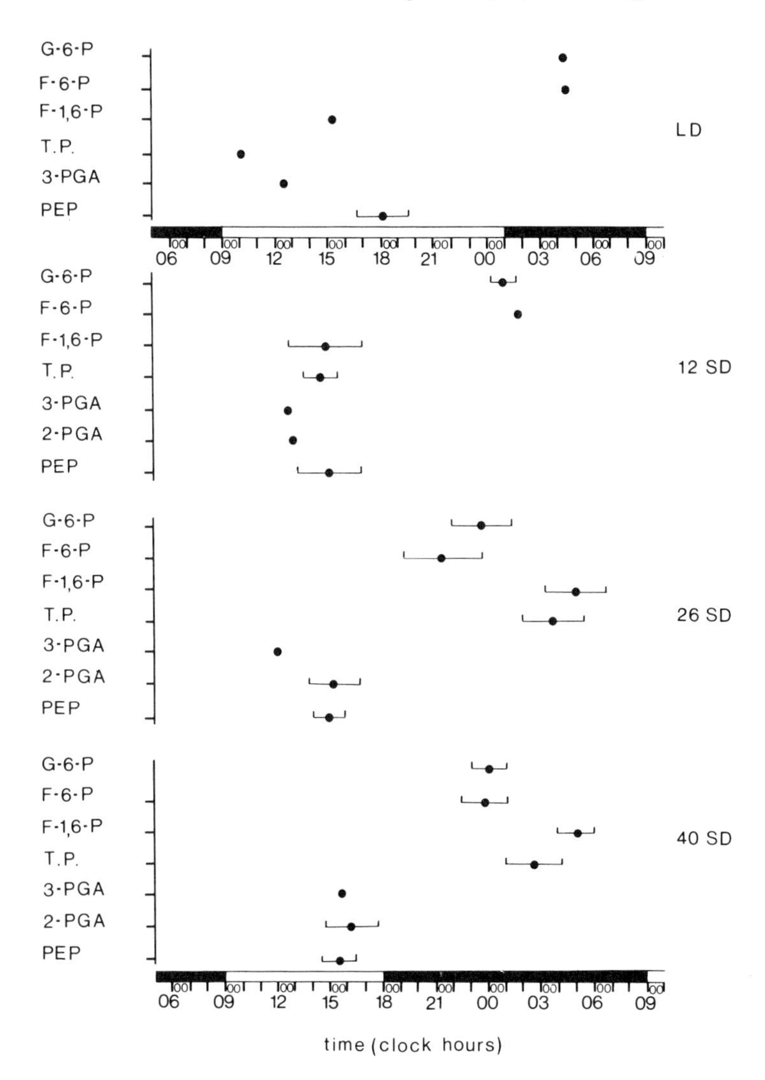

FIGURE 5.19. Acrophase of rhythms of glycolytic metabolic pools in long days (LD) and at the 12th, 26th, and 40th short days in *Kalanchoe blossfeldiana*. G-6-P, glucose-6-phosphate; F-6-P, fructose-6-phosphate; F-1,6-P, fructose-1,6-diphosphate; TP, triose phosphate; 3-PGA, 3-phosphoglyceric acid; PEP, phosphoenol pyruvate. (Reprinted from Pierre et al., 1985, with permission Casa Editrice Il Ponte S.r.l.)

5.3.1 Specificity and Anti-Specificity Factors in Time Metering

The fact that eukaryotic microorganisms tend to divide at more or less constant intervals suggests that they (as well as prokaryotes) have accurate clocks at the cellular level (see Chapter 3). Thus, at 37°C the rate at which RNA polymerase molecules move along the DNA templates of the chromosome is approximately 30 to 40 nucleotides per sec. At this rate the total possible transcription time for all the chromosomes of *Escherichia coli* would be 33 h (assuming that several operons are not transcribed simultaneously). Consequently, only 1% of the total genome would be needed to separate periodic events occurring once every 20-min cell cycle, and even the 3 to 4 h necessary for spore germination could easily be directly timed by the DNA tape. Furthermore, these longer intervals do not necessarily have to be measured by the transcription of a single contiguous piece of DNA; several operons, located on one or more chromosomes, could each code for a unique specificity factor (a subunit of RNA polymerase that recognizes specific sites on DNA for initiation of RNA synthesis, such as the σ factors in *E. coli*) that is necessary for the reading of the subsequently transcribed operon (Watson, 1976, pp. 509–510).

Further regulatory control and delay could be built into biological clocks by the presence of antispecificity factors (also coded for on the genome), which specifically prevent certain specificity factors from functioning and provide a convenient way to limit synthesis of particular proteins to restricted portions of the CDC. These are already known to exist in T4 phages, in which a viral gene product inhibits the functioning of an *E. coli* σ factor. The cell cycles of higher organisms (which can undergo circadian modulation) (see Section 3.2), though much longer, have correspondingly greater amounts of DNA. The haploid content of human genes would require an estimated 1000 days at 37°C to be completely transcribed by a single molecule of RNA polymerase (Watson, 1976, p. 510). Perhaps 24 h is not so long a time span after all!

5.3.2 The Chronon Model

Perhaps the most explicit, and consequently most speculative, transcriptional model that has appeared in the marketplace of ideas, however, is the chronon model proposed by Ehret and Trucco (1967). In essence, this theory for circadian timekeeping postulates that within every cell there exist hundreds of replicons of a special type, called chronons, each of which comprises a polycistronic strand of nuclear (or organellar) DNA some 200 to 2000 cistrons in length (Fig. 5.20). These chronon replicons are the longest and, therefore, are the rate-limiting components of the basic transcriptional cycle, represented by the sequence:

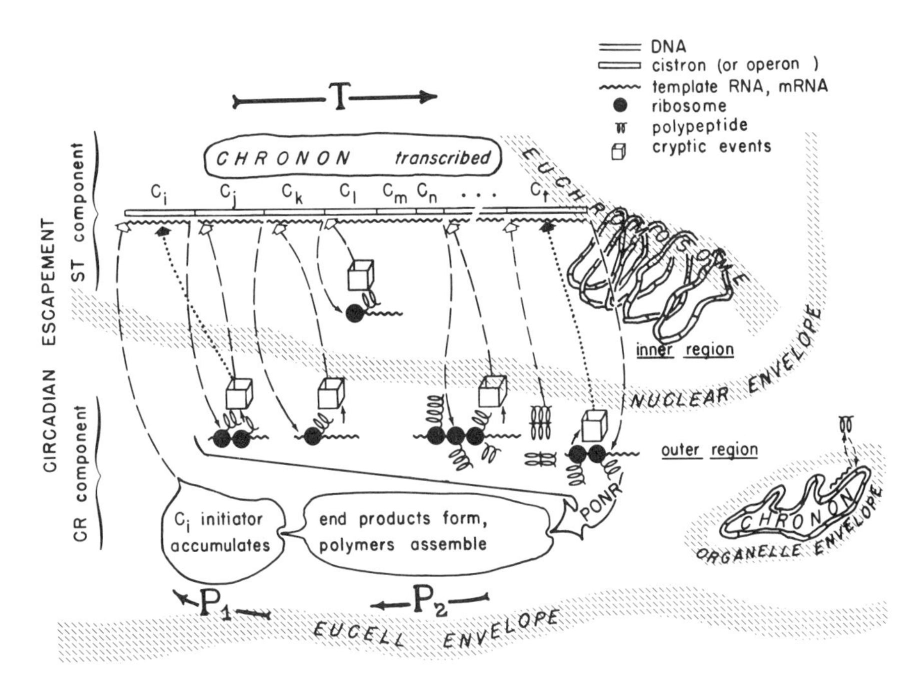

FIGURE 5.20. The chronon model for a circadian clock. The circadian cycle consists of pretranscriptional (P_1), transcriptional (T), and posttranscriptional (P_2) phases, which are analogs of G_1, S, and G_2 phases of the cell mitotic cycle. The sequential transcription (ST) component is the chronon, one of the hundreds of very long DNA polycistron complexes on a single eukaryotic chromosome (or in some cases in a cell organelle). Transcription of template RNA proceeds from left to right, starting at the initiator cistron (C_i). The overall rate of this process is slow and relatively temperature-independent because mRNA formed by each cistron must diffuse out to cytoplasmic ribosomes and then direct protein synthesis, whereupon some of the ribosomal products diffuse back to the chronon and initiate transcription of the next cistron in the sequence. This cycling process continues until the terminator cistron (C_t) is reached. An initiator substance then accumulates and restarts the transcription cycle again some 24 h later. (Reprinted from Ehret and Trucco, 1967, with permission of Academic Press.)

$$P_1 \rightarrow T \rightarrow P_2$$

where P_1, T, and P_2 designate the pretranscriptional, transcriptional, and posttranscriptional phases, respectively [note the analogy to the G_1, S, and G_2 phases of the typical CDC (see Fig. 3.1). These chronons, comprise the sequential transcription component of the circadian escapement and would have achieved circadian length through natural selection over the eons:

$$\frac{\text{Chronon DNA template distance (nm)}}{\text{RNA transcription rate (nm/h)}} = 24 \text{ h}$$

The transcription rate would be limited by time-consuming, diffusion feedback loops in which a given cistron would not be transcribed before the synthesis of a specific precursor catalyst or enzyme whose synthesis, in turn, depends on the presence of an RNA transcript from the most recently read cistron (Fig. 5.20). Thus, the entire cycle is itself composed of a series of nested do-loops, to borrow from computer jargon.

When the terminator cistron is reached after almost 24 h of sequential transcription has taken place, an initiator substance is synthesized on the ribosomes, accumulates, and then diffuses back through the nuclear membrane to the initiator cistron, whereupon the entire circadian transcriptional cycle begins anew. This second key process constitutes the chronon recycling and initiation component of the circadian escapement (Ehret and Trucco, 1967). A measure of temperature compensation would be afforded by the diffusion steps. Light sensitivity could be incorporated easily into the model via photoreceptor molecules (or organs) or photoactivable metabolic reaction sequences.

Because of the richness of detail in the chronon model, a number of predictions—some relatively easily testable, others untestable until a better understanding of the structure and function of eukaryotic DNA is attained—have emerged, (Ehret and Trucco, 1967; Ehret, 1974; Sargent et al., 1976): (1) The longest replicons (chronons) of all plant and animal circadian systems should be of approximately equal length. Unfortunately, this challenging consequence of the chronon theory is difficult to confirm (or better yet, to disconfirm), since the chronon DNA has not yet been isolated. (2) Circadian clocks should be confined to eukaryotic cells. Although this prediction is not necessarily demanded by the model in one of its more relaxed forms, it would seem to be a reasonable assumption because of the importance of the nuclear envelope in providing a compartment for the diffusion circuitry. At the time the model was first proposed (1967), no circadian rhythms had been reported and substantiated in bacteria or other prokaryotes, and most workers consider this still to be the case, although there have been recent reports of circadian rhythmicity in nitrogen-fixing cyanobacteria (Grobbelaar et al., 1986; Mitsui et al., 1986, 1987; see Section 6.1.1). (3) The approximate temperature independence of the escapement is a consequence primarily of the diffusion steps; impediments to diffusion, therefore, should slow down the sequential transcription component considerably. In this regard, it is interesting to note that deuterated circadian systems display almost universally a lengthening of the free-running period (Enright, 1971), but these findings are consistent with several other models for circadian timekeeping as well. The diffusion circuits, on the other hand, could easily be replaced by temperature-compensated enzyme systems or by some other appropriate mechanism (Pavlidis and Kauzmann, 1969; see Section 5.2.1). (4) Different mRNAs should be synthesized at different times of the circadian cycle. This consequence is one of the most fascinating and has undergone at least a degree of empirical verification. Cultures of the ciliates *Paramecium* and *Tetrahymena*, slowly growing in the infradian mode, were synchronized by an LD cycle ($g > 24$ h) and then pulse-labeled with [^{3}H]uridine or ^{32}P at different CTs to label the RNA molecules being synthesized at those times. The RNA was next extracted from the cells and mixed with previously isolated, single-stranded, native DNA that had been immobilized on filters. Finally, the degree of molecular hybridization resulting from this annealing reaction was determined by measuring the radioactivity on the washed filters. In a number of experiments comparing reaction kinetics, saturation, and competition capacities of the different RNAs (Barnett et al., 1971a,b; Ehret, 1974), there was convincing evidence for temporally characteristic RNA species (circadian transcriptotypes). Indeed, in competition-hybridization experiments in which an unlabeled RNA competitor from a given time point was included in the annealing mixture with the DNA filter and labeled RNA preparation, unlabeled circadian t_{24} RNA not only competed best with its labeled homolog stock for the available template sites against unlabeled t_6, t_{12}, t_{18}, and t_{24} RNAs (as might have been anticipated), but also proved to be the best competitor against all labeled RNAs used. (5) Enzymatic activities and metabolically related functions should also display a temporal phenotype or chronotype. This they definitely do, in both single cells and in higher organisms [see, e.g., the acrophase chart for *Euglena* (Fig. 2.4)], but the fact that they do so is certainly not uniquely predicted by the chronon model. (6) One might expect to find correlations between the events and mechanisms that regulate the circadian cycle, on the one hand, and the cell (mitotic) cycle on the other. Circadian modulation of (coupling to) the CDC has already been discussed in Section 3.2, and the similarity between the $P_1 \rightarrow T \rightarrow P_2$ and $G_1 \rightarrow S \rightarrow G_2 \rightarrow M$ cycles has been noted. Conceptually, the same chronons that limit T in the circadian cycle might also limit *S* in the CDC. Indirect support for this hypothesis is afforded by the phenomenon called by its acronym the G-E-T effect (*Gonyaulax, Euglena, Tetrahymena*). Cell cultures in the infradian growth mode, regardless of their generation time, are synchronizable by diurnal LD cycles or L-D transitions and are capable of circadian outputs (Ehret and Wille, 1970; see Section 3.2.2).

The single experimental finding that most discredits the chronon theory is the observation that circadian rhythms can continue in the apparent absence of DNA transcription (Sargent et al., 1976, pp. 296–299). Thus, the rhythm of photosynthetic activity persists unabatedly in enucleated *Acetabularia* (see Section 2.2.1), although the participation of long-lived, stable mRNAs in this rather unique algal cell could be invoked. Inhibitors of transcription (such as dactinomycin, aflatoxin B1, camptothecin, and rifampicin) have been tested in a variety of circadian systems (see Section 4.3.2), and the results in general indicate that neither continuous daily transcription from nuclear DNA nor from organellar DNA is essential for the functioning of the oscillator. Once again, there are the usual caveats:

Is the process assumed to have been inhibited truly blocked? Even if such inhibition is verified, in most cases it is incomplete. It may be, however, that RNA synthesis influences a parameter of the clock (such as availability of RNA for protein synthesis) but does not affect directly the state variables characterizing the pacemaker.

In attempting to evaluate the chronon hypothesis, one must consider such provocative recent findings as of Horseman and Will (1984), who tested circadian cycles at various levels of transcriptional regulation by assaying DNA-dependent RNA polymerases I and II and chromatin structure (as revealed by DNase I and micrococcal nuclease digestion) in liver nuclei from rats free-running in LL. Both polymerases exhibited out-of-phase circadian rhythms in their activities and a circadian variation in the sensitivity of chromatin to DNase I and to nuclease digestion. These findings suggest potential mechanisms for the regulation of circadian changes in gene action. Similarly, Wainright and Wainright (1986) hypothesized that excision repair of DNA may be a major component in the mechanism of the chick pineal clock controlling the NAT cycle in vitro (see Section 4.1.1), and that a circadian clock represents a truncated cell cycle in which DNA synthesis is restricted to excision repair of the genome. Their studies reinforce the caveat that the role of gene transcription in the control of the NAT cycle as inferred from the use of inhibitors is not necessarily that of the formation of the cognate mRNA.

5.3.3 Chronogenes and Cell Division Cycles

In Section 3.2.2, it was argued that the sequential events of the CDC, although in themselves constituting a timer of sorts, cannot be the chronometer controlling the temporal spacing of these events by merely stopping and restarting the sequence, either in response to environmental shifts or to commands from an endogenous (circadian) oscillator. The CDC is not the clock. Rather, a separate programmable entity, termed the cytochron, was hypothesized (Edmunds and Adams, 1981; Adams et al., 1984), which would interpret environmental signals and insert appropriate time loops of specified sign and duration (selected from a finite library of available loops) into the CDC or delete them from it and thereby control the time at which each change in CDC state was triggered (see Fig. 3.48). This cytochron also would be functionally independent of the circadian clock(s), although not necessarily an entirely separate mechanism. Under certain circumstances the two can be uncoupled (Edmunds et al., 1976).

But what is the immediate molecular basis whereby such time dilation or contraction is achieved? Edmunds and Adams (1981) have suggested one possibility that would implicate the direct readout of a nucleotide sequence as a basis for both the timing and the programming functions of the cytochron. This rather speculative model (Fig. 5.21) hypothesizes a small segment of chromosomal DNA, embracing a hundred or so transcriptional units, that is folded into loops by bridging crosslinks (or protein or DNA) at genetically defined loci to form a three-dimensional network of anastomosing loops. The entire complex would constitute a giant functional gene (a chronogene) of up to 3000 kilobase pairs that is capable of metering periods as long as 24 h by the incorporation of 20 to 40 nucleotides per second, as originally proposed in the chronon model of Ehret and Trucco (1967).

The chronogene model differs radically from the chronon model, however, in proposing that (1) the transcription of one loop in the sequence triggers transcription of another specific loop somewhere else within the giant gene without the involvement of a translation step, and (2) the sequence of transcription of the units is programmable. As a transcription complex nears the end of a transcription unit (Fig. 5.21), the initiator site on a loop joined to it by a crosslink opens, and transcription of this second unit begins in response to a signal (possibly torsional) transmitted across the link. To permit programming of the network, each bridging crosslink (perhaps a protein dimer linked to the hairpins of two inverted-repeat cruciforms) would exist in a passive (flop) or an active (flip) mode. In the former, the bridge could not transmit a signal across to a distant loop; instead, as the transcription complex nears the end of the unit and approaches the link, an initiator site opens on the next unit in tandem sequence along the DNA tape. When the bridge switches to the flip mode, however, the completion of transcription of a unit triggers transcription of the distant unit coupled to the other side of the bridge instead of the one in genetic sequence. Modulation of the program by accessory proteins that change modes would permit time-metering loops (groups of transcriptional units) to be added to or deleted from the cyclic program, thus generating instant advances or delays or serving to compensate the cycle time for thermal effects on transcription rate. Some of the bridges might be attached to complexes embedded in the nuclear envelope and could be switched from the flop to the flip mode by the collapse of a membrane potential generated across the nuclear membrane. This collapse could occur in response to the opening of ion gates linked to light-sensitive pigments and temperature sensors in the envelope (see Section 5.4.1), the magnitude of the resulting advance or delay being dependent on the position of the time-metering transcriptional complexes in the transcription circuit at the time of the collapse and on the position of the next available bridge in the programmed topology of the network.

In this model, most of the RNA transcripts have no coding function and are never capped or processed to serve as message, being produced solely to meter time in between structural genes and possibly derive from some of the highly mutated, rusting hulks of DNA that persist in eukaryotic genomes (Edmunds and Adams, 1981). Small RNA segments transcribed directly from critically spaced loops in the temporal program would serve to trigger CDC events, such as S or M, at the appropriate times, either

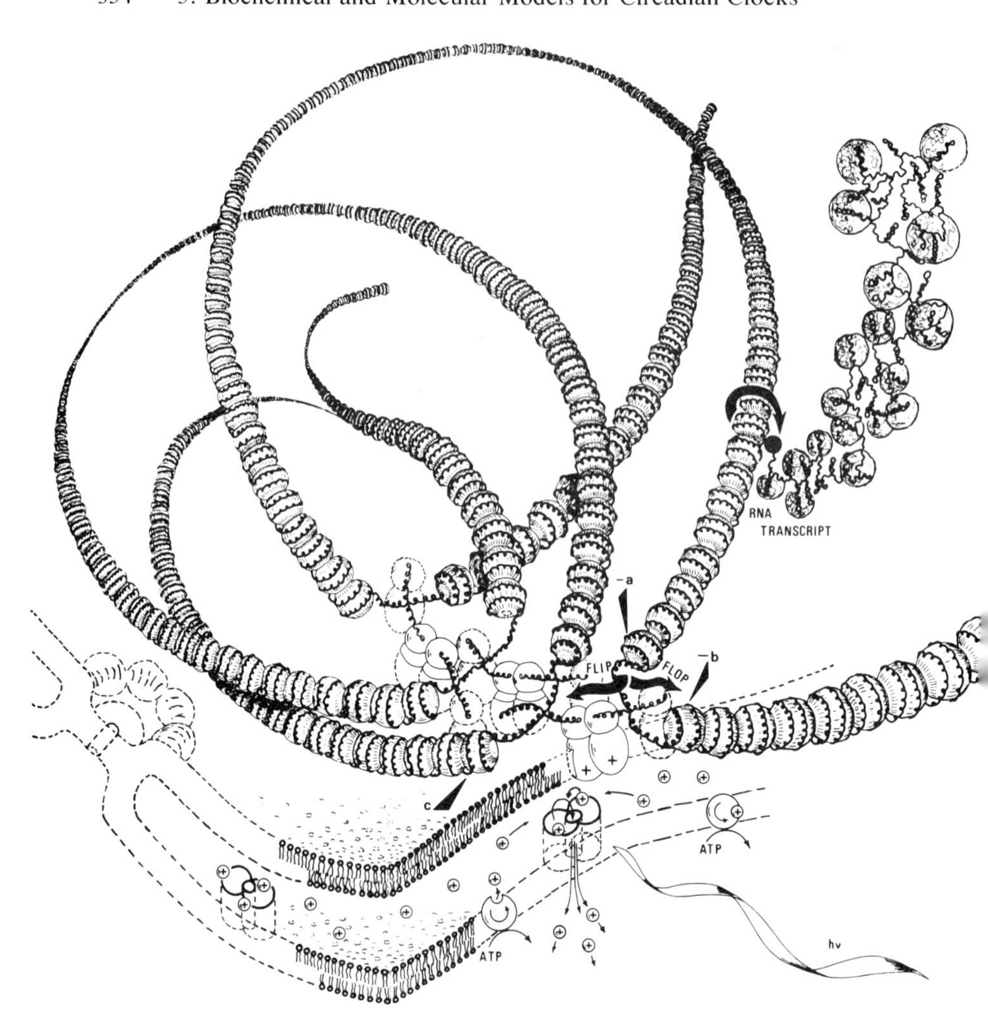

FIGURE 5.21. Diagram of a chronogene segment, one possible molecular basis for metering time in longer-period cellular oscillations. Transcriptional units of chromosomal DNA, wound around their nucleosomes, loop out from protein complexes that anchor them by their inverted-repeat cruciforms to the nuclear envelope and crosslink them to other units at paired genetic loci. At dawn, light-absorbing pigments in the membrane trigger the opening of ion gates that collapse a membrane potential accumulated by ATP (adenosine triphosphate)-driven pumps. The resulting transient change in electric field reprograms the time-metering transcriptional sequence by switching all the protein links to the flip mode. Upon completion of transcription of the top loop (a), transcription is initiated on the adjacent unit (c) in response to a (torsional?) signal transmitted across the link rather than on the next unit (b) in the tandem sequence (as it would in the flop mode). As the membrane potential restabilizes, the flip mode decays so that the only effective switching is mediated by the link at the end of the transcriptional unit being actively transcribed at the time that the transitions between light and dark occur. In this

by being processed and translated into an enzyme or in a more subtle way, so that splicing enzymes can process mRNA precursors. Chronogenes could even be modulated by the products of successive cycles to generate long-term programs that control sequential development in multicells.

The molecular chronogene model, originally developed to account for the cytochron, might also account for the complex light-resetting and temperature-resetting patterns (embodied in their PRCs) of circadian clocks (Pittendrigh, 1965; Pavlidis et al., 1968; Winfree, 1970). This type of model, involving a discrete clock gene, is consistent with (1) observations that double clock mutants exhibit a rather remarkable additivity and map at a small number of genetic loci in *Neurospora* and *Chlamydomonas* (see Section 4.4.1) and (2) the demonstration of protein cross-linked, 30- to 90-kilobase-pair DNA loops, apparently radially arranged, in eukaryotic chromosomes. On the other hand, this class of transcriptional model—as for the chronon model—appears to be at variance with the results obtained with enucleated *Acetabularia* and those with dactinomycin and rifampicin, which suggest that neither nuclear nor chloroplastic DNA is essential for the expression of a circadian rhythm or its resetting by light or dark (see Sections 4.3.2, 5.3.2), although counterarguments are possible (Edmunds and Adams, 1981; Sargent et al., 1976). Alternatively, these conflicting results could be interpreted as indirectly supporting a nucleic-acid independent, membrane-based clock, Perhaps cytochrons rely on chronogenes, whereas circadian clocks rely on membrane-based devices requiring the nucleus only for a supply of parts. At present, one must consider the chronogene construct as a thought model having several possible solutions, arising from the need to explain the insertion and deletion of finite time segments in CDCs formally demanded by their empirically observed variability.

5.4 Membrane Models

The models for circadian clocks considered thus far embody two different conceptual approaches: the feedback-loop models (see Section 5.2) conceive of the clock as a biochemical network with self-sustained oscillations arising from feedback control circuitry within the biochemical system; at

◁

way, long segments of tandemly arranged units are either inserted into or are deleted from a coupled transcriptional circuit at dusk or dawn to generate the advancing or delaying adjustments that serve to synchronize the clock with the earth's rotation. The genetically programmed siting of the links ensures that the precise loop lengths required to effect these phase shifts occur at the right places in the multiple-path transcriptional circuit. (Reprinted from Edmunds and Adams, 1981, with permission of American Association for the Advancement of Science.)

the other pole, the transcriptional tape-reading models envisage time-metering as arising from sequential gene expression. It is difficult for either of these two categories of clock models to stand alone; in fact, they do tend to merge. What is required is a unifying scheme that will provide a structural home for diverse biochemical species. Several membrane models address themselves precisely to this need.

Given the explosive growth of the field of membrane biology, it is not surprising that membranes and cellular compartments have been causally implicated in the generation of circadian rhythmicity (reviewed by Edmunds and Cirillo, 1974; Sweeney, 1976b; Sweeney and Herz, 1977; Edmunds, 1980c; Engelmann and Schrempf, 1980). Several models for circadian oscillators have been proposed that derive from experimental results showing that a number of agents that affect ion transport and other membrane properties also alter period and phase (see Section 4.3.3) and that incorporate as well the requirement for protein synthesis in clock function (see Section 4.3.2). In a sense, it is almost a truism to invoke membranes in circadian timekeeping. It is difficult, although not impossible to imagine a biological process not involving membranes in some manner. What is important is to identify roles of membranes that are essential to the mechanism—the familiar problem of distinguishing between gears and hands (see Section 4.3). Thus, membranes could play a role in clocks in two ways: (1) the membrane function, although not a part of the clock itself and thus not necessarily oscillatory, may control τ by controlling the velocities of parts of the oscillatory process, or (2) the membrane function may constitute an element of the clock and, therefore, obligatorily undergo rhythmic change. The membrane function in either role could reside in the lipid (or other) composition of the membranes, in the activity of membrane-bound (transport) proteins or enzymes, in membrane potential, or in the configuration and conformation of membranes of any or all cell compartments. Let us now examine several models in which membranes and their properties are considered to be clock elements in themselves (Edmunds, 1980c; Engelmann and Schrempf, 1980).

5.4.1 Early membrane models

Two important and similar models for circadian clocks arose approximately at the same time (both published in 1974) and served to focus attention on the key role of membranes. The more general model, in some respects, was proposed by Sweeney (1974b), who derived it from a consideration of the paradoxical role of the nucleus in the circadian rhythm of photosynthesis in *Acetabularia* (see Section 5.4.1): (1) the rhythm persisted for months in the absence of the nucleus, yet the phase in transplantation experiments was determined by that of the donor of the nucleus, and (2) dactinomycin inhibited rhythmicity in intact cells but not in anucleate ones. According to Sweeney's model, circadian rhythms are generated by a feedback mechanism in which some molecule X (perhaps a protein or an ion, such as K^+) is actively transported into organelles until some critical concentration of X is reached within. Passive diffusion then would re-establish an even distribution of X between organelle (e.g., nucleus, chloroplast) and cytoplasm, and the cycle would repeat. Diffusion across relatively long distances within the cell would account for the long period. Light pulses would reset rhythms by affecting transport rate through the membranes, the sign and magnitude of the engendered $\Delta\phi$ being dependent on the phase of the oscillator. During active pumping, light would induce a greater membrane leakage rate, prolonging the active transport phase, whereas during the passive phase, light-enhanced leakage would accelerate active pumping and lead to a phase advance. Finally, temperature compensation of the period would be the result of a higher leakage rate at higher temperatures, necessitating increased active transport to attain the critical concentration of X at which active transport is terminated. The first paradox could be explained if it were assumed that nuclear volume served nonobligatorily as a pool for X, which would be taken up or dissipated depending on the phase, thus generating delays or advances. The second paradox might be resolved by assuming that certain proteins in the *Acetabularia* stalk associated with the transport of X are stable in anucleate cells, but unstable in intact ones.

Sweeney's model (1974b) made several testable predictions, the two most important being that (1) analysis of cellular organelles and their membranes by such sophisticated techniques as electron microprobe examination or freeze-fracture or freeze-etch electron microscopy should reveal circadian changes in their composition (with respect to ions or other molecules) and structural arrangement, and (2) short perturbations of cells by specific ionophores should phase shift their overt rhythms. Although there is much circumstantial evidence in support of both predictions in *Acetabularia* (see Table 2.3) and *Gonyaulax* (see Table 2.5), for example, the data on the effects of pulses of membrane-active agents (see Section 4.3.3) are sometimes contradictory (e.g., gramacidin, A23187, did not shift the phase of the rhythm of stimulated bioluminescence in *Gonyaulax*). There is no reason, moreover, to propose that the oscillator is located in either nucleus or chloroplast, since circadian rhythms have been reported in organisms lacking one or the other of these two types of organelle. The evidence at best is consistent with the model but by no means demands it.

The second membrane model for circadian timekeeping, put foward by Njus et al. (1974), is richer in detail. Their model incorporates the network concept, but identifies ions and membrane-bound ion transport elements with the biochemical clock, and thus with the primary oscillations; the two variables would interact in a limit cycle system. Feedback control would arise in the system by the effect of ion concentrations (implying changes in transmembrane ion gradients) on the functioning of the ion transport structures themselves, which, in turn, would affect the distribution of those ions. As will be seen, temperature compensation would derive from the lipids within the membrane (known to be capable of adapting to environmental temperature fluctuations by changing their physical state), which would determine the rate-limiting kinetics of transport activity. Finally, this membrane clock would be coupled biochemically to the diverse overt rhythms surveyed in earlier chapters. Some could be driven directly (e.g., leaf movement or phototaxis), some perhaps by ion-mediated activation and inactivation of enzymes, and some by ion-stimulated inducers or hormones.

Light would affect the clock by causing an ion gate in the membrane to open that may be either photosensitive itself or coupled hormonally or neurally to an anatomically distinct photoreceptor. Now, what might be the effect of a perturbing light pulse on such a system? For the sake of illustration (Fig. 5.22), let us assume that a rhythm involves just one ion and only two membrane states. The transmembrane ionic gradient will oscillate as shown in the top panel of Figure 5.22, and membrane transport will fluctuate as depicted in the bottom panel. The impinging light pulse would open the ion gate and thus lead to the depletion of the ionic gradient (to x_o). The consequence of this drop in the gradient, however, depends on the state of the membrane *(y)* at the time of the pulse. Applied before CT 18, it will generate a phase delay because the membrane is in the active mode and the ion gradient *(x)* will build back up again; but imposed just after CT 18, it will cause a phase advance, since the membrane is in its passive mode and x would proceed further into its state of discharge.

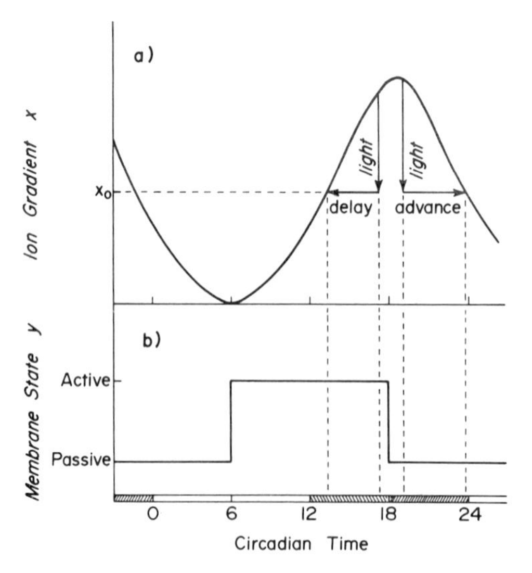

FIGURE 5.22. Hypothetical oscillations in a simple one-ion system (a) with only two membrane states (b). A light pulse applied just before CT 18 results in a phase delay, since it brings the ion membrane system to a state equivalent to that at about CT 13. Similarly, an identical light pulse impinging just after CT 18 generates a phase advance. Maximum phase shifts are obtained during the subjective night. (Reprinted from Njus et al., 1974, with permission of Macmillan Journals Limited.)

(Note the intended analogy with the breakpoint of the typical PRC for light pulses.)

If, then, the basic circadian clock is a feedback oscillator comprising ions and membranes, the consequent transmembrane ion fluxes might be expected to be mediated primarily by either the synthesis and degradation or the activation and inhibition of membrane proteins. Since protein synthesis was thought probably not to be required to drive the oscillation, the latter control process seemed to be more likely. It is plausible that the kinetics of the resulting oscillation depend, at least partially, on the physicochemical properties of the lipids that compose the membrane lipid bilayer into which these proteins are embedded and are relatively free to migrate about. A simplistic, structural representation of the way in which this fluid mosaic membrane complex might serve as the circadian clock is given in Figure 5.23. This diagram shows changes both in the arrangement of the intercalated particles and in their sizes caused by the varying distributions of K^+ ions. In turn, the state of these proteins determines the direction and activity of membrane transport. A photoreceptor that serves as a K^+ gate or ion channel is depicted also. Translational diffusion of the membrane particles, coupled with time-dependent cooperative phe-

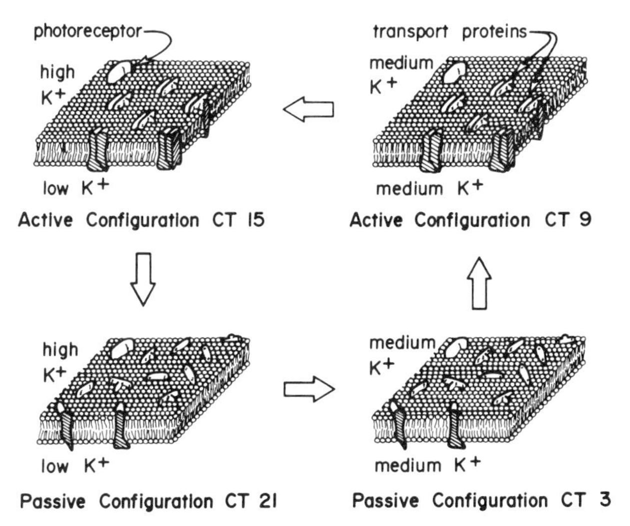

FIGURE 5.23. A membrane model of the circadian clock. Schematic repesentation of the way in which a fluid mosaic membrane might keep time. Changes in both particle arrangements and sizes caused by ion distributions are shown. The state of these particles (hypothesized to be membrane proteins intercalated into the lipid bilayer that are capable of migration within the plane of the membrane), in turn, determines the direction and activity of membrane transport. A photoreceptor that serves as a K^+ gate is included. This ionic clock would couple biochemically to the diverse overt circadian rhythms. (Reprinted from Njus et al., 1974, with permission of Macmillan Journals Limited.)

nomena, supposedly would yield the long circadian period (although this postulate is less fact than act of faith and, in fact, may be contradicted by observations of rapid lateral movement of proteins in cellular membranes).

Now it is necessary to explain the effects of temperature on circadian systems, one of the more attractive features of the model. For temperature steps-up, known to cause phase advances in overt rhythms, the lipid (fatty acid) composition would change over the course of a few hours to compensate for the temporarily increased rate of ion transport, and the initial free-running period would be restored (Njus et al., 1974). The opposite adaptive change in the membrane lipids would soon negate the delaying effects of a temperature step-down. Since a temperature pulse or cycle can be treated as a series of step-changes, a similar analysis would hold for the resultant period in the entrained steady state. Such changes in the physical state of membranes during temperature adaptation are well known in such diverse organisms as goldfish and bears (Wisnieski and Fox, 1976).

This membrane model, like that of Sweeney (1974b), provides several avenues for experimental verification (Njus et al., 1974, 1976; Njus, 1976): (1) Clock mutants might well phenotypically exhibit altered ion transport capacities or differences in membrane lipid composition [as has been found in the *zonata* strain of the fungus *Podospora anserina* (Lysek, 1976) and in the *cel*$^-$ mutant of *Neurospora* that is defective in temperature compensation (Coté and Brody, 1987; see Section 4.4.3)]. (2) Changes in lipid composition should be detectable (perhaps by gas chromatography) during adaptation to changing temperatures [as has been seen in *Neurospora* (Vokt and Brody, 1985)]. (3) Perturbation of lipid composition by chemicals or nutritional changes should be reflected in the output period length of the system [as demonstrated by Brody and Martin's manipulation (1979) of the period of *Neurospora* by the addition of unsaturated fatty acids to the medium (see Section 4.3.3, Fig. 4.29)]. (4) Phase shifting by light in unicellular organisms should cause a change in the intracellular ionic distributions [as observed in the pulvinus of many leguminous plants during the resetting of the leaf movement rhythm, or in the eye of *Bulla* or *Aplysia* (see Section 4.3.3). (5) Ferritin labeling and freeze-fracture electron microscopy might permit one to directly observe circadian oscillations in the arrangement of the postulated membrane-intercalated particles [as shown for *Gonyaulax* by Sweeney (1976a)]. (6) Electrical events in the cell may be correlated (causally?) with the changes in K^+ flux, which, in turn, lead to a change in transmembrane potential [as found in the pulvinus of *Samanea* (Racusen and Satter, 1975) or in the white clover, *Trifolium repens* (Scott et al., 1977)]. Most of this evidence, along with all of that implicating ions in some manner in timekeeping (see Section 4.3.3), however, is correlative rather than causal. Membrane function is so pervasive in biological processes that it is difficult to design an experiment that will falsify explicitly this or any membrane model for a circadian oscillator.

5.4.2 Coupled Translation–Membrane Model

In developing this model for a circadian oscillator, Schweiger and Schweiger (1977) not only incorporated many of the features of the earlier membrane models (see Section 5.4.1) but also took into account the large body of data on the effects of various inhibitors of macromolecular synthesis on circadian rhythms in *Acetabularia* and other microorganisms and higher systems that implicates the essential role of protein synthesis on 80S ribosomes in clock functioning (see Section 4.3.2, Fig. 4.17). The central components of the oscillator were considered to be membranes (in particular, those of chloroplasts for *Acetabularia*) and essential membrane proteins, whose synthesis is feedback regulated. These changes in the protein composition of the membrane (both qualitative and quantitative) would require assembly (synthesis on 80S ribosomes and their loading, or insertion, into the thylakoid membrane) and then disassembly (unloading from the membrane and subsequent degradation), as illustrated schematically in Figure 5.24A. Loading would change the functional state of the thylakoid membrane, which would lead to changes in ion concentration that, in turn, would inhibit either the de novo synthesis of the essential proteins or their transport from the site of synthesis to the site of consumption in the membrane. The rate of disassembly could be constant, with the rate of assembly undergoing circadian oscillations, or the reverse situation could hold true. The slow, time-consuming processes of assembly would generate the long circadian period (at least in theory). As for the model of Njus et al. (1974), light and dark would affect rhythmicity by opening ion gates in the thylakoid membrane. Perturbations given during the loading phase would retard loading and generate a phase delay of the rhythm, whereas those imposed during unloading would accelerate the process and yield a phase advance. Temperature compensation would derive from the fact that the processes of transport and synthesis, on the one hand, and insertion (loading), on the other, were assumed to have counterbalancing temperature coefficients (high and low values of Q_{10}, respectively), leading to an overall net Q_{10} of about 1 (Fig. 5.24B), although some observations have challenged these assumptions (Fowler and Branton, 1977).

An obvious line of experimental verification of this model would be to identify and demonstrate the cyclic synthesis of the limited number of hypothesized proteins essential for clock function. Indeed, it is quite satisfying that since the model was put foward, Hartwig et al. (1985, 1986) have detected one (and only one) such 80S protein (p230) having a high molecular weight (230,000) in the chloroplast fraction of both nucleate and anucleate *Acetabularia* that fulfills these requirements (see Section 4.3.2). Its rate of synthesis in LL exhibited highly significant circadian oscillations, its synthesis was inhibited by cycloheximide, and 8-h pulses of the inhibitor phase-shifted the rhythm of p230 synthesis (see Fig. 4.19). Schweiger et

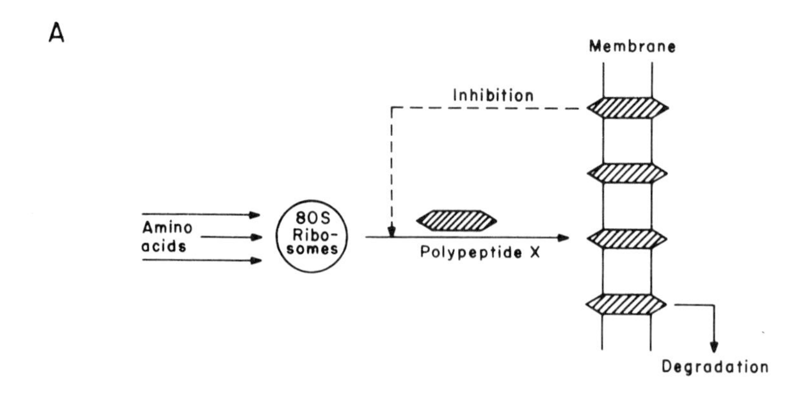

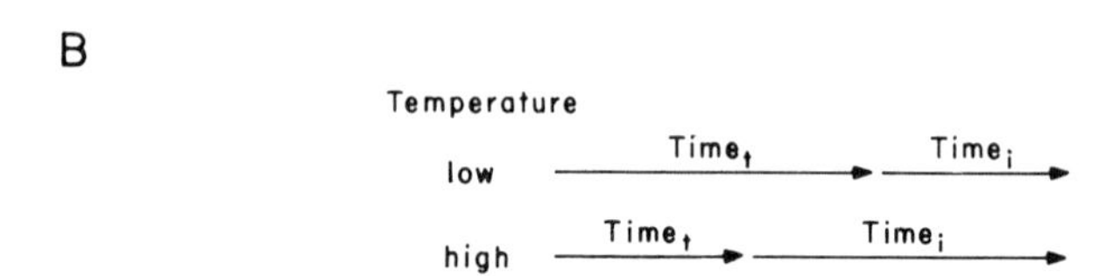

FIGURE 5.24. The coupled translation–membrane model. (A) The model postulates the existence of a polypeptide that is synthesized on 80S ribosomes and then integrated into a membrane. This integrated polypeptide affects membrane function, resulting in inhibition of the production of the polypeptide. Owing to degradation of the essential protein, the inhibition is suspended, and the polypeptide is synthesized again. (B) Temperature compensation of τ is based on the fact that at low temperature the time needed for translation (Time$_t$) is longer, whereas the time needed for integration (Time$_i$) is shorter (perhaps due to the fact that the degree of order in the membrane also is higher at lower temperatures). At high temperature, the situation would be reversed. (Reprinted from Schweiger et al., 1986, with permission of The Company of Biologists Limited.)

al. (1986) conclude that translation of the message for p230 is rhythmically switched on and off, a result in accord with the inhibitor studies (see Fig. 4.17), perhaps as the result of a specific selection of the p230 message from a cytoplasmic storage pool and the transport of this species to the polysomes, where it would be bound and immediately translated.

DNA sequences homologous to those in *per* probes from *Drosophila* (see Section 4.4.2) have been found in *Acetabularia*, suggesting that the essential protein is ubiquitous and that its primary structure has been conserved, but in the chloroplast rather than in the nuclear genome (Li-Weber et al., 1987). This seemingly contradictory discovery led to the hypothesis that in *Acetabularia* the activity of p230 (having a highly specialized membrane function) would be affected by the product of the *per* locus, which would be a near-neighbor protein (but not an essential clock element) also encoded in the chloroplast genome. This situation perhaps reflects an ancient phase of symbiosis between a eukaryotic host cell and a prokaryotic, photosynthetic endosymbiont in which the relevant genes have not yet been translocated to the nuclear genome, as they have been in the fully integrated chloroplasts of higher plants (Schweiger et al., 1986).

5.4.3 Monovalent Ion-Mediated Translational Control Model

This membrane model (Burgoyne, 1978) for circadian rhythmicity is essentially a hybrid of those of Njus et al. (1974) and Schweiger and Schweiger (1977) discussed in the preceding two sections. The former had some difficulty in accounting for the long time constant and the demonstrable requirement for protein synthesis (see Section 4.3.2), whereas the latter did not address itself in detail to the known effects of altered levels of monovalent ions (especially K^+) on circadian rhythms (see Section 4.3.3).

In Burgoyne's model (1978) of ion-mediated translation control, the rate of synthesis of a membrane protein involved in ion transport (perhaps the Na^+/K^+-ATPase, or a modulator of this activity) is feedback regulated by the intracellular monovalent ion concentration. Thus, as the membrane protein(s) is synthesized and inserted into the membrane, its activity would lead to a change in the monovalent ion concentration that would inhibit its own synthesis. Eventually, due to normal turnover of the membrane protein and passive diffusion of the ion, the ion concentration would change in the reverse direction, and a critical point would be reached at which synthesis of the membrane protein would be switched on again. As with the coupled translation–membrane model, the circadian period could be attributed to the long half-life of the membrane protein and the time lag between its synthesis and subsequent transport to and insertion into the membrane. Phase-shifting by light would take place via photosensitive gates in the plasma membrane (Njus et al., 1974), and temperature compensation presumably would have an explanation similar to that given by Schweiger and Schweiger (1977).

In addition to predicting that there would be circadian variations in the amount in the membrane of a protein involved in ion transport, in the rate of synthesis of this membrane protein, and in the levels of monovalent ions within the cell—all anticipated in the earlier membrane models—Burgoyne's model (1978) uniquely and critically predicts that the translation of the mRNA coding for the membrane protein would be sensitive to the postulated changes in monovalent ion concentration during the circadian cycle. A test of this prediction, however, awaits the identification of the putative membrane protein(s), a good candidate for which is the p230 protein recently described by Hartwig et al. (1985, 1986) (see Section 4.3.2).

5.4.4 Other Variations on the Membrane Model Theme

Some other membrane-type models for circadian oscillators are worthy of brief consideration (Table 5.1).

For example, Konopka and Orr (1980) have proposed a model for the oscillator in *Drosophila* that formally is similar to those of Sweeney (1974b) and Njus et al. (1974), which propose active and passive membrane processes as the mechanism. In this model, the oscillation consists of a buildup of a gradient across the membrane during the subjective day and a dissipation of the gradient during the subjective night (in the earlier models the active and passive phases began, respectively, at the midpoints of the subjective day and night). The gradient would be established by the operation of a light-insensitive ion pump; when the gradient reached it maximum value, the pump would cease to operate and light-sensitive ion gates would open to allow the gradient to dissipate (in earlier membrane models, light closes the gates). Light pulses during the subjective night would cause the gates to close and the pump to start up again, generating either net phase delays or advances depending on whether the perturbation occurred in the earlier or later part of the night. Konopka and Orr (1980) hypothesize that the per^S gene product (see Sections 4.4.1, 4.4.2) increases the activity of the ion pump, resulting in a shorter subjective day. Indeed, this is precisely what is observed experimentally. The daytime interval when the fly is active and the portion of the PRC for light signals corresponding to subjective day are both shortened in per^S flies. Long-period mutants (per^L) would be obtained by decreasing pump activity, raising the gradient's maximum value, or decreasing the number of channels. Other loci even might code for different pump subunits or channel proteins. Thus there would appear to be separate molecular processes characterizing subjective day and night.

Satter and Galston (1981) and Iglesias and Satter (1983a,b) have detailed a model for coupled K^+/H^+ fluxes to account for the direction and pathway of K^+, Cl^-, and H^+ movements and the role of membrane potential underlyng circadian leaf (sleep) movements in leguminous plant, such as *Samana* (see Section 4..3). Active extrusion of protons by an electrogenic proton pump was postulated to energize K^+ accumulation and the resultant changes in turgor in the flexor and extensor cells of the pulvinus. As a consequence, cytoplasmic pH might also vary, potentially affecting a multitude of cellular processes.

This model involving proton extrusion is particularly interesting in light of the feedback model of Chay (1981) for biological oscillations, in which changes in H^+ transport, cytoplasmic pH, and pH-sensitive, H^+-producing or H^+-transporting enzymes play an essential role. The oscillator would have the following three features: (1) a carrier embedded in a membrane removes H^+ from the inner compartment to the outside by active transport, (2) a key membrane-bound enzyme or one localized within the compartment translocates H^+ from outside to inside or produces H^+ as a product, and (3) the activity of this key enzyme depends on the pH of the inner compartment. When the two competing processes become counterbalanced (i.e., when the rate of removal of H^+ from the compartment equals the rate of production or translocation of H^+ by the enzyme), the system may break into sustained oscillations. With this model, oscillations having a very long period, as observed in mitochondrial and circadian rhythms, could be simulated quantitatively by appropriately adjusting the parameters K_S (where S is the substrate concentration) and pK_Q (the acid dissociation constant of the reduced state of a carrier protein); τ would be inversely proportional to the value of V_{max}/K_S. Thus, not only does Chay's model (1981) accommodate the oscillations of enzyme and substrate observed in some circadian rhythms (see Sections 4.5.1, 5.2.4), but also it can generate a 24-h period with the requirement of slow lateral diffusion of proteins (as called for by Njus et al., 1974).

A mathematical model of periodic processes in membranes, particularly of the regulation of the CDC by the plasma membrane, has been developed by Chernavskii et al. (1977). A relatively simple autonomous cell clock is assumed, in which membrane processes (as opposed to energy metabolism; Sel'kov, 1970, 1975) play a leading role. These processes entail specific membrane changes in chemical (oxidation) and physical (phase-transition between solid and fluid states) properties, and a dynamic, oscillating system can result under certain circumstances. Conditions of transition between the cellular proliferative state and the resting state were investigated and modeled.

The possibility that the phosphorylation and dephosphorylation of proteins, many of which are membrane bound, may play a key role in timekeeping and perhaps even be part of the molecular clock mechanism itself has received increasing attention (Sweeney, 1983). For example, the regulatory properties of PEP carboxylase, which undergo cyclic changes at the posttranslational level in many CAM plants (see Section 5.1), has been attributed to a reversible phosphorylation process that actually has been observed in *Bryophyllum* (Nimmo et al., 1984, 1986, 1987); and *Kalenchoe* (Brulfert et al., 1986). Similarly, the rate of protein phosphorylation, especially of a 95-kDa species, shows circadian changes in cell-free extracts of *G. polyedra* (Schröder-Lorenz and Rensing, 1987). Finally, serotonin (5-HT) increases the phosphorylation of two low molecular weight (23 and 15 kd) phosphoproteins and decreases the phosphorylation of a 20-kd species in the isolated *Aplysia* eye; this effect is mimicked by the cAMP analog 8-benzylthio-cAMP (Lotshaw and Jacklet, 1987). These phosphoproteins, therefore, may mediate the phase shift of the circadian rhythm of CAP induced by 5-HT pulses (see Section 4.2.2). The kinases and phosphatases responsible for reversible phosphorylation of proteins are under the control of a variety of factors, such as cAMP, Ca^{2+}, calmodulin, or the plastoquinone pool. Some of these elements have already been incorporated into a model for a circadian oscillator (Goto et al., 1985; Fig. 5.16). Because phosphorylation reactions are rapid, they would require additional time-delaying elements in order to generate circadian periods. In fact, there are recent reports of Ca^{2+}-calmodulin-dependent protein kinases undergoing autophosphorylation and acting as molecular switches (Miller and Kennedy, 1986). Incorporation of 3 to 12 of a possible total of 30 phosphate groups per holoenzyme caused kinase activity toward exogenous substrates, as well as autophosphorylation itself, to then become Ca^{2+} independent. Thus, kinase activity could be prolonged beyond the duration of an initial activating Ca^{2+} signal (even more so if dephosphorylation by cellular phosphatases were also opposed), thereby introducing considerable time delay into the system. Phosphorylation reactions are attractive as a general control mechanism, since they affect so many diverse cellular processes. Whether they constitute clock gears or merely coupling elements, however, remains to be determined. It is also interesting to note that the phosphatidylinositol cycle has been hypothesized to mediate the effects of light on leaflet movement in *Samanea saman* pulvini; hydrolysis of membrane-localized phosphoinositides, accompanied by an increase in cytosolic free Ca^{2+}, would provide a mechanism for phototransduction in the motor cells (Morse et al., 1987b,c).

5.5 Problems and Prospects

The various models proposed for circadian clocks (Table 5.1) sometimes overlap, for they often incorporate several different notions from each other. This is not at all surprising, since each has strengths and shortcomings for which it attempts to compensate by hybridization (Edmunds, 1983). For example, although the network models deal with known biochemical oscillations, they have difficulty in accounting for the long period of circadian rhythms and their temperature compensation (notwithstanding appeals to deposition effects and the fiddling with parameter values). Similarly, the chronon hypothesis (Ehret and Trucco, 1967) and the chronogene model (Edmunds and Adams, 1981) combine gene action and cell cycle controls but are weakened by the demonstration that DNA-dependent RNA synthesis does not seem to be important in the clock mechanism of *Acetabularia*. Enucleated cells continue to exhibit a rhythm of photosynthesis, and inhibiting extranuclear RNA synthesis with rifampicin does not stop the clock. Finally, the earlier membrane models (e.g., that of Njus et al., 1974), although neatly accounting for temperature compensation, had difficulty in accounting for the circadian period. This deficiency was addressed by Schweiger and Schweiger's coupled translation–membrane model (1977), invoking the time-consuming processes of assembly, transport, and loading of essential proteins into membranes. It incorporated also the recent reemphasis on the necessary role of protein synthesis in the maintenance (if not the generation) of persisting circadian rhythms in *Acetabularia*, *Gonyaulax*, and *Neurospora*.

Although the search for essential proteins may be very time-consuming with no guarantee of immediate success, it seems to have been rewarded by the identification of a cyclically appearing clock species (p230) in *Acetabularia* (Hartwig et al., 1985) and of a 34-kDa protein in *Aplysia* that Yeung and Eskin (1987) have nominated as a worthy candidate for a component of the circadian oscillator. The advent of the powerful techniques of recombinant DNA research and gene cloning, may offer some solace, as indeed they have in the case of the identification of the gene product (a proteoglycan) of the *per* locus in *Drosophila* (Jackson et al., 1986; Reddy et al., 1986; see Section 4.4.2). Li-Weber et al. (1987) have reported provocatively that the chloroplast genome of *Acetabularia* which is assumed to encode the p230 clock protein, shares a homology with the *per* locus of *Drosophila*. But what does the *per* protein do? Paradoxically, it is not cyclic in itself, normally a formal requirement for a putative clock gear. Does it only secondarily affect the rhythmicity generated by some other oscillator, or is it a component of an intercellular clock, forming or maintaining communication among rhythmic cells that would lead to mutual synchronization (Bargiello et al., 1987; Dowse and Ringo, 1987). Perhaps the *per* gene product is essential for the modulation (frequency multiplication) of another ultradian oscillator located in a membrane or neural net.

One must note that the search for the elusive clock per se could be doomed to failure if circadian timekeeping is not attributable to any one entity or subset of reactions in a cell, or it could be exceedingly difficult if "all these aspects of cell chemistry (soluble enzyme kinetics, nuclear message transcription, membranes and the second messenger) are susceptible to circadian modification of their regulatory dynamics, and that the clock, volatile as a ghost, lurks now in one room, now in another, in different cell types" (Hastings and Schweiger, 1976, pp. 54–55).

No matter whether or not our taste runs to ghosts, let us end on a note of cautious optimism: ". . . We will all have some fun in any case" (C.S. Pittendrigh, in a rather heated exchange with F.A. Brown Jr., Cold Spring Harbor Symp. Quant. Biol. Vol. 25, p. 183, 1960).

6 General Considerations and Conclusions

Now that we have completed our discussion of experimental approaches to circadian clock mechanisms and have examined various biochemical and molecular models for biological oscillators, let us note some of the more provocative, ancillary unsolved problems and applications of cellular circadian rhythmicity in the hope of enticing others into the field. Obviously, these topics, many embracing entire disciplines in themselves, can be treated only in the most superficial manner—tastebud-titillating tidbits, as it were—and others necessarily will be slighted.

6.1 Evolution of Circadian Rhythmicity

Whence came circadian clocks? Circadian rhythmicity is part of a continuum of biological oscillations of varying periodicity (see Fig. 1.1) and was probably derived through the modification of existing cellular oscillations early in the evolution of eukaryotes. The use of a circadian oscillator to measure day-length precisely (photoperiodism) then developed independently in the animal and plant kingdoms (Bünning, 1986). In order to answer this question, an armchair evolutionary biologist must attempt to understand the selective forces that may have been operating at that time.

6.1.1 Early Selection for Biological Clocks

Pittendrigh long ago speculated (1965, 1966) that the daily LD cycle was the historical selective agent of circadian oscillations and that these periodicities reflected an adaptation of the primitive cell to cope with the deleterious effects of solar irradiation on cellular processes, such as gene induction and DNA replication. He reasoned that the strong ultraviolet component of light (not screened by the then underdeveloped ozone layer) would have caused the formation of thymine dimers in the DNA backbone. Thus, an organism that relegated its replicative processes to nighttime darkness would have had an advantage, and this would probably have led to its concentrating its growth and synthetic processes in the day. In this manner, the first circadian clock may have had its genesis.

Supportive arguments for this ingenious hypothesis include the observations that (1) rhythmic gating of cell division to the dark periods in present-day diurnal LD cycles (see Sections 3.2.1, 3.2.2) minimizes the inhibitory effects of light, and (2) rhythmic regulation of photosensitive membrane components minimizes the photo-induced loss of membrane function (Epel, 1973; Woodward et al., 1978; Ułaszewki et al., 1979; Edmunds, 1980a). Furthermore, Paietta (1982) proposed that one of the original selective forces involved in the evolution of circadian rhythmicity was the joint effect of the LD cycle and the increasing level of free oxygen early in eukaryotic evolution. Circadian periodicity would have provided a protective mechanism for minimizing the deleterious effects of diurnally imposed photooxidation by preventing photooxidative destruction of sensitive components through excited-state quenching by pigments and by restricting the highly photosensitive processes to darkness.

Kippert (1985a), however, has argued that the supposed selective advantage of adaptation to the daily LD cycle (whatever be its length) should be doubted on at least two grounds. First, if the LD cycle were important, it should have been so for prokaryotes also, and thus far there are few reports of bona fide prokaryotic circadian rhythms, even in the cyanobacteria (Bünning, 1986). Second, the first eukaryotes are believed to have been nonautotrophic, with no need to couple their energy metabolism to the periodic energy supply provided by the environment. Rather, Kippert argues that the symbiosis between bacteria and host cell by which the first eukaryotic organisms are thought to have arisen would have required a metabolic coordination and synchronization of the CDC among the partners and that this temporal organization would have been achieved best by the coevolution of endogenous oscillators and the calcium system of intracellular signaling (Kippert, 1985a, 1986; see Levandowsky, 1981; Paietta, 1982; see Section 5.2.4). Adaptive synchronization to external *Zeitgeber*, such as LD or temperature cycles, would have evolved at a later date, as would have temperature compensation.

There have been recent reports of circadian rhythmicity in nitrogen-fixing cyanobacteria, which are unique in their ability to carry out photosynthetic O_2 evolution and O_2-labile nitrogen fixation within the same organism (Grobbelaar et al., 1986; Mitsui et al., 1986, 1987). These seemingly incompatible reactions take place in heterocystous cyanobacteria by spatial separation of the sites for the two processes. In nonheterocystous species, however, other mechanisms must be invoked. For example, in diurnal LD cycles temporal separation of photosynthesis and nitrogen fixation into the light and dark intervals of growth, respectively, may occur. Mitsui et al. (1986) found, however, that the two processes can continue in LL for at least 3 days in certain species of *Synechococcus* sp. (Miami BG43511 and 43522), the growth of which had been synchronized by LD:

12,12. The circadian rhythm in O_2 evolution and nitrogenase activity were approximately 180° out of phase. Maximum activity in net O_2 evolution occurred during the subjective day, just before cell division, whereas peak activity of nitrogenase was found at night. If O_2 evolution was inhibited with diuron (DCMU), acetylene reduction was not affected, although in nonheterocystous species blockage enhances reduction (indicating that photosynthetic O_2 evolution has adverse effects on nitrogen fixation). Thus, a circadian oscillator has enabled *Synechococcus* sp. to achieve a functional separation of two otherwise incompatible reactions during the CDC, with the attendant implication that circadian rhythmicity (in this case of all three processes) can be expressed in a prokaryote.

6.1.2 Ultradian Clocks: Small Steps Toward Circadian Oscillators?

As Klevecz (1984; see Klevecz et al., 1984) has stressed, however, the search for selective forces is inextricably intertwined with the primitive environment. The evolution of the earth–moon system and the consequent changes in day length through geological time must be taken into account. During prebiotic evolution the solar day was quite short—probably less than 10 h, and perhaps only 4 to 5 h (Fig. 6.1)—and was punctuated by boiling tides and intense ultraviolet irradiation. If a primitive oscillator with the capacity to entrain to sunlight or temperature changes had evolved at that time, it would have had a relatively short free-running period. Klevecz speculates, therefore, that modern circadian rhythms ($\tau \cong 24$ h) have evolved by organismic fusion and coupling of systems with short-period oscillators that first emerged when the earth's rotation was rapid and the moon was in close proximity. Thus, cellular oscillators as we now see them would represent vestiges of a primitive circadian clock having a much shorter period, perhaps even that of the quantal cycle, G_q (see Section 3.1.2), which Klevecz and his coworkers have observed in cultured mammalian cells (Klevecz, 1976; Klevecz et al., 1984).

Perhaps existing ultradian oscillators should receive increased emphasis as possible mechanisms for generating circadian rhythmicity (see Sections 5.2.1 and 5.2.3). Based on their studies and those of others on leaf movements, circumnutations of shoots, levels of enzyme activities, and metabolites, Koukkari et al. (1985) have identified what they consider to be a special group of ultradian rhythms having periods ranging from approximately 60 to 240 min, although the patterns of these rhythms seldom exhibit strict periodicity or constant waveform. For the moment, however, neither the significance of these rhythms nor the reality of the boundaries of this ultradian domain is clear. Likewise, Lloyd et al. (1982a) have argued that temperature-compensated epigenetic oscillations in energy metabolism ($\tau \cong 30$ min; $Q_{10} \cong 1.12$) may serve a dual timekeeping role for both the CDC and circadian rhythmicity, although the mechanism by which this

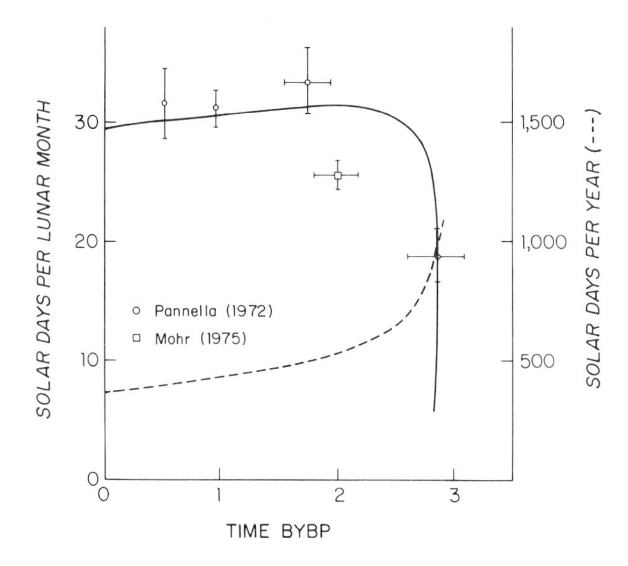

FIGURE 6.1. Cellular oscillators as vestiges of a primitive circadian clock. Data points shown are for the estimated number of solar days per lunar month as determined from depositional periodicities in stromatolites by Pannella (1972) and Mohr (1975). The solid and dotted lines are the calculated change in solar days per lunar month and per year, caused by tidal dissipation of kinetic and potential energy and the changing distribution of angular momentum in the earth–moon system, respectively, as calculated by Turcotte et al. (1977). (Reprinted from Klevecz, 1984, with permission of Marcel Dekker, Inc.)

might be accomplished is not known (see Section 5.2.2). An ancillary prediction of this interesting proposition is that the CDC would be quantized, as indeed it is in *Tetrahymena, Paramecium,* and mammalian cells (see Sections 2.1.2, 3.1.2).

Finally, although the period of ultradian rhythms is temperature compensated in many instances (Klevecz and King, 1982; Lloyd et al., 1982a; Michel and Hardeland, 1985; see Fig. 5.6), this is not necessarily the case. Analogously, the rhythmic change between resting and actively moving phases in the marine plasmodial rhizopod, *Thalassomyxa australis,* though characterized by a 24-h period at 22°C, shows marked temperature dependence (Q_{10} = 2.7 between 12 and 24°C) in contrast to most circadian rhythms and appears to be unaffected by a LD or temperature cycle (Silyn-Roberts et al., 1986). We have seen (see Section 4.4.1) that many of the *frq* circadian clock mutants of *Neurospora* have altered temperature compensation, a complete loss of which is found in *frq-9* (Loros and Feldman, 1986). These comparative observations not only raise the question whether there is a multiplicity or uniformity of cellular temperature-compensation mechanisms (Hardeland et al., 1986) but also suggest that there may exist currently halfway houses between ultradian and circadian clocks that perhaps reflect the evolution of fully formed, present-day circadian clocks (Kippert, 1985a, 1986). On this notion, motility rhythm of *Thalassomyxa* would represent a type of periodicity that is considerably longer than most metabolic rhythms but that has not yet developed temperature compensation (Silyn-Roberts and Engelmann, 1986), whereas the temperature-compensated ultradian oscillators constitute precocious anlagen for circadian clocks (but see Jerebzoff, 1987).

6.2 Multiple Cellular Oscillators

6.2.1 Intracellular Clockshops

A basic but unanswered question in the field of circadian rhythms concerns the number of clocks that might exist in a single cell or microorganism. A large part of the difficulty in addressing this question lies in our ignorance of the mechanism of any clock. Probably the preponderance of opinion is that there is but one central oscillator within the cell—or that it is the cell itself or most of it (see Section 4.1.3)—and that the numerous overt rhythms observed as circadian outputs are merely different hands of this single driving entity. The observations of McMurry and Hastings (1972a) that four separate rhythms in *Gonyaulax polyedra* (see Section 2.2.4, Table 2.5) all seem to have the same free-running period, and thus do not dissociate in LL) (Fig. 6.2A), with similar temperature coefficients (provocatively, $Q_{10} \cong 0.85$ to 0.90) and that they all respond identically when perturbed by a resetting dark pulse (Fig. 6.2B) give strong support to this notion. The simultaneous, long-term recording of the circadian rhythms of oxygen evolution, chloroplast migration, and electric potential in individual *Acetabularia* cells did not reveal any significant divergence in their respective periods despite the fact that τ is not as stable as it is in multicellular organisms. The phase relation among the rhythms was retained after a phase-shifting dark pulse or temperature shift (Broda and Schweiger, 1981b).

The results of some work (Laval-Martin et al., 1979; Edmunds and Laval-Martin, 1981) with *Euglena* maintained in higher-frequency LD cycles (see Section 2.2.3) have revealed that the phase relationship between the rhythm of photosynthetic capacity and that of chlorophyll content varied, suggesting the possibility of desynchronization among circadian rhythms in a multioscillator, unicellular organism. Schweiger et al. (1986) noted that occasionally a dark pulse shifted only one of two simultaneously monitored rhythms in an individual *Acetabularia* cell, demonstrating that one is dealing with at least two different clocks whose phases are closely coupled to each other. Hoffmans-Hohn et al. (1984) discovered multiple periodicities for each of three circadian rhythms (autokinesis, intracellular pH, and CO_2 and O_2 partial pressures) in long-term (> 15 days) cultures of *Chlamydomonas reinhardtii* (strain 11-32/89), *Chlorella fusca* (strain

211-8b), and *Euglena gracilis* (strain 1224/5-9) (Fig. 6.3A,B). Time-series analysis of some 40 data series revealed three circadian periods for each rhythm, two symmetrically placed, equal-energy side peaks flanking a center peak. There are several possible explanations of these multiple periodicities: (1) one master oscillator, with time-varying frequency (nonstationary times series, or oscillatory free-run), (2) multiple oscillators, with stable frequencies, (3) additive superposition of three different independent oscillators (due either to subpopulations of cells in the test cultures, or three physiological processes with different frequencies), and (4) multiplicative coupling of two oscillators (amplitude-modulated oscillations). Since there was no evidence of an oscillatory free-run (such as seen in higher animals), the origin of the multiple periodicities in the various measured rhythms must have been due to the existence of different oscillators with stable periods. Based on their analysis, Hoffmans-Hohn et al. (1984) favor a model of multiplicative coupling between a circadian oscillator and an unknown, endogenous, low-frequency oscillator (with τ ranging between 100 and 750 h) that modulates the amplitude of the circadian rhythm.

Dharmananda and Feldman (1979) reported a spatial distribution of circadian clock phase for the conidiation rhythm aging cultures of *Neurospora crassa*. A gradient of phases was found that was a function of the length of time the clock had been free-running in a given part of the culture. These phase differences within a single interconnected mycelium demonstrated the absence of total internal synchronization between adjacent regions of the hyphae even though these regions are replete with cellular junctions (which presumably would permit intercellular communication, yet obviously did not allow effective transfer of phase information).

6.2.2 Intercellular Communication and Coupled Oscillators

Long-term persisting circadian rhythms in populations of microorganisms raise another interesting question. Why does not the synchrony decay in the absence of a synchronizing *Zeitgeber* in the same manner that cell division synchrony, for example, decays in bacterial and mammalian cell cultures (usually within three to four cycles)? Assuming that no subtle geophysical factors are phasing the population, we are left with two alternatives. First, the free-running period of the oscillator must be exceedingly precise and almost identical with those of other cells, or second, some sort of intercellular communication must occur that maintains synchrony within the population, or network, of self-sustaining oscillators (Edmunds and Funch, 1969b; Edmunds, 1971).

With regard to the first alternative, much depends on what assumptions are made about the nature of the variance. Thus, by the random walk model (in which fast-running or slow-running cells do not necessarily

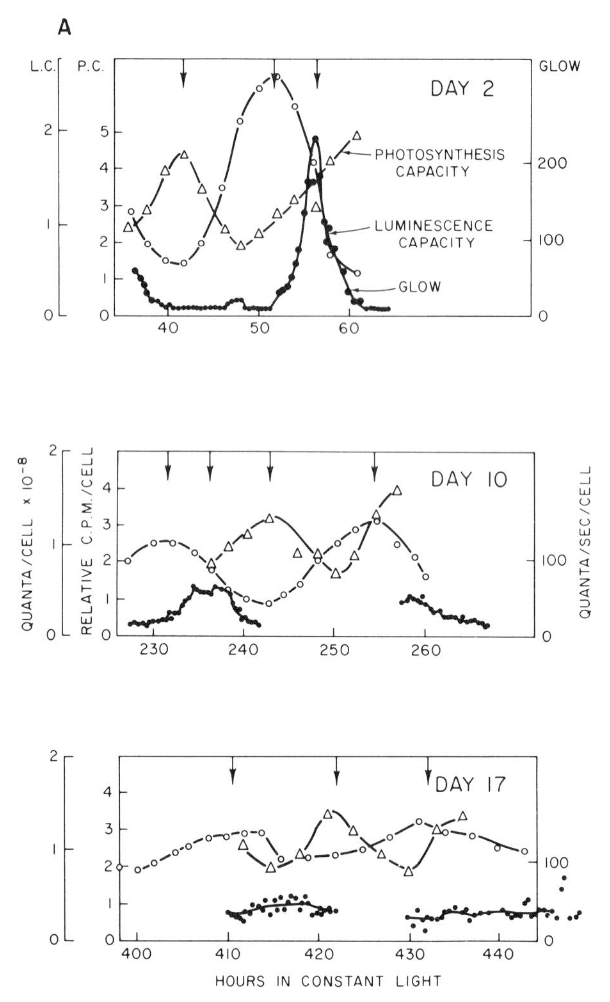

FIGURE 6.2. No desynchronization among four circadian rhythms in the unicellular alga *Gonyaulax polyedra*. (A) Luminescence capacity (L.C.), glow, and photosynthesis capacity (P.C.) rhythms after 2, 10, and 17 days in constant light (200 footcandles, FT.C., cool-white fluorescent). Arrows locate rhythm maxima, estimated visually. A flask containing 1200 ml of culture was transferred from an LD: 12,12 cycle at the end of a light period (time 0) into constant light. At the times shown, samples for assay of all three rhythms were withdrawn. Glow samples were placed on the scintillation counter turntable 32 min apart (if three replicates) or 24 min apart (if four replicates), and the glow intensity was thus measured every 32 or 24 min between times of sampling the flask (every 2 to 3 h). Three replicates for each rhythm were assayed for the day 2 and day 17 measurements; four were done for the day 10 measurements. Some glow points are missing because of turntable malfunction. The culture increased linearly in cell number from 4500 cells per ml on day 2 to 16,000 cells per ml on day 17. The cell division rhythm was not measured but in a similar experiment was not detectable after 17 days

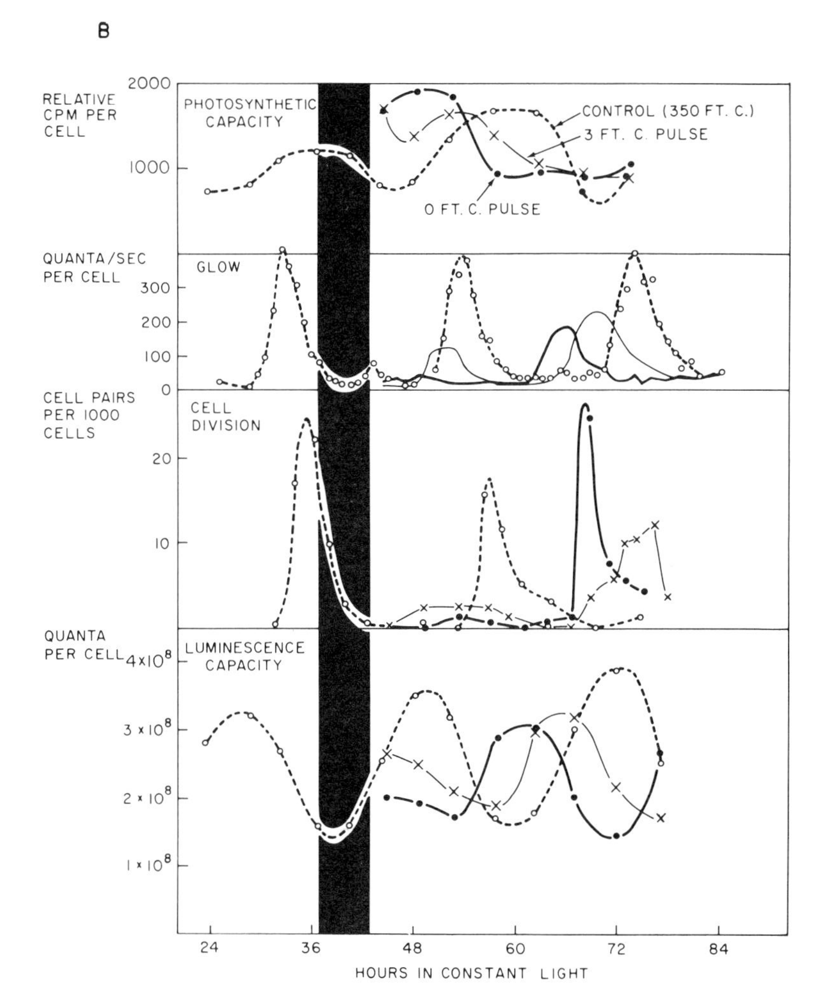

after time 0. (B) Phase shift of four rhythms caused by a 6-h dark interruption. The dark interruption is represented by a vertical black swath. Ten-milliliter samples from an LD: 12,12, culture were pipetted into vials during the light-on phase. At the end of the next light-on phase, the vials were placed on the turntable in constant light of 350 FT.C., and the turntable was started. Sampling began as shown after 23 h. The exposure to a lower light intensity was given to appropriate vials by putting them into transparent containers wrapped with the appropriate number of white towels (3 FT.C.) or by putting them in a light-tight box (0 FT.C). In assaying for the rhythms, samples were removed from the vials in a rotating schedule; only those vials used for measurement of glow intensity were undisturbed. All determinations except those of the number of cell pairs were made in triplicate from vials spaced around the turntable and, except for the glow intensity values, were averaged. (Reprinted from McMurry and Hastings, 1972a, with permission of American Association for the Advancement of Science.)

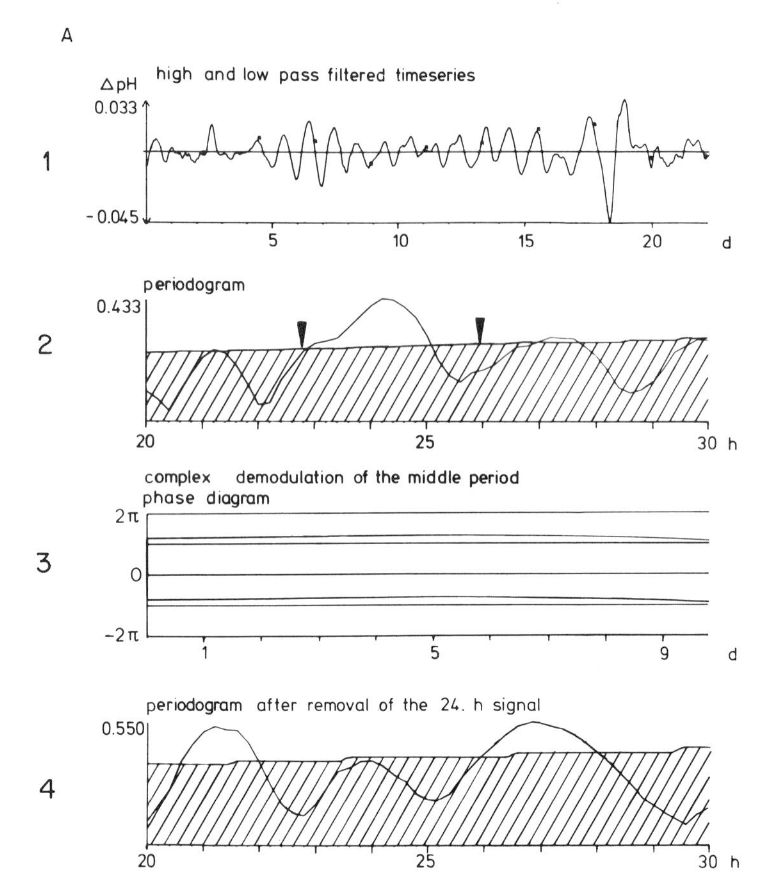

FIGURE 6.3. Multiple periodicities in the circadian system of unicellular algae: a cellular clockshop? (A) Long-time record of pH in a population of *Chlamydomonas reinhardii* after filtering. (1) A beatlike, varying amplitude, due to the three periods in the critical range, can be recognized. (2) The periodogram gives the location of the three peaks. The arrows indicate the location of the side lobe frequencies due to filter error. (3) Complex demodulation at 24 h (pass periods 23.3 and 24.74 h and stop periods 21.9 and 26.55 h) reveals the stability of the major middle peak. (Although the side peaks at 21.1 and 27 h have only low energy, they are also stable in time, not shown here.) (4) Periodogram of the long-time record after removal of the major middle peak signal by bandpass filtering proves the existence of the side peaks. Shaded area of the periodogram gives the 95% F-test threshold for each frequency in the circadian range from 20 to 30 h. (B) Multiple periodicities and amplitude relation, as derived from the periodograms, in different experiments and for different organisms. Indicated period lengths (in hours) are drawn on a frequency scale to show the symmetrical location. According to the multiplicative coupling model, the carrier periodicity and the modulation periodicity are given. (Reprinted from Hoffmans-Hohn et al., 1984, with permission of Verlag der Zeitschrift für Naturforschung, Tübingen.)

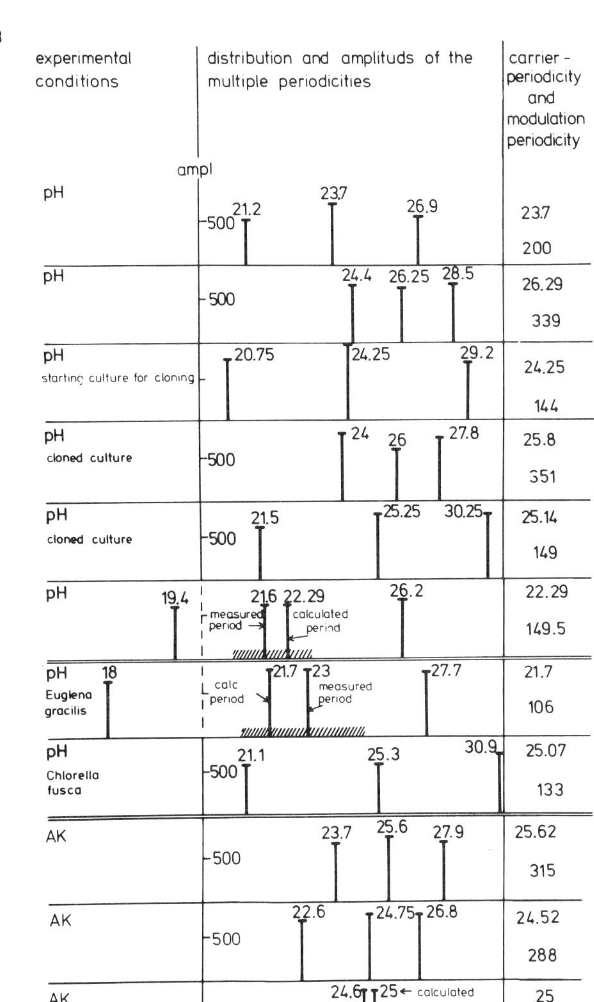

transmit this property to their progeny or even from cycle to cycle), the rate of dispersion could be quite slow, perhaps requiring several weeks before the peak of some rhythm became so spread out that it was obliterated (damped out). Recent work on the rhythm of glow luminescence in *Gonyaulax* (see Section 2.2.4) suggests that gradual decay does occur over such an extended timespan, but that it is very slow because the rhythm is accurate to within 2 min a day and the variance in circadian period among individual cells is about 18 min (Njus et al., 1981).

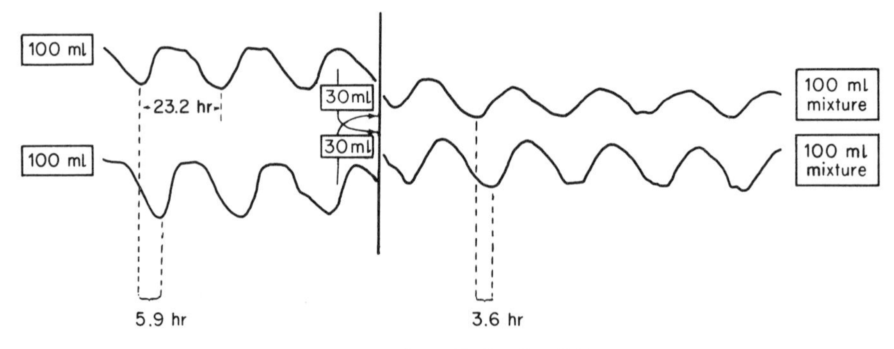

FIGURE 6.4. Mixing experiment designed to test for intercellular communication in *Euglena gacilis*. Aliquots (30 ml) were taken simultaneously from two 100-ml autotrophic cultures 5.9 h (90°) out-of-phase with respect to their free-running rhythms (τ = 23.2 h) of random motility and reciprocally mixed with the opposite culture. The resulting phase angle difference of 3.6 h between the rhythms in the two mixed cultures was in close agreement with the theoretical prediction (3.1 h) on the hypothesis that no cell-cell interaction occurred. (Reprinted from Brinkmann in Edmunds, 1982, with permission of Academic Press.)

FIGURE 6.5. Mixing experiments designed to test for intercellular communication in *Gonyaulax polyedra*. (A) Comparison (exp. 139) of circadian rhythm of glow luminescence (ordinate) in a culture comprising a mixture [Panel (b), solid line, No. 9 = No. 7 + No. 11] of two out-of-phase cultures (No. 7 and No. 11) with the expected luminescence [Panel (b), dotted line] based on the behavior of the parent cultures [Panels (a) and (c)] maintained in LL (abscissa) under the same conditions. For the first 4 days, the vials whose traces are reflected in Panel (a) contained No. 7 cells, and those for Panel (c) contained No. 11 cells. On day 5 the contents of these vials were evenly mixed and divided back into the two vials. The corresponding calculated trace [Panel (b), No. 9 = (No. 7 + No. 11)/2], representing the arithmetic sum of traces (a) and (c), is plotted with open circles and superimposed on the actual trace of the mix. Panel (d) [No. 9-(No. 7 + No. 11)] shows the result of subtracting the actual mix (No. 9) from the calculated trace [No. 7 + No. 11)/2]. The increasing amplitude after day 9 is indicative of the growing phase difference. This is not the case for two vials (No. 7, No. 13) containing the same population, as indicated by the traces given in Panel (e). Subtraction (No. 7 − No. 13) yields only very small differences. Scaling for Panels (a, b, c) is identical; the calculated subtractions are scaled for maximum amplitude for each panel. (B) Phase plots (Exp. 152) of normalized raw data from a mixed (AB) culture [Panel (a), No. 17) compared with the calculated summation [Panel (b)] of the two parent cultures maintained separately (A + B, No. 15 and No. 19). Note the apparent merging of peaks in the mixed culture. From plots of maxima, as determined by a peakfinder program, values for the free running periods of the different components were calculated (Panels c, d) and are noted on the graphs. (Reprinted from Broda et al., 1986a, with permission of Humana Press, Inc.; see Hastings et al., 1985.)

A

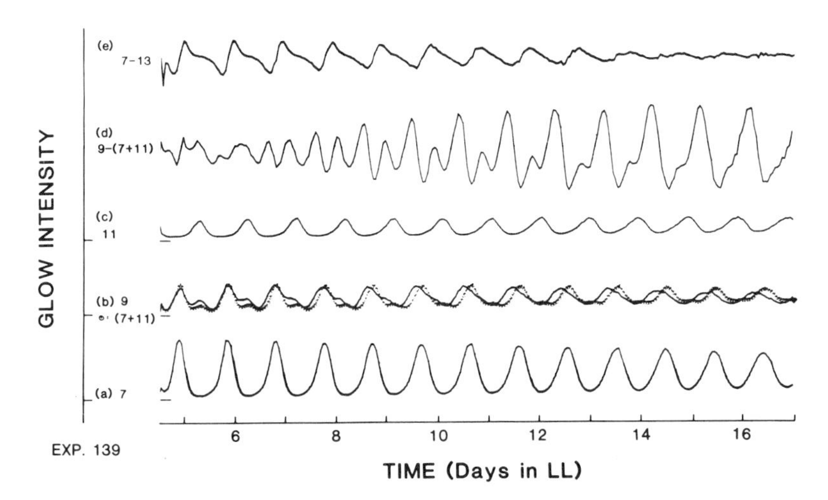
GLOW INTENSITY
(e) 7-13
(d) 9-(7+11)
(c) 11
(b) 9
(7+11)
(a) 7
EXP. 139
TIME (Days in LL)

B

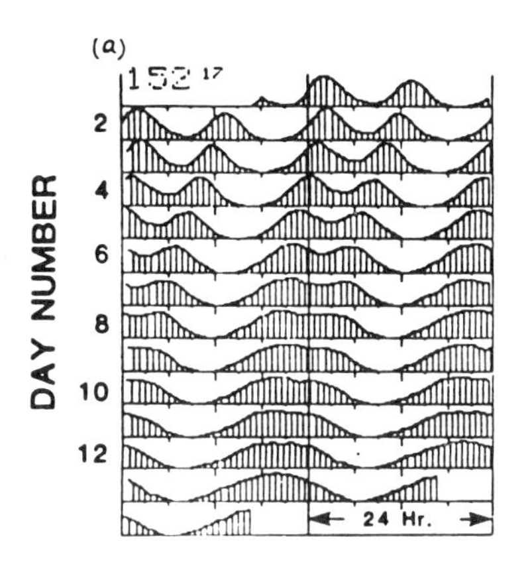
(a)
152 17
DAY NUMBER
24 Hr.

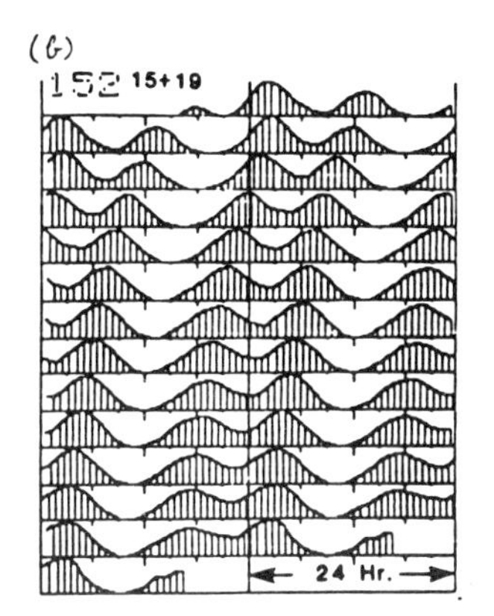
(b)
152 15+19
24 Hr.

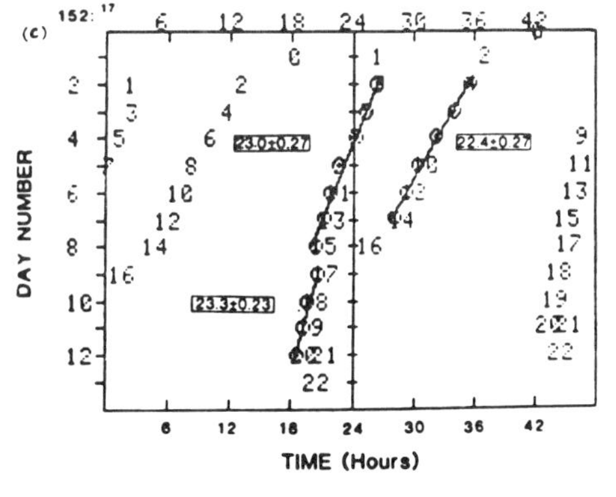
(c)
152: 17
23.0±0.27
22.4±0.27
23.3±0.23
DAY NUMBER
TIME (Hours)

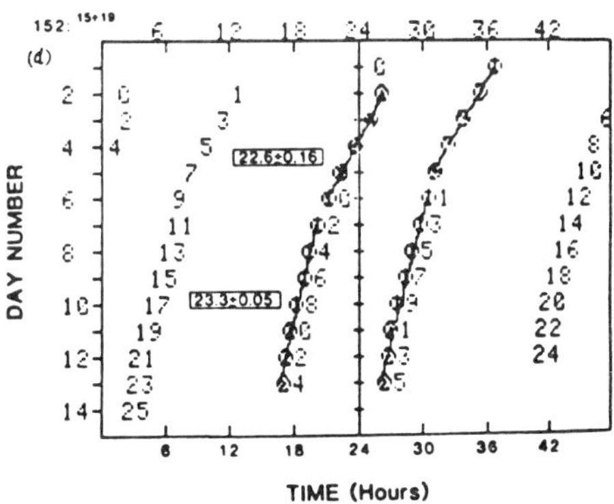
(d)
152: 15+19
22.6±0.16
23.3±0.05
DAY NUMBER
TIME (Hours)

The alternative hypothesis of intercellular crosstalk among coupled oscillators (see Section 5.2.3) has been examined theoretically from both a mathematical and a biochemical standpoint (Goodwin, 1963; Winfree, 1967; Pavlidis, 1969, 1971). Experimental tests of this hypothesis in microorganisms, however, almost uniformly have been inconclusive or negative. Thus, Brinkmann (1966) found no evidence for intercellular communication in *Euglena* when out-of-phase cultures were mixed and the resultant phase of the mobility rhythm examined (Fig. 6.4). Similarly, mixing experiments with synchronously dividing populations of this flagellate did not support this notion (Edmunds, 1971). Mixing experiments for the rhythm of glow bioluminescence in *Gonyaulax* (Hastings and Sweeney, 1958; Sulzman et al., 1982), experiments with *Acetabularia* cells whose rhythms of photosynthetic O_2 evolution were out-of-phase (Mergenhagen and Schweiger (1974), and the results for the condiation rhythm in *Neurospora* (Dharmananda and Feldman, 1979) discussed in the preceding section all indicated the absence of internal communication and supported the cellular autonomy of circadian clocks. Nevertheless, a recent reexamination of the question in *Gonyaulax* (Broda et al., 1986a), in which out-of-phase cultures were mixed and then carefully monitored over longer timespans than in the previous work (Sulzman et al., 1982), has indicated an effect on period by cell conditioning of the medium (Fig. 6.5A,B) that suggests that circadian communication among unicells does occur after all. In this regard, it is provocative that a substance *("Uhrstoff")* that shortens the period of the glow rhythm has been isolated from *Gonyaulax* (as well as from extracts of several animal tissues). Pulses of *Uhrstoff* have long-lasting effects, which are a function of both concentration and cell density and which can be expressed only under white or blue background light (Roenneberg et al., 1987).

6.3 The Breakdown of Temporal Organization at the Cellular Level

The major thrust of this book has been to document organization along the time axis at the cellular and molecular levels and to extol its virtues for the organism. The development of the concept of open systems enabled one to predict in principle the spontaneous appearance (or self-organization) of temporal dissipative structures, including autonomous oscillations, in certain nonlinear systems in which there is positive or negative feedback, sometimes combined with cross-catalysis (Nicolis and Portnow, 1973; Nicolis and Progogine, 1977; see Section 5.2). But what of the inverse: Is the breakdown of temporal organization deleterious in its consequences, just as chaotic fluctuations and futile cycles might be destabilizing in complex metabolic networks, such as the glycolytic cycle (see

Sections 5.2.1 and 5.2.2)? Let us briefly speculate on this possibility in the areas of malfunctioning cellular pacemakers and cell cycle oscillators.

6.3.1 Dysfunction of Cellular Pacemakers in Pathology and Disease

In recent years, there has been a burgeoning interest in oscillatory neural networks (reviewed by Selverston and Moulins, 1985) and cellular and neural pacemakers (Berridge et al., 1979; Carpenter, 1981, 1982; Connor, 1985; Torras, 1985; Jacklet, 1987). These neural networks are ensembles of neurons responsible for a wide variety of periodic behavior patterns—perhaps even underlying certain types of learning at cellular sites—and are capable of generating oscillatory electrical output without requiring synaptic or bloodborne input, although many do need tonic excitation of some sort.

There are two ways that such neuronal oscillating systems could work (Selverston and Moulins, 1985). One or more cells embedded within the network would have the property of endogenous bursting (cell-driven oscillators, or pacemakers), or else the network itself would produce bursts as a result of synaptic interactions and the intrinsic membrane properties of the individual neurons, with no single cell's being capable of producing bursts (network oscillators). The nine-cell cardiac ganglion of the lobster, which produces bursts of motor impulses that drive the heart muscle at a frequency of 20 to 50 per min, is an example of a cell-driven oscillator, with primary control's being exerted by the four smaller interneurons, or pacemaker cells. The swim oscillator of the sea slug *Tritonia diomedia*, which fires antagonist bursts of impulses to the dorsal and ventral musculature and thereby generates alternating flexion movements that enable the mollusc to escape from predators, exemplifies the network-type of oscillator. The circuit that controls the leech heartbeat is a good example of a mixed oscillator, in which the properties of both other types of oscillator are present.

We introduced in Section 1.3.3 the synchronous, rhythmic, flashing of populations of fireflies as an example of a population of interacting, coupled neural oscillators (see Section 5.2.3), replete with the properties of entrainment, phase-shiftability, and free-run. Furthermore, Enright (1980) has theorized that the temporal precision inherent in circadian systems could be achieved in a neuronal clock comprising intrinsically unreliable, sloppy component oscillators by mutual coupling: sufficient conditions would be that the output of the individual elements be summed and that mutual, reciprocal triggering be mediated by a nonlinear phenomenon, such as a threshold. Carpenter and Grossberg (1983) have suggested a neural theory of circadian rhythms—the gated pacemaker—in which the positive feedback signals of competing on-cell and off-cell populations are gated by a slowly accumulating, chemical transmitter substance, and the clocklike nature would be a mathematical property. Honerkamp et al. (1985) have outlined mathematically a general mechanism underlying autonomous bursting, in which two nonlinear oscillators with different frequencies are coupled so that the slower oscillator alternatively switches the fast one on and off.

At this juncture it is relevant, therefore, to consider some examples of diseases that are associated with and even possibly due to abnormalities in the function of normal cellular pacemakers (Carpenter, 1982).

The Cardiac Pacemaker

The coordination of pumping activity of the four heart chambers is maintained by the passage through the cardiac tissue of a wave of electrical impulses that travels rapidly from cell to cell. In isolated right atrial preparations, electrical activity normally originates in a cluster of pacemaker cells in the sinus node. Activation then spreads preferentially toward the atrioventricular (AV) node, a second pacemaker lying between the right atrium and right ventricle, as well as through the remainder of the atrium. The AV node gives a physiological delay to impulse propagation, and then rapid propagation resumes down the Purkinje fibers to the ventricular myocardium, accompanied by a wave of contraction. Although the sinus node is the most important pacemaker unit, with the highest intrinsic firing rate, several other heart areas have been shown both in vivo and in vitro to possess the property of self-excitation and, therefore, to serve as potential pacemaker sites. These subsidiary pacemakers, including cells in the right atrium, the mitral and tricuspid valves, the lower part of the AV node, and the His-Purkinje system, assume an important role in many pathological situations, such as the failure of pacemaker activity in the sinus node or the interruption of impulse transmission from atria to ventricles.

Dysfunction of the cardiac pacemaker itself, though primarily a disease of the elderly, occurs frequenctly in younger age groups. Electrocardiographic manifestations of the sick sinus syndrome (Kerr et al., 1982) include sinus bradycardia, sinus arrest or pause, sinoatrial exit block, chronic atrial fibrillation, inability to resume the sinus rhythm following cardioversion, and bradycardia-tachycardia syndrome. Treatment is aimed at control of these symptoms, often by the implantation of an artificial pacemaker into the ventricle and by pharmacological manipulation. Provocatively, Winfree (1983) has suggested that fibrillation and disruption of the coordinated contraction of the heart muscle fibers, which cause many sudden cardiac deaths, are the result of an inadvertent, critical (strength, duration) electrical stimulus impinging on the system at a critical phase point of the pacemaker oscillation, the singularity point (Winfree, 1980). Whereas electrical stimuli applied to the heart normally phase-advance or delay the next heartbeat, such a critical stimulus given at the singularity point disperses the normal wave of contraction and leads to arrhythmia in a manner analogous to that observed (see Section 3.2.1) for the suppression of the circadian rhythmicity of *Drosophila* eclosion (Winfree, 1970) or cell division in *Euglena* (Malinowski et al., 1985).

Tremor

Tremor is small-amplitude rhythmic movement, usually dominated by one or more frequencies ranging from 3 to 30 Hz and occurring in a limb that is either at rest or engaged in other motor activity (see review by Wolpaw, 1982). It may originate at a number of points within the motor system (including the central and peripheral nervous system and the mechanical apparatus comprising muscles and bones) or in the interaction between points. Both normal, 8 to 12 Hz physiological postural tremor (e.g., of the outstretched hand) and pathological parkinsonian resting tremor (3 to 7 Hz) are due to rhythmic change in the force applied to a limb by muscle contraction, reflecting the firing of the relevant alpha-motorneurons. These rhythmic electrical impulses at least in part are due to intrinsic CNS pacemaker activity. Parkinsonian resting tremor does not require peripheral input at all and seems to be the result of rhythmic neuronal firing in specific regions of the ventrolateral thalamic mass and possibly the primary motor cortex, which, in turn, is supported by a thalamic or thalamocortical neuronal loop.

Epilepsy

Epilepsy, a pathological condition of the human central nervous system that affects millions of people, might be described as a condition of the CNS in which groups of neurons generate abnormal bursts of action potentials (Schwartzkroin, 1982). The disease, in fact, a complicated set of electrical and behavorial symptoms, that may result from diverse causes, is the product of relatively large cell aggregates, in which the bursting activities of individual cells are more or less synchronized. The underlying cellular mechanism could entail an abnormal epileptic neuron that might be considered to be a pacemaker cell, intrinsically capable of burst generation independent of extrinsic input, which would recruit other neurons in surrounding tissue into an abnormal burst discharge; or it could comprise an abnormal aggregate, a system in which the effectiveness of excitatory synaptic contacts, or the potency of excitatory circuitry has been enhanced. Experimental data suggest that although individual neurons within chronic foci do have intrinsic burst-generating capabilities, these pacemakers also are highly controlled by extrinsic factors (Schwartzkroin, 1982).

Aging clocks?

As opposed to the genetically programmed theories of aging (see Section 6.5.3), Walker and Timiras (1982) have proposed that aging is the post-maturational extension of pacemaker-regulated events essential for development (Samis and Capobianco, 1978). In the adult, cellular pacemakers would mediate physiological decline and promote pathology by affecting those control systems that regulate the quality of the internal environment. Progressive impairment of homeostatic competence could take place either directly, by an influence of pacemakers on processes that at first promote growth and development but that subsequently cause aging, or indirectly, by a decline (resulting from these processes) in the ability of the pacemakers to temporally coordinate multiple, interdependent rhythmic systems. Although it is difficult to distinguish between cause and effect, the latter hypothesis was favored: aging is the process of increasing temporal disorganization among complex physiological axes (see Brock, 1985). Walker and Timiras (1982) further suggest that thyroid hormone and serotonin represent endogneous compounds whose general function and wide phylogenetic distribution make them potential regulators, or pacemakers, for the ontogeny and expression of biological rhythms. The obvious predictions of this hypothesis—borne out by some experimental data—are that normal aging would be accompanied by alterations in thyroid-serotonergic interactions and that senile-type syndromes might be induced prematurely in animals whose serotonin metabolism in brain pacemaker areas had been manipulated.

6.3.2 Cancer: The Malignant Transformation

Cell cycle clocks and the interaction of circadian oscillators with the CDC are treated in detail in Chapter 3. Is it possible that the breakdown of temporal order could lead to certain types of cancer (Edmunds, 1978)?

Sel'kov (1970), for example, has considered the CDC to be a type of limit cycle oscillation in which the variables are the concentrations of sulfhydryl (—SH) and disulfide (—SS—) groups (see Section 3.1.2,Fig. 3.17). Under certain conditions this —SH-controlling system would have alternative stationary states, and the transition of the system from one state to another, induced by some environmental or extrinsic perturbation, could cause profound changes in cellular physiology, among which would be malignant growth and carcinogenesis (Gilbert, 1974b, 1977, 1978a, 1980).

Another conceptual approach is concerned with the relationships among cellular communication, contact inhibition, and cell division (mitotic) clocks (Burton, 1971, 1975; Burton and Canham, 1973; Loewenstein, 1975). Three general postulates were made: (1) Every cell throughout its life retains the capacity to divide, although it may be suppressed by intercellular factors, and possesses an intrinsic, autonomous oscillation in the concentration of some key substance(s) that regulates cell division (formally analogous to the mitogen relaxation and limit cycle oscillators discussed in Section 3.1.2). (2) These endogenous oscillations of individual cells have periods scattered around a mean cycle time with a considerable coefficient of variation, and hence are asynchronous and out-of-phase with each other. (3) Normal cells of a tissue freely communicate with cells in contact with them via tight junctions; the key mitotic substance *(X)* would move from cell to cell by passive diffusion down a concentration gradient. Because of the asynchrony of the constitutent individual oscillators, the resultant amplitude of the oscillation in the level of *X* does not reach the threshold value for further cell division to ensue, and contact inhibition occurs. Cancer would result from an excessive intrinsic oscillation in *X* (despite asynchrony and intercellular communication) that leads to insufficient contact inhibition, from some perturbation that brings the cells into synchrony, or from the lack of sufficient intercellular communication. Obviously, this is a very general hypothesis, difficult to completely falsify, but it does stress the importance of the temporal coordination and phasing of intrinsic cellular oscillations.

Wille and Scott (1984) have discussed CDC-dependent, integrated control of cell proliferation and differentiation in normal and in neoplastic mammalian cells. Particularly germane is their speculation that neoplastic transformation might be attributable to defective control of the CDC. They suggest that the two-stage model of carcinogenesis could be explained by specific alterations in the kinetic parameters controlling the dynamics of a nonlinear, dynamic biochemical system comprising both a high-amplitude and a low-amplitude, stable, limit cycle oscillation nested within the mitotic oscillator (Wille, 1979).

6.4 Cellular Aspects of Chronopharmacology and Chronotherapy

Circadian and cell cycle clocks may play a profound role in the treatment of disease (McGovern et al., 1977; Reinberg and Halberg, 1979; Reinberg and Smolensky, 1983; Lemmer, 1984). In Chapter 3 we documented much of the experimental evidence for viewing the cell cycle as an oscillatory system. In this short synopsis, the role of cell cycle clocks and circadian rhythmicity in the treatment of cancer is considered.

The importance of chronobiology for basic cancer research and for chemotherapy has been stressed by Scheving (1984) and by Møller (1984). There is no lack of awareness that the CDC is most significant for cancer treatment, if for no other reason than the fact that studies of synchronized mammalian cells have shown that their sensitivity to many cytostatic drugs (such as cytosine arabinoside), as well as to ionizing radiation, is highly dependent on the stage of the CDC. Some of these drugs also may lead to a partial synchronization of the CDC and have been so employed to obtain synchronized cell populations both in vitro and in vivo. This strategy can maximize the chance for survival and cure by applying the minimum dose of drug necessary to kill the phased, malignant cells at the time of their CDC when they are most susceptible. It is obvious, therefore, that the circadian rhythmicities observed in cell flux and in the distribution

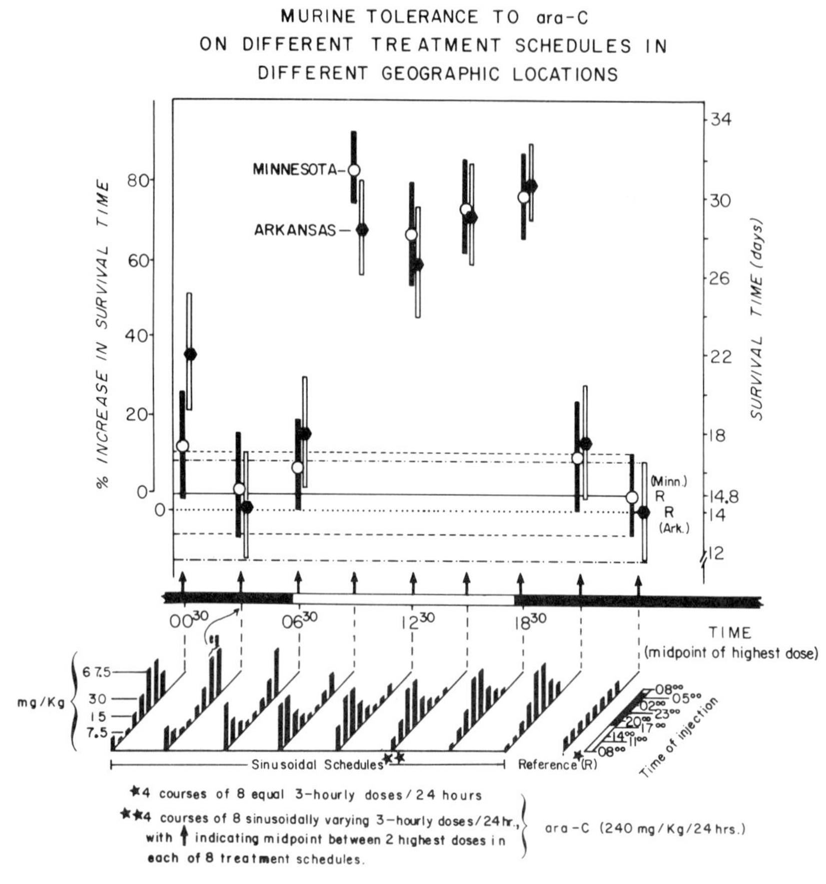

FIGURE 6.6. Chronochemotherapy: Survival times of $CD2F_1$ mice on different drug administration schedules (top) and timing of doses of 1-β-D-arabinofuranosylcytosine (ara-c) (bottom). All treatment schedules comprise four courses, each consisting of a total of 240 mg/kg per 24 h. When the same total dose of ara-C is given, certain sinusoidal drug admininstration schedules definitely are better tolerated by mice than are the other sinusoidal schedules or a reference schedule [at the time the study was done the reference treatment schedule of eight equal doses over a 24-h span was considered the best experimental treatment schedule]. Also note the unequivocal reproducibility of chronotoxicity of the ara-C from experiments done on the same day in different laboratories in different geographic locations. (Reprinted from Scheving et al., 1976, with permission of Cancer Research, Inc.)

and duration of CDC phases must be taken into account in the design of an appropriate chemotherapeutic regimen, as has been suggested recently by Klevecz et al. (1987), who found using flow cytometry that the timing of the phase of DNA synthesis in human ovarian cancer is nearly 12 h out of phase with that of normal cells.

However, there is another, equally important consideration for maximizing the results of radiotherapy and chemotherapy: host tolerance. It is unfortunate that in most of the earlier cancer work there is little mention of the role of rhythmic variations (particularly circadian periodicities) in the susceptibility of the whole organism to the toxicity of the drug(s) being used for treatment. Hrushesky (1983), Scheving (1984), and Møller (1984) haved reviewed the experimental evidence indicating that properly designed protocols (such as sinusoidally varying drug courses during the 24-h day) can dramatically enhance survival and cure rates by concommitantly maximizing the tolerance of the host to the drug (such as 1-β-D-arabinofuranosylcytosine or cisplatin) through a temporal shielding of normal, healthy tissues (Fig. 6.6). Thus, when to treat must assume importance together with the what and where (Haus et al., 1974; Halberg, 1975; Hrushesky, 1983).

6.5 Cellular Clocks in Development and Aging

It seems appropriate that in this last section we deal briefly with the role of cellular clocks in development and in aging—the α and the ω of biological existence.

6.5.1 Timing of Developmental Events

The fact that the development of the eggs of most higher organisms takes place in a relatively constant, nurturing environment suggests that these cells must depend on endogenous signals for timing their developmental stages. But what are these cues? Is the temporal ordering of developmental processes merely a causal chain of events wherein each successive event is cued by the completion of the preceding one, or are there also long-term mechanisms (clocks, for example) which trigger the events at the proper time? These alternatives are formally analogous to the dependent and independent pathways considered in Section 3.1.2 for the regulation of the cell division (developmental) cycle. Satoh (1984) reviewed the evidence that CDCs serve as timing mechanisms in early embryonic development of animals. He concludes that the developmental clock that determines the time of cellular differentiation is closely associated with the DNA replication cycle (appearing to be independent of the cytoplasmic clock that times early morphogenetic events) and outlines a model based on the demethylation of DNA for counting cell divisions and for the establishment and maintenance of the differentiated state (Fig. 6.7).

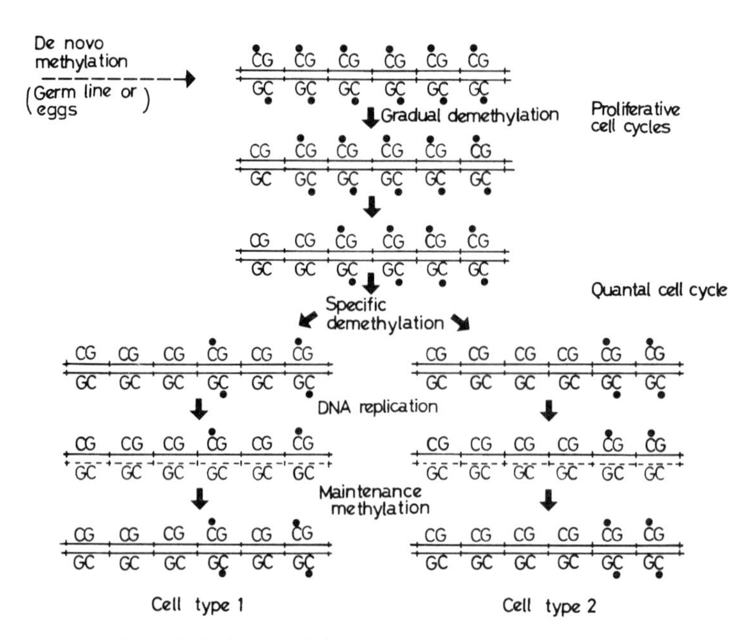

FIGURE 6.7. A demethylation model of DNA for counting cell divisions and establishment and maintenance of a differentiated state, based on the proposals of Holliday and Pugh (1975) and Razin and Riggs (1980). Closed circles represent methyl group. C, cytosine; G, guanine. (Reprinted from Satoh, 1984, with permission of Marcel Dekker, Inc.)

At the same time, there is considerable evidence for the existence of autonomous clocks in the cytoplasm (Satoh, 1984). Mitchison (1984) has noted that one line of evidence that supports the existence of independent-timer periodic controls (which continue to function even if growth or division is blocked or perturbed) is afforded by the development of eggs and early embryos. As he states, these systems are particularly interesting in that "they break many of the normal rules of the cell cycle." Thus, Yoneda et al. (1978) discovered that enucleated merogones of sea urchin eggs undergo rhythmic increases and decreases of cortical stiffness, and Hara et al. (1980) found that *Xenopus* eggs prevented from cleaving with antimitotic drugs undergo a sequence of periodic surface contraction waves (SCWs) timed with the division cycle in untreated eggs (Fig. 6.8). Both sets of observations implicate cytoplasmic clocks; their mechanism(s) is the subject of investigation.

Recent studies on *Xenopus* have revealed three timing mechanisms based on different principles (reviewed by Kirschner et al., 1985). The early synchronous cleavage cycle (the first 12 cleavages) is timed by a cytoplasmic oscillator present in each blastomere, which can operate independently of the nuclear events that it controls. The cessation of synchronous cleavage at the midblastula transition then engages a new de-

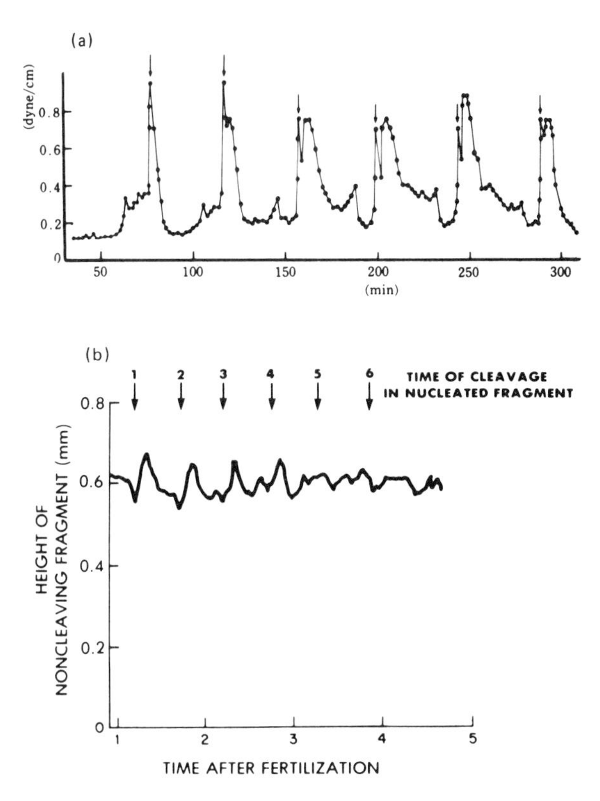

FIGURE 6.8. Evidence for cytoplasmic clocks. (a) Changes in tension at the surface of activated nonnucleate fragment of a sea urchin egg. The egg of *Hemicentrotus pulcherrimus* was separated into two (with and without nucleus) fragments by centrifugal force. Subsequently, the nonnucleate fragment was activated by treatment with butyric acid, and the surface tension was measured continuously by a compression method. Timings of sharp peaks in each cycle (arrows) almost coincide with divisions of normal eggs (Yoneda et al., 1978). (b) Periodic surface contraction waves in a nonnucleate fragment obtained by constriction of a fertilized *Xenopus* egg. The height of the egg fragment measured from the projected image is plotted versus the time after fertilization. Arrows 1, 2, 3, 4, 5, 6 indicate the time of onset of cleavage in the cleaving (nucleate) fragment (Kirschner et al., 1980). (Reprinted from Satoh, 1984, with permission of Marcel Dekker, Inc.)

velopmental program, whose timing depends on the ratio of nuclear to cytoplasmic volume, which is sensed in each cell by the titration of some material by the nucleus. Finally, the timing of the ensuing gastrulation appears to be executed by the measurement of time elapsed since fertilization. It is not clear how many other developmental timers may be implicated during embryogenesis. Although the increase in spatial and temporal heterogeneity that arises as differentiation progresses may be the result of cell–cell interactions, in the early embryo it must depend on the use of intercellular pathways, some of which may be peculiar to a single developmental stage while others may be elaborations of some generalized eukaryotic regulatory circuits.

Indeed, Shinagawa (1985) has noted the similarities between the maturation promoting factor (MPF)—the cytoplasmic signal that has been implicated in the transition from the S to M phase in cleaving embryos (see Kirschner et al., 1985)—and cytoplasmic factors thought to be responsible for triggering the periodic SCWs and cyclic rounding-up and flattening activities associated with the cleavage cycle. Each factor is present in the cytoplasm and is translocatable or transplantable; each causes changes with a periodicity corresponding to the cleavage cycle; each is capable of affecting nuclear morphology; and each might derive at least in part from germinal vesicle materials. Although the nature of the factors is unknown, they may be related to a complex of calcium and the calcium-sequestering system. Free calcium, known to affect contractile systems, cytoskeletons, and the mitotic apparatus, might participate in inducing the autogenous oscillation (Shinagawa, 1985), and may be an element of the developmental timer itself—a role suggested for calcium in circadian oscillators (see Section 4.3.3 and 5.2.5). It is important to note that these cell cycle and cytoplasmic clocks models are not mutually exclusive. Thus, Belisle et al. (1984) have given experimental evidence implicating a central timer mechanism in the regulation of early division cycles during the development of sea urchin embryos. These authors propose a sequential interactive network (SIN) model that couples (by gene action) a semi-autonomous chromosome cycle with an independent metabolic cycle that would regulate the former (Fig. 6.9).

Finally, a set of methods has been developed for investigating the relationships of developmental times (Soll, 1979, 1983). This methdology depends upon the differential sensitivities of timers to small changes in a single environmental variable (such as temperature) and can be applied to any discrete developmental sequence. By measuring the duration of developmental stages at different temperatures and the intervals between any two stages after reciprocal shifts between low and high temperatures at the time the first stage is formed, single, sequential, and parallel timer models can be distinguished. When these methods were applied to morphogenesis in *Dictyostelium discoideum*, a minimum of six independent timers (the majority of which appeared to function in parallel) were delineated for consecutive developmental stages. Recently, a number of unique timing mutants have been isolated, which progress normally through the entire sequence of morphologic stages, but which exhibit selective alterations in the lengths of individual rate-limiting components (Soll and Finney, 1986). One of these (FM-1), which lacks the first ratelimiting component of the pre-aggregative period and begins to aggregate 4 h earlier than the wild type, "switches" at relatively high frequency to several other timing phenotypes with extended pre-aggregative periods

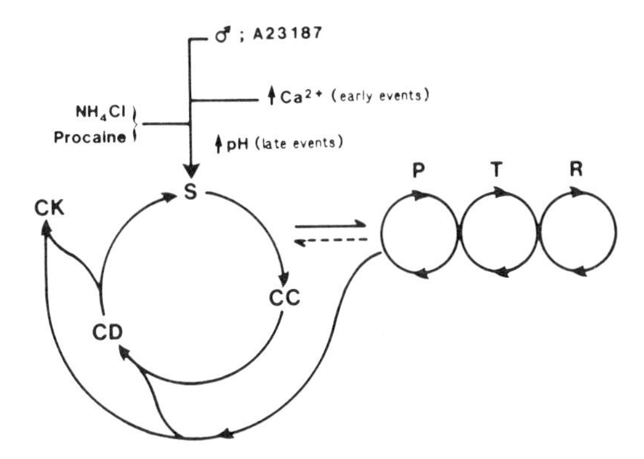

FIGURE 6.9. Sequential interactive network (SIN) model for regulation of early division cycles in sea urchin embryos. Fertilization or egg activation results in an alteration of intracellular ionic composition and reactivation of a start event (S). The chromosome condensation–decondensation loop cycles semiautonomously but is influenced by G_2-phase events occurring on an independent metabolic loop (outer loop). Coupling between the chromosome loop and the metabolic loop is mediated at two sites: (1) a pre-CC event (→) that switches on protein synthesis and that, in turn, supplies gene products necessary for switching on the chromosome loop (←-), and (2) a G_2 event associated with gene products that supply the chromosome loop with proteins needed for chromosome decondensation. CC, chromosome condensation; CD, chromosome decondensation; CK, cytokinesis; P, protein synthesis; R, replication events; T, transcription events. (Reprinted from Belisle et al., 1984, with permission of Marcel Dekker.)

and back again to that of FM-1, a finding that suggests mobile genetic elements may be involved (Soll et al., 1987).

6.5.2 Spatial and Temporal Morphogenetic Fields

A key problem in embryology and development is to explain how gene activity is converted into morphology, that is, to elucidate the coordination of both the spatial and temporal organization of developing systems (Jeffery and Raff, 1983). As Goodwin and Cohen (1969) have noted, it is as if every cell has access to and can read a clock and a map. These authors proposed that the map arises from wavelike propagation of activity from localized clocks, or pacemakers, the essential features of which would include intercellular signalling, entrainment of all cells in the tissue by the fastest cells in the pacemaker region, and a refractory period to ensure unidirectionality. The morphogenetic field thus established by these populations of coupled oscillators (see Sections 5.2.3 and 6.2.2) bears a formal resemblance to the dynamic, spatiotemporal patterns, or dissipative structures (Prigogine, 1961) encountered, for example, during the aggregation of cellular slime molds (see Section 5.2.1).

The graded distribution of molecules in the control of morphogenesis of a cell, such as *Acetabularia*, on the one hand, and their temporal distribution, on the other, would provide the organism with a powerful and versatile tool for obtaining threshold levels in particular regions and at particular times, effectively combining temporal phase (environmentally controlled) and local concentration and giving rise to polarity. Electrical fields (transcellular currents) and specific ion species almost certainly play a major role in this process (Goodwin and Pateromichelakis, 1979; Dazy et al., 1986; Vanden Driessche and Cotton, 1986; see Section 4.3.3).

6.5.3 Aging: Life Cycle Clocks?

At the other end of biological existence, one might consider cell cycle clocks and aging: Is there a life cycle clock? The viewpoint that aging can be viewed as the loss of temporal organization has been treated comprehensively in Samis and Capobianco (1978), but what does this mean at the level of the cell? Theories of cellular aging can be divided into two general categories: the genetic program theories, in which aging would be an event caused by the active expression of specific aging genes or cell-clock longevity genes or by the passive exhaustion of vital genetic information, and cumulative error, wear and tear theories, in which senescence would result from the progressive damage to organelles or to

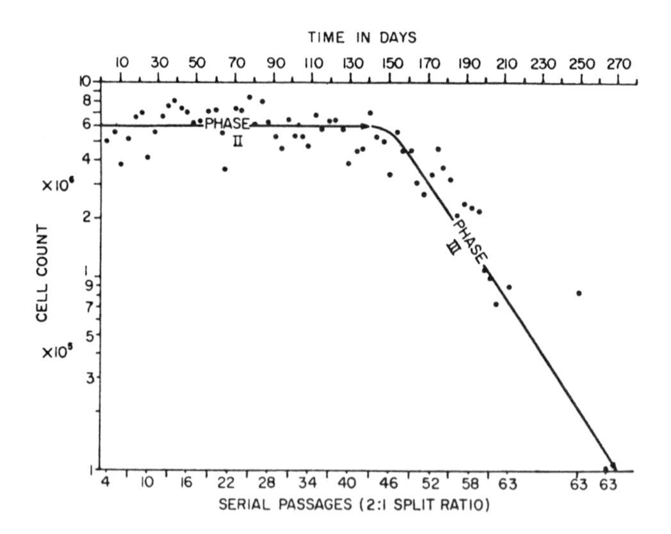

FIGURE 6.10. Limitation in proliferative capacity of populations of normal human diploid cells (strain WI-44) cultured in vitro. Cell counts are given for each passage or subcultivation. The initial plateau (phase II), with no apparent loss of biological function as measured by constant doubling time, is followed by phase III, where doubling time increases exponentially until the culture is ultimately lost. (Reprinted from Hayflick, 1965, with permission of Academic Press.)

the accumulating error in cytoplasmic, information-carrying molecules (Finch and Hayflick, 1977). Ever since Hayflick and Moorhead's report (1961) that normal human embryonic fibroblasts cultured in vitro were limited in their capacity to replicate (50 ± 10 population doublings), there has been considerable controversy about whether the so-called phase III phenomenon (the period of greatly reduced replicative capacity) is reflective of cellular aging in vivo, and, therefore, whether the study of cellular senescence in vitro has any relevance to the normal mammalian lifespan (Fig. 6.10). Many investigators believe, nevertheless, that the biological changes preceding both phase III in vitro and mortality in vivo may be at least indirectly responsible for both phenomena.

Zorn and Smith (1984) are among this group and have reported recent experiments with reconstructed cells (in which old fibroblast nuclei, or karyoplasts, were fused with young cytoplasms, or cytoplasts, and vice versa) designed to determine the relative contributions of the nucleus and cytoplasm to the phase III phenomenon. Their results, supporting a predominantly nuclear role in cellular senescence, provide evidence that the expression of the phase III clock is encoded in the genetic material, presumably on one or more chromosomes. If so, the authors note that it should be feasible to transfer this ticking clock from an old cell to another younger one that has gone through fewer population doublings (as they did with whole-cell hybrids) and then identify by karyotype analysis those chromosomes responsible for phase III—that is, the senescence clock could be localized. Finally, they note the intriguing possibility that cellular

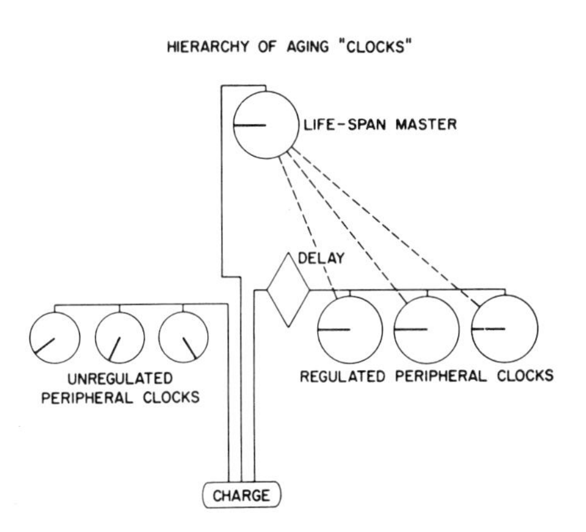

FIGURE 6.11. Life cycle clocks. The determinant of lifespan normally is the master, which fires first, either directly or via regulated secondary clocks with varying intervals after firing. Unregulated clocks are set slow and do not normally fire within the lifespan. If the master were to be artificially slowed, one such unregulated clock might be able to detonate eventual aging. (Reprinted from Comfort, 1979, with permission of Elsevier/North Holland.)

aging and cellular transformation might map on the same chromosome(s) and be related by differential expression of the same genes.

Lengthening (or shortening) a putative life cycle clock may not be as speculative as it might seem. Haga and Karino (1986) report that the diminished sexual activity observed in aged *Paramecium* that had undergone about 700 fissions could be restored to the level of that in presenescent cells by the microinjection of a soluble cytoplasmic fraction (containing the protein factor immaturin) prepared from young cells. Likewise, the synergistic interaction of two linked genes, *incoloris* and *vivax*, inhibits senescence in the ascomycete *Podospora anserina*, prolonging the onset of senescence from 26 days (in the wild type) to 42 and 66 days, respectively, in the mutant strains, whereas the double mutant showed no signs of aging after more than 1 year in culture (Esser and Keller, 1976). Tudzynski et al. (1980) have identified a plasmidlike DNA, identical with the infective principle discovered earlier and present exclusively in aging mycelia of this fungus, that can transform juvenile protoplasts to senescence. Could these gene products that enhance aging be formally equivalent to the detonators envisaged by Comfort (1979), or perhaps be part of the lifespan master clock (Fig. 6.11)?

6.6 Epilogue

A picture thus begins to emerge of the mutual interactions among cell cycles and cellular clocks and such seemingly diverse phenomena as cancer, aging, and pharmacological effectiveness of drugs. Because I have a predilection for pentacles, these speculative formal relationships are summarized in the pentagram illustrated in Figure 6.12. Each element on the perimeter is related to the other four elements, yielding a total of some 10 reciprocal relationships.

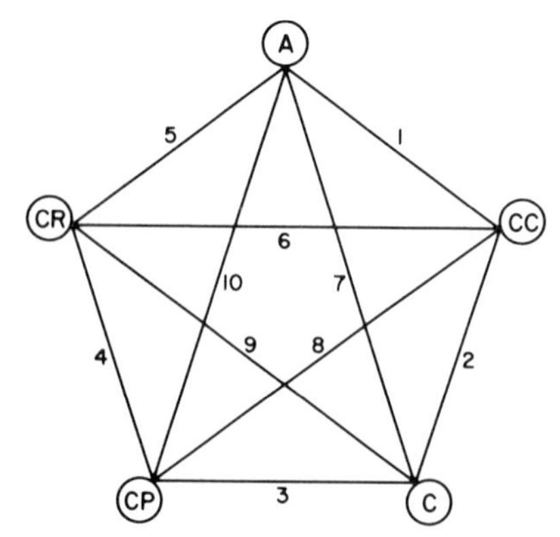

FIGURE 6.12. A pentagram showing the formal relationships existing among aging (A), cancer (C), cell cycles (CC), chronopharmacology (CP) and circadian rhythms (CR). (Reprinted from Edmunds, 1978, with permission of Plenum Press.)

Thus, circadian rhythms (CR) have been shown (1) to modulate cell division cycles (CC), (2) perhaps to play a role in runaway cancer cells (C), and (3) to underlie rhythmic sensitivity and susceptibility to pharmacological agents (CP), while at the same time perhaps being affected by these drugs, and possibly play a role in longevity and aging (A), even as their amplitude and phasing undergo age-related changes. The same type of analysis can be carried out for the other elements: C, CC, and cancer treatment (CP) were all found to be intimately related, the efficacy of the chemotherapeutic drug treatment being dependent on the cell cycle stage at which it was administered even as it synchronized the CCs of the tumor cells. Cellular senescence and aging (A) were related to the CC via spanning and the Hayflick phenomenon, and the correlation between A and the incidence of cancer is well known.

Perhaps biological timekeeping, can serve as a unifying theme at the interfaces among the fields of oncology, gerontology, pharmacology, and cell cycle regulation (Edmunds, 1978).

References

Adamich, M. and Sweeney, B.M. (1976) The preparation and characterization of *Gonyaulax* spheroplasts. Planta **130**: 1–6

Adamich, M., Laris, P.C. and Sweeney, B.M. (1976) In vivo evidence for a circadian rhythm in membranes in *Gonyaulax*. Nature **261**: 583–585

Adams, K.J., Weiler, C.S. and Edmunds, L.N. Jr. (1984) The photoperiodic control of cell division in *Euglena* and *Ceratium*. In: Cell Cycle Clocks. L.N. Edmunds Jr., ed. Marcel Dekker, New York, pp. 395–429

Akitaya, A., Ohsaka, S., Ueda, T. and Kobatake, Y. (1985) Oscillations in intracellular ATP, cAMP and cGMP concentration in relation to rhythmical sporulation under continuous light in the myxomycete *Physarum polycephalum*. J. Gen. Microbiol. **131**: 195–200

Anderson, R.W., Laval-Martin, D.L. and Edmunds, L.N. Jr. (1985) Temperature compensation of the circadian rhythm of cell division in *Euglena*. Exp. Cell Res. **157**: 144–158

Andrews, R.V. (1968) Daily variation in membrane flux of cultured hamster adrenals. Comp. Biochem. Physiol. **26**: 469–488

Andrews, R.V. (1971) Circadian rhythms in adrenal organ cultures. Gegenbaurs Morphol. Jahrb. **117**: 89–98

Andrews, R.V. and Folk, G.E. Jr. (1964) Circadian metabolic patterns in cultured hamster adrenals. Comp. Biochem. Physiol. **11**: 393–410

Aschoff, J., ed. (1965) Circadian Clocks. North-Holland Publishing Co., Amsterdam

Aschoff, J., ed. (1981) Biological Rhythms. Handbook of Behavorial Neurobiology. (F.A. King, series ed.). Plenum Press, New York, Vol. 4

Aschoff, J., Daan, S. and Groos, G.A., eds. (1982) Vertebrate Circadian Systems: Structure and Physiology. Springer-Verlag, Berlin, Heidelberg

Ashkenazi, I.E., Hartman, H., Strulowitz, B. and Dar, O. (1975) Activity rhythms of enzymes in human red blood cell suspensions. J. Interdisc. Cycle Res. **6**: 291–301

Atkinson, D.E. (1968) The energy charge of the adenylate pool as a regulatory parameter: Interaction with feedback modifiers. Biochemistry **7**: 4030–4034

Atkinson, D.E. (1969) Regulation of enzyme function. Annu. Rev. Microbiol. **23**: 47–68

Bachmann, B.J. and Low, K.B. (1980) Linkage map of *Escherichia coli* K-12, ed, 6. Microbiol. Rev. **44**: 1–56

Ball, N.G. and Dyke, I.J. (1957) The effects of decapitation, lack of oxygen, and low temperature on the endogenous 24-hour rhythm in the growth-rate of the Avena coleoptile. J. Exp. Bot. **8**: 323–338

Balzer, I.U. and Hardeland, R. (1981) Advance shifts of the bioluminescence rhythms in *Gonyaulax polyedra* by pharmaca potentially acting on membranes. J. Interdisc. Cycle Res. **12**: 29–34

Barbacka-Surowiak, G. (1986) Existence of an endogenous circadian rhythm in fluorescent dye-labelled cells in hypothalamus culture. Folia Biol. (Krakow) **34**: 73–83

Bargiello, T.A. and Young, M.W. (1984) Molecular genetics of a biological clock in *Drosophila*. Proc. Natl. Acad. Sci. USA **81**: 2142–2146

Bargiello, T.A., Jackson, F.R. and Young, M.W. (1984) Restoration of circadian behavioural rhythms by gene transfer in *Drosophila*. Nature **312**: 752–754

Bargiello, T.A., Saez, L., Baylies, M.K., Gasic, G., Young, M.W. and Spray, D.C. (1987) The *Drosophila* clock gene *per* affects intercellular junctional communication. Nature **328**: 686–691

Barlow, R.B. Jr., Bolanowski, S.J. Jr. and Brachman, M.L. (1987) Efferent optic nerve fibers mediate circadian rhythms in the *Limulus* eye. Science **197**: 86–89

Barlow, R.B. Jr., Kaplan, E., Renninger, G.H. and Saito, T. (1987) Circadian rhythms in *Limulus* photoreceptors. J. Gen. Physiol. **89**: 353–378

Barnett, A. (1959) The effect of continuous light and darkness on the mating type cycle in syngen 2 of *Paramecium multimicronucleatum*. J. Protozool. **6** [Suppl.]: 22

Barnett, A. (1961) Inheritance of mating type and cycling in *Paramecium multimicronucleatum*. Am. Zool. **1**: 341

Barnett, A. (1965) A circadian rhythm of mating type reversals in *Paramecium multimicronucleatum*. In: Circadian Clocks. J. Aschoff, ed. North-Holland Publishing Co., Amsterdam, pp. 305–308

Barnett, A. (1966) A circadian rhythm of mating type reversals in *Paramecium multimicronucleatum*, syngen 2, and its genetic control. J. Cell. Physiol. **67**: 239–270

Barnett, A. (1969) Cell division: A second circadian clock system in *Paramecium multimicronucleatum*. Science **164**: 1417–1419

Barnett, A., Ehret, C.F. and Wille, J.J. Jr. (1971a) Testing the chronon theory of circadian timekeeping. In: Biochronometry. M. Menaker, ed. National Academy of Sciences, Washington, DC, pp. 637–650

Barnett, A., Wille, J.J. and Ehret, C.F. (1971b) Resolution of some component classes of complex RNA by molecular hybridization in the eukaryote *Tetrahymena pyriformis*. Biochim. Biophys. Acta **247**: 243–261

Baylies, M.K., Bargiello, T.A., Jackson, F.R. and Young, M.W. (1987) Changes in abundance or structure of the *per* gene product can alter periodicity of the *Drosophila* clock. Nature **326**: 390–392

Belisle, B.W., Wille, J.J. Jr. and Byrd, E.W. (1984) Central timer versus dependent pathway regulation of cell division in sea urchin cell cycles. In: Cell Cycle Clocks. L.N. Edmunds Jr., ed. Marcel Dekker, New York, pp. 539–555

Benson, J.A. and Jacklet, J.W. (1977a) Circadian rhythm of output from neurones in the eye of *Aplysia*. I. Effects of deuterium oxide and temperature. J. Exp. Biol. **70**: 151–166

Benson, J.A. and Jacklet, J.W. (1977b) Circadian rhythm of output from neurones in the eye of *Aplysia*. II. Effects of cold pulses on a population of coupled oscillators. J. Exp. Biol. **70**: 167–181

Benson, J.A. and Jacklet, J.W. (1977c) Circadian rhythm of output from neurones in the eye of *Aplysia*. IV. A model of the clock: Differential sensitivity to light and low temperature pulses. J. Exp. Biol. **70**: 195–211

Berridge, M.J., Rapp, P.E. and Treherne, J.E., eds. (1979) Cellular Oscillators. Cambridge University Press, Cambridge, England

Bierhuizen, J.F., ed. (1972) Circadian Rhythmicity. Proc. Int. Symp. Circadian Rhythmicity (Wageningen, 1971). Centre for Agricultural Publishing and Documentation, Wageningen, The Netherlands

Binkley, S. (1983) Rhythms in ocular and pineal *N*-acetyltransferase: A portrait of an enzyme clock. Comp. Biochem. Physiol. **75A**: 123–129

Binkley, S. and Geller, E.B. (1975) Pineal *N*-acetyltransferase in chickens: Rhythm persists in constant darkness. J. Comp. Physiol. **99**: 67–70

Binkley, S., MacBride, S.E., Klein, D.C. and Ralph, C.L. (1973) Pineal enzymes: Regulation of avian melatonin synthesis. Science **181**: 273–275

Binkley, S.A., Riebman, J.B. and Keilly, K.B. (1978) The pineal gland: A biological clock *in vitro*. Science **202**: 1198–1201

Bittman, E.L., Dempsey, R.J. and Karsch, F.J. (1983) Pineal melatonin secretion drives the reproductive response to daylength in the ewe. Endocrinology **113**: 2276–2283

Bitz, D.M. and Sargent, M.L. (1974) A failure to detect an influence of magnetic fields on the growth rate and circadian rhythm of *Neurospora crassa*. Plant Physiol. **53**: 154–157

Blessing, J. and Lempp, H. (1978) An immunological approach to the isolation of factors with mitotic activity from the plasmodial stage of the Myxomycete *Physarum polycephalum*. Exp. Cell Res. **113**: 435–438

Block, G.D. and McMahon, D.G. (1984) Cellular analysis of the *Bulla* ocular circadian pacemaker sytem. III. Localization of the circadian pacemaker. J. Comp. Physiol. A **155**: 387–395

Block, G.D. and Page, T.L. (1978) Circadian pacemakers in the nervous system. Annu. Rev. Neurosci. **1**: 19–34

Block, G.D. and Wallace, S.F. (1982) Localization of a circadian pacemaker in the eye of a mollusc, *Bulla*. Science **217**: 155–157

Block, G.D., McMahon. D.G., Wallace, S.F. and Friesen, W.O. (1984) Cellular analysis of the *Bulla* ocular circadian pacemaker system. I. A model for retinal organization. J. Comp. Physiol. A **155**: 365–378

Blum, J.J. (1967) An adrenergic control system in *Tetrahymena*. Proc. Natl. Acad. Sci. USA **58**: 81–88

Bode, V.C., De Sa, R. and Hastings, J.W. (1963) Daily rhythm of luciferin activity in *Gonyaulax polyedra*. Science **141**: 913–915

Boiteux, A., Goldbeter, A. and Hess, B. (1975) Control of oscillating glycolysis of yeast by stochastic, periodic, and steady source of substrate: A model and experimental study. Proc. Natl. Acad. Sci. USA, **72**: 3829–3833

Bollig, I.C., Mayer, K., Mayer, W.-E. and Engelmann, W. (1978) Effects of cAMP, theophylline, imidazole, and 4-(3,4-dimethyoxybenzyl)-2-imidazolidone on the leaf movement rhythm of *Trifolium repens*—A test of the cAMP-hypothesis of circadian rhythms. Planta **141**: 225–230

Borghi, H., Puiseux-Dao, S. and Dazy, A.-C. (1986) The effects of blue and red light on *Acetabularia mediterranea* after a long dark period: Recovery of the endogenous rhythms of transcellular electric potential and chloroplast velocity. Can. J. Bot. **64**: 1134–1137

Bose, R.N., Rajasekar, N., Thompson, D.M. and Gould, E.S. (1986) Electron transfer. 78. Reduction of carboxylato-bound chromium(V) with bisulfite. A "clock reaction" involving chromium(IV). Inorg. Chem. **25**: 3349–3353

Boulos, Z. and Rusak, B. (1982) Circadian phase response curves for dark pulses in the hamster. J. Comp. Physiol. **146**: 411–417

Bradbury, E.M., Inglis, R.J., Matthews, H.R. and Langan, T.A. (1974) Molecular basis of control of mitotic cell division in eukaryotes. Nature **249**: 553–555

Brady,J., ed. (1982) Biological Timekeeping. Soc. Exp. Biol. Seminar Series, No. 14. Cambridge University Press, Cambridge

Brain, R.D., Freeberg, J., Weiss, C.W. and Briggs, W.R. (1977a) Blue light-induced absorbance changes in membrane fraction from corn and *Neurospora*. Plant Physiol. **59**: 948–952

Brain, R.D., Woodward, D.O. and Briggs, W.R. (1977b) Correlative studies of light sensitivity and cytochrome content in *Neurospora crassa*. Carnegie Inst. Wash. Yearbook **1976**: 295–299

Brand, L.E. (1982) Persistent diel rhythms in the chlorophyll fluorescence of marine phytoplankton species. Marine Biol. **69**: 253–262

Brandt, W.H. (1953) Zonation in a prolineless strain of *Neurospora*. Mycologia, **45**: 194–208

Bré, M.H., El Ferjani, E. and Lefort-Tran, M. (1981) Sequential protein dependent steps in the cell cycle. Initiation and completion of division in vitamin B_{12}-replenished *Euglena gracilis*. Protoplasma **108**: 301–318

Brewer, E.N. and Rusch, H.P. (1968) Effect of elevated temperature shocks on mitosis and on the initiation of DNA replication in *Physarum polycephalum*. Exp. Cell Res. **49**: 79–86

Briggs, W.R. (1976) The nature of the blue light photoreceptor in higher plants and fungi. In: Light and Plant Development. Proc. 22nd Univ. Nottingham Easter School in Agric. Sci. H. Smith, ed. Butterworths, London and Boston, pp. 7–18

Brinkmann, K. (1966) Temperatureinflüsse auf die circadiane Rhythmik von *Euglena gracilis* bei Mixotrophie und Autotrophie. Planta **70**: 344–389

Brinkmann, K. (1971) Metabolic control of temperature compensation in the circadian rhythm of *Euglena gracilis*. In: Biochronometry. M. Menaker, ed. National Academy of Sciences, Washington, DC, pp. 567–593

Brinkmann, K. (1976a) The influence of alcohols on the circadian rhythm and metabolism of *Euglena gracilis*. J. Interdisc. Cycle Res. **7**: 149–170

Brinkmann, K. (1976b) Circadian rhythm in the kinetics of acid denaturation of cell membranes of *Euglena gracilis*. Planta **129**: 221–227

Britz, S.J., Schrott, E., Widell, S., Brain, R.D. and Briggs, W.R. (1977) Methylene blue-mediated red light photoreduction of cytochromes in particulate fractions of corn and *Neurospora*. Carnegie Inst. Wash. Yearbook **1976**: 289–293

Brock, M.A. (1985) Biological clocks and aging. In: Review of Biological Research in Aging, vol. 2. M. Rothstein, ed. Alan R. Liss, Inc., New York, pp. 445–462

Broda, H. and Schweiger, H.-G. (1981a) Long-term measurement of endogenous diurnal oscillations of the electric potential in an individual Acetabularia cell. Eur. J. Cell Biol. **26**: 1–4

Broda, H. and Schweiger, H.-G. (1981b) Long-term measurements of three rhythmic-parameters in *Acetabularia*. Protoplasma **105**: 352–353

Broda, H., Schweiger, G., Koop, H.-U., Schmid, R. and Schweiger, H.-G. (1979) Chloroplast migration: A method for continuously monitoring a circadian rhythm in a single cell of *Acetabularia*. In: Developmental Biology of *Acetabularia*. S. Bonotto, V. Kefeli and S. Puiseux-Dao, eds. Elsevier/North-Holland Biomedical Press, Amsterdam, pp. 163–168

Broda, H., Dirk, J. and Schweiger, H.-G. (1983) On-line data acquisition and evaluation of long-term measurements of circadian rhythms in individual cells of *Acetabularia*. Cell Biophys. **5**: 43–59

Broda, H., Brugge, D., Homma, K. and Hastings, J.W. (1986a) Circadian communication between unicells? Effects on period by cell-conditioning of medium. Cell Biophys., **8**: 47–67

Broda, H., Gooch, V.D., Taylor, W., Aiuto, N. and Hastings, J.W. (1986b) Acquisition of circadian bioluminescence data in *Gonyaulax* and an effect of the measurement procedure on the period of the rhythm. J. Biol. Rhythms **1**: 251–263

Brody, S. (1981) Oligomycin-resistant mutations in the DCCD binding protein of the mitochondrial ATPase affect the circadian rhythm of *Neurospora*. Fed. Proc. **40**: 1734

Brody, S. and Forman, L. (1980) Interactions between exogenous unsaturated fatty acids and mitochondria in *Neurospora crassa* (Abstr.) Annu. Meet. Am. Soc. Microbiol., K185

Brody, S. and Harris, S. (1973) Circadian rhythms in *Neurospora*: Spatial differences in pyridine nucleotide levels. Science **180**: 498–500

Brody, S. and Martins, S.A. (1979) Circadian rhythms in *Neurospora crassa*: Effects of unsaturated fatty acids. J. Bacteriol. **137**: 912–915

Brody, S., Dieckmann, C. and Mikolajczyk, S. (1985) Circadian rhythms in *Neurospora crassa*: The effects of point mutations on the proteolipid portion of the mitochondrial ATP synthetase. Mol. Gen. Genet. **200**: 155–161

Bronk, B.V., Dienes, G.J. and Pastkin, A. (1968) The stochastic theory of cell proliferation. Biophys. J. **8**: 1353–1338

Brooks, R.F. (1981) Variability in the cell cycle and the control of proliferation. In: The Cell Cycle. P.C.L. John, ed. Cambridge University Press, Cambridge, pp. 35–61

Brooks, R.F., Bennett, D.C. and Smith, J.A. (1980) Mammalian cell cycles need two random transitions. Cell **19**: 493–504

Bruce, V.G. (1960) Environmental entrainment of circadian rhythms. Cold Spring Harbor Symp. Quant. Biol. **25**: 29–48

Bruce, V.G. (1965) Cell division rhythms and the circadian clock. In: Circadian Clocks. J. Aschoff, ed. North-Holland Publishing Company, Amsterdam, pp. 125–138

Bruce, V.G. (1970) The biological clock in *Chlamydomonas reinhardi*. J. Protozool. **17**: 328–333

Bruce, V.G. (1972) Mutants of the biological clock in *Chlamydomonas reinhardi*. Genetics **70**: 537–548

Bruce, V.G. (1974) Recombinants between clock mutants of *Chlamydomonas reinhardi*. Genetics **77**: 221–230

Bruce, V.G. (1976) Clock mutants. In: The Molecular Basis of Circadian Rhythms. J.W. Hastings and H.-G. Schweiger, eds. Dahlem Konferenzen, Berlin, pp. 339–351

Bruce, V.G. and Bruce, N.C. (1981) Circadian clock-controlled growth cycle in *Chlamydomonas reinhardi*. In: International Cell Biology 1980–1981. Proc. 2nd Int. Cong. Cell Biol., Berlin, 1980. H.G. Schweiger, ed. Springer-Verlag, Berlin, pp. 823–830

Bruce, V.G. and Minis, D.H. (1969) Circadian clock action spectrum in a photoperiodic moth. Science **163**: 583–585

Bruce, V.G. and Pittendrigh, C.S. (1956) Temperature independence in a unicellular "clock". Proc. Natl. Acad. Sci. USA **42**: 676–681

Bruce, V.G. and Pittendrigh, C.S. (1957) Endogenous rhythms in insects and microorganisms. Am. Nat. **91**: 179–195

Bruce, V.G. and Pittendrigh, C.S. (1958) Resetting the *Euglena* clock with a single light stimulus. Am. Nat. **92**: 294–306

Bruce, V.G. and Pittendrigh. C.S. (1960) An effect of heavy water on the phase and period of the circadian rhythm in *Euglena*. J. Cell Comp. Physiol. **56**: 25–31

Brulfert, J., Vidal, J., Gadal, P. and Queiroz, O. (1982) Daily rhythm of phosphoenolpyruvate carboxylase. Immunological evidence for the absence of a rhythm in protein synthesis. Planta **156**: 92–94

Brulfert, J., Vidal, J., Le Marechal, P., Gadal, P., Queiroz, O., Kluge, M. and Kruger, I. (1986) Phosphorylation–dephosphorylation as a probable mechanism for the diurnal regulatory changes of phosphoenolpyruvate carboxylase in CAM plants. Biochem. Biophys. Res. Commun. **136**: 151–159

Buck, J. (1937) Studies on the firefly. I. The effect of light and other agents on flashing in *Photinus pyralis*, with special reference to periodicity and diurnal rhythm. Physiol. Zool. **10**: 45–58

Buck, J. and Buck, E. (1968) Mechanism of rhythmic synchronous flashing of fireflies. Science **159**: 1319–1327

Buck, J. and Buck, E. (1978) Toward a functional interpretation of synchronous flashing by fireflies. Am. Natur. **112**: 471–492

Buck, J., Buck, E., Case, J.F. and Hanson, F.E. (1981a) Control of flashing in fireflies. V. Pacemaker synchronization in *Pteroptyx cribellata*. J. Comp. Physiol. **144**: 287–298

Buck, J., Buck, E., Hanson, F.E., Case, J.F., Mets, L. and Atta, G.J. (1981b) Control of flashing in fireflies. IV. Free run pacemaking in a synchronic *Pteroptyx*. J. Comp. Physiol. **144**: 277–286

Buetow, D.E., ed. (1968a,b; 1982) The Biology of Euglena. Academic Press, New York, Vols. I, II, and III

Bünning, E. (1973) The Physiological Clock. Springer-Verlag, Berlin and New York

Bünning, E. (1986) Evolution der circadianen Rhythmik und ihrer Nutzung zur Tageslangenmessung. Naturwissenshaften **73**: 70–77

Bünning, E. and Lörcher, L. (1957) Regulierung und Auslösung endogenetagesperiodischer Blattbewegungen durch verschiedener Lichtqualitäten. Naturwissenschaften **44**: 472

Bünning, E. and Moser, I. (1966) Response curves of the circadian rhythm in *Phaseolus*. Planta **69**: 101–110

Bünning, E. and Moser, I. (1972) Influence of valinomycin on circadian leaf movements of *Phaseolus*. Proc. Natl. Acad. Sci. USA **69**: 2732–2733

Bünning, E. and Moser, I. (1973) Light-induced phase shifts of circadian leaf movements of *Phaseolus*: Comparison with the effects of potassium and ethyl alcohol. Proc. Natl. Acad. Sci. USA **70**: 3387–3389

Bünning, E., Kurras, S. and Vielhaben, V. (1965) Phasenverschiebungen der endogenen Tagesrhythmik durch Reduktion der Atmung. Planta **64**: 291–300

Burgoyne, R.D. (1978) A model for the molecular basis of circadian rhythms involving monovalent ion-mediated translational control. FEBS Lett. **94**: 17–19

Burns, F.J. and Tannock, I.F. (1970) On the existence of a G0-phase in the cell cycle. Cell Tissue Kinet. **3**: 321–334

Burton, A.C. (1971) Cellular communication, contact inhibition, cell clocks, and cancer: The impact of the work and ideas of W.R. Loewenstein. Perspect. Biol. Med. **14**: 301–318

Burton, A.C. (1975) The role of biochemical rhythms in contact inhibition of cellular division. In: Cellular Membranes and Tumor Cell Behavior. Williams & Wilkins Co., Baltimore, pp. 249–266

Burton, A.C. and Canham, P.B. (1973) The behavior of coupled biochemical oscillators as a model of contact inhibition of cellular division. J. Theor. Biol. **39**: 555–580

Bush, K. and Sweeney, B.M. (1972) The activity of ribulose diphosphate carboxylase in extracts of *Gonyaulax polyedra* in the day and night phases of short period mutant of *Chlamydomonas reinhardii*. Eur. J. Cell Biol. **33**: 13–18

Cameron, I.L. and Padilla, G.M., eds. (1966) Cell Synchrony: Studies in Biosynthetic Regulation. Academic Press, New York and London

Campbell, A. (1957) Synchronization of cell division. Bacteriol. Rev. **21**: 263–271

Campbell, A. (1964) The theoretical basis of synchronization by shifts in environmental conditions. In: Synchrony in Cell Division and Growth. E. Zeuthen, ed. Wiley-Interscience, New York, pp. 469–484

Campbell, N., Satter, R.L. and Garber, R.C. (1981) Apoplastic transport of ions in the motor organ of *Samanea*. Proc. Natl. Acad. Sci. USA **78**: 2981–2984

Cardinali, D.P., Larin, F. and Wurtman, R.J. (1972) Control of the rat pineal gland by light spectra. Proc. Natl. Acad. Sci. USA **69**: 2003–2005

Carpenter, D.O., ed. (1981) Cellular Pacemakers. Vol. 1: Mechanisms of Pacemaker Generation. John Wiley & Sons, New York

Carpenter, D.O., ed. (1982) Cellular Pacemakers. Vol. 2: Function in Normal and Disease States. John Wiley & Sons, New York

Carpenter, G.A. and Grossberg, S. (1983) A neural theory of circadian rhythms: The gated pacemaker. Biol. Cybern. **48**: 35–59

Carrino, J.J. and Laffler, T. (1985) The effect of heat shock on the cell cycle regulation of tubulin expression in *Physarum polycephalum*. J. Cell Biol. **100**: 642–647

Carrino, J.J. and Laffler, T.G. (1986) Transcription of α-tubulin and histone H4 genes begins at the same point in the *Physarum* cell cycle. J. Cell Biol. **102**: 1666–1670

Carter, B.L.A. and Halvorson, H.O. (1973) Periodic changes in rate of amino acid uptake during yeast cell cycle. J. Cell Biol. **58**: 401–409

Cassone, V.M., Chesworth, M.J. and Armstrong, S.M. (1986) Entrainment of rat circadian rhythms by daily injection of melatonin depends upon the hypothalamic suprachiasmatic nuclei. Physiol. Behav. **36**: 1105–1110

Castor, L.N. (1980) A G_1 rate model accounts for cell-cycle kinetics attributed to 'transition probability.' Nature **287**: 857–859

Chance, B., Pye, K. and Higgins, J. (1967) Waveform generation by enzymatic oscillators. IEEE Spectrum **4**: 79–86

Chance, B., Pye, E.K., Ghosh, A.K. and Hess, B., eds. (1973) Biological and Biochemical Oscillators. Academic Press, New York and London

Chandler, M., Bird, R.E. and Caro, L. (1975) The replication time of the Escherichia coli K 12 chromosome as a function of cell doubling time. J. Mol. Biol. **94**: 127–132

Chandrashekaran, M.K. and Engelmann, W. (1973) Early and late subjective night phases of the *Drosophila pseudoobscura* circadian rhythm require different energies of blue light for phase shifting. Z. Naturforsch. **28C**: 750–753

Chay, T.R. (1981) A model for biological oscillations. Proc. Natl. Acad. Sci. USA **78**: 2204–2207

Chen, P.-C. and Lorenzen, H. (1986a) Changes in the productivity and nuclear divisions in synchronous *Chlorella* and circadian rhythm. Plant Cell Physiol. **27**: 1423–1427

Chen, P.-C. and Lorenzen, H. (1986b) Changes of enzyme activities and isozyme patterns in synchronous *Chlorella* due to circadian rhythm. J. Plant Physiol. **125**: 87–94

Chernavskii, D.S., Palamarchuk, E.K., Polezhaev, A.A., Solyanik, G.I. and Burlakova, E.B. (1977) A mathematical model of periodic processes in membranes (with application to cell cycle regulation. Biosystems **9**: 187–193

Cheung, W.Y. (1980) Calmodulin plays a pivotal role in cellular regulation. Science **207**: 19–27

Cheung, W.Y., ed. (1981) Calcium and Cell Function. Academic Press, New York, Vol. 1

Chin, B., Friedrich, P.D. and Bernstein, I.A. (1972) Stimulation of mitosis following fusion of plasmodia in the myxomycete *Physarum polycephalum*. J. Gen. Microbiol. **71**: 93–101

Chisholm, S.W. (1981) Temporal patterns of cell division in unicellular algae. In: Physiological Bases of Phytoplankton Ecology. T. Platt, ed. Canad. Bull. Fish. Aquatic Sci. **210**: 150–181

Chisholm, S.W. and Brand, L.E. (1981) Persistence of cell division phasing in marine phytoplankton in continuous light after entrainment to light:dark cycles. J. Exp. Mar. Biol. Ecol. **51**: 107–118

Chisholm, S.W., Morel, F.M.M. and Slocum, W.S. (1980) The phasing and distribution of cell division cycles in marine diatoms. In: Primary Productivity in the Sea. Brookhaven Symp. in Biology., No. 31). P. Falkowski, ed. Plenum Press, New York, pp. 281–300

Chisholm, S.W., Vaulot, D. and Olson, R.J. (1984) Cell cycle controls in phytoplankton: Comparative physiology and ecology. In: Cell Cycle Clocks. L.N. Edmunds Jr., ed. Marcel Dekker, New York, pp. 365–394

Christianson, R. and Sweeney, B.M. (1972) Sensitivity to stimulation, a component of the circadian rhythm in luminescence in *Gonyaulax*. Plant Physiol. **49**: 994–997

Citri, Y., Colot, H.V., Jacquier, A.C., Yu, Q., Hall, J.C., Baltimore, D. and Rosbash, M. (1987) A familly of unusually spliced biologically active transcripts encoded by a *Drosophila* clock gene. Nature **326**: 42–47

Codd, G.A. and Merrett, M.J. (1971) Photosynthetic products of division synchronized cultures of *Euglena*. Plant Physiol. **47**: 635- 639

Comfort, A. (1979) The Biology of Senescence, 3rd ed. Elsevier/North Holland, New York

Connor, J.A. (1985) Neural pacemakers and rhythmicity. Annu. Rev. Physiol. **47**: 17–28

Cook, J.R. (1961a) *Euglena gracilis* in synchronous division. I. Dry mass and volume characteristics. Plant Cell Physiol. **2**: 199–202

Cook, J.R. (1961b) *Euglena gracilis* in synchronous division. II. Biosynthetic rates over the light cycle. Biol. Bull. **121**: 277–289

Cook, J.R. (1966) Photosynthetic activity during the division cycle in synchronized *Euglena gracilis*. Plant Physiol. **41**: 821–825

Cook, J.R. and Cook, B. (1962) Effect of nutrients on the variation of individual generation times. Exp. Cell Res. **28**: 524–530

Cook, J.R. and James,T.W. (1960) Light-induced division synchrony in *Euglena gracilis* var. *bacillaris*. Exp. Cell Res. **21**: 583–589

Cooper, S. (1979) A unifying model for the G1 period in prokaryotes and eukaryotes. Nature **280**: 17–19

Cooper, S. (1982a) The continuum model. Application to G1-arrest and G(0). In: Cell Growth. C. Nicolini, ed. Plenum Press, New York, pp. 315–336

Cooper, S. (1982b) The continuum model. Statistical implications. J. Theor. Biol. **94**:783–800

Cooper, S. (1984a) Application of the continuum model to the clock model of the cell division cycle. In: Cell Cycle Clocks. L.N. Edmunds Jr., ed. Marcel Dekker, New York, pp. 209–218

Cooper, S. (1984b) The continuum model as a unified description of the division cycle of eukaryotes and prokaryotes. In: The Microbial Cell Cycle. P. Nurse and E. Streiblová, eds. CRC Press, Boca Raton, FL, pp. 7–18

Cooper, S. and Helmstetter, C.E. (1968) Chromosome replication and the division cycle of *Escherichia coli* B/r. J. Mol. Biol. **31**: 519–540

Cornélissen, G., Broda, H. and Halberg, F. (1986) Does *Gonyaulax polyedra* measure a week? Cell Biophys. **8**: 69–85

Cornelius, G. (1980) On the existence of circadian rhythms in human erythrocyte suspensions. Enzyme activities and membrane functions. J. Interdisc. Cycle Res. **11**: 55–68

Cornelius, G. and Rensing, L. (1976) Daily rhythmic changes in Mg^{2+}-dependent ATPase activity in human red blood cell membranes in vitro. Biochem. Biophys. Res. Commun. **71**: 1269–1272

Cornelius, G. and Rensing, L. (1982) Can phase response curves of various treatments of circadian rhythms be explained by effects on protein synthesis and degradation? Biosystems **15**: 35–47

Cornelius, G. and Rensing, L. (1986) Circadian rhythm of heat shock protein synthesis of *Neurospora crassa*. Eur. J. Cell Biol. **40**: 130–132

Cornelius, G., Schroeder-Lorenz, A. and Rensing, L. (1985) Circadian-clock control of protein synthesis and degradation in *Gonyaulax polyedra*. Planta **166**: 365–370

Corrent, G., McAdoo, D.J. and Eskin, A. (1978) Serotonin shifts the phase of the circadian rhythm from the *Aplysia* eye. Science **202**: 977–979

Corrent, G., Eskin, A. and Kay, I. (1982) Entrainment of the circadian rhythm from the eye of *Aplysia*: Role of serotonin. Am. J. Physiol. **242**: R326–R332

Coté, G.G. and Brody, S. (1986) Circadian rhythms in *Drosophila melanogaster*: Analysis of period as a function of gene dosage at the *per* (period) locus. J. Theor. Biol. **121**: 487–503

Coté, G.G. and Brody, S. (1987) Circadian rhythms in *Neurospora crassa*: Membrane composition of a mutant defective in temperature compensation. Biochim. Biophys. Acta, **898**: 23–36

Cotton, G. and Vanden Driessche, T. (1987) Identification of calmodulin in *Acetabularia*: its distribution and physiological significance. J. Cell Sci. **87**: 337–347

Craigie, R.A. and Cavalier-Smith, T. (1982) Cell volume and the control of the *Chlamydomonas* cell cycle. J. Cell Sci. **54**: 173–191

Cramer-Herold, R., Lysek, G. and Varchmin-Fuchs, B. (1986) Induction by sodium dodecylsulfate of the circadian rhythm of conidiation in *Neurospora crassa*. Experientia **42**: 184–185

Creanor, J. and Mitchison, J.M. (1986) Nucleoside diphosphokinase, an enzyme with step changes in activity during the cell cycle of the fission yeast *Schizosaccharomyces pombe*. I. Persistence of steps after a block to the DNA–division cycle. J. Cell Sci. **86**: 207–215

Cummings, F.W. (1975) A biochemical model of the circadian clock. J. Theor. Biol. **55**: 455–470

Das, J. and Busse, H.-G. (1985) Long term oscillation in glycolysis. J. Biochem. **97**: 719–727

Das, J., Busse, H.-G. and Havsteen, B.H. (1982) Long-period glycolytic oscillations. Hoppe-Seyler's Z. Physiol. Chem. **363**: 952

Dazy, A.-C., Borghi, H., Garcia, E. and Puiseux-Dao, S. (1986) Transcellular current and morphogenesis in *Acetabularia mediterranea* grown in white, blue and red light. In: Ionic Currents in Development. Alan R. Liss, New York, pp. 123–130

DeCoursey, P., ed. (1976) Biological Rhythms in the Marine Environment. University of South Carolina Press, Columbia

DeCoursey, P.J. and Buggy, J. (1986) Restoration of locomotor rhythmicity in SCN-lesioned golden hamsters by transplantation of fetal SCN. Soc. Neurosci. Abst. **12**: 210 (No. 61.4)

Decroly, O. and Goldbeter, A. (1982) Birhythmicity, chaos, and other patterns of temporal self-organization in a multiply regulated biochemical system. Proc. Natl. Acad. Sci. USA **79**: 6917–6921

Deguchi, T. (1979a) Circadian rhythm of serotonin *N*-acetyltransferase activity in organ culture of chicken pineal gland. Science **203**: 1245–1247

Deguchi, T. (1979b) A circadian oscillator in cultured cells of chicken pineal gland. Nature **282**: 94–96

Deitzer, G.F., Kempf, O., Fischer, S. and Wagner, E. (1974) Endogenous rhythmicity and energy transduction. IV. Rhythmic control of enzymes involved in the tricarboxylic acid cycle and the oxidative pentose phosphate pathway in *Chenopodium rubrum*. Planta **117**: 29–41

Delmer, D.P. and Brody, S. (1975) Circadian rhythms in *Neurospora crassa*: Oscillation in the level of an adenine nucleotide. J. Bacteriol. **121**: 548–553

Demets, R., Tomson, A.M., Stegwee, D. and Van den Ende, H. (1987) Control of the mating competence rhythm in *Chlamydomonas eugametos*. J. Gen. Microbiol. **133**: 1081–1088

Dennis, P. and Bremer, J. (1974) Macromolecular composition during steady-state growth of *Escherichia coli* B/r. J. Bacteriol. **119**: 270–281

Dharmananda, S. and Feldman, J.F. (1979) Spatial distribution of circadian clock phase in aging cultures of *Neurospora crassa*. Plant Physiol. **63**: 1049–1054

Dieckmann, C.L. (1980) Circadian rhythms in *Neurospora crassa*: A biochemical and genetic study of the involvement of mitochondrial metabolism in periodicity. PhD thesis, University of California, San Diego

Dieckmann, C.L. and Brody, S. (1980) Circadian rhythms in *Neurospora crassa*: Oligomycin-resistant mutations affect periodicity. Science **207**: 896–898

Dobra, K.W. and Ehret, C.F. (1977) Circadian regulation of glycogen, tyrosine aminotransferase, and several respiratory parameters in solid agar cultures of *Tetrahymena pyriformis*. In: Proceedings of the XII International Conference of the International Society for Chronobiology, Washington, DC, 1975. Publishing House Il Ponte, Milano, pp. 589–594

Donachie, W.D. (1968) Relationship between cell size and time of initiation of DNA replication. Nature **219**: 1077–1079

Donachie, W.D. (1973) Regulation of cell division in bacteria. Br. Med. Bull. **29**: 203–207

Donachie, W.D. (1974) Cell division in bacteria. In: Mechanism and Regulation of DNA Replication. A.R. Kolber and M. Kohiyama, eds. Plenum Press, New York, pp. 431–445

Donachie, W.D. (1979) The cell cycle of *Escherichia coli*. In: Developmental Biology of Prokaryotes. J.H. Parish, ed. Studies in Microbiology, Vol. 1. N.G. Carr, J.L. Ingraham and S.C. Rittenberg, Series eds. Blackwell Scientific, Oxford, pp. 11–35

Donachie, W.D. (1981) The cell cycle of *Escherichia coli*. In: The Cell Cycle. P.C.L. John, ed. Cambridge University Press, Cambridge, pp. 63–83

Donachie, W.D. and Masters, M. (1969) Temporal control of gene expression in bacteria. In: The Cell Cycle: Gene–Enzyme Interactions. G.M. Padilla, G.L. Whitson and I.L. Cameron, eds. Academic Press, New York and London, pp. 37–76

Donachie, W.D., Jones, N.C. and Teather, R.M. (1973) The bacterial cell cycle. Symp. Soc. Gen. Microbiol. **23**: 9–44

Donachie, W.D., Begg, K.J. and Vicente, M. (1976) Cell length, cell growth and cell division. Nature **264**: 328–333

Donnan, L. and John, P.C.L. (1983) Cell cycle control by timer and sizer in *Chlamydomonas*. Nature **304**: 630–633

Donnan, L. and John, P.C.L. (1984) Timer and sizer controls in the cell cycles of *Chlamydomonas* and *Chlorella*. P. Nurse and E. Streiblová, eds. CRC Press, Boca Raton, FL pp. 231–251

Donnan, L., Carvill, E.P., Gilliland, T.J. and John, P.C.L. (1985) The cell cycles of *Chlamydomonas* and *Chlorella*. New Phytol. **99**: 1–40

Donner, B., Helmboldt-Caesar, U. and Rensing, L. (1985) Circadian rhythm of total protein synthesis in the cytoplasm and chloroplasts of *Gonyaulax polyedra*. Chronobiol. Int. **2**: 1–9

Dowse, H.B. and Ringo, J.M. (1987) Further evidence that the circadian clock in *Drosophila* is a population of coupled ultradian oscillators. J. Biol. Rhythms **2**: 65–76

Dowse, H.B., Hall, J.C. and Ringo, J.M. (1987) Circadian and ultradian rhythms in *period* mutants of *Drosophila melanogaster*. Behav. Gen. **17**: 19–35

Dreisig, H. (1978) The circadian rhythms of bioluminescence in the glowworm, *Lampyris noctiluca* L. (Coleoptera, Lampyridae). Behav. Ecol. Sociobiol. **3**: 1–18

Drescher, K., Cornelius, G. and Rensing, L. (1982) Phase response curves obtained by perturbing different variables of a 24 hr model oscillator based on translational control. J. Theor. Biol. **94**: 345–353

Dunlap, J.C. and Feldman, J. (1982) *Frq-7*, a circadian clock mutant whose clock is insensitive to cycloheximide. Neurospora Newslett. **29**: 12

Dunlap, J.C. and Hastings, J.W. (1981) The biological clock in *Gonyaulax* controls luciferase activity by regulating turnover. J. Biol. Chem. **256**: 10509–10518

Dunlap, J., Taylor, W. and Hastings, J.W. (1980) The effects of protein synthesis inhibitors on the *Gonyaulax* clock. I. Phase shifting effects of cycloheximide. J. Comp. Physiol. **138**: 1–8

Dunlap, J., Taylor, W. and Hastings, J.W. (1981) The control and expression of bioluminescence in dinoflagellates. In: Bioluminescence: Current Perspectives. K.H. Nealson, ed. Burgess Publishing Company, Minneapolis, pp. 108–124

Dunlap, J., McClung, C.R. and Fox, B.A. (1987a) Molecular characteristics of the *frequency* and *formate* loci in *N. crassa*. (In preparation)

Dunlap, J., Loris, J.J. and Denome, S. (1987b) Molecular cloning of genes under control of the biological clock in *N. crassa*. (In preparation)

Earnest, D.J. and Sladek, C.D. (1985) Circadian rhythms of vasopressin release from perifused rat suprachiasmatic explants in vitro. Soc. Neurosci. Abst. **11**: 385

Earnest, D.J. and Sladek, C.D. (1986) Mechanisms underlying the acute and circadian release of vasopressin from perifused rat suprachiasmatic explants *in vitro*. Soc. Neurosci. Abst. **12**: 845 (No. 233.11)

Edmunds, L.N. Jr. (1964) Replication of DNA and cell division in synchronously dividing cultures of *Euglena gracilis*. Science **145**: 266–268

Edmunds, L.N. Jr. (1965a) Studies on synchronously dividing cultures of *Euglena gracilis* Klebs (strain Z.). I. Attainment and characterization of rhythmic cell division. J. Cell. Comp. Physiol. **66**: 147–158

Edmunds, L.N. Jr. (1965b) Studies on synchronously dividing cultures of *Euglena gracilis* Klebs (strain Z). II. Patterns of biosynthesis during the cell cycle. J. Cell. Comp. Physiol. **66**: 159–182

Edmunds, L.N. Jr. (1966) Studies on synchronously dividing cultures of *Euglena gracilis* Klebs (strain Z). III. Circadian components of cell division. J. Cell. Physiol. **67**: 35–44

Edmunds, L.N. Jr. (1971) Persisting circadian rhythm of cell division in *Euglena*: Some theoretical considerations and the problem of intercellular communication. In: Biochronometry. M. Menaker, ed. National Academy of Science, Washington, DC, pp. 594–611

Edmunds, L.N. Jr. (1974) Phasing effects of light on cell division in exponentially increasing cultures of *Tetrahymena* grown at low temperatures. Exp. Cell Res. **83**: 367–379

Edmunds, L.N. Jr. (1975) Temporal differentiation in *Euglena*: Circadian phenomena in non-dividing populations and in synchronously dividing cells. In: Les Cycles Cellulaires et leur Blocage chez Plusieurs Protistes. M. Lefort-Tran and R. Valencia, eds. Colloques Int. C.N.R.S., No. 240, Centre National de la Recherche Scientifique, Paris, pp. 53–67

Edmunds, L.N. Jr. (1976) Models and mechanisms for circadian timekeeping. In: An Introduction to Biological Rhythms. J.D. Palmer, F.A. Brown Jr. and L.N. Edmunds Jr. Academic Press, New York and London, pp. 280–361

Edmunds, L.N. Jr. (1978) Clocked cell cycle clocks: Implications toward chronopharmacology and Aging. In: Aging and Biological Rhythms. H.V. Samis Jr. and S. Capobianco, eds. Plenum, New York, pp. 125–184

Edmunds, L.N. Jr. (1980a) Blue-light photoreception in the inhibition and synchronization of growth and transport in the yeast *Saccharomycs*. In: The Blue Light Syndrome. H. Senger, H., ed.Springer Verlag, Berlin, pp. 584–596

Edmunds, L.N. Jr. (1980b) Long-term chloroplast culture and the genesis of circadian photosynthetic rhythms in *Euglena*: Problem and prospects. In: Endocytobiology: Endosymbiosis and Cell Biology, A Synthesis of Recent Research. W. Schwemmler and H.E.A. Schenk, eds. Walter de Gruyter, Berlin, pp. 685–702

Edmunds, L.N. Jr. (1980c) Current concepts of membrane control of circadian rhythms. In: Chronobiology: Principles and Applications to Shifts in Schedules. Proc. NATO Advanced Study Institute, Hannover, 1979. L.E. Scheving and F. Halberg, eds. Sijthoff and Noordhoff, Alphen aan den Rijn, The Netherlands, pp. 205–228

Edmunds, L.N. Jr. (1981) Clocked cell cycle clocks: Ultradian, circadian, and infradian interfaces. In: International Cell Biology 1980–1981. Proc. 2nd Int. Cong. Cell Biol., Berlin, 1980. H.G. Schweiger, ed. Springer-Verlag, Berlin and Heidelberg, pp. 831–845

Edmunds, L.N. Jr. (1982) Circadian and infradian rhythms. In: The Biology of *Euglena*. D.E. Buetow, ed. Academic Press, New York, Vol. III, pp. 53–142

Edmunds, L.N. Jr. (1983) Chronobiology at the cellular and molecular levels: models and mechanisms for circadian timekeeping. Am. J. Anat. **168**: 389–431

Edmunds, L.N. Jr., ed. (1984a) Cell Cycle Clocks. Marcel Dekker, New York

Edmunds, L.N. Jr. (1984b) Circadian oscillators and cell cycle controls in algae. In: The Microbial Cell Cycle. P. Nurse and E. Streiblova, eds. CRC Press, Boca Raton, FL, pp. 209–230

Edmunds, L.N. Jr. (1984c) Physiology of circadian rhythms in microorganisms. In: Advances in Microbial Physiology. A.H. Rose and D.W. Tempest, eds. Academic Press, London, Vol. 25, pp. 61–148

Edmunds, L.N. Jr. (1987) Interaction of circadian oscillators and the cell developmental cycle in *Euglena*. In: Algal Development. Molecular and Cellular Aspects. W. Wiessner, D.G. Robinson, and R.C. Starr, eds. Springer-Verlag, Berlin and Heidelberg, pp. 1–8

Edmunds, L.N. Jr. and Adams, K.J. (1981) Clocked cell cycle clocks. Science **211**: 1002–1013

Edmunds, L.N. Jr. and Cirillo, V.P. (1974) On the interplay among cell cycle, biological clock and membrane transport control systems. Int. J. Chronobiol. **2**: 233–246

Edmunds, L.N. Jr. and Funch, R.R. (1969a) Circadian rhythm of cell division in *Euglena*: Effects of a random illumination regimen. Science **165**: 500–503

Edmunds, L.N. Jr. and Funch, R.R. (1969b) Effects of 'skeleton' photoperiods and high frequency light-dark cycles on the rhythm of cell division in synchronized cultures of *Euglena*. Planta **87**: 134–163

Edmunds, L.N. Jr. and Halberg, F. (1981) Circadian time structure of *Euglena*: A model system amenable to quantification. In: Neoplasms—Comparative Pathology of Growth in Animals, Plants and Man. H.E. Kaiser, ed. Williams & Wilkins, Baltimore, pp. 105–134

Edmunds, L.N. Jr. and Laval-Martin, D.L. (1981) 'Free-running' circadian rhythms of photosynthesis elicited by short-period cycles of light and darkness in synchronously dividing and nondividing *Euglena*. In: Photosynthesis. G. Akoyunoglou, ed. Proc. 5th Intl. Cong. on Photosynthesis, Halkidiki, Greece, 1980. Balaban International Science Services, Philadelphia, Vol. VI, pp. 313–322

Edmunds, L.N. Jr. and Laval-Martin, D.L. (1984) Cell division cycles and circadian oscillators. In: Cell Cycle Clocks. L.N. Edmunds Jr., ed. Marcel Dekker, New York, pp. 295–324

Edmunds, L.N. Jr. and Laval-Martin, D.L. (1986) Chronopharmacological analysis of an autonomously oscillatory circadian clock in the algal flagellate *Euglena*. In: Annual Review of Chronopharmacology. A. Reinberg, M. Smolensky and G. Labrecque, eds. Pergamon Press, Oxford, Vo. 3, pp. 165–168

Edmunds, L.N. Jr., Chuang, L., Jarrett, R.M. and Terry, O.W. (1971) Long- term persistence of free-running rhythms of cell division in *Euglena* and the implication of autosynchrony. J. Interdisc. Cycle Res. **2**: 121–132

Edmunds, L.N. Jr., Sulzman, F.M. and Walther, W.G. (1974) Circadian oscillations in enzyme activity in *Euglena* and their relation to the circadian rhythm of cell division. In: Chronobiology. L.E. Scheving, F. Halberg and J.E. Pauly, eds. Igaku Shoin, Tokyo, pp. 61–66

Edmunds, L.N. Jr., Jay, M.E., Kohlmann, A., Liu, S.C., Merriam, V.H. and Sternberg, H. (1976) The coupling effects of some thiol and other sulfur-containing compounds on the circadian rhythm of cell division in photosynthetic mutants of *Euglena*. Arch. Microbiol. **108**: 1–8

Edmunds, L.N. Jr., Apter, R.I., Rosenthal, P.J., Shen, W.-K. and Woodward, J.R. (1979) Light effects in yeast: Persisting oscillations in cell division activity and amino acid transport in cultures of *Saccharomyces* entrained by light-dark cycles. Photochem. Photobiol. **30**: 595–601

Edmunds, L.N. Jr., Tay, D.E. and Laval-Martin, D.L. (1982) Cell division cycles and circadian clocks: Phase-response curves for light perturbations in synchronous cultures of *Euglena*. Plant Physiol. **70**: 297–302

Edmunds, L.N. Jr., Laval-Martin, D.L. and Goto, K. (1987) Cell division cycles and circadian clocks: Modeling a metabolic oscillator in the algal flagellate *Euglena*. In: Endocytobiology III. J.J. Lee and J.F. Fredrick, eds. Ann. N.Y. Acad. Sci. **503**: 459–475

Ehret, C.F. (1951) The effects of visible, ultraviolet, and x-irradiation on the mating reaction in *Paramecium bursaria*. Anat. Rec. **111**: 112

Ehret, C.F. (1959) Photobiology and biochemistry of circadian rhythms in nonphotosynthesizing cells. Fed. Proc. **18**: 1232–1240

Ehret, C.F. (1960) Action spectra and nucleic acid metabolism in circadian rhythms at the cellular level. Cold Spring Harbor Symp. Quant. Biol. **25**: 149–158

Ehret, C.F. (1974) The sense of time: Evidence for its molecular basis in the eukaryotic gene-action system. Adv. Biol. Med. Phys. **15**: 47–77

Ehret, C.F. and Dobra, K.W. (1977) The infradian eukaryotic cell: A circadian energy-reserve escapement. In: Proceedings of the XII International Conference of the International Society for Chronobiology, 1975, Washington, DC Publishing House Il Ponte, Milano, pp. 563–570

Ehret, C.F. and Potter, V.R. (1974) Circadian chronotypic induction of tyrosine aminotransferase and depletion of glycogen by theophylline in the rat. Int. J. Chronobiol. **2**, 321–326

Ehret, C.F. and Trucco, E. (1967) Molecular models for the circadian clock. I. The chronon concept. J. Theor. Biol. **15**: 240–262

Ehret, C.F. and Wille, J.J. (1970) The photobiology of circadian rhythms in protozoa and other eukaryotic microorganisms. In: Photobiology of Microorganisms. P. Halldal, ed. Wiley (Interscience), New York, pp. 369–416

Ehret, C.F., Barnes, J.H. and Zichal, K.E. (1974) Circadian parameters of the infradian growth mode in continuous cultures: Nucleic acid synthesis and oxygen induction of the ultradian mode. In: Chronobiology. L.E. Scheving, F. Halberg and J.E. Pauly, eds. Igaku Shoin, Tokyo, pp. 44–50

Ehret, C.F., Potter, V.R. and Dobra, K.W. (1975) Chronotypic action of theophylline and of pentobarbital as circadian Zeitgebers in the rat. Science **188**: 1212–1215

Ehret, C.F., Meinert, J.C., Groh, K.R. and Antipa, G.A. (1977) Circadian regulation: Growth kinetics of the infradian cell. In: Growth Kinetics and Biochemical Regulation of Normal and Malignant Cells. B. Drewinko and R.M. Humphrey, eds. Williams & Wilkins, Baltimore, pp. 49–76

Ehrhardt, V., Krug, H.F. and Hardeland, R. (1980) On the role of gene expression in the circadian oscillator mechanism. J. Interdisc. Cycle Res. **11**: 257–276

Ellis, G.B., McKlveen, R.E. and Turek, F.W. (1982) Dark pulses affect the circadian rhythms of activity in hamsters kept in constant light. Am. J. Physiol. **242**: R44–R50

Engelberg, J. (1964) The decay of synchronization of cell division. Exp. Cell Res. **36**: 647–662

Engelberg, J. (1968) On the deterministic origins of mitotic variability. J. Theor. Biol. **20**: 249–259

Engelmann, W. and Mack, J. (1978) Different oscillators control the circadian rhythm of eclosion and activity in Drosophila. J. Comp. Physiol. **127**: 229–237

Engelmann, W. and Schrempf, M. (1980) Membrane models for circadian rhythms. Photochem. Photobiol. Rev. **5**: 49–86

Engelmann, W., Johnsson, A., Karlsson, H.G., Kobler, R. and Schimmel, M.-L. (1978) Attenuation of the petal movement rhythm in *Kalenchoë* with light pulses. Physiol. Plant. **43**: 68–76

Enright, J.T. (1971) Heavy water slows biological timing processes. Z. Vergl. Physiol. **72**: 1–16

Enright, J.T. (1980) Temporal precision in circadian systems: A reliable neuronal clock from unreliable components. Science **209**: 1542–1545

Eon-Gerhardt, R., Tollon, Y., Chraibi, R. and Wright, M. (1981) Regulation of thymidine kinase synthesis during the cell cycle of *Physarum polycephalum*: The effects of two microtubule inhibitors. Cytobios **32**: 47–62

Epel, B.L. (1973) Inhibition of growth and respiration by visible and near visible light. In: Photophysiology. A.C. Giese, ed. Academic Press, New York, Vol VIII, pp. 209–229

Eppley, R.W., Holm-Hansen, O. and Strickland, J.D.H. (1968) Some observations on the vertical migration of Dinoflagellates. J. Phycol. **4**: 333–340

Eskes, G.A. and Rusak, B. (1985) Horizontal knife cuts in the suprachiasmatic area prevent hamster gonadal responses to photoperiod. Neurosci. Lett. **61**: 261–266

Eskes, G.A., Wilkinson, M., Moger, W.H. and Rusak, B. (1984) Periventricular and suprachiasmatic lesion effects on photoperiodic responses of the hamster hypophyseal–gonadal axis. Biol. Reprod. **30**: 1073–1081

Eskin, A. (1972) Phase shifting a circadian rhythm in the eye of *Aplysia* by high potassium pulses. J. Comp. Physiol. **80**: 353–376

Eskin, A. (1977) Neurophysiological mechanisms involved in photoentrainment of the circadian rhythm from the *Aplysia* eye. J. Neurobiol. **8**: 273–299

Eskin, A. (1979a) Identification and physiology of circadian pacemakers. Fed. Proc. **38**: 2570–2572

Eskin, A. (1979b) Circadian system of the *Aplysia* eye: Properties of the pacemaker and mechanisms of its entrainment. Fed. Proc. **38**: 2573–2579

Eskin, A. (1982a) Differential effects of amino acids on the period of the circadian rhythm from the *Aplysia* eye. J. Neurobiol. **13**: 233–239

Eskin, A. (1982b) Increasing external K^+ blocks phase shift in a circadian rhythm produced by serotonin or 8-benzylthio-cAMP. J. Neurobiol. **13**: 241–249

Eskin, A. and Corrent, G. (1977) Effects of divalent cations and metabolic poisons on the circadian rhythm of the *Aplysia* eye. J. Comp. Physiol. **117**: 1–21

Eskin, A. and Takahashi, J.S. (1983) Adenylate cyclase activation shifts the phase of a circadian pacemaker. Science **220**: 82–84

Eskin, A., Corrent, G., Lin, C.-Y. and McAdoo, D.J. (1982) Mechanism for shifting the phase of a circadian rhythm by serotonin: Involvement of cAMP. Proc. Natl. Acad. Sci. USA **79**: 660–664

Eskin, A., Takahashi, J.S., Zatz, M. and Block, G.D. (1984a) Cyclic guanosine 3′,5′-monophosphate mimics the effects of light on a circadian pacemaker in the eye of *Aplysia*. J. Neurosci. **4**: 2466–2471

Eskin, A., Yeung, S.J. and Klass, M.R. (1984b) Requirement for protein synthesis in the regulation of a circadian rhythm by melatonin. Proc. Natl. Acad. Sci. USA **81**: 7637–7641

Esser, K. and Keller, W. (1976) Genes inhibiting senescence in the ascomycete *Podospora anserina*. Mol. Gen. Genet. **144**: 107–110

Estabrook, R.W. and Srere, P., eds. (1981) Biological Cycles. Current Topics in Cellular Regulation. Academic Press, New York and London, Vol. 18

Evans, L.T. and Allaway, W.G. (1972) Action spectrum for the opening of *Albizzia julibrissin* pinnules, and the role of phytochrome in the closing movements of pinnules and of stomata of *Vicia faba*. Aust. J. Biol. Sci. **25**: 885–893

Fantes, P.A. (1984a) Temporal control of the *Schizosaccharomyces pombe* cell cycle. In: Cell Cycle Clocks. L.N. Edmunds Jr., ed. Marcel Dekker, New York, pp. 233–252

Fantes, P.A. (1984b) Cell cycle control in *Schizosaccharomyces pombe*. In: The Microbial Cell Cycle. P. Nurse and E. Streiblová, eds. CRC Press, Boca Raton, FL pp. 109–125

Fantes, P.A. and Nurse, P. (1977) Control of cell size at division in fission yeast by a growth-modulated size control over nuclear division. Exp. Cell Res. **107**: 377–386

Fantes, P.A., Grant, W.D., Pritchard, R.H., Sudbery, P.E. and Wheals, A.E. (1975) The regulation of cell size and the control of mitosis. J. Theor. Biol. **50**: 213–244

Feldman, J.F. (1967) Lengthening the period of a biological clock in *Euglena* by cycloheximide, an inhibitor of protein synthesis. Proc. Natl. Acad. Sci. USA **57**: 1080–1087

Feldman, J.F. (1968) Circadian rhythmicity in amino acid incorporation in *Euglena gracilis*. Science **160**: 1454–1456

Feldman, J.F. (1975) Circadian periodicity in *Neurospora*: Alteration by inhibitors of cyclic AMP phosphodiesterase. Science **190**: 789–790

Feldman, J.F. (1982) Genetic approaches to circadian clocks. Annu. Rev. Plant Physiol. **33**: 583–608

Feldman, J.F. (1983) Genetics of circadian clocks. BioScience **33**: 426–431

Feldman, J.F. and Atkinson, C.A. (1978) Genetic and physiological characteristics of a slow-growing circadian clock mutant of *Neurospora crassa*. Genetics **88**: 255–265

Feldman, J.F. and Bruce, V.G. (1972) Circadian rhythm changes in autotrophic *Euglena* induced by organic carbon sources. J. Protozool. **19**: 370–372

Feldman, J.F. and Dunlap, J.C. (1983) *Neurospora crassa*: A unique system for studying circadian rhythms. Photochem. Photobiol. Rev. **7**: 319–368

Feldman, J.F. and Hoyle, M.N. (1973) Isolation of circadian clock mutants of *Neurospora crassa*. Genetics **75**: 605–613

Feldman, J.F. and Hoyle, M.N. (1974) A direct comparison between circadian and non-circadian rhythms in *Neurospora crassa*. Plant Physiol. **53**: 928–930

Feldman, J.F. and Hoyle, M.N. (1976) Complementation analysis of linked circadian clock mutants in *Neurospora*. Genetics **82**: 9–17

Feldman, J.F. and Widelitz, R. (1977) Manipulation of circadian periodicity in cysteine auxotrophs of *Neurospora crassa*. (Abstr.). Annu. Meet. Am. Soc. Microbiol., p. 125

Feldman, J.F., Gardner, G.F. and Denison, R.A. (1979) Genetic analysis of the circadian clock of *Neurospora*. In: Biological Rhythms and Their Central Mechanism. M. Suda, O. Hayaishi and H. Nakagawa, eds. Elsevier/North-Holland Biomedical Press, Amsterdam, pp. 56–66

Feldman, J.F., Pierce, S.B. and Brown, D. (1986) Temperature compensation in circadian clock mutants of *Neurospora*. Plant Physiol. **80**[Suppl.]: 92 (Abstr. 487)

Finch, C.E. and Hayflick, L., eds. (1977) Handbook on the Biology of Aging. Van Nostrand-Reinhold, New York

Fitt, W.K., Chang, S.S. and Trench, R.K. (1981) Motility patterns of different strains of the symbiotic dinoflagellate *Symbiodinium* (= *Gymnodinium*) *microadriaticum* (Freudenthal) in culture. Bull. Mar. Sci. **31**: 436–443

Follett, B.K. and Follett, D.E., eds. (1981) Biological Clocks in Seasonal and Reproductive Cycles. Proc. 32nd Symp. Colston Res. Soc., Bristol, 1980. John Wiley & Sons (Halsted Press), New York

Foward, R.B. Jr. and Davenport, D. (1970) Circadian rhythm of a behavioural photoresponse in the dinoflagellate *Gyrodinium dorsum*. Planta **92**: 259–266

Fowler, V. and Branton, D. (1977) Lateral mobility of human erythrocyte intregral membrane proteins. Nature **268**: 23–26

Fox, T.O. and Pardee, A.B. (1970) Animal cells: Noncorrelation of length of G_1 phase with size after mitosis. Science **167**: 80–82

Francis, C.D. and Sargent, M.L. (1979) Effects of temperature perturbations on circadian conidiation in *Neurospora*. Plant Physiol. **64**: 1000–1004

Frank, K.D. and Zimmerman, W.F. (1969) Action spectra for phase shifts of a circadian rhythm in *Drosophila*. Science **163**: 688–689

Frisch, L., ed. (1960) Biological Clocks. Cold Spring Harbor Symposia on Quantitative Biology, Cold Spring Harbor Laboratory, NY, Vol. 25

Gardner, G.F. and Feldman, J.F. (1980) The *frq* locus in *Neurospora crassa*: A key element in circadian clock organization. Genetics **96**: 877–886

Gardner, G.F. and Feldman, J.F. (1981) Temperature compensation of circadian periodicity in clock mutants of *Neurospora crassa*. Plant Physiol. **68**: 1244–1288

Gardner, G.F. and Galston, A.W. (1977) Fatty acids and circadian rhythms in *Phaseolus* (Abstr. 168). Plant Physiol. **59**[Suppl.]: 30

Gaston, S. and Menaker, M. (1968) Pineal function: The biological clock in the sparrow? Science **160**: 1125–1127

Gerisch, G. (1971) Periodische Signale steuern die Musterbildung in Zellverbänden. Naturwissenschaften **58**: 430–438

Gerisch, G. (1976) Cyclic-AMP oscillation and signal transmission in aggregating *Dictyostelium* cells. In: The Molecular Basis of Circadian Rhythms. J.W.Hastings and H.-G. Schweiger, eds. Dahlem Konferenzen, Berlin, pp. 433–440

Gerisch, G. (1986) *Dictyostelium discoideum*: A eukaryotic microorganism that develops by cell aggregation from a unicellular to a multicellular stage. In: Cellular and Molecular Aspects of Developmental Biology. M. Fougereau and R. Stora, eds. Elsevier Science Publ., Amsterdam, pp. 47–66

Gilbert, D.A. (1974a) The nature of the cell cycle and the control of cell proliferation. Biosystems **5**: 197–206

Gilbert, D.A. (1974b) The temporal response of the dynamic cell to disturbances and its possible relationship to differentiation and cancer. S. Afr. J. Sci. **70**: 234–244

Gilbert, D.A. (1977) Density dependent limitation of growth and the regulation of cell replication by changes in the triggering level of the cell cycle switch. Biosystems **9**: 215–228

Gilbert, D.A. (1978a) The malignant transformation: The nature of its effects on cell replication characteristics. S. Afr. J. Sci. **74**: 48–49

Gilbert, D.A. (1978b) The relationship between the transition probability and oscillator concepts of the cell cycle and the nature of the commitment to replication. Biosystems **10**: 235–240

Gilbert, D.A. (1978c) Feedback quenching as a means of effectively increasing the period of biochemical and biological oscillations. Biosystems **10**: 241–245

Gilbert, D.A. (1980) Mathematics and cancer. In: Mathematical Modelling in Biology and Ecology, Lecture Notes in Biomathematics. W. Getz, ed. Springer-Verlag, Berlin, Vol. 33, pp. 97–115

Gilbert, D.A. (1981) The cell cycle: One or more limit cycle oscillations? S. Afr. J. Sci. **77**: 541–546

Gilbert, D.A. (1982a) Cell cycle variability: The oscillator model of the cell cycle yields transition probability alpha and beta type curves. Biosystems **15**: 317–330

Gilbert, D.A. (1982b) An oscillator cell cycle model needs no first or second chance event. Biosystems **15**: 331–339

Gillette, M.U. (1986) The suprachiasmatic nuclei: Circadian phase-shifts induced at the time of hypothalamic slice preparation are preserved *in vitro*. Brain Res. **379**: 176–181

Gillette, M.U., Reiman, A.M. and Lipeski, L.E. (1986) Circadian protein and phosphoprotein changes in the suprachiasmatic nuclei: The difference between night and day. Soc. Neurosci. Abst. **12**: 845 (No. 233.12)

Goldbeter, A. (1975) Mechanism for oscillatory synthesis of cyclic AMP in *Dictyostelium discoideum*. Nature **253**: 540–542

Goldbeter, A. and Caplan, S.R. (1976) Oscillatory enzymes. Annu. Rev. Biophys. Bioeng. **5**: 449–476

Goldbeter, A. and Nicolis, G. (1976) An allosteric enzyme model with positive feedback applied to glycolytic oscillations. In: Progress in Theoretical Biology. R. Rosen, ed. Academic Press, New York, Vol. 4, pp. 65–160

Goldman, B.D. and Darrow, J.M. (1983) The pineal gland and mammalian photoperiodism. Neuroendocrinology **37**: 386–396

Goodenough, J.E. and Bruce, V.G. (1980) The effects of caffeine and theophylline on the phototactic rhythm of *Chlamydomonas reinhardii*. Biol. Bull. **159**: 649–655

Goodwin, B.C. (1963) Temporal Organization in Cells. Academic Press, London

Goodwin, B.C. (1965) Oscillatory behaviour in enzymatic control processes. Adv. Enzyme Regul. **3**: 425–438

Goodwin, B.C. (1966) An entrainment model for timed enzyme syntheses in bacteria. Nature **209**: 479–481

Goodwin, B.C. (1976) Analytical Physiology of Cells and Developing Organisms. Academic Press, London

Goodwin, B.C. and Cohen, M.H. (1969) A phase-shift model for the spatial and temporal organization of developing systems. J. Theor. Biol. **25**: 49–108

Goodwin, B.C. and Pateromichelakis, S. (1979) The role of electrical fields, ions and the cortex in the morphogenesis of *Acetabularia*. Planta **145**: 427–436

Gorman, J., Tauro, P., LaBerge, M. and Halvorson, H.O. (1964) Timing of enzyme synthesis during synchronous division in yeast. Biochem. Biophys. Res. Commun. **15**: 43–49

Gorton, H.L. and Satter, R.L. (1983) Circadian rhythmicity in leaf pulvini. Bioscience **33**: 451–457

Goshima, K. (1979) Rhythmic and arrhythmic contraction of cultured heart muscle cells. In: Biological Rhythms and Their Central Mechanism. M. Suda, O. Hayaishi and H. Nakagawa, eds. Elsevier/North-Holland Biomedical Press, Amsterdam, pp. 67–76

Goto, K. (1984) Causal relationships among metabolic circadian rhythms in *Lemna*. Z. Naturforsch. **39c**: 73–84

Goto, K., Laval-Martin, D. and Edmunds, L.N. Jr. (1985) Biochemical modeling of an autonomously oscillatory circadian clock in *Euglena*. Science **228**: 1284–1288

Govindjee, Wong, D., Prézelin, B.B. and Sweeney, B.M. (1979) Chlorophyll *a* fluorescence of *Gonyaulax polyedra* grown on a light-dark cycle and after transfer to constant light. Photochem. Photobiol. **30**: 405–411

Green, D.J. and Gillette, R. (1982) Circadian rhythm of firing rate recorded from single cells in the rat suprachiasmatic brain slice. Brain Res. **245**: 198–200

Grobbelaar, N., Huang, T.C., Lin, H.Y. and Chow, T.J. (1986) Dinitrogen-fixing endogenous rhythm in *Synechococcus* RF-1. FEMS Microbiol. Lett. **37**: 173–178

Groos, G. and Hendriks, J. (1982) Circadian rhythms in electrical discharge of rat suprachiasmatic neurones recorded *in vitro*. Neurosci. Lett. **34**: 283–288

Grover, N.B., Woldringh, C.L., Zaritsky, A. and Rosenberger, R.F. (1977) Elongation of rod-shaped bacteria. J. Theor. Biol. **67**: 181–193

Guiguet, M., Kupiec, J.-J. and Valleron, A,-J. (1984) A systematic study of the variability of cell cycle phase durations in experimental mammalian systems. In: Cell Cycle Clocks. L.N. Edmunds Jr., ed. Marcel Dekker, New York, pp. 97–112

Gurel, O. and Gurel, D. (1983) Types of oscillations in chemical reactions. Top. Curr. Chem. **118**: 1–74

Gwinner, E. (1978) Effects of pinealectomy on circadian locomotor activity rhythms in European starlings, *Sturnus vulgaris*. J. Comp. Physiol. **126**: 123–129

Gwinner, E. and Benzinger, I. (1978) Synchronization of a circadian rhythm in pinealectomized European starlings by daily injections of melatonin. J. Comp. Physiol. **127**: 209–214

Haga, N. and Karino, S. (1986) Microinjection of immaturin rejuvenates sexual activity of old *Paramecium*. J. Cell Sci. **86**: 263–271

Halaban, R. (1969) Effects of light quality on the circadian rhythm of leaf movement of a short-day plant. Plant Physiol. **44**: 973–977

Halberg, F. (1960) The 24-hour scale: A time dimension of adaptive functional organization. Perspect. Biol. Med. **3**: 491–527

Halberg, F. (1975) When to treat. Ind. J. Cancer **12**: 1–20

Hall, J.C. (1986) Learning and rhythms in courting, mutant *Drosophila*. Trends Neurosci. **9**: 414–418

Hall, J.C. and Rosbash, M. (1987) Genes and biological rhythms. Trends Genet. **3**: 185–191

Halvorson, H.O. (1977) A review of current models on temporal gene expression in *Saccharomyces cerevisiae*. In: Cell Differentiation in Microorganisms, Plants and Animals. L. Nover and K. Mothes, eds. VEB Gustav Fischer Verlag, Jena, pp. 361–376

Halvorson, H.O., Carter, B.L.A. and Tauro, P. (1971) Synthesis of enzymes during the cell cycle. Adv. Microbial Physiol. **6**: 47–106

Hamblen, M., Zehring, W.A., Kyriacou, C.P., Reddy, P., Yu, Q., Wheeler, D.A., Zwiebel, L.J., Konopka, R.J., Robash, M. and Hall, J.C. (1986) Germ-line transformation involving DNA from the *period* locus in *Drosophila melanogaster*: Overlapping genomic fragments that restore circadian and ultradian rhythmicity to per^o and per^- mutants. J. Neurogenet. **3**: 249–291

Handler, A.M. and Konopka, R.J. (1979) Transplantation of a circadian pacemaker in *Drosophila*. Nature **279**: 236–238

Hanes, S.D., Koren, R. and Bostian, K.A. (1986) Control of cell growth and division in *Saccharomyces cerevisiae*. CRC Crit. Rev. Biochem. **21**: 153–223

Hanson, F.E. (1978) Comparative studies of firefly pacemakers. Fed. Proc. **37**: 2158–2164

Hanson, F.E. (1982) Pacemaker control of rhythmic flashing of fireflies. In: Cellular Pacemakers. Function in Normal and Disease States. D.O. Carpenter, ed. John Wiley & Sons, New York, Vol. 2, pp. 81–100

Hara, K., Tydeman, P. and Kirschner, M. (1980) A cytoplasmic clock with the same period as the division cycle in *Xenopus* eggs. Proc. Natl. Acad. Sci. USA **77**: 462–466

Hardeland, R. (1972) Circadian rhythmicity of tyrosine aminotransferase in suspensions of isolated rat liver cells. J. Interdisc. Cycle Res. **3**: 109–114

Hardeland, R. (1973a) Circadian rhythmicity in cultured liver cells. I. Rhythms in tyrosine aminotransferase activity and inducibility and in [^{3}H]leucine incorporation. Int. J. Biochem. **4**: 581–590

Hardeland, R. (1973b) Circadian rhythmicity in cultured liver cells. II. Reinduction of rhythmicity in tyrosine aminotransferase activity. Int. J. Biochem. **4**: 591–595

Hardeland, R. (1982) Circadian rhythms of bioluminescence in two species of *Pyrocystis* (Dinophyta): Measurements in cell populations and in single cells. J. Interdisc. Cycle Res. **13**: 49–54

Hardeland, R. and Nord, P. (1984) Visualization of free-running circadian rhythms in the dinoflagellate *Pyrocystis noctiluca*. Mar. Behav. Physiol. **11**: 199–207

Hardeland, R., Harnau, G., Rüsenberg and Balzer, I. (1986) Multiplicity or uniformity of cellular temperature compensation mechanisms? J. Interdisc. Cycle Res. **17**: 121–123

Harding, L.W. Jr., Meeson, B.W., Prézelin, B.B. and Sweeney, B.M. (1981) Diel periodicity of photosynthesis in marine phytoplankton. Mar. Biol. **61**: 95–105

Harris, P.J.C. and Wilkins, M.B. (1978) Evidence of phytochrome involvement in the entrainment of the circadian rhyhthm of carbon dioxide metabolism in *Bryophyllum*. Planta **138**: 271–278

Hartman, H., Ashkenazi, I. and Epel, B.L. (1976) Circadian changes in membrane properties of human red blood cells in vitro, as measured by a membrane probe. FEBS Lett. **67**: 161–163

Hartwell, L.H. (1971) Genetic control of the cell division cycle in yeast. II. Genes controlling DNA replication and its initiation. J. Mol. Biol. **59**: 183–194

Hartwell, L.H. (1974) *Saccharomyces cerevisiae* cell cycle. Bacteriol. Rev. **38**: 164–198

Hartwell, L.H. (1976) Sequential function of gene products relative to DNA synthesis in the yeast cell cycle. J. Mol. Biol. **104**: 803–817

Hartwell, L.H. (1978) Cell division from a genetic perspective. J. Cell Biol. **77**: 627–637

Hartwell, L.H., Culotti, J. and Reid, B. (1974) Genetic control of the cell division cycle in yeast. Science **183**: 46–51

Hartwig, R., Schweiger, M., Schweiger, R. and Schweiger, H.-G. (1985) Identification of a high molecular weight polypeptide that may be part of the circadian clockwork in *Acetabularia*. Proc. Natl. Acad. Sci. USA **82**: 6899–6902

Hartwig, R., Schweiger, R. and Schweiger, H.-G. (1986) Circadian rhythm of the synthesis of a high molecular weight protein in anucleate cells of the green alga *Acetabularia*. Eur. J. Cell Biol. **41**: 139–141

Hasegawa, K. and Tanakadate, A. (1984) Circadian rhythm of locomotor behavior in a population of *Paramecium multimicronucleatum*: Its characteristics as derived from circadian changes in the swimming speeds and the frequencies of avoiding response among individual cells. Photochem. Photobiol. **40**: 105–112

Hasegawa, K., Katakura, T. and Tanakadate, A. (1984) Circadian rhythm in the locomotor behavior in a population of *Paramecium multimicronucleatum*. J. Interdisc. Cycle Res. **15**: 45–56

Hastings, J.W. (1959) Unicellular clocks. Annu. Rev. Microbiol. **13**: 297–312

Hastings, J.W. (1960) Biochemical aspects of rhythms: Phase-shifting by chemicals. Cold Spring Harbor Symp. Quant. Biol. **25**: 131–143

Hastings, J.W. and Bode, V.C. (1962) Biochemistry of rhythmic systems. Ann. NY Acad. Sci. **98**: 876–889

Hastings, J.W. and Krasnow, R. (1981) Temporal regulation in the individual *Gonyaulax* cell. In: International Cell Biology 1980–1981. Proc. 2nd Int. Cong. Cell Biol., Berlin, 1980. H.-G. Schweiger, ed. Springer-Verlag, Berlin, pp. 815–822

Hastings, J.W. and Schweiger, H.-G., eds. (1976) The Molecular Basis of Circadian Rhythms. Dahlem Konferenzen, Berlin

Hastings, J.W. and Sweeney, B.M. (1958) A persistent diurnal rhythm of luminescence in *Gonyaulax polyedra*. Biol. Bull. **115**: 440–458

Hastings, J.W. and Sweeney, B.M. (1959) The *Gonyaulax* clock. In: Photoperiodism and Related Phenomena in Plants and Animals. R.B. Withrow, ed. American Association for the Advancement of Science, Washington, DC, pp. 567–584

Hastings, J.W. and Sweeney, B.M. (1960) The action spectrum for shifting the phase of the rhythm of luminescence in *Gonyaulax polyedra*. J. Gen. Physiol. **43**: 697–706

Hastings, J.W. and Sweeney, B.M. (1964) Phased cell division in marine dinoflagellates. In: Synchrony in Cell Division and Growth. E. Zeuthen, ed. Wiley-Interscience, New York, pp. 307–321

Hastings, J.W., Astrachan, L. and Sweeney, B.M. (1961) A persistent daily rhythm in photosynthesis. J. Gen. Physiol. **45**: 69–76

Hastings, J.W., Broda, H. and Johnson, C.H. (1985) Phase and period effects of physical and chemical factors. Do cells communicate? In: Temporal Order. L. Rensing and N.I. Jaeger, eds. Springer-Verlag, Berlin and Heidelberg, pp. 213–221

Hastings, J.W., Johnson, C. and Kondo, T. (1987) Action spectrum for phase shifting of the circadian rhythm of phototaxis in *Chlamydomonas*. 15th Annual Meeting, American Society for Photobiology, 21-25 June 1987, Bal Harbour, FL (Abstract #WPM-C46)

Hasunuma, K. (1984a) Rhythmic conidiation and light sensitivity of mutants in orthophosphate repressible cyclic phosphodiesterase in *Neurospora crassa*. Proc. Japan Acad. [Ser. B.] **60**: 260–264

Hasunuma, K. (1984b) Circadian rhythm in *Neurospora* includes oscillation of cyclic 3′,5′-AMP level. Proc. Japan Acad. [Ser. B] **60**: 377–380

Hasunuma, K. and Shinohara, Y. (1985) Characterization of *cpd-1* and *cpd-2* mutants which affect the activity of orthophosphate regulated cyclic phosphodiesterase in *Neurospora*. Curr. Genet. **10**: 197–203

Haus, E. and Halberg, F. (1966) Persisting circadian rhythm in hepatic glycogen of mice during inanition and dehydration. Experientia **22**: 113–114

Haus, E., Halberg, F., Kühl, J.F.W. and Lakatua, D.J. (1974) Chronopharmacology in animals. Chronobiologia, suppl. 1, **1**: 122–156

Hayflick, L. (1965) The limited in vitro lifetime of human diploid cell strains. Exp. Cell Res. **37**: 614–636

Hayflick, L. and Moorehead, P.S. (1961) The serial culture of human diploid cell strains. Exp. Cell Res. **25**: 585–621

Hayles, J. and Nurse, P. (1986) Cell cycle regulation in yeast. J. Cell Sci. [Suppl.] **4**: 155–170

Heath, M.R. and Spencer, C.P. (1985) A model of the cell cycle and cell division phasing in a marine diatom. J. Gen. Microbiol. **131**: 411–425

Heide, O.M. (1977) Photoperiodism in higher plants: An interaction of phytochrome and circadian rhythms. Plant Physiol. **39**: 25–32

Hellebust, J.A., Terborgh, J. and McLeod, G.C. (1967) The photosynthetic rhythm of *Acetabularia crenulata*. II. Measurements of carbon dioxide and the activities of enzymes of the reductive pentose cycle. Biol. Bull. **133**: 670–678

Helmstetter, C., Cooper, S., Pierucci, O. and Revelas, E. (1968) On the bacterial life sequence. Cold Spring Harbor Symp. Quant. Biol. **33**: 809–822

Helmstetter, C.E., Pierucci, O., Weinberger, M., Holmes, M. and Tang, M.-S. (1979) Control of cell division in *Escherichia coli*. In: The Bacteria: A Treatise on Structure and Function, Mechanisms of Adaptation. J. R. Sokatch and L.N. Ornston, eds. Academic Press, New York, Vol. 7, pp. 517–519

Herman, E. and Sweeney, B.M. (1975) Circadian rhythm of chloroplast ultrastructure in *Gonyaulax polyedra*. Concentric organization around a central cluster of ribosomes. J. Ultrastruct. Res. **50**: 347–354

Hess, B. (1976) Oscillations in biochemical systems. In: The Molecular Basis of Circadian Rhythms. J.W. Hastings and H.-G. Schweiger, eds. Dahlem Konferenzen, Berlin, pp. 175–191

Hess, B. and Boiteux, A. (1971) Oscillatory phenomena in biochemistry. Annu. Rev. Biochem. **40**: 237–258

Hesse, M. (1972) Endogene Rhythmik der Produktionsfähigkeit bei *Chlorella* und ihre Beeinflussung durch Licht. Z. Pflanzenphysiol. **67**: 58–77

Higgins, J. (1964) A chemical mechanism for oscillation of glycolytic intermediates in yeast cells. Proc. Natl. Acad. Sci. USA **51**: 989–994

Higgins, J. (1967) The theory of oscillating reactions. Ind. Eng. Chem. **59**: 18–62

Hillman, W.S. (1971) Entrainment of *Lemna* CO_2 output through phytochrome. Plant Physiol. **48**: 770–774

Hillman, W.S. and Koukkari, W.L. (1967) Phytochrome effects in the nyctinastic leaf movements of *Albizza julibrissin* and some other legumes. Plant Physiol. **42**: 1413–1418

Hiroshige, T. and Honma, K., eds. (1985) Circadian Clocks and Zeitgebers. Hokkaido University Press, Sapporo, Japan

Hobohm, U., Cornelius, G., Taylor, W. and Rensing, L. (1984) Is the circadian clock of *Gonyaulax* held stationary after a strong pulse of anisomycin? Comp. Biochem. Physiol. **79A**: 371–378

Hochberg, M.L. and Sargent, M.L. (1974) Rhythms of enzyme activity associated with circadian conidiation in *Neurospora crassa*. J. Bacteriol. **120**, 1164–1175

Hoffmann, K. (1965) Overt circadian frequencies and circadian rule. In: Circadian Clocks. J. Aschoff, ed. North-Holland Publ. Co., Amsterdam, pp. 87–94

Hoffmans, M. and Brinkmann, K. (1979) Circadian rhythms of external pH and photosynthesis in 'Chlamydomonas reinhardii'. Chronobiologia **6**: 111

Hoffmans-Hohn, M., Martin. W. and Brinkmann, K. (1984) Multiple periodicities in the circadian system of unicellular algae. Z. Naturforsch. **39c**: 791–800

Holland, I.B. (1987) Genetic analysis of the E. coli division clock. Cell **48**: 361–362

Holliday, R. and Pugh, J.E. (1975) DNA modification mechanisms and gene activity during development. Science **187**: 226–232

Homma, K. (1987) Circadian control of cell division in *Gonyaulax polyedra*. Ph.D. Thesis, Harvard University, Cambridge

Honerkamp, J., Mutschler, G. and Seitz, R. (1985) Coupling of a slow and a fast oscillator can generate bursting. Bull. Math. Biol. **47**: 1–21

Hopkins, J.T. (1965) Some light induced changes in behaviour and cytology of an estuarine mud-flat diatom. In: Light as an Ecological Factor. R. Bainbridge, G.C. Evans and O. Rockam, eds. John Wiley & Sons, New York, pp. 335–358

Horseman, N.D. and Will, C.L. (1984) Circadian regulation of RNA polymerase and nuclease sensitivity in rat liver nuclei. J. Interdisc. Cycle Res. **15**: 169–178

Howard, A. and Pelc, S.R. (1953) Synthesis of desoxyribonucleic acid in normal and irradiated cells and its relation to chromosome breakage. Heredity (London) [Suppl.] **6**: 261–273

Howell, S.H. (1974) An analysis of cell cycle controls in temperature sensitive mutants of *Chlamydomonas reinhardi*. In: Cell Cycle Controls. G. M. Padilla, I.L. Cameron and A. Zimmerman, eds. Academic Press, New York and London, pp. 235–249

Howell, S.H. and Naliboff, J.A. (1973) Conditional mutants in *Chlamydomonas reinhardtii* blocked in the vegetative cell cycle. I. An analysis of cell cycle block points. J. Cell Biol. **57**: 760–772

Hrushesky, W. (1983) The clinical applications of chronobiology to oncology. Am. J. Anat. **168**: 519–542

Hyver, C., Jerebzoff, S. and Nguyen, V.H. (1979) An attempt to establish a model for the rhythmic phenomena affecting the growth or sporulation of certai fungi. Chronobiologia **6**: 213–228

Iglesias, A. and Satter, R.L. (1983a) H^+ fluxes in excised *Samanea* motor tissue. I. Promotion by light. Plant Physiol. **72**: 564–569

Iglesias, A. and Satter, R.L. (1983b) H^+ fluxes in excised *Samanea* motor tissue. II. Rhythmic properties. Plant Physiol. **72**: 570–572

Inouye, S.T. (1984) Light responsiveness of the suprachiasmatic nucleus within the island with the retino-hypothalamic tract spared. Brain Res. **294**: 263–268

Inouye, S.T. and Kawamura, H. (1979) Persistence of circadian rhythmicity in a mammalian hyptothalamic "island" containing the suprachiasmatic nucleus. Proc. Natl. Acad. Sci. USA **76**: 5962–5966

Inouye, S.T. and Kawamura, H. (1982) Characteristics of a circadian pacemaker in the suprachiasmatic nucleus. J. Comp. Physiol. A **146**: 153–160

Jacklet, J.W. (1969) Circadian rhythm of optic nerve impulses recorded in darkness from isolated eye of *Aplysia*. Science **164**: 562–563

Jacklet, J.W. (1977) Neuronal circadian rhythm: Phase-shifting by a protein synthesis inhibitor. Science **198**: 69–71

Jacklet, J.W. (1980a) Circadian rhythm from the eye of *Aplysia*: Temperature compensation of the effects of protein synthesis inhibitors. J. Exp. Biol. **84**: 1–15
Jacklet, J.W. (1980b) Protein synthesis requirements of the *Aplysia* circadian clock. J. Exp. Biol. **85**: 33–42
Jacklet, J. (1981) Circadian timing by endogenous oscillators in the nervous system: Toward cellular mechanisms. Biol. Bull. **160**: 99–227
Jacklet, J.W. (1984) Neural organization and cellular mechanisms of circadian pacemakers. Int. Rev. Cytol. **89**: 251–294
Jacklet, J.W. (1985) Neurobiology of circadian rhythms generators.Trends Neurosci. **8**: 69–73
Jacklet, J.W., ed. (1987) Cellular and Neuronal Oscillators. Cellular Clocks, L.N. Edmunds Jr., series ed. Marcel Dekker, New York, Vol. 2, in press
Jacklet J. and Colquhoun, W. (1983) Ultrastructure of photoreceptors and circadian pacemaker neurons in the eye of a gastropod, *Bulla gouldiana*. J. Neurocytol. **12**: 373–396
Jacklet, J.W. and Geronimo, J. (1971) Circadian rhythm: Population of interacting neurons. Science **174**: 299–302
Jacklet, J.W. and Lotshaw, D. (1981) Light and high potassium cause similar phase shifts of the *Aplysia* eye circadian rhythm. J. Exp. Biol. **94**: 345–349
Jackson, F.R., Bargiello, T.A., Yun, S.-H. and Young, M.W. (1986) Product of *per* locus of *Drosophila* shares homology with proteoglycans. Nature **320**: 185–187
Jaffe, M.J. and Galston, A.W. (1967) Phytochrome control of rapid nyctinastic movements and membrane permeability in *Albizzia julibrissin*. Planta **77**: 135–141
James, A.A., Ewer, J., Reddy, P., Hakk, J.C. and Rosbash, M. (1986) Embryonic expression of the *period* clock gene in the central nervous system of *Drosophila melanogaster*. EMBO J. **5**: 2313–2320
Janakidevi, K., Dewey, V.C. and Kidder, G.W. (1966a) The biosynthesis of catecholamines in two genera of protozoa. J. Biol. Chem. **241**: 2576–2578
Janakidevi, K., Dewey, V.C. and Kidder, G.W. (1966b) Serotonin in protozoa. Arch. Biochem. Biophys. **113**: 758–759
Jarrett, R.M. and Edmunds, L.N. Jr. (1970) Persisting circadian rhythm of cell division in a photosynthetic mutant of *Euglena*. Science **167**: 1730–1733
Jeffery, W.R. and Raff, R.A., eds. (1983) Time, Space, and Pattern in Embryonic Development. Alan R. Liss, New York
Jerebzoff, S. (1979) Systemes métaboliques oscillants chez les végétaux inférieurs. Bull. Soc. Bot. Fr. Actual. Bot. **126**: 23–38
Jerebzoff, S. (1986) Cellular circadian rhythms in plants: Recent approaches to their molecular bases. Physiol. Vég. **24**: 367–376
Jerebzoff, S. (1987) Are there ultradian rhythms at the molecular level? J. Interdisc. Cycle Res. **18**: 9–16
Jerebzoff, S. and Jerebzoff-Quintin, S. (1983) Control of enzyme activity rhythms in *Leptosphaeria michotii*. I. Effects of cycloheximide. Physiol. Vég. **21**: 233–245
Jerebzoff, S. and Jerebzoff-Quintin, S. (1984) Cyclic activity of L-asparaginase through reversible phosphorylation in *Leptosphaeria michotii*. FEBS Lett. **171**: 67–71

Jerebzoff-Quintin, S. and Jerebzoff, S. (1980) Métabolisme de l'asparagine et rythme endogène de sporulation chez le *Leptosphaeria michotii* (West) Sacc. Physiol. Vég. **18**: 147–156

Jerebzoff-Quintin, S. and Jerebzoff, S. (1985) L-Aspariginase activity in *Leptosphaeria michotii*. Isolation and properties of two forms of the enzyme. Physiol. Plant. **64**: 74–80

John, P.C.L., ed. (1981) The Cell Cycle. Cambridge University Press, Cambridge

John, P.C.L. (1984) Control of the cell division cycle in *Chlamydomonas*. Microbial. Sci. **1**: 96–101

Johnson, C.H. (1983) Changes in intracellular pH are not correlated with the circadian rhythm of *Neurospora*. Plant Physiol. **72**: 129–133

Johnson, C.H. and Hastings, J.W. (1986) The elusive mechanism of the circadian clock. Am. Sci. **74**: 29–36

Johnson, C.H., Roeber, J.F. and Hastings, J.W. (1984) Circadian changes in enzyme concentration account for rhythm of enzyme activity in *Gonyaulax*. Science **223**: 1428–1430

Johnston, G.C. and Singer, R.A. (1985) Novel cell cycle regulation in the yeast *Schizosaccharomyces pombe*: The DNA–division sequence modulates mass accumulation. Exp. Cell Res. **158**: 544–553

Johnston, G.C., Pringle, J.R. and Hartwell, L.H. (1977) Coordination of growth with cell division in the yeast *Saccharomyces cerevisiae*. Exp. Cell Res. **105**: 79–98

Jones, N.C. and Donachie, W.D. (1973) Chromosome replication, transcription and control of cell division in *Escherichia coli*. Nature New Biol. **243**: 100–103

Kadle, R. and Folk, G.E., Jr. (1983) Importance of circadian rhythms in animal cell cultures. Comp. Biochem. Physiol. **76**A: 773–776

Kämmerer, J. and Hardeland, R. (1982) On the chronobiology of *Tetrahymena*. J. Interdisc. Cycle Res. **13**: 297–302

Karakashian, M.W. (1965) The circadian rhythm of sexual reactivity in *Paramecium aurelia*, syngen 3. In: Circadian Clocks. J. Aschoff, ed. North-Holland Publishing Company, Amsterdam, pp. 301–304

Karakashian, M.W. (1968) The rhythm of mating in *Paramecium aurelia*, syngen 3. J. Cell Physiol. **71**: 197–209

Karakashian, M.W. and Hastings, J.W. (1962) Inhibition of a biological clock by actinomycin D. Proc. Natl. Acad. Sci. USA **48**: 2130–2137

Karakashian, M.W. and Hastings, J.W. (1963) The effects of inhibitors of macromolecular biosynthesis upon the persistent rhythm of bioluminescence in *Gonyaulax*. J. Gen. Physiol. **47**: 1–12

Karakashian, M.W. and Schweiger, H.-G. (1976a) Circadian properties of the rhythmic system in individual nucleated and anucleated cells of *Acetabularia mediterranea*. Exp. Cell Res. **97**: 366–377

Karakashian, M.W. and Schweiger, H.-G. (1976b) Evidence for a cycloheximide sensitive component in the biological clock of *Acetabularia mediterranea*. Exp. Cell Res. **98**: 303–312

Karakashian, M.W. and Schweiger, H.-G. (1976c) Temperature dependence of cycloheximide sensitive phase of circadian cycle in *Acetabularia mediterranea*. Proc. Natl. Acad. Sci. USA **73**: 3216–3219

Karvé, A., Engelmann, W. and Schoser, G. (1961) Initiation of rhythmical petal movements in *Kalanchoe blossfeldiana* by transfer from continuous darkness to continuous light or vice versa. Planta **56**: 700–711

Kasal, C. and Perez-Polo, R. (1980) In vitro evidence of photoreception in the chick pineal gland and its interaction with the circadian clock controlling *N*-acetyltransferase (NAT). J. Neurosci. Res. **5**: 579–585

Kasal, C., Menaker, M. and Perez-Polo, R. (1979) Circadian clock in culture: *N*-acetyltransferase activity of chick pineal glands oscillates *in vitro*. Science **203**: 656–658

Kauffman, S.A. and Wille, J.J. (1975) The mitotic oscillator in *Physarum polycephalum*. J. Theor. Biol. **55**: 47–93

Kauffman, S.A. and Wille, J.J. (1976) Evidence that the mitotic "clock" in *Physarum polycephalum* is a limit cycle oscillator. In: The Molecular Basis of Circadian Rhythms. J.W. Hastings and H.-G. Schweiger, eds. Dahlem Konferenzen, Berlin, pp. 421–431

Kawamura, H. and Nihonmatsu, I. (1985) The suprachiasmatic nucleus as a circadian rhythm generator: Immunocytochemical identification of the suprachiasmatic nucleus within the transplanted hypothalamic tissue. In: Circadian Clocks and Zeitgebers. T. Hiroshige and K. Honma, eds. Hokkaido University Press, Sapporo, Japan, pp. 55–63

Keiding, N., Hartmann, N.R. and Møller, U. (1984) Diurnal variation in influx and transition intensities in the S phase of hamster cheek pouch epithelium cells. In: Cell Cycle Clocks. L.N. Edmunds Jr., ed. Marcel Dekker, New York, pp. 135–159

Keller, S. (1960) Über die Wirkung chemischer Faktoren auf die tagesperiodischen Blattbewegungen von *Phaseolus multiflorus*. Z. Bot. **48**: 32–57

Kerr, C.R., Strauss, H.C., Wallace, A.G. and Scheinman, M.M. (1982) Disease of the human cardiac pacemaker. In: Cellular Pacemakers. Function in Normal and Disease States. D.O. Carpenter, ed. John Wiley & Sons, New York, Vol. 2, pp. 229–259

Khalsa, S.B.S. and Block, G.D. (1986) The *Bulla* ocular circadian pacemaker is phase shifted by pentylenetetrazole independently of extracellular calcium concentration. Soc. Neurosci. Abst. **12**: 596 (No.166.3)

Khalsa, S.B.S. and Block, G.D. (1987) Light-induced phase shifts of the *Bulla* ocular pacemaker are not blocked by calmodulin antagonists. Soc. Neurosci. Abst., **13** (in press)

Kiefner, G., Schliessmann, F. and Engelmann, W. (1974) Methods for recording circadian rhythms in *Euglena*. Int. J. Chronobiol. **2**: 189–195

Kippert, F. (1985a) On the origin of circadian rhythms. J. Interdisc. Cycle Res. **16**: 77–84

Kippert, F. (1985b) Evidence for the concept of quantized cell cycles: The cell cycle of individual *Paramecium* cells is a multiple of ultradian subcycles. Eur. J. Cell Biol. **38** [Suppl. 9]: 16

Kippert, F. (1986) Endocytobiotic coordination: Intracellular calcium signalling and the origin of endogenous rhythms. In: Endocytobiology III. J.J. Lee and J.F. Fredrick, eds. Ann. NY Acad. Sci. **503**: 476–495

Kirschner, M., Gerhart, L.C., Hara, K. and Ubbels, G.A. (1980) Initiation of the cell cycle and establishment of bilateral symmetry in *Xenopus* eggs. In: The Cell Surface: Mediator of Developmental Processes. S. Subtelny and N.K. Wessels, eds. Academic Press, New York, pp. 187–215

Kirschner, M., Newport, J. and Gerhart, J. (1985) The timing of early developmental events in *Xenopus*. Trends Genet. **1**: 41–47
Kirschstein, M. (1969) Das rhythmische Verhalten einer farblosen Mutante von *Euglena gracilis*. Planta **85**: 126–134
Klein, D.C. and Weller, J. (1970) Indole metabolism in the pineal gland: A circadian rhythm in *N*-acetyltransferase. Science **169**: 1093–1095
Klein, D.C., Smoot, R., Weller, J.L., Higa, S., Markey, S.P., Creed, G.J. and Jacobowitz, D.M. (1983) Lesions of the paraventricular nucleus area of the hypothalamus disrupt the suprachiasmatic-spinal cord circuit in the melatonin rhythm generating system. Brain Res. Bull. **10**: 647–652
Klemm, E. and Ninnemann, H. (1976) Detailed action spectrum for the delay shift in pupal emergence of *Drosophila pseudoobscura*. Photochem. Photobiol. **24**: 369–371
Klemm, E. and Ninnemann, H. (1979) Nitrate reductase—A key enzyme in blue light-promoted conidiation and absorbance change of *Neurospora*. Photochem. Photobiol. **29**: 629–632
Klevecz, R.R. (1969a) Temporal order in mammalian cells. I. The periodic synthesis of lactate dehydrogenase in the cell cycle. J. Cell Biol. **43**: 207–219
Klevecz, R.R. (1969b) Temporal coordination of DNA replication with enzyme synthesis in diploid and heteroploid cells. Science **166**: 1536–1538
Klevecz, R.R. (1976) Quantized generation time in mammalian cells as an expression of the cellular clock. Proc. Natl. Acad. Sci. USA **73**: 4012–4016
Klevecz, R.R. (1978) The clock in animal cells is a limit cycle oscillator. In: Cell Reproduction. E.R. Dirksen, D.M. Prescott and C.F. Fox, eds. Academic Press, New York, pp. 139–146
Klevecz, R.R. (1984) Cellular oscillators as vestiges of a primitive circadian clock. In: Cell Cycle Clocks. L.N. Edmunds Jr., ed. Marcel Dekker, New York, pp. 47–61
Klevecz, R.R. and King, G.A. (1982) Temperature compensation in the mammalian cell cycle. Exp. Cell Res. **140**: 307–313
Klevecz, R.R., Kros, J. and Gross, S.D. (1978) Phase response versus positive and negative division delay in animal cells. Exp. Cell Res. **116**: 285–290
Klevecz, R.R., King, G.A. and Shymko, R.M. (1980a) Mapping the mitotic clock by phase perturbation. J. Supramol. Struct. **14**: 329–342
Klevecz, R.R., Kros, J. and King, G.A. (1980b) Phase response to heat shock as evidence for a timekeeping oscillator in synchronous animal cells. Cytogenet. Cell Genet. **26**: 236–243
Klevecz, R.R., Kauffman, S.A. and Shymko, R.M. (1984) Cellular clocks and oscillators. Int. Rev. Cytol. **86**: 97–128
Klevecz, R.R., Shymko, R.M., Blumenfeld, D. and Braly, P.S. (1987) Circadian timing of DNA synthesis is aberrant in human ovarian cancer. Cancer Res. (in press)
von Klitzing, L. and Schweiger, H.-G. (1969) A method for recording the circadian rhythm of the oxygen balance in a single cell of *Acetabularia mediterranea*. Protoplasma **67**: 327–332
Klotter, K. (1960) General properties of oscillating systems. Cold Spring Harbor Symp. Quant. Biol. **25**: 185–187
Knobil, F. (1980) The neuroendocrine control of the menstrual cycle. Rec. Prog. Hormone Res. **36**: 53–88

Koch, A.L. (1977) Does the initiation of chromosome replication regulate cell division? Adv. Microb. Physiol. **16**: 49–98

Koch, A.L. (1980) Does the variability of the cell cycle result from one or many events? Nature **286**: 80–82

Koch, A.L. and Schaechter, M. (1962) A model for statistics of the cell division process. J. Gen. Microbiol. **29**: 435–454

Kondo, T. (1983) Phase shift in the potassium uptake rhythm of the duckweed *Lemna gibba* G3 caused by an azide pulse. Plant Physiol. **73**: 605–608

Kondo, T. (1984a) The period of circadian rhythm in *Lemna gibba* G3 is influenced by the substitution of rubidium for potassium. Plant Cell Physiol. **25**: 1313–1317

Kondo, T. (1984b) Removal by a trace of sodium of the period lengthening of the potassium uptake rhythm due to lithium in *Lemna gibba* G3. Plant Cell Physiol. **75**: 1071–1074

Konopka, R.J. (1979) Genetic dissection of the *Drosophila* circadian system. Fed. Proc. **38**: 2602–2605

Konopka, R.J. (1981) Genetics and development of circadian rhythms in invertebrates. In: Biological Rhythms. Handbook of Behavorial Neurobiology. J. Aschoff, ed. Plenum Press, New York, Vol. 4, pp. 173–181

Konopka, R.J. (1987) Neurogenetics of *Drosophila* circadian rhythms. In: Evolutionary Genetics of Invertebrate Behavior. M.D. Heuttel, ed. Plenum Press, New York (in press)

Konopka, R.J. and Benzer, S. (1971) Clock mutants of *Drosophila melanogaster*. Proc. Natl. Acad. Sci. USA **68**: 2112–2116

Konopka, R.J. and Orr, D. (1980) Effects of a clock mutation on the subjective day—Implications for a membrane model of the *Drosophila* circadian clock. In: Development and Neurobiology of *Drosophila*. O. Siddiqu, P. Babu, L.M. Hall and J.C. Hall, eds. Plenum Press, New York, pp. 409–416

Konopka, R.J. and Wells, S. (1980) *Drosophila* clock mutations affect the morphology of a brain neurosecretory cell group. J. Neurobiol. **11**: 411–415

Konopka, R., Wells, S. and Lee, T. (1983) Mosaic analysis of a *Drosophila* clock mutant. Mol. Gen. Genet. **190**: 284–288

Koop, H.-U., Schmid, R., Heunert, H.-H. and Milthaler, B. (1978) Chloroplast migration: A new circadian rhythm in *Acetabularia*. Protoplasma **97**: 301–310

Koukkari, W.L. and Hillman, W.S. (1968) Pulvini as the photoreceptors in the phytochrome effect on nyctinasty in *Albizzia julibrissin*. Plant Physiol. **43**: 698–704

Koukkari, W.L., Bingham, C. and Duke, S.H. (1985) A special group of ultradian oscillations. Chronobiologia, **12**: 253 (Abstr. 51).

Krasnow, R., Dunlap, J., Taylor, W., Hastings, J.W., Vetterling, W. and Haas, E. (1981) Measurements of *Gonyaulax* bioluminescence, including that of single cells. In: Bioluminescence: Current Perspectives. K.H. Nealson, ed. Burgess Publishing Co., Minneapolis, MN, pp. 52–63

Kreuels, T. and Brinkmann, K. (1979) The versatility of D_2O effects on biological and chemical oscillations. Chronobiologia **6**: 121–122

Kreuels, T., Martin, W. and Brinkmann, K. (1979) Influence of D_2O on the Belousov-Zhabotinsky reaction. In: Proceedings of the Discussion Meeting on the Kinetics of Physiochemical Oscillations, Aachen, 1979, Vol. 1, pp. 51–60

Kreuels, T., Jörres, R., Martin, W. and Brinkmann, K. (1984) System analysis of the circadian rhythm of *Euglena gracilis*. II. Masking effects and mutual interactions of light and temperature responses. Z. Naturforsch. **39c**: 801–811

Krug, H.F. and Hardeland, R. (1985) Diurnal rhythmicity in cytosolic control of cell-free hepatic protein synthesis. J. Interdisc. Cycle Res. **16**: 193–201

Kubitschek, H.E. (1962) Normal distribution of cell generation rate. Exp. Cell Res. **26**: 439–450

Kubitschek, H.E. (1971) The distribution of cell generation times. Cell Tissue Kinet. **4**: 113–122

Kubitschek, H.E. and Freedman, M.L. (1971) Chromosome replication and the division cycle of *Escherichia coli* B/r. J. Bacteriol. **107**: 95–99

Kyriacou, C.P. and Hall, J.C. (1980) Circadian rhythm mutations in *Drosophila melanogaster* affect short-term fluctuations in the male's courtship song. Proc. Natl. Acad. Sci. USA **77**: 6729–6733

Kyriacou, C.P. and Hall, J.C. (1985) Action potential mutations stop a biological clock in *Drosophila*. Nature **314**: 171–173

Kyriacou, C.P. and Hall, J.C. (1986) Interspecific genetic control of courtship song production and reception in *Drosophila*. Science **232**: 494–497

Lachney, C.L. and Lonergan, T.A. (1985) Regulation of cell shape in *Euglena gracilis*. III. Involvement of stable microtubules. J. Cell Sci. **74**: 219–237

Laffler, T.G. and Carrino, J.J. (1986) The tubulin and histone genes of *Physarum polycephalum*: Models for cell cycle-regulated gene expression. Bioessays, **5**: 62–65

Laffler, T., Chang, M.T. and Dove, W.F. (1981) Periodic synthesis of microtubular proteins in the cell cycle of *Physarum*. Proc. Natl. Acad. Sci. USA **78**: 5000–5004

Lakin-Thomas, P.L. (1985) Biochemical genetics of the circadian rhythm in *Neurospora crassa*: Studies on the *cel* strain. Ph.D. Thesis, University of California, San Diego

Lakin-Thomas, P.L. and Brody, S. (1985) Circadian rhythms in *Neurospora crassa*: Interactions between clock mutations. Genetics **109**: 49–66

Laval-Martin, D.L., Shuch, D.J. and Edmunds, L.N. Jr. (1979) Cell cycle-related and endogenously controlled circadian photosynthetic rhythms in *Euglena*. Plant Physiol. **63**: 495–502

Ledoigt, G. and Calvayrac, R. (1979) Phénomènes périodiques, métaboliques et structuraux chez un protiste, *Euglena gracilis*. J. Protozool. **26**: 632–643

Lehman, M.N., Bittman, E.L. and Newman, S.W. (1984) Role of the hypothalamic paraventricular nucleus in neuroendocrine responses to daylength in the golden hamster. Brain Res. **308**: 25–32

Lehman, M.N., Silver, R., Gladstone, W.R., Kahn, R.M., Gibson, M. and Bittman, E.L. (1987a) Circadian rhythmicity restored by neural transplant: Immunocytochemical characterization of the graft and its integration with the host brain. J. Neurosci. **6**: 1626–1638

Lehman, M.N., Silver, R., Gibson, M. and Bittman, E.L. (1987b) Dispersed cell suspensions of fetal suprachiasmatic nucleus (SCN) restore circadian locomotor rhythms in SCN-lesioned hamsters. Soc. Neurosci. Abst., **13** (in press)

Lemmer, B. (1984) Chronopharmakologie: Tagesrhythmen und Arzneimittelwirkung, 2nd ed. Wissenschaftliche Verlagsgesellschaft, Stuttgart

Leong, T.-Y. and Schweiger, H.-G. (1978) The role of chloroplast membrane protein synthesis in the circadian clock. Occurrence of a polypetide which is tentatively involved in the clock. In: Chloroplast Development. Proc. Int. Symp. on Chloroplast Development, Spetsai, Greece, 1978. G. Akoyunoglou and J.H. Argyroudi-Akoyunoglou, eds. Elsevier/North-Holland Biomedical Press, Amsterdam, pp. 323–332

Leong, T.-Y. and Schweiger, H.-G. (1979) The role of chloroplast-membrane-protein synthesis in the circadian clock. Purification and partial characterization of a polypeptide which is suggested to be involved in the clock. Eur. J. Biochem. **98**: 187–194

Levandowsky, M. (1981) Endosymbionts, biogenic amines and a heterodyne hypothesis for circadian rhythms. Ann. NY Acad. Sci. **361**: 369–375

Livingstone, M.S. (1981) Two mutations in *Drosophila* affect the synthesis of octopamine, dopamine, and serotonin by altering the activities of two amino-acid decarboxylases. Neurosci. Abstr. **7**: 351

Li-Weber, M., de Groot, E.J. and Schweiger, H.-G. (1987) Sequence homology to the *Drosophila per* locus in higher plant nuclear DNA and in *Acetabularia* chloroplast DNA. Mol. Gen. Genet. **209**: 1–7

Lloyd, D. and Edwards, S.W. (1986) Temperature-compensated ultradian rhythms in lower eukaryotes: Periodic turnover coupled to a timer for cell division. J. Interdisc. Cycle Res., **17**: 321–326

Lloyd, D. and Edwards, S.W. (1987) Temperature-compensated ultradian rhythms in lower eukaryotes: Timers for cell cycles and circadian events? In: Advances in Chronobiology, Pt. A. J.E. Pauly and L.E. Scheving, eds. Alan R. Liss, Inc., New York, pp. 131–151

Lloyd, D., Edwards, S.W. and Fry, J.C. (1982a) Temperature-compensated oscillations in respiration and cellular protein content in synchronous cultures of *Acanthamoeba castellanii*. Proc. Natl. Acad. Sci. USA **79**: 3785–3788

Lloyd, D., Poole, R.K. and Edwards, S.W. (1982b) The Cell Division Cycle: Temporal Organization and Control of Cellular Growth and Reproduction. Academic Press, London

Loewenstein, W.R. (1975) Intercellular communication in normal and neoplastic tissues. In: Cellular Membranes and Tumor Cell Behavior. Williams & Wilkins, Baltimore, pp. 239–248

Loidl, P. and Gröbner, P. (1982) Acceleration of mitosis induced by mitotic stimulators of *Physarum polycephalum*. Exp. Cell Res. **137**: 469–472

Loidl, P. and Sachsenmaier, W. (1982) Control of mitotic synchrony in *Physarum polycephalum*. Phase shifting by fusion of heterophasic plasmodia contradicts a limit cycle oscillator model. Eur. J. Cell Biol. **28**: 175–179

Lonergan, T.A. (1983) Regulation of cell shape in *Euglena gracilis*. I. Involvement of the biological clock, respiration, photosynthesis, and cytoskeleton. Plant Physiol. **71**: 719–730

Lonergan, T.A. (1984) Regulation of cell shape in *Euglena gracilis*. II. The effects of altered extra- and intracellular Ca^{2+} concentrations and the effect of calmodulin antagonists. J. Cell Sci. **71**: 37–50

Lonergan, T.A. (1985) Regulation of cell shape in *Euglena gracilis*. IV. Localization of actin, myosin and calmodulin. J. Cell Sci. **77**: 197–208

Lonergan, T.A. (1986a) A possible second role for calmodulin in biological clock-controlled processes of *Euglena*. Plant Physiol. **82**: 226–229

Lonergan, T.A. (1986b) The photosynthesis and cell shape rhythms can be naturally uncoupled from the biological clock in *Euglena gracilis*. J. Exp. Bot. **37**: 1334–1340

Lonergan, T.A. and Sargent, M.L. (1978) Regulation of the photosynthesis rhythm in *Euglena gracilis*. I. Carbonic anhydrase and glyceraldehyde-3-phosphate dehydrogenase do not regulate the photosynthesis rhythm. Plant Physiol. **61**: 150–153

Lonergan, T.A. and Sargent, M.L. (1979) Regulation of the photosynthesis rhythm in *Euglena gracilis*. II. Involvement of electron flow through both photosystems. Plant Physiol. **64**: 99–103

Lörcher, L. (1958) Die Wirkung verschiedener Lichtqualitäten auf die endogene Tagesrhythmik von *Phaseolus*. Z. Bot. **46**: 209–241

Lorenzen, H. (1980) Time measurements in unicellular algae and its influence on productivity. In: Algae Biomass. G. Shelef and C.J. Soeder, eds. Elsevier/North-Holland Biomedical Press, Amsterdam, pp. 411–419

Lorenzen, H. and Albrodt, J. (1981) Timing and circadian rhythm and their importance for metabolic regulation in *Chlorella*. Ber. Deutsch. Bot. Ges. **94**: 347–355

Lorenzen, H., Chen, P.-C. and Tischner, R. (1985) Zeitmessung und Circadiane Rhythmik bei zwei Stämmen von *Chlorella*. Biochem. Physiol. Pflanzen. **180**: 149–156

Loros, J.J. and Feldman, J.F. (1986) Loss of temperature compensation of circadian period length in the *frq-9* mutant of *Neurospora crassa*. J. Biol. Rhythms **1**: 187–198

Loros, J.J., Richman, A. and Feldman, J.F. (1986) A recessive circadian clock mutant at the *frq* locus in *Neurospora crassa*. Genetics **114**: 1095–1110

Lotshaw, D. and Jacklet, J.W. (1981) Kinetics of phase shifts induced by protein synthesis inhibitors in the circadian rhythm of neuronal activity from the eye of *Aplysia*. Soc. Neurosci. Abstr. **7**: 218 (Abstr. 1654)

Lotshaw, D. and Jacklet, J.W. (1983) Serotonin stimulates protein phosphorylation in the eye of *Aplysia*, an endogenous circadian oscillator. Fed. Proc. **42**: 575 (Abstr. 229.11)

Lotshaw, D.P and Jacklet, J.W. (1986) Involvement of protein synthesis in circadian clock of *Aplysia* eye. Am. J. Physiol. **250**: R5–R17

Lotshaw, D.P. and Jacklet, J.W. (1987) Serotonin induced protein phosphorylation in the *Aplysia* eye. Comp. Biochem. Physiol. **86C**: 27–32

Lövlie, A. and Farfaglio, G. (1965) Increase in photosynthesis during the cell cycle of *Euglena gracilis*. Exp. Cell Res. **39**: 418–434

Lysek, G. (1976) Alterations of the phospholipids in a rhythmically growing mutant of *Podospora anserina* (Ascomycetes). Biochem. Physiol. Pflanzen **169**: 207–212

Lysek, G. and Von Witsch, H. (1974a) Lichtabhängige Zonirerungen bei Pilzen und ihre physiologischen Grundlagen. Ber. Deutsch. Bot. Ges., **87**: 207–213

Lysek, G. and Von Witsch, H. (1974b) Rhythmisches Mycelwachstum bei *Podospora anserina*. 7. Der Einfluss oberflächenaktiver Substanzen und Antibiotika im Dunkeln und im Licht. Arch. Mikrobiol. **97**: 227–237

Mabood, S.F., Newman, P.F.J. and Nimmo, I.A. (1978) Circadian rhythms in the activity of acetylcholinesterase of human erythrocytes incubated in vitro. Biochem. Soc. Trans. **6**: 305–308

de Mairan, M. (1729) Observation botanique. Histoire Acad. Roy. Sci. Paris, p. 35

Malinowski, J.R., Laval-Martin, D.L. and Edmunds, L.N., Jr. (1985) Circadian oscillators, cell cycles, and the singularity: Light perturbations of the free-running rhythm of cell division in *Euglena*. J. Comp. Physiol. **B 155**: 257–267

Mandell, A.J. and Russo, P.V. (1981) Striatal tyrosine hydroxylase activity: Multiple conformational kinetic oscillators and product concentration frequencies. J. Neurosci. **1**: 380–389

Mano, Y. (1970) Cytoplasmic regulation and cyclic variation in protein synthesis in the early cleavage stage of the sea urchin embryo. Dev. Biol. **22**: 433–460

Mano, Y. (1971) Cell-free cyclic variation of protein synthesis associated with the cell cycle in sea urchin embryos. J. Biochem. (Tokyo) **69**: 11–25

Mano, Y. (1975) Systems constituting the metabolic sequence in the cell cycle. Biosystems **7**:51–65

Marques, N., Edwards, S.W., Fry, J.C., Halberg, F. and Lloyd, D. (1987) Temperature-compensated ultradian variation in cellular protein content of *Acanthamoeba castellanii* revisited. In: Advances in Chronobiology, Pt. A. J.E. Pauly and L.E. Scheving, eds. Alan R. Liss, Inc., New York, pp. 105–119

Martens, C.L. and Sargent, M.L. (1974) Conidiation rhythms of nucleic acid metabolism in *Neurospora crassa*. J. Bacteriol. **117**: 1210–1215

Martin, W., Jörres, Kreuels, T., Lork, W. and Brinkmann, K. (1985) Systemanalyse der circadianen Rhythmik von *Euglena gracilis*: Linearitäten und Nichtlinearitäten in der Reaktion auf Temperatursignale. Ber. Deutsch. Bot. Ges. **98**: 173–186

Masters, M. and Donachie, W.E. (1966) Repression and control of cyclic enzyme synthesis in *Bacillus subtilis*. Nature **209**: 476–479

Masters, M. and Pardee, A.B. (1965) Sequence of enzyme synthesis and gene replication during the cell cycle of *Bacillus subtilis*. Proc. Natl. Acad. Sci. USA **54**: 64–70

Mattern, D.L (1985a) Unsaturated fatty acid isomers: Effects on the circadian rhythm of a fatty acid-deficient *Neurospora crassa* mutant. Arch. Biochem. Biophys. **237**: 402–407

Mattern, D.L. (1985b) Preparation of functional group analogs of unsaturated fatty acids and their effects on the circadian rhythm of a fatty acid- deficient mutant of *Neurospora crassa*. Chem. Phys. Lipids **37**: 297–306

Mattern, D. and Brody, S. (1979) Circadian rhythms in *Neurospora crassa*: Effects of saturated fatty acids. J. Bacteriol. **139**: 977–983

Mattern, D.L., Forman, L.R. and Brody, S. (1982) Circadian rhythms in *Neurospora crassa*: A mutation affecting temperature compensation. Proc. Natl. Acad. Sci. USA **79**: 825–829

Matthews, H.R., Hardie, D.G., Inglis, R.J. and Bradbury, E.M. (1976) The molecular basis of control of mitotic cell division. In: The Molecular Basis of Circadian Rhythms. J.W. Hastings and H.-G. Schweiger, eds. Dahlem Konferenzen, Berlin, pp. 395–408

Mayer, W.-E. (1981) Energy-dependent phases of the circadian clock and the clock-controlled leaf movement in *Phaseolus coccineus* L. Planta **152**: 292–301

Mayer, W.-E. and Knoll, U. (1981) Temperature compensation of cycloheximide-sensitive phases in the *Phaseolus* pulvinus. Z. Pflanzenphysiol. **103**: 413–425

Mayer, W. and Scherer, I. (1975) Phase shifting effect of caffeine in the circadian rhythm of *Phaseolus coccineus* L. Z. Naturforsch. **30c**: 855–856

Mayer, W.-E., Gruner, R. and Strubel, H. (1975) Periodenverlängerung und Phasenverschiebungen der circadianen Rhythmik von *Phaseolus coccineus* L. durch Theophyllin. Planta **125**: 141–148

von Mayersbach, H., Scheving, L.E. and Pauly, J.E., eds. (1981) Biological Rhythms in Structure and Function. Eleventh International Congress of Anatomy, Part C. Alan R. Liss, New York

McAteer, M., Donnan, L. and John, P.C.L. (1985) The timing of division in *Chlamydomonas*. New Phytol. **99**: 41–56

McDaniel, M., Sulzman, F.M. and Hastings, J.W. (1974) Heavy water slows the *Gonyaulax* clock: A test of the hypothesis that D_2O affects circadian oscillations by diminishing the apparent temperature. Proc. Natl. Acad. Sci. USA **71**: 4389–4391

McGovern, J.P., Smolensky, M.H. and Reinberg, A., eds. (1977) Chronobiology in Allergy and Immunology. Charles C. Thomas, Springfield, IL

McMahon, D.G. and Block, G.D. (1986) Light-induced phase shifts of the *Bulla* ocular circadian pacemaker are calcium-dependent. Soc. Neurosci. Abst. **12**: 596 (No.166.2)

McMahon, D.G., Wallace, S.F. and Block, G.D. (1984) Cellular analysis of the *Bulla* ocular circadian pacemaker system. II. Neurophysiological basis of circadian rhythmicity. J. Comp. Physiol. A **155**: 379–385

McMillan, J.P. (1972) Pinealectomy abolishes the circadian rhythm of migratory restlessness. J. Comp. Physiol. **79**: 105–112

McMurry, L. and Hastings, J.W. (1972a) No desynchronization among four different circadian rhythms in the unicellular alga, *Gonyaulax polyedra*. Science, **175**: 1137–1139

McMurry, L. and Hastings, J.W. (1972b) Circadian rhythms: mechanism of luciferase activity changes in *Gonyaulax*. Biol. Bull. **143**: 196–206

Meinert, J.C., Ehret, C.F. and Antipa, G.A. (1975) Circadian chronotypic death in heat-synchronized infradian mode cultures of *Tetrahymena pyriformis*. Microbial Ecol. **2**: 201–214

Menaker, M. (1968) Extraretinal light perception in the sparrow, I. Entrainment of the biological clock. Proc. Natl. Acad. Sci. USA **59**: 414–421

Menaker, M., ed. (1971) Biochronometry. National Academy of Sciences, Washington, DC

Menaker, M. and Eskin, A. (1967) Circadian clock in photoperiodic time measurement: A test of the Bünning hypothesis. Science **157**: 1182–1185

Menaker, M. and Wisner, S. (1983) Temperature-compensated circadian clock in the pineal of *Anolis*. Proc. Natl. Acad. Sci. USA **80**: 6119–6121

Mergenhagen, D.M. (1976) Gene expression and its role in rhythms. In: The Molecular Basis of Circadian Rhythms. J.W. Hastings and H.-G. Schweiger, eds. Dahlem Konferenzen, Berlin, pp. 353–359

Mergenhagen, D. (1980) Circadian rhythms in unicellular organisms. Curr. Top. Microbiol. **1980**: 123–147

Mergenhagen, D. (1984) Circadian clock: Genetic characterization of a phase and period of the circadian rhythm in Euglena. J. Cell. Comp. Physiol. **56**: 25–31

Mergenhagen, D. (1986) The circadian rhythm in *Chlamydomonas reinhardii* in a zeitgeber-free environment. Naturwissenschaften **73**: 410–412

Mergenhagen, D.M. and Hastings, J.W. (1977) The circadian rhythm in metabolic mutant strains of *Chlamydomonas reinhardi*. In: Proceedings of the XII Int. Conference of the International Society for Chronobiology, 1975, Washington, DC. Publishing House Il Ponte, Milano, pp. 735–740

Mergenhagen, D. and Mergenhagen, E. (1987) The biological clock of *Chlamydomonas reinhardii* in space. Eur. J. Cell Biol., **43**: 203–207

Mergenhagen, D. and Schweiger, H.-G. (1973) Recording the oxygen production of a single *Acetabularia* cell for a prolonged period. Exp. Cell Res. **81**: 360–364

Mergenhagen, D. and Schweiger, H.-G. (1974) Circadian rhythmicity: Does intercellular synchronization occur in *Acetabularia*? Plant Sci. Lett. **3**: 387–389

Mergenhagen, D. and Schweiger, H.-G. (1975a) Circadian rhythm of oxygen evolution in cell fragments of *Acetabularia mediterranea*. Exp. Cell Res. **92**: 127–130

Mergenhagen, D. and Schweiger, H.-G. (1975b) The effect of different inhibitors of transcription and translation on the expression and control of circadian rhythm in individual cells of *Acetabularia*. Exp. Cell Res. **94**: 321–326

Michel, U. and Hardeland, R. (1985) On the chronobiology of *Tetrahymena*. III. Temperature compensation and temperature dependence in the ultradian oscillation of tyrosine aminotransferase. J. Interdisc. Cycle Res. **16**: 17–23

Miller, S.G. and Kennedy, M.B. (1986) Regulation of brain type II Ca^{2+}/calmodulin-dependent protein kinase by autophosphorylation: A Ca^{2+}-triggered molecular switch. Cell **44**: 861–870

Millet, B., Melin, D., Bonnet, B., Ibrahim, C.A. and Mercier, J. (1984) Rhythmic circumnutation movement of the shoots in *Phaseolus vulgaris* L. Chronobiol. Int. **1**: 11–19

Milos, P., Roux, E., Alexander, D., Johnson, C. and Hastings, J.W. (1987) Isolation of a cDNA for *Gonyaulax* luciferase: rhythmic protein levels are not mediated by rhythmic mRNA levels. (In preparation)

Mitchell, J.L.A. (1971) Photoinduced division synchrony in permanently bleached *Euglena gracilis*. Planta **100**: 244–257

Mitchison, J.M. (1969) Enzyme synthesis in synchronous cultures. Science **165**: 657–663

Mitchison, J.M., ed. (1971) The Biology of the Cell Cycle. Cambridge University Press, London

Mitchison, J.M. (1974) Sequences, pathways and timers in the cell cycle. In: Cell Cycle Controls. G.M. Padilla, I.L. Cameron and A. Zimmerman, eds. Academic Press, New York, pp. 125–142

Mitchison, J.M. (1977) Enzyme synthesis during the cell cycle. In: Cell Differentiation in Microorganisms, Plants and Animals. L. Nover and K. Mothes, eds. VEB Gustav Fischer Verlag, Jena, pp. 377–401

Mitchison, J.M. (1984) Dissociation of cell cycle events. In: Cell Cycle Clocks. L.N. Edmunds Jr., ed. Marcel Dekker, New York, pp. 163–171

Mitra, B. and Sen, S.P. (1976) Evidence of the involvement of membrane-bound steroids in the photoperiodic induction of flowering in *Xanthium*. Experientia **33**: 316–317

Mitsui, A., Kumazawa, S., Takahashi, A., Ikemoto, H., Cao, S. and Arai, T. (1986) Strategy by which nitrogen-fixing unicellular cyanobacteria grow photoautotrophically. Nature **323**: 720–722

Mitsui, A., Cao, S., Takahashi, A. and Arai, T. (1987) Growth synchrony and cellular parameters of unicellular nitrogen-fixing marine cyanobacterium, Synechococcus sp. strain Miami BG 043511 under continuous illumination. Physiol. Plant. **69**: 1–8

Miwa, I., Kondo, T., Johnson, C. and Hastings, J.W. (1986) Circadian rhythm of photoaccumulation in *Paramecium bursaria*. Zool. Sci. **3**: 1101 (Abstr.).

Miwa, I., Nagatoshi, H. and Horie, T. (1987) Circadian rhythmicity within single cells of *Paramecium bursaria*. J. Biol. Rhythms **2**: 57–64

Miyamoto, H., Rasmussen, L. and Zeuthen, E.G. (1973) Studies of the effect of temperature shocks on preparation for cell division in mouse fibroblast cells (L cells). J. Cell Sci. **13**: 889–900

Mohr, R.E. (1975) In: Growth Rhythms and the History of the Earth Rotation. G.D. Rosenberg and S.K. Runcorn, eds. John Wiley & Sons, London

Møller, U. (1984) Diurnal rhythmicity used in experimental cancer research. In: Cell Cycle Clocks. L.N. Edmunds Jr., ed. Marcel Dekker, New York, pp. 501–524

Monod, J. and Jacob, F. (1961) Teleonomic mechanisms in cellular metabolism, growth and differentiation. Cold Spring Harbor Symp. Quant. Biol. **26**: 389–401

Moore, R.Y. (1979) The anatomy of central mechanisms regulating endocrine rhythms. In: Endocrine Rhythms. D.T. Krieger, ed. Raven Press, New York, pp. 63–87

Moore, R.Y. (1983) Organization and function of a central nervous system circadian oscillator. Fed. Proc. **42**: 2783–2789

Moore, R.Y. and Eichler, V.B. (1972) Loss of circadian adrenal corticosterone rhythm following suprachiasmatic lesion in rat. Brain Res. **42**: 201–206

Moore-Ede, M.C., Sulzman, F.M. and Fuller, C.A. (1982) The Clocks That Time Us: Physiology of the Circadian Timing System. Harvard University Press, Cambridge, MA

Morse, D., Milos, P., Roux, E. and Hastings, J.W. (1987a) Both luciferase binding protein and luciferase exhibit a circadian rhythm, but their mRNA's do not. (In preparation)

Morse, M.J., Crain, R.C. and Satter, R.L. (1987b) Phosphatidylinositol cycle metabolites in *Samanea saman*. Plant Physiol. **83**: 640–644

Morse, M.J., Crain, R.C. and Satter, R.L. (1987c) Light-stimulated phosphatidylinositol turnover in *Samanea saman*. Proc. Natl. Acad. Sci. USA (in press)

Muñoz, V. and Butler, W.L. (1975) Photoreceptor pigment for blue light in *Neurospora crassa*. Plant Physiol. **55**: 421–426

Nakashima, H. (1981) A liquid culture method for the biochemical analysis of the circadian clock of *Neurospora crassa*. Plant Cell Physiol. **22**: 231–238

Nakashima, H. (1982a) Effects of membrane ATPase inhibitors on light-induced phase shifting of the circadian clock in *Neurospora crassa*. Plant Physiol. **69**: 619–623

Nakashima, H. (1982b) Phase shifting of the circadian clock by diethylstilbestrol and related compounds in *Neurospora crassa*. Plant Physiol. **70**: 982–986

Nakashima, H. (1983) Is the fatty acid composition of phospholipids important for the function of the circadian clock in *Neurospora crassa*? Plant Cell Physiol. **24**: 1121–1127

Nakashima, H. (1984a) Calcium inhibits phase shifting of the circadian conidiation rhythm of *Neurospora crassa* by the calcium ionophore A23187. Plant Physiol. **74**: 268–271

Nakashima, H. (1984b) Effects of respiratory inhibitors on respiration, ATP contents, and the circadian onidiation rhythm of *Neurospora crassa*. Plant Physiol. **76**: 612–614

Nakashima, H. (1985) Biochemical and genetic aspects of the conidiation rhythm in *Neurospora crassa*: Phase shifting by metabolic inhibitors. In: Circadian Clocks and Zeitgebers. T. Hiroshige and K. Honma, eds. Hokkaido University Press, Sapporo, Japan, pp. 35–44

Nakashima, H. (1986) Phase shifting of the circadian conidiation rhythm in *Neurospora crassa* by calmodulin antagonists. J.Biol. Rhythms **1**: 163–169

Nakashima, H. (1987) Comparison of phase shifting by temperature of wild type *Neurospora crassa* and the clock mutant, *frq-7*. J. Interdisc. Cycle Res. **18**: 1–8

Nakashima, H. and Feldman, J.F. (1980) Temperature sensitivity of light-induced phase shifting of the circadian clock of *Neurospora*. Photochem. Photobiol. **32**: 247–252

Nakashima, H. and Fujimura, Y. (1982) Light-induced phase shifting of the circadian clock in *Neurospora crassa* requires ammonium salts at high pH. Planta **155**: 431–436

Nakashima, H., Perlman, J. and Feldman, J.F. (1981a) Cycloheximide-induced phase shifting of the circadian clock of *Neurospora*. Am. J. Physiol. **241**: R31–R35

Nakashima, H., Perlman, J. and Feldman, J.F. (1981b) Genetic evidence that protein synthesis is required for the circadian clock of *Neurospora*. Science **212**: 361–362

Naylor, E. and Hartnoll, R.G., eds. (1979) Cyclic Phenomena in Marine Plants and Animals. Proc. 13th Eur. Mar. Biol. Symp., 1978, Isle of Man. Pergamon Press, Oxford

Newman, G.C. and Hospod, F.E. (1986) Rhythm of suprachiasmatic nucleus 2-deoxyglucose uptake in vitro. Brain Res. **381**: 345–350

Nicolis, G. and Portnow, J. (1973) Chemical oscillations. Chem. Rev. **73**: 365–384

Nicolis, G. and Prigogine, I. (1977) Self-Organization in Nonequilibrium Systems. John Wiley & Sons, New York

Nimmo, G.A., Nimmo, H.G., Fewson, C.A. and Wilkins, M.B. (1984) Diurnal changes in the properties of phosphoenolpyruvate carboxylase in *Bryophyllum* leaves: A possible covalent modification. FEBS Lett. **178**: 199–203

Nimmo, G.A., Nimmo, H.G., Hamilton, I.D., Fewson, C.A. and Wilkins, M.B. (1986) Purification of the phosphorylated night form and dephosphorylated day form of phosphoenolpyruvate carboxylase from *Bryophyllum fedtschenkoi*. Biochem. J. **239**: 213–220

Nimmo, G.A., Wilkins, M.B., Fewson, C.A. and Nimmo, H.G. (1987) Persistent circadian rhythms in the phosphorylation state of phosphoenolpyruvate carboxylase from *Bryophyllum fedtschenkoi* leaves and its sensitivity to inhibition by malate. Planta **170**: 408–415

Ninnemann, H. (1979) Photoreceptors for circadian rhythms. Photochem. Photobiol. Rev. **4**: 207–265

Njus, D. (1976) Experimental approaches to membrane models. In: The Molecular Basis of Circadian Rhythms. J.W. Hastings and H.-G. Schweiger, eds. Dahlem Konferenzen, Berlin, pp. 283–294

Njus, D., Sulzman, F.M. and Hastings, J.W. (1974) Membrane model for the circadian clock. Nature **248**: 116–120

Njus, D., Gooch, V.D., Mergenhagen, D., Sulzman, F.M. and Hastings, J.W. (1976) Membranes and molecules in circadian systems. Fed. Proc. **35**: 2353–2357

Njus, D., McMurry, L. and Hastings, J.W. (1977) Conditionality of circadian rhythmicity: Synergistic action of light and temperature. J. Comp. Physiol. **117**, 335–344

Njus, D., Gooch, V.D. and Hastings, J.W. (1981) Precision of the *Gonyaulax* circadian clock. Cell Biophys. **3**: 223–231

Novák, B. and László, E. (1986) A limit cycle type oscillation model for the cell cycle regulated by carbon dioxide. J. Theor. Biol. **120**: 309–320

Novák, B. and Mitchison, J.M. (1986) Change in the rate of CO_2 production in synchronous cultures of the fission yeast *Schizosaccharomyces pombe*: A periodic cell cycle event that persists after the DNA division cycle has been blocked. J. Cell Sci. **86**: 191–206

Novák, B. and Mitchsion, J.M. (1987) Periodic cell cycle changes in the rate of CO_2 production in the fission yeast *Schizosaccharomyces pombe* persist after a block to protein synthesis. J. Cell Sci. **87**: 323–325

Novák, B. and Sironval, C. (1976) Circadian rhythm of the transcellular current in regenerating enucleated posterior stalk segments of *Acetabularia mediterranea*. Plant Sci. Lett. **6**: 273–283

Nover, L., ed. (1984) Heat shock response of eukaryotic cells. Springer-Verlag, Berlin, Heidelberg

Nurse, P. (1975) Genetic control of cell size at cell division in yeast. Nature **256**: 547–551

Nurse, P. and Fantes, P.A. (1981) Cell cycle controls in fission yeast: A genetic analysis. In: The Cell Cycle. P.C.L. John, ed. Cambridge University Press, London, pp. 85–98

Nurse, P. and Streiblová, E., eds. (1984) The Microbial Cell Cycle. CRC Press, Boca Raton, FL

Nurse, P. and Thuriaux, P. (1977) Controls over the timing of DNA replication during the cell cycle of fission yeast. Exp. Cell Res. **107**: 365–375

Nurse, P., Thuriaux, P. and Nasmyth. K. (1976) Genetic control of the cell division cycle in the fission yeast *Schizosaccharomyces pombe*. Mol. Gen. Genet. **146**: 167–178

O'Hare, M.J. and Hornsby, P.J. (1975) Absence of circadian rhythm of corticosterone secretion in monolayer cultures of adult rat adrenocortical cells. Experientia **31**: 378–380

Olson, R.J. and Chisholm, S.W. (1986) Effects of light and nitrogen limitation on the cell cycle of the dinoflagellate *Amphidinium carteri*. J. Plank. Res. **8**: 785–793

Oppenheim, A. and Katzir, N. (1971) Advancing the onset of mitosis by cell-free preparations of *Physarum polycephalum*. Exp. Cell. Res. **68**: 224–226

Østgaard, K. and Jensen, A. (1982) Diurnal and circadian rhythms in the turbidity of growing *Skeletonema costatum* cultures. Mar. Biol. **66**: 261–268

Østgaard, K., Jensen, A. and Johnsson, A. (1982) Lithium ions lengthen the period of growing cultures of the diatom *Skeletonema costatum*. Physiol. Plant. **55**: 285–288

Padilla, G.M., Whitson, G.L. and Cameron, I.L., eds. (1969) The Cell Cycle: Gene–Enzyme Interactions. Academic Press, New York and London

Padilla, G.M., Cameron, I.L. and Zimmerman, A., eds. (1974) Cell Cycle Controls. Academic Press, New York and London

Page, T.L. (1982) Transplantation of the cockroach circadian pacemaker. Science **216**: 73–75

Paietta, J. (1982) Photooxidation and the evolution of circadian rhythmicity. J. Theor. Biol. **97**: 77–82

Paietta, J. and Sargent, M.L. (1981) Photoreception in *Neurospora crassa*: Correlation of reduced light sensitivity with flavin deficiency. Proc. Natl. Acad. Sci. USA **78**: 5573–5577

Paietta, J. and Sargent, M.L. (1982) Blue light responses in nitrate reductase mutants of *Neurospora crassa*. Photochem. Photobiol. **35**: 853–855

Paietta, J. and Sargent, M.L. (1983) Isolation and characterization of light-insensitive mutants of *Neurospora crassa*. Genetics **104**: 11–21

Palmer, J.H. and Asprey, G.F. (1958) Studies in the nyctinastic movement of the leaf pinnae of *Samanea saman* (Jacq) Merrill. II. The behaviour of the upper and lower half pulvini. Planta **51**: 770–785

Palmer, J.D. and Round, F.E. (1967) Persistent, vertical-migration rhythms in benthic microflora. VI. The tidal and diurnal nature of the rhythm in the diatom *Hantzschia virgata*. Biol. Bull., **132**: 44–55

Palmer, J.D., Livingston, L. and Zusy, F.D. (1964) A persistent diurnal rhythm in photosynthetic capacity. Nature **203**: 1087–1088

Palmer, J.D., Brown, F.A. Jr. and Edmunds, L.N. Jr. (1976) An Introduction to Biological Rhythms. Academic Press, New York and London

Pannella, G. (1972) Paleontological evidence on the earth's rotational history since early precambrian. Astrophys. Space Sci. **16**: 212–237

Pardee, A.B. (1974) A restriction point for control of normal animal cell proliferation. Proc. Natl. Acad. Sci. USA **71**: 1286–1290

Pardee, A.B., Dubrow, R., Hamlin, J.L. and Kleitzen, R.F.A. (1978) Animal cell cycle. Annu. Rev. Biochem. **47**: 715–750

Pavlidis, T. (1968) Studies on biological clocks: A model for the circadian rhythms of nocturnal organisms. In: Lectures in Mathematics in the Life Sciences. M. Gerstenhaber, ed. American Mathematical Society, Providence, RI, Vol. 1, pp. 88–112

Pavlidis, T. (1969) Populations of interacting oscillators and circadian rhythms. J. Theor. Biol. **22**: 418–436

Pavlidis, T. (1971) Populations of biochemical oscillators as circadian clocks. J. Theor. Biol. **33**: 319–338

Pavlidis, T. (1973) Biological Oscillators: Their Mathematical Analysis. Academic Press, New York and London

Pavlidis, T. and Kauzmann, W. (1969) Toward a quantitative biochemical model for circadian oscillators. Arch. Biochem. Biophys. **132**: 338–348

Pavlidis, T., Zimmerman, W.F. and Osborn, J. (1968) A mathematical model for temperature effects on circadian rhythms. J. Theor. Biol. **18**: 210–221

Peleg, L., Dotan, A. and Ashkenazi, I.E. (1979) Biological oscillations in human red blood cell (RBC) suspensions and in human red blood cell-ghosts suspensions (RBCG). Chronobiologia **6**: 142

Pengelley, E.T., ed. (1974) Circannual Clocks. Academic Press, New York and London

Perlman, J. (1981) Physiological and biochemical studies of circadian rhythmicity in *Neurospora crassa*. Ph.D. Thesis, University of California, Santa Cruz

Perlman, J., Nakashima, H. and Feldman, J.F. (1981) Assay and characteristics of circadian rhythmicity in liquid cultures of *Neurospora crassa*. Plant Physiol. **67**: 404–407

Petersen, D.F. and Anderson, E.C. (1964) Quantity production of synchronized mammalian cells in suspension culture. Nature **203**: 642–643

Peterson, E.L. (1980) Phase-resetting a mosquito circadian oscillator. I. Phase-resetting surface. J. Comp. Physiol. **138**: 201–211

Petrovic, A.G., Oudet, C.L. and Stutzmann, J.J. (1984) Temporal organization of rat and human skeletal cells. In: Cell Cycle Clocks. L.N. Edmunds Jr., ed. Marcel Dekker, New York, pp. 325–349

Pickard, G.E. and Turek, F.W. (1983) The hypothalamic paraventricular nucleus mediates the photoperiodic control of reproduction but not the effects of light on the circadian rhythm of activity. Neurosci. Lett. **43**: 67–72

Pickard, G.E. and Zucker, I. (1986) Influence of deuterium oxide on circadian activity rhythms of hamsters: Role of the suprachiasmatic nuclei. Brain Res. **376**: 149–154

Pierre, J.-N., Celati, C. and Queiroz, O. (1984) Control, entrainment and coordination by photoperiodism of 24h-period enzyme rhythms (phosphofructokinase and PEP carboxylase) in Kalanchoe. J. Interdisc. Cycle Res. **15**: 267–279

Pierre, J.-N., Celati, C. and Queiroz, O. (1985) 24-h organization of glycolysis and control by photoperiodism. Chronobiologia **12**: 1–9

Pierucci, O. and Helmstetter, C.E. (1969) Chromosome replication, protein synthesis and cell division in *Escherichia coli*. Fed. Proc. **28**: 1755–1760

Piggott, J.R., Rai, R. and Carter, B.L.A. (1982) A bifunctional gene product involved in two phases of the yeast cell cycle. Nature **298**: 391–393

Pirson, A. and Lorenzen, H.G. (1958) Ein endogener Zeitfaktor bei der Teilung von *Chlorella*. Z. Bot. **46**: 53–66

Pittendrigh, C.S. (1960) Circadian rhythms and the circadian organization of living systems. Cold Spring Harbor Symp. Quant. Biol. **25**: 159–184

Pittendrigh, C.S. (1961) On temporal organization in living systems. Harvey Lect. **56**: 93–125

Pittendrigh, C.S. (1965) On the mechanism of the entrainment of a circadian clock by light cycles. In: Circadian Clocks. J. Aschoff, ed. North-Holland Publ. Co., Amsterdam, pp. 277–297

Pittendrigh, C.S. (1966) Biological clocks: The functions, ancient and modern, of circadian oscillations. In: Sciences and the Sixties. Proc. Cloudcroft Symp., Albuquerque, 1965. D.L. Arm, ed. University of New Mexico, Albuquerque, pp. 96–111

Pittendrigh, C.S. (1974) Circadian oscillations in cells and the circadian organization of multicellular systems. In: The Neurosciences: Third Study Program. F.O. Schmitt and F.G. Worden, eds. MIT Press, Cambridge, MA, pp. 437–458

Pittendrigh, C.S. (1981) Circadian organization and the photoperiodic phenomena. In: Biological Clocks in Seasonal Reproductive Cycles. B.K. Follett and D.E. Follett, eds. John Wiley & Sons (Halsted Press), New York, pp. 1–35

Pittendrigh, C.S. and Daan, S. (1976) A functional analysis of circadian pacemakers in nocturnal rodents. V. Pacemaker structure: A clock for all seasons. J. Comp. Physiol. A**106**: 333–355

Pittendrigh, C.S., Bruce, V.G., Rosenzweig, N.S. and Rubin, M.L. (1959) A biological clock in *Neurospora*. Nature **184**: 169–170

Pittendrigh, C.S., Eichhorn, J.H., Minis, D.H. and Bruce, V.G. (1970) Circadian systems. VI. Photoperiodic time measurement in *Pectinophora gossypiella*. Proc. Natl. Acad. Sci. USA **66**: 758–764

Pittendrigh, C.S., Caldarola, P.C. and Cosbey, E.S. (1973) A differential effect of heavy water on temperature-dependent and temperature-compensated aspects in the *Drosophila pseudoobscura* circadian system. Proc. Natl. Acad. Sci. USA **70**: 2037–2041

Pittendrigh, C.S., Elliott, J. and Takamura, T. (1984) The circadian component in photoperiodic induction. In: Photoperiodic Regulation of Insect and Molluscan Hormones. Ciba Foundation Symposium 104. R. Porter and G.M. Collins, eds. Pitman Publishing Ltd., London, pp. 26–47

Pohl, R. (1948) Tagesrhythmus in phototakischen verhalten der *Euglena gracilis*. Z. Naturforsch. **3b**: 367–374

Polanshek, M.M. (1977) Effects of heat shock and cycloheximide on growth and division of the fission yeast, *Schizosaccharomyces pombe*. J. Cell Sci. **23**: 1–23

Prescott, D.M., ed. (1976) Reproduction of Eukaryotic Cells. Academic Press, New York

Prézelin, B.B. and Sweeney, B.M. (1977) Characterization of photosynthetic rhythms in marine dinoflagellates. II. Photosynthesis-irradiance curves in *in vitro* chlorophyll *a* fluorescence. Plant Physiol. **60**: 388–392

Prézelin, B.B., Meeson, B.W. and Sweeney, B.M. (1977) Characterization of photosynthetic rhythms in marine dinoflagellates. I. Pigmentation, photosynthetic capacity and respiration. Plant Physiol. **60**: 384–387

Prigogine, I. (1961) Introduction to the Thermodynamics of Irreversible Processes. Wiley-Interscience, New York

Pringle, J.R. (1981) The genetic approach to the study of the cell cycle. In: Mitosis/Cytokinesis. A.M. Zimmerman and A. Forer, eds. Academic Press, New York and London, pp. 3–28

Pringle, J.R. and Hartwell, L.H. (1981) The *Saccharomyces cerevisiae* cell cycle. In: The Molecular Biology of the Yeast *Saccharomyces*. J.N. Strathern, E.W. Jones and J.R. Broach, eds. Cold Spring Harbor Laboratory, Cold Spring Harbor, NY, pp. 97–142

Pritchard, R.H. (1984) Control of DNA replication in bacteria. In: The Microbial Cell Cycle. P. Nurse and E. Streiblová, eds. CRC Press, Boca Raton, FL, pp. 19–27

Proceedings of the International Conferences of the International Society for Chronobiology: XI (Little Rock, 1971), Igaku Shoin Ltd., Tokyo, 1974 (Chronobiology, L.E. Scheving, F. Halberg and J.E. Pauly, eds.); XII (Washington, D.C.,1975), Publishing House Il Ponte, Milano, 1977; XIII (Pavia, Italy, 1977), Publishing House Il Ponte, Milano, 1981 (F. Halberg, L.E. Scheving, E.W.

Powell and D.K. Hayes, eds.); XV (Minneapolis, 1981), S. Karger Publishers, New York, 1984 (Chronobiology 1982–1983, E. Haus and H. Kabat, eds.); XVI (Dublin, 1983), Chronobiol. Int., **1**, 1984; XVII (Little Rock, 1985), Alan R. Liss, New York, 1987 (Advances in Chronobiology, Parts A and B, J.E. Pauly and L.E. Scheving, eds.); XVIII (Leiden, 1987), Pergamon Press (W.J. Rietveld, G.A. Kerkhof and W. Th.J.M., eds.), in press

Prosser, R.A. and Gillette, M.U. (1986) cAMP analogs reset the circadian oscillator of rat SCN *in vitro*. Soc. Neurosci. Abst. **12**: 211 (No.61.8)

Puiseux-Dao, S. (1984) Environmental signals and rhythms on the order of hours. Role of cellular membranes and compartments and the cytoskeleton. In: Cell Cycle Clocks. L.N. Edmunds Jr., ed. Marcel Dekker, New York, pp. 351–363

Pye, K. (1969) Biochemical mechanisms underlying the metabolic oscillations in yeast. Can. J. Bot. **47**: 271–285

Queiroz, O. (1983) An hypothesis on the role of photoperiodism in the metabolic adaptation to drought. Physiol. Vég. **23**: 577–588

Queiroz-Claret, C. and Queiroz, O. (1981) Rythmes circadiens spontanés d'activité enzymatique (PEP carboxylase et malate déshydrogénase de *Kalenchoe*) dans des extraits maintenus en conditions constantes. C. R. Acad. Sc. Paris **292**: 1237–1240

Queiroz-Claret, C., Girard, Y., Girard, B. and Queiroz, O. (1985) Spontaneous long-period oscillations in the catalytic capacity of enzymes in solution. J. Interdisc. Cycle Res. **16**: 1–9

Quentin, E. and Hardeland, R. (1986a) Circadian rhythmicity of protein synthesis and translational control in *Euglena gracilis*. J. Interdisc. Cycle Res. **17**: 197–205

Quentin, E. and Hardeland, R. (1986b) Stimulation of 80S protein synthesis by heat-stable cytosolic extracts in *Euglena gracilis*. Comp. Biochem. Physiol. **85B**: 89–92

Racusen, R. and Satter, R.L. (1975) Rhythmic and phytochrome-regulated changes in transmembrane potential in *Samanea* pulvini. Nature **255**: 408–410

Radha, E., Hill, T.D., Rao, G.H.R. and White, J.G. (1985) Glutathione levels in human platelets display a circadian rhythm in vitro. Thromb. Res. **40**: 823–831

Rapkine, L. (1931) Sur les processus chimiques au cours de la division cellulaire. Ann. Physiol. Physicochim. Biol. **7**: 382–418

Rapp, P.E. (1979) An atlas of cellular oscillators. In: Cellular Oscillators. M.J. Berridge, P.E. Rapp and J.E. Treherne, eds. Cambridge University Press, Cambridge, pp. 281–306

Rapp, P.E. and Berridge, M.J. (1977) Oscillations in calcium-cyclic AMP control loops form the basis of pacemaker activity and other high frequency biological rhythms. J. Theor. Biol. **66**: 497–525

Raschke, K. (1975) Stomatal action. Annu. Rev. Plant Physiol. **26**: 309–340

Razin, A. and Riggs, A.D. (1980) DNA methylation and gene function. Science **210**: 604–610

Readey, M.A. (1987) Ultradian photosynchronization in *Tetrahymena pyriformis* GLC is related to modal cell generation time: Further evidence for a common timer model. Chronobiol. Int. **4**: 195–208

Readey, M.A. and Groh, K.R. (1986) Polymodal distribution of cell generation times in *Tetrahymena*. (In preparation)

Reddy, P., Zehring, W.A., Wheeler, D.A., Pirrotta, V., Hadfield, C., Hall, J.C.

and Rosbash, M. (1984) Molecular analysis of the *period* locus in Drosophila melanogaster and identification of a transcript involved in biological rhythms. Cell **38**: 701–710

Reddy, P., Jacquier, A.C., Abovich, N., Petersen, G. and Rosbash, G. (1986) The *period* clock locus of D. melanogaster codes for a proteoglycan. Cell **46**: 53–61

Reich, J.G. and Sel'kov, E.E. (1981) Energy Metabolism of the Cell: A Theoretical Treatise. Academic Press, London and New York

Reinberg, A. and Halberg, F., eds. (1979) Chronopharmacology. Pergamon Press, Oxford

Reinberg, A. and Smolensky, M. (1983) Biological Rhythms and Medicine. Springer-Verlag, New York

Rensing, L. (1969) Tagesperiodik von Zellfunktionen und Strahlenempfindlichkeit in Normal- und Tumorgewebe. Naturwissenschaften **22**: 390–396

Rensing, L. and Goedeke, K. (1976) Circadian rhythm and cell cycle: Possible entraining mechanisms. Chronobiologia **3**: 53–65

Rensing, L. and Jaeger, N.I. (1985) Temporal Order. Springer-Verlag, Berlin, Heidelberg

Rensing, L. and Schill, W. (1985) Perturbation by single and double pulses as analytical tool for analyzing oscillatory mechanisms. In: Temporal Order. L. Rensing and N.I. Jaeger, eds. Springer-Verlag, Berlin, Heidelberg, pp. 226–231

Rensing, L., Bos, A., Kroeger, J. and Cornelius, G. (1986) Possible link between circadian rhythm and heat shock response in *Neurospora crassa*. Chronobiol. Int. (in press)

Reppert, S.M. and Gillette, M.U. (1985) Circadian patterns of vasopressin secretion and SCN neuronal activity in vitro. Soc. Neurosci. Abst. **11**: 385

Rinnan, T. and Johnsson, A. (1986) Effects of alkali ions on the circadian leaf movements of *Oxalis regnellii*. Physiol. Plant. **66**: 139–143

Roberts, M.H. and Block, G.D. (1985) Analysis of mutual circadian pacemaker coupling between the two eyes of *Bulla*. J. Biol. Rhythms **1**: 55–75

Robertson, L.M. and Takahashi, J.S. (1986) Photic pulses perturb pineal pacemaker: *in vitro* entrainment of a cellular circadian oscillator. Soc. Neurosci. Abst. **12**: 211 (No.61.9)

Robinson, G.A., Butcher, R.W. and Sutherland, E.W. (1971) Cyclic AMP. Academic Press, New York

Roeder, P.E., Sargent, M.L. and Brody, S. (1982) Circadian rhythms in *Neurospora crassa*: Oscillations in fatty acids. Biochemistry **21**: 4909–4916

Roenneberg, T., Hastings, J.W. and Krieger, M.R. (1987) A period-shortening substance in 'Gonyaulax polyedra'. Chronobiologia **14**: 228 (Abst. No. 189)

Rosbash, M. and Hall, J.C. (1985) Biological clocks in *Drosophila*: Finding the molecules that make them tick. Cell **43**: 3–4

Rothman, B.S. and Strumwasser, F. (1976) Phase-shifting the circadian rhythm of neuronal activity in the isolated *Aplysia* eye with puromycin and cycloheximide. J. Gen. Physiol. **68**: 359–384

Round, F.E. and Palmer, J.D. (1966) Persistent, vertical-migration rhythms in benthic microflora. II. Field and laboratory studies on diatoms from the banks of the River Avon. J. Mar. Biol. Assoc. UK **46**: 191–214

Roux, S.J., McEntire, K., Slocum, R.D., Cedel, T.E. and Hale, C.C. II. (1981) Phytochrome induced photoreversible calcium fluxes in a purified mitochondrial fraction from oats. Proc. Natl. Acad. Sci. USA **78**: 283–287

Rusak, B. (1980) Suprachiasmatic lesions prevent an antigonadal effect of melatonin. Biol. Reprod. **22**: 148–154

Rusak, B. (1982) Physiological models of the rodent circadian system. In: Vertebrate Circadian Systems: Structure and Physiology. J. Aschoff, S. Daan and G.A. Groos, eds. Springer-Verlag, Berlin and Heidelberg, pp. 62–74

Rusak, B. and Groos, G. (1982) Suprachiasmatic stimulation phase shifts rodent circadian rhythms. Science **215**: 1407–1409

Rusak, B. and Morin, L.P. (1976) Testicular responses to photoperiod are blocked by lesions of the suprachiasmatic nuclei in golden hamsters. Biol. Reprod. **15**: 366–374

Rusak, B. and Zucker, I. (1979) Neural regulation of circadian rhyhthms. Physiol. Rev. **59**: 449–526

Rusch, H.P. (1970) Some biochemical events in the life cycle of *Physarum polycephalum*. Adv. Cell Biol. **1**: 297–327

Rusch, H.P., Sachsenmaier, W., Behrens, K. and Gruter, V. (1966) Synchronization of mitosis by the fusion of the plasmodia of *Physarum polycephalum*. J. Cell Biol. **31**: 204–209

Sachsenmaier, W. (1976) Control of synchronous nuclear mitosis in *Physarum polycephalum*. In: The Molecular Basis of Circadian Rhythms. J.W. Hastings and H.-G. Schweiger, eds. Dahlem Konferenzen, Berlin, pp. 409–420

Sachsenmaier, W. (1981) The mitotic cycle of *Physarum*. In: The Cell Cycle. P.C.L. John, ed. Cambridge University Press, London, pp. 139–160

Sachsenmaier, W., Donges, K.H., Rupff, H. and Czihak, G. (1970) Advanced initiation of synchronous mitosis in *Physarum polycephalum* following UV- irradiation. Z. Naturforsch. **25b**: 866–871

Sachsenmaier, W., Remy, U. and Plattner-Schobel, R. (1972) Initiation of synchronous mitosis in *Physarum polycephalum*. A model of the control of cell division in eukaryotes. Exp. Cell Res. **73**: 41–48

Samis, H.V. Jr. and Capobianco, S., eds. (1978) Aging and Biological Rhythms. Plenum Publ. Corp., New York

Samuelsson, G., Sweeney, B.M., Matlick, H.A. and Prézelin, B.B. (1983) Changes in photosystem II account for the circadian rhythm in photosynthesis in *Gonyaulax polyedra*. Plant Physiol. **73**: 329–331

Sargent, M.G. (1979) Surface extension and the cell cycle in prokaryotes. In: Advances in Microbial Physiology. A.H. Rose and J.G. Morris, eds. Academic Press, London, Vol. 18, pp. 105–176

Sargent, M.L. (1969) Response of *Neurospora* to various antibiotics and other chemicals. Neurospora Newslett. **5**: 17

Sargent, M.L. and Briggs, W.R. (1967) The effects of light on a circadian rhythm of conidiation in *Neurospora*. Plant Physiol. **42**: 1504–1510

Sargent, M.L. and Kaltenborn, S.H. (1972) Effects of medium composition and carbon dioxide on circadian conidiation in *Neurospora*. Plant Physiol. **50**: 171–175

Sargent, M.L. and Woodward, D.O. (1969) Genetic determinants of circadian rhythmicity in *Neurospora*. J. Bacteriol. **97**: 861–866

Sargent, M.L., Briggs, W.R. and Woodward, D.O. (1966) The circadian nature of a rhythm expressed by an invertaseless strain of *Neurospora crassa*. Plant Physiol. **41**: 1343–1349

Sargent, M.L., Ashkenazi, I.E., Bradbury, E.M., Bruce, V.G., Ehret, C.F., Feldman, J.F., Karakashian, M.W., Konopka, R.J., Mergenhagen, D., Schütz,

G.A., Schweiger, H.-G. and Vanden Driessche, T.E.A. (1976) The role of genes and their expression: Group report. In: The Molecular Basis of Circadian Rhythms. J.W. Hastings and H.-G. Schweiger, eds. Dahlem Konferenzen, Berlin, pp. 295–310

Sato, R., Kondo, T. and Miyoshi, Y. (1985) Circadian rhythms of potassium and sodium contents in the growing front of *Neurospora crassa*. Plant Cell Physiol. **26**: 447–453

Satoh, N. (1984) Cell division cycles as the basis for timing mechanisms in early embryonic development of animals. In: Cell Cycle Clocks. L.N. Edmunds Jr., ed. Marcel Dekker, New York, pp. 527–538

Satter, R.L. and Galston, A.W. (1971a) Potassium flux: A common feature of *Albizzia* leaflet movement controlled by phytochrome or endogenous rhythm. Science **174**: 518–520

Satter, R.L. and Galston, A.W. (1971b) Interaction between an endogenous rhythm and phytochrome in control of potassium flux and leaflet movement. Plant Physiol. **48**: 740–746

Satter, R.L. and Galston, A.W. (1973) Leaf movements: Rosetta stone of plant behavior? BioScience **23**: 407–416

Satter, R.L. and Galston, A.W. (1981) Mechanisms of control of leaf movements. Annu. Rev. Plant Physiol. **32**: 83–110

Satter, R.L., Marinoff, P. and Galston, A.W. (1970) Phytochrome-controlled nyctinasty in *Albizzia julibrissin*. II. Potassium flux as a basis for leaflet movement. Am. J. Bot. **57**: 916–926

Satter, R.L., Geballe, G.T., Applewhite, P.B. and Galston, A.W. (1974a) Potassium flux and leaf movement in *Samanea saman*. I. Rhythmic movement. J. Gen. Physiol. **64**: 413–430

Satter, R.L., Geballe, G.T. and Galston, A.W. (1974b) Potassium flux and leaf movement in *Samanea saman*. II. Phytochrome controlled movement. J. Gen. Physiol. **64**: 431–442

Satter, R.L., Schrempf, M., Chaudhri, J. and Galston, A.W. (1977) Phytochrome and circadian clocks in *Samanea*. Rhythmic redistribution of potassium and chloride within the pulvinus during long dark periods. Plant Physiol. **59**: 231–235

Satter, R.L., Garber, R.C., Khairallah, L. and Cheng, Y.-S. (1982) Elemental analysis of freeze-dried thin sections of *Samanea* motor organs: Barriers to ion diffusion through the apoplast. J. Cell Biol. **95**: 893–902

Saunders, D.S. (1976) Insect Clocks. Pergamon Press, Oxford

Saunders, D.S. (1978) An experimental and theoretical analysis of photoperiodic induction in the flesh-fly, *Sarcophaga argyrostoma*. J. Comp. Physiol. **124**: 75–95

Sawaki, Y., Nihonmatsu, I. and Kawamura, H. (1984) Transplantation of the neonatal suprachiasmatic nuclei into rats with complete bilateral suprachiasmatic lesions. Neurosci. Res. **1**: 67–72

Schedl, T., Burland, T., Gull, K. and Dove, W.F. (1984) Cell cycle regulation of tubulin RNA level, tubulin protein synthesis, and assembly of microtubules in *Physarum*. J. Cell Biol. **99**: 155–165

Scheffey, C. and Wille, J.J. (1978) Cycloheximide-induced mitotic delay in *Physarum polycephalum*. Exp. Cell Res. **113**: 259–262

Scheving, L.E. (1981) Circadian rhythms of cell proliferation: Their importance when investigating the basic mechanism of normal vs. abnormal growth. In:

Biological Rhythms in Structure and Function. H. von Mayersbach, L.E. Scheving and J.E. Pauly, eds. Alan R. Liss, New York, pp. 39–79

Scheving, L.E. (1984) Chronobiology of cell proliferation in mammals: Implications for basic research and cancer chemotherapy. In: Cell Cycle Clocks. L.N. Edmunds Jr., ed. Marcel Dekker, New York, pp. 455–500

Scheving, L.E. and Pauly, J.E. (1973) Cellular mechanisms involving biorhythms with emphasis on those rhythms associated with the S and M stages of the cell cycle. Int. J. Chronobiol. **1**: 269–283

Scheving, L.E., Haus, E., Kühl, J.F.W., Pauly, J.E., Halberg, F. and Cardoso, S. (1976) Close reproduction by different laboratories of characteristics of circadian rhythm in 1-beta-D-arabinofuranosylcytosine tolerance by mice. Cancer Res. **36**: 1133–1137

Scheving, L.E., Pauly, J.E., T.H. Tsai and Scheving, L.A. (1983) Chronobiology of cellular proliferation: Implication for cancer chemotherapy. In: Topics in Environmental Physiology and Medicine. A. Reinberg and M. Smolensky, eds. Springer-Verlag, New York, pp. 79–130

Schmid, R. (1986) Twofold effect of blue light on a circadian rhythm in *Acetabularia*. J. Interdisc. Cycle Res. **17**: 99–107

Schmid, R. and Koop, H.-U. (1983) Properties of the chloroplast movement during the circadian chloroplast migration in *Acetabularia mediterranea*. Z. Pflanzenphysiol. **112**: 351–357

Schmidt, R.R. (1966) Intracellular control of enzyme synthesis and activity during synchronous growth of *Chlorella*. In: Cell Synchrony—Studies in Biosynthetic Regulation. I.L. Cameron nd G.M. Padilla, eds. Academic Press, New York, pp. 189–235

Schmit, J.C. and Brody, S. (1976) Biochemical genetics of *Neurospora crassa* conidial germination. Bacteriol. Rev. **40**: 1–41

Schnabel, G. (1968) Der Einflüss von Licht auf die circadiane Rhythmik von *Euglena gracilis* bei Autotrophie und Mixotrophie. Planta **81**: 49–63

Scholübbers, H. G., Taylor, W. and Rensing, L. (1984) Are membrane properties essential for the circadian rhythm of *Gonyaulax*? Am. J. Physiol. **247**: R250–R256

Schrempf, M. (1975) Eigenschaften und Lokalisation des Photorezeptors für phasenverschiebendes Störlicht bei der Blüttenblattbewegung von *Kalanchoe blossfeldiana* (v. *Poelin*). Ph.D. Dissertation, University Tübingen

Schröder-Lorenz, A. and Rensing, L. (1986) Circadian clock mechanism and synthesis rates of individual protein species in *Gonyaulax polyedra*. Comp. Biochem. Physiol. **85B**: 315–323

Schröder-Lorenz, A. and Rensing, L. (1987) Circadian changes in protein-synthesis rate and protein phosphorylation in cell-free extracts of *Gonyaulax polyedra*. Planta **170**: 7–13

Schulz, R., Pilatus, U. and Rensing, L. (1985) On the role of energy metabolism in *Neurospora* circadian clock function. Chronobiol. Int. **2**: 223–233

Schwartz, W.J., Davidsen, L.C. and Smith, C.B. (1980) *In vivo* metabolic activity of a putative circadian oscillator, the rat suprachiasmatic nuclei. J. Comp. Neurol. **189**: 157–167

Schwartzkroin, P.A. (1982) Epilepsy: A result of abnormal pacemaker activity in central nervous system neurons? In: Cellular Pacemakers. Function in Normal and Disease States. D.O. Carpenter, ed. John Wiley & Sons, New York, Vol. 2, pp. 323–343

Schweiger, E., Wallraff, H.G. and Schweiger, H.-G. (1964a) Über tagesperiodische Schwankungen der Sauerstoffbilanz kernhaltiger und kernloser *Acetabularia mediterranea*. Z. Naturforsch. **19C**: 499–505

Schweiger, E., Wallraff, H.G. and Schweiger, H.-G. (1964b) Endogenous circadian rhythm in cytoplasm of *Acetabularia*: Influence of the nucleus. Science **146**: 658–659

Schweiger, H.-G. (1977) Die biologische Uhr, zircadianne Organisation der Zelle. Arzneim. Forsch. **27**: 202–208

Schweiger, H.-G. and Schweiger, M. (1977) Circadian rhythms in unicellular organisms: An endeavour to explain the molecular mechanism. Int. Rev. Cytol. **51**: 315–342

Schweiger, H.-G., Hartwig, R. and Schweiger, M. (1986) Cellular aspects of circadian rhythms. J. Cell Sci. **4**[suppl.]: 181–200

Scott, B.I.H., Gulline, H.F. and Robinson, G.R. (1977) Circadian electrochemical changes in the pulvinules of *Trifolium repens* L. Aust. J. Plant Physiol. **4**: 193–206

Sel'kov, E.E. (1968) Self oscillations in glycolysis. Eur. J. Biochem. **4**: 79–86

Sel'kov, E.E. (1970) Two alternative, self-oscillating stationary states in thiol metabolism—two alternative types of cell division: Normal and malignant ones. Biophysika **15**: 1065–1073 (in Russian)

Sel'kov, E.E. (1975) Stabilization of energy charge, generation of oscillations and multiple steady states in energy metabolism as a result of purely stoichiometric regulation. Eur. J. Biochem. **59**: 151–157

Sel'kov, E.E. (1979) A unifying theory of the cell mechanism. Chronobiologia **6**: 155

Selverston, A.I. and Moulins, M. (1985) Oscillatory neural networks. Annu. Rev. Physiol. **47**: 29–48

Senger, H., ed. (1980) The Blue Light Syndrome. Springer-Verlag, New York

Shibata, S. and Oomura, Y. (1985) Observations of suprachiasmatic neuronal activity in relation to circadian rhythm in rat hypothalamic slice preparations. In: Circadian Clocks and Zeitgebers. T. Hiroshige and K. Honma, eds. Hokkaido University Press, Sapporo, Japan, pp. 64–74

Shibata, S., Newman, G.C. and Moore, R.Y. (1986) Effects of calcium on 2-deoxyglucose uptake in the suprachiasmatic nucleus (SCN) in vitro. Soc. Neurosci. Abst. **12**: 845 (No.233.10)

Shields, R. (1977) Transition probability and the origin of variation in the cell cycle. Nature **267**: 704–707

Shields, R. (1978) Further evidence for a random transition in the cell cycle. Nature **273**: 755–758

Shields, R., Brooks, R.F., Riddle, P.N., Capellaro, D.F. and Delia, D. (1978) Cell size, cell cycle and transition probability in mouse fibroblasts. Cell **15**: 469–474

Shilo, B., Shilo, V.and Simchen, G. (1976) Cell cycle initiation in yeast follows frist-order kinetics. Nature **264**: 767–770

Shin, H.-S., Bargiello, T.A., Clark. B.T., Jackson, R.J. and Young, M.W. (1985) An unusual coding sequence from *Drosophila* clock gene is conserved in vertebrates. Nature **317**: 445–448

Shinagawa, A. (1985) Localization of the factors producing the periodic activities responsible for synchronous cleavage in *Xenopus* embryos. J. Embryol. Exp. Morph., **85**: 33–46

Shiotsuka, R., Jovonovich, J. and Jovonovich, J.A. (1974) Circadian and ultradian rhythms in adrenal organ cultures. Chronobiologia **1**[Suppl. 1]: 109

Shnoll, S.E. and Chetverikova, E.P. (1975) Synchronous reversible alterations in enzymatic activity (conformational fluctuations) in actomyosin and creatine kinase preparations. Biochim. Biophys. Acta **403**: 89–97

Shymko, R.M. and Klevecz, R.R. (1981) Cell division gated by oscillatory timekeeping and critical size. In: Biomathematics and Cell Kinetics. M. Rotenberg, ed. Elsevier/North-Holland Biomedical Press, Amsterdam, pp. 329–348

Shymko, R.M., Klevecz, R.R. and Kauffman, S.A. (1984) The cell cycle as an oscillatory system. In: Cell Cycle Clocks. L.N. Edmunds Jr., ed. Marcel Dekker, New York, pp. 273–293

Silyn-Roberts, H. and Engelmann, W. (1986) *Thalassomyxa australis*, a model organism for the evolution of circadian rhythms? Endocytobios. Cell Res. **3**: 239–242

Silyn-Roberts, H., Engelmann, W. and Grell, K.G. (1986) *Thalassomyxa australis* rhythmicity. 1, Temperature dependence. J. Interdisc. Cycle Res. **16**: 77–84

Simon, E., Satter, R.L. and Galston, A.W. (1976a) Circadian rhythmicity in excised *Samanea* pulvini. I. Sucrose-white light interactions. Plant Physiol. **58**: 417–420

Simon, E., Satter, R.L. and Galston, A.W. (1976b) Circadian rhythmicity in excised *Samanea* pulvini. II. Resetting the clock by phytochrome conversion. Plant Physiol. **58**: 421–425

Sisken, J.E. and Morasca, L. (1965) Intrapopulation kinetics of the mitotic cycle. J. Cell. Biol. **25** (part 2): 179–189

Slater, M. and Schaechter, M. (1974) Control of cell division in bacteria. Bacteriol. Rev. **38**: 199–221

Smith, H.T.B. and Mitchison, J.M. (1976) Anaesthetics delay and accelerate division in the fission yeast *Schizosaccharomyces pombe*. Exp. Cell Res. **99**: 432–435

Smith, J.A. and Martin, L. (1973) Do cells cycle? Proc. Natl. Acad. Sci. USA **70**: 1263–1267

Smith, J.A. and Martin, L. (1974) Regulation of cell proliferation. In: Cell Cycle Controls. G.M. Padilla, I.L. Cameron and A. Zimmerman, eds. Academic Press, New York, pp. 43–60

Smith, R.F. and Konopka, R.J. (1981) Circadian clock phenotypes of chromosomal aberrations with a breakpoint at the *per* locus. Mol. Gen. Genet. **183**: 243–251

Smith, R.F. and Konopka, R.J. (1982) Effects of dosage alterations at the *per* locus on the period of the circadian clock of *Drosophila*. Mol. Gen. Genet. **185**: 30–36

Snaith, P.J. and Mansfield, T.A. (1985) Responses of stomata to IAA and fusicoccin at the opposite phases of an entrained rhythm. J. Exp. Bot. **36**: 937–944

Snaith, P.J. and Mansfield, T.A. (1986) The circadian rhythm of stomatal opening: Evidence for the involvement of potassium and chloride fluxes. J. Exp. Bot. **37**: 188–199

Soll, D.R. (1979) Timers in developing systems. Science **203**: 841–849

Soll, D.R. (1983) A new method for examining the complexity and relationships of "timers" in developing systems. Dev. Biol. **95**: 73–91

Soll, D.R. and Finney, R. (1986) Regulation of timing during the foward program of development and the reverse program of dedifferentiation in *Dictyostelium discoideum*. In: The Genetic Regulation of Development. W.F. Loomis, ed. 45th Symp. Soc. Dev. Biol. Academic Press, New York (in press)

Soll, D.R., Mitchell, L., Kraft, B., Alexander, S., Finney, R. and Varnum- Finney, B. Characterization of a timing mutant of *Dictyostelium discoideum* which exhibits "high frequency switching." Dev. Biol. **120**: 25–37

Sonnenborn, T.M. and Sonnenborn, D. (1958) Some effects of light on the rhythm of mating type changes in stock 232–35, of syngen 2 of *Paramecium multimicronucleatum*. Anat. Rec. **131**: 601

Spudich, J and Sager R. (1980) Regulation of *Chlamydomonas* cell cycle by light and dark. J. Cell Biol. **85**: 136–145

Stadler, D.R. (1959 Genetic control of a cyclic growth pattern in *Neurospora*. Nature **184**: 170–171

Stephan, F. and Zucker, I. (1972) Circadian rhythms in drinking behavior and locomotor activity are eliminated by hypothalamic lessions. Proc. Natl. Acad. Sci. USA **69**: 1583–1586

Straley, S.C. and Bruce, V.G. (1979) Stickiness to glass. Circadian changes in the cell surface of *Chlamydomonas reinhardi*. Plant Phyiol. **63**: 1175–1181

Stutzmann, J. and Petrovic, A. (1979) Persistence in organ culture of a growth rate circadian rhythm. In: Chronopharmacology, Advances in Biosciences. A. Reinberg and F. Halberg, eds. Pergamon Press, Oxford, Vol. 19, pp. 89–100

Subbaraj, R. and Chandrashekaran, M.K. (1978) Pulses of darkness shift the phase of a circadian rhythm in an insectivorous bat. J. Comp. Physiol. **127**: 239–243

Suda, M., Hayaishi, O. and Nakagawa, H., eds. (1979) Biological Rhythms and Their Central Mechanism. Proc. Naito Symp. on Biorhythm and Its Central Mechanism, Japan, 1980. Elsevier/North-Holland Biomedical Press, Amsterdam

Sudbery, P.E. and Grant, W.D. (1975) The control of mitosis in *Physarum polycephalum*. The effect of lowering the DNA:mass ratio by UV-irradiation. Exp. Cell Res. **95**: 405–415

Sulzman, F.M. and Edmunds, L.N. Jr. (1972) Persisting circadian oscillations in enzyme activity in non-dividing cultures of *Euglena*. Biochem. Biophys. Res. Commun. **47**: 1338–1344

Sulzman, F.M. and Edmunds, L.N. Jr. (1973) Characterization of circadian oscillations in alanine dehydrogenase activity in non-dividing populations of *Euglena gracilis* (Z). Biochim. Biophys. Acta **320**: 594–609

Sulzman, F.M., Krieger, N.R., Gooch, V.D. and Hastings, J.W. (1978) A circadian rhythm of the luciferin binding protein from *Gonyaulax polyedra*. J. Comp. Physiol. **128**: 251–257

Sulzman, F.M., Gooch, V.D., Homma, K. and Hastings, J.W. (1982) Cellular autonomy of the *Gonyaulax* circadian clock. Cell Biophys. **4**: 97–103

Sulzman, F.M., Ellman, D., Fuller, C.A., Moore-Ede, M.C. and Wassmer, G. (1984) *Neurospora* rhythms in space: A reexamination of the endogenous–exogenous question. Science **225**: 232–234

Sundararajan, K.S., Subbaraj, R., Chandrashekaran, M.K. and Shanmugasundaram, S. (1978) Influence of fusaric acid on circadian leaf movement of the cotton plant, *Gossypium hirsutum*. Planta **144**: 111–112

Sussman, A.S., Lowry, R.J. and Durkee, T. (1964) Morphology and genetics of a periodic colonial mutant of *Neurospora crassa*. Am. J. Bot. **51**: 243--252

Svetina, S. (1977) An extended transition probability model of the variability of cell cycle generation times. Cell Tissue Kinet. **10**: 575–581

Sweeney, B.M. (1960) The photosynthetic rhythm in single cells of *Gonyaulax polyedra*. Cold Spring Harbor Symp. Quant. Biol. **25**: 145–148

Sweeney, B.M. (1963) Resetting the biological clock in *Gonyaulax* with ultraviolet light. Plant Physiol. **38**: 704–708

Sweeney, B.M. (1965) Rhythmicity in the biochemistry of photosynthesis in *Gonyaulax*. In: Circadian Clocks. J. Aschoff, ed. North-Holland Publishing Company, Amsterdam, pp. 190–194

Sweeney, B.M. (1969a) Rhythmic Phenomena in Plants. Academic Press, New York

Sweeney, B.M. (1969b) Transducing mechanisms between circadian clock and overt rhythms in *Gonyaulax*. Can. J. Bot. **47**: 299–308

Sweeney, B.M. (1972) Circadian rhythms in unicellular organisms. In: Circadian Rhythmicity. Proc. Int. Symp. on Circadian Rhythmicity, Wageningen, 1971. J.F. Bierhuizen, chmn. Centre for Agricultural Publishing and Documentation, Wagingen, The Netherlands, pp. 137–156

Sweeney, B.M. (1974a) The potassium content of *Gonyaulax polyedra* and phase changes in the circadian rhythm of stimulated bioluminescence by short exposures to ethanol and valinomycin. Plant Physiol. **53**: 337–347

Sweeney, B.M. (1974b) A physiological model for circadian rhythms derived from the *Acetabularia* rhythm paradoxes. Int. J. Chronobiol. **2**: 25–33

Sweeney, B.M. (1976a) Freeze-fracture studies of thecal membranes of *Gonyaulax polyedra*. Circadian changes in the particles of one membrane face. J. Cell Biol. **68**: 451–461

Sweeney, B.M. (1976b) Evidence that membranes are components of circadian oscillators. In: The Molecular Basis of Circadian Rhythms. J.W. Hastings and H.-G. Schweiger, eds. Dahlem Konferenzen, Berlin, pp. 267–281

Sweeney, B.M. (1979) Bright light does not immediately stop the circadian clock of *Gonyaulax*. Plant Physiol. **64**: 341–344

Sweeney, B.M. (1981) Circadian timing in the unicellular autotrophic dinoflagellate, *Gonyaulax polyedra*. Ber. Deutsch. Bot. Ges. **94**: 335–345

Sweeney, B.M. (1982) Interaction of the circadian cycle with the cell cycle in *Pyrocystis fusiformis*. Plant Physiol. **70**: 272–276

Sweeney, B.M. (1983) Circadian time-keeping in eukaryotic cells, models and hypotheses. In: Progress in Phycological Research. F.E. Round and D. J. Chapman, eds. Elsevier Science Publishers, Amsterdam, Vol. 2, pp. 189–225

Sweeney, B.M. (1986) The loss of the circadian rhythm in photosynthesis in an old strain of *Gonyaulax polyedra*. Plant Physiol. **80**: 978–981

Sweeney, B.M. and Folli, S.I. (1984) Nitrate deficiency shortens the circadian period in *Gonyaulax*. Plant Physiol. **75**: 242–245

Sweeney, B.M. and Hastings, J.W. (1957) Characteristics of the diurnal rhythm of luminescence in *Gonyaulax polyedra*. J. Cell. Comp. Physiol. **49**: 115–128

Sweeney, B.M. and Hastings, J.W. (1958) Rhythmic cell division in populations of *Gonyaulax polyedra*. J. Protozool. **5**: 217–224

Sweeney, B.M. and Hastings, J.W. (1960) Effects of temperature upon diurnal rhythms. Cold Spring Harbor Symp. Quant. Biol. **25**: 87–104

Sweeney, B.M. and Haxo, F.T. (1961) Persistence of a photosynthetic rhythm in enucleated *Acetabularia*. Science **134**: 1361–1363

Sweeney, B.M. and Herz, J.M. (1977) Evidence that membranes play an important role in circadian rhythms. In: Proceedings of the XII International Symposium of the International Society for Chronobiology, Washington, D.C., 1975 Publishing House Il Ponte, Milano, pp. 751–761

Sweeney, B.M., Haxo, F.T. and Hastings, J.W. (1959) Action spectra for two effects of light on luminescence in *Gonyaulax polyedra*. J. Gen. Physiol. **43**: 285–299

Sweeney, B.M., Tufli, C.F. Jr. and Rubin, R.H. (1967) The circadian rhythm in photosynthesis in *Acetabularia* in the presence of actinomycin D, puromycin, and chloramphenicol. J. Gen. Physiol. **50**: 647–659

Sweeney, B.M., Prézelin, B.B., Wong, D. and Govindjee (1979) *In vivo* chlorophyll *a* fluorescence transients and the circadian rhythm of photosynthesis in *Gonyaulax polyedra*. Photochem. Photobiol. **30**: 309–311

Swift, E. and Taylor, W.R. (1967) Bioluminescence and chloroplast movement in the dinoflagellate *Pyrocystis lunula*. J. Phycol. **3**: 77–81

Szyszko, A.H., Prazak, B.L., Ehret, C.F., Eisler, W.J. and Wille, J.J. (1968) A multi-unit sampling system and its use in the characterization of ultradian and infradian growth in *Tetrahymena*. J. Protozool. **15**: 781–785

Takahashi, J.S. and Menaker, M. (1979a) Physiology of avian circadian pacemakers. Fed. Proc. **38**: 2583–2588

Takahashi, J.S. and Menaker, M. (1979b) Brain mechanisms in avian circadian systems. In: Biological Rhythms and Their Central Mechanism. M. Suda, O. Hayaishi and H. Nakagawa, eds. Elsevier/North-Holland Biomedical Press, Amsterdam, pp. 95–109

Takahashi, J.S. and Menaker, M. (1982a) Entrainment of the circadian system of the house sparrow: A population of oscillators in pinealectomized birds. J. Comp. Physiol. **146**: 245–253

Takahashi, J.S. and Menaker, M. (1982b) Role of the suprachiasmatic nuclei in the circadian system of the house sparrow, *Passer domesticus*. J. Neurosci. **2**: 815–828

Takahashi, J.S. and Menaker, M. (1984) Multiple redundant circadian oscillators within the isolated avian pineal gland. J. Comp. Physiol. **154**: 435–440

Takahashi, J.S. and Zatz, M. (1982) Regulation of circadian rhythmicity. Science **217**: 1104–1111

Takahashi, J.S., Hamm, H. and Menaker, M. (1980) Circadian rhythms of melatonin release from individual superfused chicken pineal glands *in vitro*. Proc. Natl. Acad. Sci. USA **77**: 2319–2322

Takahashi, J.S., DeCoursey, P.J., Baumann, L. and Menaker, M. (1984) Spectral sensitivity of a novel photoreceptive system mediating entrainment of mammalian circadian rhythms. Nature **308**: 186–188

Tanakadate, A. and Ishikawa, H. (1985) Microcomputerized measurement of the circadian locomotor rhythm in microorganisms. Physiol. Behav. **34**: 241–248

Tauro, P., Halvorson, H.L. and Epstein, R.L. (1968) Time of gene expression in relation to centromere distance during the cell cycle of *Saccharomyces cerevisiae*. Proc. Natl. Acad. Sci. USA **59**: 277–284

Taylor, W. and Hastings, J.W. (1979) Aldehydes phase shift the *Gonyaulax* clock. J. Comp. Physiol. **130**: 359–362

Taylor, W. and Hastings, J.W. (1982) Minute-long pulses of anisomycin phase-shift the biological clock in *Gonyaulax* by hours. Naturwissenschaften **69**: 94–96

Taylor, W., Gooch, V.D. and Hastings, J.W. (1979) Period shortening and phase shifting effects of ethanol on the *Gonyaulax* glow rhythm. J. Comp. Physiol. **130**: 355–358

Taylor, W.R., Dunlap, J.C. and Hastings, J.W. (1982a) Inhibitors of protein synthesis on 80S ribosomes phase shift the *Gonyaulax* clock. J. Exp. Biol. **97**: 121–136

Taylor, W., Krasnow, R., Dunlap, J.C., Broda, H. and Hastings, J.W. (1982b) Critical pulses of anisomycin drive the circadian oscillator in *Gonyaulax* towards its singularity. J. Comp. Physiol. **148**: 11–25

Temin, H.M. (1971) Stimulation by serum of multiplication of stationary chicken cells. J. Cell. Physiol. **78**: 161–170

Terborgh, J. and McLeod, G.C. (1967) The photosynthetic rhythm of *Acetabularia crenulata*. I. Continuous measurements of oxygen exchange in alternating light-dark regimes and in constant light of different intensities. Biol. Bull. **133**: 659–669

Terry, O. and Edmunds, L.N., Jr. (1969) Semi-continuous culture and monitoring system for temperature-synchronized *Euglena*. Biotechnol. Bioeng. **11**: 745–756

Terry, O.W. and Edmunds, L.N., Jr. (1970a) Phasing of cell division by temperature cycles in *Euglena* cultured autotrophically under continuous illumination. Planta **93**: 106–127

Terry, O.W. and Edmunds, L.N., Jr. (1970b) Rhythmic settling induced by temperature cycles in *Euglena* cultured autotrophically under continuous illumination. Planta **93**: 128–142

Tharp, G.D. and Folk, G.E. Jr. (1965) Rhythmic changes in rate of the mammalian heart and heart cells during prolonged isolatioin. Comp. Biochem. Physiol. **14**: 255–273

Thiel, G., Hardeland, R. and Michel, U. (1985) On the chronobiology of *Tetrahymena*. II. Further evidence for the persistence of ultradian rhythmicity in the absence of protein synthesis and cell growth. J. Interdisc. Cycle Res. **16**: 11–16

Thompson, G.A. Jr. (1980) The Regulation of Membrane Lipid Metabolism. CRC Press, Boca Raton, FL

Thorud, E., Clausen, O.P.F. and Laerum, O.D. (1984) Circadian rhythms in cell population kinetics of self-renewing mammalian tissues. In: Cell Cycle Clocks. L.N. Edmunds Jr., ed. Marcel Dekker, New York, pp. 113–133

Ting, I.P. and Gibbs, M., eds. (1982) Crassulacean Acid Metabolism. American Society of Plant Physiologists, Rockville, MD

Tomson, A., Demets, R., Homan, W., Stegwee, D. and Van den Ende, H. (1987) A circadian rhythm in the concentration of mating receptors in *Chlamydomonas eugametos*. Manuscript in preparation.

Töpperwien, F. and Hardeland, R. (1980) Free-running circadian rhythm of plastid movements in individual cells of *Pyrocystis lunula* (Dinophyta). J. Interdisc. Cycle Res. **11**: 325-329

Tormo, A., Martínez-Salas, E. and Vincente, M. (1980) Involvement of the *ftsA* gene product in late stages of the *Escherichia coli* cell cycle. J. Bacteriol. **141**: 806–813

Torras, C. (1985) Modelling and Simulation of Neurons and Neural Networks with

Temporal-Pattern Learning Capability. Lecture Notes in Biomathematics. Springer-Verlag, Berlin and Heidelberg, Vol. 63

Truman, J. (1972) Physiology of insect rhythms. II. The silkmoth brain as the location of the biological clock controlling eclosion. J. Comp. Physiol. **881**: 99–114

Tudzynski, P., Stahl, U. and Esser, K. (1980) Transformation to senescence with plasmid-like DNA in the ascomycete *Podospora anserina*. Curr. Genet., **2**: 181–184

Turcotte, D.L., Cisne, J.L. and Nordmann, J.C. (1977) On the evolution of the lunar orbit. Icarus **30**: 254–266

Turek, F.W. (1985) Circadian neural rhythms in mammals. Annu. Rev. Physiol. **47**: 49–64

Turek, F.W., McMillan, J.P. and Menaker, M. (1976) Melatonin: Effects on the circadian locomotor rhythm of sparrows. Science **194**: 1441–1443

Tyson, J.J. (1976) The Belousov-Zhabotinskii Reaction. Lecture Notes in Biomathematics. S. Levin, ed. Springer-Verlag, Berlin, Vol. 10

Tyson, J.J. (1979) Periodic enzyme synthesis: Reconsideration of the theory of oscillatory repression. J. Theor. Biol. **80**: 27–38

Tyson, J.J. (1982) Periodic phenomena in *Physarum*. In: Cell Biology of *Physarum* and *Didymium*. H.C. Aldrich and J.W. Daniel, eds. Academic Press, New York, pp. 61–110

Tyson, J.J. (1983a) Periodic enzyme synthesis and oscillatory repression: Why is the period of oscillation close to the cell cycle time? J. Theor. Biol. **103**: 313–328

Tyson, J.J. (1983b) Unstable activator models for size control of the cell cycle. J. Theor. Biol. **104**: 617–631

Tyson, J.J. (1984) The control of nuclear division in *Physarum polycephalum*. In: The Microbial Cell Cycle. P. Nurse and E. Streiblová, eds. CRC Press, Boca Raton, FL, pp.175–190

Tyson, J.J. (1985) The coordination of cell growth and division—Intentional or incidental? Bioessays, **2**: 72–77

Tyson, J.J. (1987) Size control of cell division. J. Theor. Biol. **126**: 381–391

Tyson, J.J. and Diekmann, O. (1986) Sloppy size control of the cell division cycle. J. Theor. Biol. **118**: 405–426

Tyson, J.J. and Hannsgen, K.B. (1985) The distribution of cell size and generation time in a model of the cell cycle incorporating size control and random transitions. J. Theor. Biol. **113**: 29–62

Tyson, J.J. and Kauffman, S. (1975) Control of mitosis by a continuous biochemical oscillation: synchronization, spatially inhomogeneous oscillations. J. Math. Biol. **1**: 289–310

Tyson, J.J. and Sachsenmaier, W. (1978) Is nuclear division in *Physarum* controlled by a continuous limit cycle oscillator? J. Theor. Biol. **73**: 723–737

Tyson, J.J. and Sachsenmaier, W. (1984) The control of nuclear division in *Physarum polycephalum*. In: Cell Cycle Clocks. L.N. Edmunds Jr., ed. Marcel Dekker, New York, pp. 253–270

Tyson, J.J., Alivisatos, S.G.A., Grün, F., Pavlidis, T., Richter, O. and Schneider, F.W. (1976) Mathematical background: Group Report. In: The Molecular Basis of Circadian Rhythms. J.W. Hastings and H.-G. Schweiger, eds. Dahlem Konferenzen, Berlin, pp. 85–108

Tyson, J.J., Garcia-Herdugo, G. and Sachsenmaier, W. (1979) Control of nuclear division in *Physarum polycephalum*. Comparison of cycloheximide pulse treatment, UV irradiation, and heat shock. Exp. Cell Res. **119**: 87–98

Uhl, G.R. and Reppert, S.M. (1986) Suprachiasmatic nucleus vasopressin messenger RNA: Circadian variation in normal and Brattleboro rats. Science **232**: 390–393

Ułaszewski, S., Mamouneas, T., Shen, W.-K., Rosenthal, P.J., Woodward, J.R., Cirillo, V.P. and Edmunds, L.N., Jr. (1979) Light effects in yeast: Evidence for participation of cytochrbmes in photoinhibition of growth and transport in *Saccharomyces cerevisiae* cultured at low temperatures. J. Bacteriol. **138**: 523–529

Vanden Driessche, T. (1966a) Circadian rhythms in *Acetabularia*: Photosynthetic capacity and chloroplast shape. Exp. Cell Res. **42**: 18–30

Vanden Driessche, T. (1966b) The role of the nucleus in the circadian rhythms of *Acetabularia mediterranea*. Biochim. Biophys. Acta **126**: 456–470

Vanden Driessche, T. (1967) Experiments and hypothesis on the role of RNA in the circadian rhythm of photosynthetic capacity in *Acetabularia mediterranea*. Nachr. Akad. Wiss. Gottingen Math.-Phys. Kl II **10**: 108–109

Vanden Driessche, T. (1970a) Les rythmes circadiens chez les unicellulaires. J. Interdisc. Cycle Res. **1**: 21–42

Vanden Driessche, T. (1970b) Circadian variation in ATP content in the chloroplasts of *Acetabularia mediterranea*. Biochim. Biophys. Acta **205**: 526–528

Vanden Driessche, T. (1973) The chloroplasts of *Acetabularia*. The control of their multiplication and activities. Subcell. Biochem. **2**: 33–67

Vanden Driessche, T. (1974) Circadian rhythm in the Hill reaction of *Acetabularia*. In: Proceedings of the Third International Congress on Photosynthesis Rehovot, Israel, 1974. M. Avron, ed. Elsevier Scientific Publishing Company, Amsterdam, pp. 745–751

Vanden Driessche, T. (1975) Circadian modulation of structure and function in *Acetabularia*. In: Les Cycles Cellulaires et leur Blocage chez Plusieurs Protistes. M. Lefort-Tran and R. Valencia, eds. Colloq. Int. C.N.R.S., No. 240. Centre National de la Recherche Scientifique, Paris, pp. 33–40

Vanden Driessche, T. (1979) Phase-shifting effect of IAA on the photosynthetic circadan rhythm of *Acetabularia*. In: Developmental Biology of *Acetabularia*. S. Bonotto, V. Kefeli and S. Puiseux-Dao, eds. Elsevier/North-Holland Biomedical Press, Amsterdam, pp. 195–204

Vanden Driessche, T. (1980) Circadian rhythmicity: General properties—as exemplified mainly by *Acetabularia*—and hypotheses on its cellular mechanism. Arch. Biol. (Bruxelles) **91**: 49–76

Vanden Driessche, T. (1987) The molecular mechanism of circadian rhythms. News Physiol. Sci. (in press)

Vanden Driessche, T. and Bonotto, S. (1969) The circadian rhythm in RNA synthesis in *Acetabularia mediterranea*. Biochim. Biophys. Acta **179**: 58–66

Vanden Driessche, T. and Cotton, G. (1986) Polarity and circadian rhythmicity are regulatory components in *Acetabularia* morphogenesis. Endocytobiol. Cell Res. **3**: 275–297

Vanden Driessche, T. and Hars, R. (1972a) Variations circadiennes de l'ultrastructure des chloroplastes d'*Acetabularia*. I. Algues entieres. J. Microscopie **15**: 85–90

Vanden Driessche, T. and Hars, R. (1972b) Variations circadiennes de l'ultrastructure des chloroplastes d'*Acetabularia*. I. Algues anucléées. J. Microscopie **15**: 91–98

Vanden Driessche, T., Bonotto, S. and Brachet, J. (1970) Inability of rifampicin to inhibit circadian rhythmicity in *Acetabularia* despite inhibition of RNA synthesis. Biochim. Biophys. Acta **224**: 631–634

Vanden Driessche, T., Doege, K.J., Minder, C. and Cairns, W.L. (1979) Circadian rhythm in cyclic AMP content in *Acetabularia*. In: Chronopharmacology. Advances in the Biosciences. A. Reinberg and F. Halberg, eds. Pergamon Press, Oxford, England, Vol. 19, pp. 291–299

Vasquez, D. (1979) Inhibitors of protein biosynthesis. Mol. Biol. Biochem. Biophys. **30**: 1–312

Vaulot, D. and Chisholm, S.W. (1987) A simple model of the growth of phytoplankton populations in light/dark cycles. J. Plank. Res. **9**: 345–366

Vicente, M. (1984) The control of cell division in bacteria. In: The Microbial Cell Cycle. P. Nurse and E. Streiblová, eds. CRC Press, Boca Raton, FL, pp. 29–49

Vokt, J.P. and Brody, S. (1985) The kinetics of changes in the fatty acid composition of *Neurospora crassa* lipids after a temperature increase. Biochim. Biophys. Acta **835**: 176–182

Volknandt, W. and Hardeland, R. (1984a) Circadian rhythmicity of protein synthesis in the dinoflagellate, *Gonyaulax polyedra*: A biochemical and radioautographic investigation. Comp. Biochem. Physiol. **77B**: 493–500

Volknandt, W. and Hardeland, R. (1984b) Effects of aldehydes and alcohols on protein synthesis in the dinoflagellate, *Gonyaulax polyedra*: Relation to phase-shifting potency in the circadian oscillation. Comp. Biochem. Physiol. **78C**: 51–54

Volm, M. (1964) Die Tagesperiodik der Zellteilung von *Paramecium bursaria*. Z. Vergl. Physiol. **48**: 157–180

Wagner, E. (1976a) The nature of photoperiodic time measurement: Energy transduction and phytochrome action in seedlings of *Chenopodium rubrum*. In: Light and Plant Development. Proc. 22nd Nottingham Easter School in Agric. Sci., 1975. H. Smith, ed. Butterworths, London and Boston, pp. 419–443

Wagner, E. (1976b) Endogenous rhythmicity in energy metabolism: Basis for timer-photoreceptor interactions in photoperiodic control. In: The Molecular Basis of Circadian Rhythms. J.W. Hastings and H.-G. Schweiger, eds. Dahlem Konferenzen, Berlin, pp. 215–238

Wagner, E. (1977) Molecular basis of physiological rhythms. Symp. Soc. Gen. Microbiol. **31**: 33–70

Wagner, E. (1985) Membrane-oscillator hypothesis of metabolic control in photoperiodic time measurement and the temporal organization of development and behaviour in plants. In: Recent Advances in Biological Membrane Studies. L. Packer, ed. Plenum Publ. Corp., New York, pp. 525–546

Wagner, E. and Cumming, B.G. (1970) Betacyanin accumulation, chlorophyll content, and flower initiation in *Chenopodium rubrum* as related to endogenous rhythmicity and phytochrome action. Can. J. Bot. **48**: 1–18

Wagner, E. and Frosch, S. (1974) Endogenous rhythmicity and energy transduction. VI. Rhythmicity in reduced and oxidized pyridine nucleotide levels in seedlings of *Chenopodium rubrum*. J. Interdisc. Cycle Res. **5**: 231–239

Wagner, E., Frosch, S. and Deitzer, G.F. (1974a) Metabolic control of photoperiodic time measurement. J. Interdisc. Cycle Res. **5**: 240–246

Wagner, E., Frosch, S. and Deitzer, G.F. (1974b) Membrane oscillator hypothesis of photoperiodic control. In: Proceedings of the Annual European Symposium on Plant Photomorphogenesis. J.A. de Greef, ed. Campus of the State University Center, Antwerp, pp. 15–19

Wagner, E., Stroebele, L. and Frosch, S. (1974c) Endogenous rhythmicity and energy transduction. V. Rhythmicity in adenine nucleotides and energy charge in seedlings of *Chenopodium rubrum*. J. Interdisc. Cycle Res. **3**: 77–88

Wainwright, S.D. and Wainwright, L.K. (1979) Chick pineal serotonin acetyltransferase: A diurnal cycle maintained *in vitro* and its regulation by light. Can. J. Biochem. **57**: 700–709

Wainwright, S.D. and Wainwright, L.K. (1986) Effects of some inhibitors of DNA synthesis and repair upon the cycle of serotonin *N*-acetyltransferase activity in cultured chick pineal cells. Biochem. Cell Biol. **64**: 344–355

Walker, R.F. and Timiras, P.S. (1982) Pacemaker insufficiency and the onset of aging. In: Cellular Pacemakers. Function in Normal and Disease States. D.O. Carpenter, ed. John Wiley & Sons, New York, Vol. 2, pp. 345–365

Walther, W.G. and Edmunds, L.N. Jr. (1973) Studies on the control of the rhythm of photosynthetic capacity in synchronized cultures of *Euglena gracilis* (Z). Plant Physiol. **51**: 250–258

Walz, B. and Sweeney, B.M. (1979) Kinetics of the cycloheximide-induced phase changes in the biological clock in *Gonyaulax*. Proc. Natl. Acad. Sci. USA **76**: 6443–6447

Walz, B., Walz, A. and Sweeney, B.M. (1983) A circadian rhythm in RNA in the dinoflagellate *Gonyaulax polyedra*. J. Comp. Physiol. B **151**: 207–213

Watanabe, S. and Sibaoka, T. (1973) Site of photo-reception to opening response in *Mimosa* leaflets. Plant Cell Physiol. **14**: 1121–1124

Watson, J.D. (1976) Molecular Biology of the Gene, 3rd ed. W.A. Benjamin, Menlo Park, CA

Webb, G.F. (1986) Random transitions, size control, and inheritance in cell population dynamics. (In preparation)

Weiler, C.S. and Chisholm, S.W. (1976) Phased cell division in natural populations of marine dinoflagellates from shipboard cultures. J. Exp. Mar. Biol. Ecol. **25**: 239–247

Weiler, C.S. and Eppley, R.W. (1979) Temporal pattern of division in the dinoflagellate genus *Ceratium* and its application to the determination of growth rate. J. Exp. Mar. Biol. Ecol. **39**: 1–24

Weitzel, G. and Rensing, L. (1981) Evidence for cellular circadian rhythms in isolated fluorescent dye-labelled salivary glands of wild type and an arrhythmic mutant of *Drosophila melanogaster*. J. Comp. Physiol. **143**: 229–235

West, D.J. (1976) Phase shift of the circadian rhythm of conidiation in response to ultraviolet light. Neurospora Newslett. **23**: 17–18

Wetterberg, L., Geller, E. and Yuwiler, A. (1970) Harderian gland: An extraretinal photoreceptor influencing the pineal gland in neonatal rats? Science **167**: 884–885

Wever, R. (1965a) A mathematical model for circadian rhythms. In: Circadian Clocks. J. Aschoff, ed. North-Holland Publishing Co., Amsterdam, pp. 47–63

Wever, R. (1965b) Pendulum versus relaxation oscillation. In: Circadian Clocks. J. Aschoff, ed. North-Holland Publishing Co., Amsterdam, pp. 74–83

Wever, R. (1979) The Circadian System of Man. Springer-Verlag, New York

Wheals, A. and Silverman, B. (1982) Unstable activator model for size control of the cell cycle. J. Theor. Biol. **97**: 505–510

Wilkins, M.B. (1960) The effect of light upon plant rhythms. Cold Spring Harbor Symp. Quant Biol. **25**: 115–129

Wilkins, M.B. (1973) An endogenous circadian rhythm in the rate of carbon dioxide output of *Bryophyllum*. VI. Action spectrum for the induction of phase shifts by visible radiation. J. Exp. Bot. **24**: 488–496

Wilkins, M.B. and Warren, D.M. (1963) The influence of low partial pressures of oxygen on the rhythm in the growth rate of the Avena coleoptile. Planta **60**: 261–273

Wille, J.J. Jr. (1974) Light entrained circadian oscillations of growth rate in the yeast *Candida utilis*. In: Chronobiology. L.E. Scheving, F. Halberg and J.E. Pauly, eds. Igaku Shoin, Tokyo, pp. 72–77

Wille, J.J. Jr. (1979) Biological rhythms in protozoa. In: Biochemistry and Physiology of Protozoa, 2nd ed. M. Levandowsky and S.H. Hutner, eds. Academic Press, New York, Vol. 2, pp. 67–149

Wille, J.J. Jr. and Ehret, C.F. (1968a) Light synchronization of an endogenous circadian rhythm of cell division in *Tetrahymena*. J. Protozool., **15**: 785–788

Wille, J.J. Jr. and Ehret, C.F. (1968b) Circadian rhythm of pattern formation in populations of a free-swimming organism, *Tetrahymena*. J. Protozool. **15**: 789–792

Wille, J.J. Jr. and Scott, R.E. (1984) Cell cycle-dependent integrated control of cell proliferation and differentiation in normal and neoplastic mammalian cells. In: Cell Cycle Clocks. L.N. Edmunds Jr., ed. Marcel Dekker, New York, pp. 433–453

Wille, J.J. Barnett, A. and Ehret, C.F. (1972) Participation of rare templates in DNA/RNA hybridization in the eukaryote, *Tetrahymena pyriformis*. Biochem. Biophys. Res. Commun. **46**: 685–691

Wille, J.J., Scheffey, C. and Kauffman, S.A. (1977) Novel behavior of the mitotic clock in *Physarum*. J. Cell Sci. **27**: 91–104

Winfree, A.T. (1967) Biological rhythms and the behavior of populations of coupled oscillators. J. Theor. Biol. **16**: 15–42

Winfree, A.T. (1970) Integrated view of resetting a circadian clock. J. Theor. Biol. **28**: 327–374

Winfree, A.T. (1974) Rotating chemical reactions. Sci. Am. **230**: 82–95

Winfree, A.T. (1975) Unclocklike behavior of biological clocks. Nature **253**: 315–319

Winfree, A.T. (1980) The Geometry of Biological Time. Springer-Verlag, New York

Winfree, A.T. (1983) Sudden cardiac death: A problem in topology. Sci. Am. **248**: 144–160

Winfree, A.T. (1987) The Timing of Biological Clocks. W.H. Freeman, San Francisco and London

Winfree, A.T. and Gordon, H. (1977) The photosensitivity of a mutant circadian clock. J. Comp. Physiol. **122**: 87–109

Wisnieski, B.J. and Fox, C.F. (1976) Correlations between physical state and physiological activities in eukaryotic membranes, especially in response to temperature. In: The Molecular Basis of Circadian Rhythms. J.W. Hastings and H.-G. Schweiger, eds. Dahlem Konferenzen, Berlin, pp. 247–266

Wolf, R., Wick, R. and Sauer, H. (1979) Mitosis in *Physarum polycephalum*: Analysis of time-lapse films and DNA replication of normal and heat-shocked macroplasmodia. Eur. J. Cell Biol. **19**: 49–59

Wolpaw, J.R. (1982) Tremor: Extrinsic and intrinsic origins. In: Cellular Pacemakers. Function in Normal and Disease States. D.O. Carpenter, ed. John Wiley & Sons, New York, Vol. 2, pp. 299–319

Woodward, D.O. and Sargent, M.L. (1973) Circadian rhythms in *Neurospora*. In: Behaviour of Microorganisms. A. Perez-Miravete, ed. Plenum Press, New York, pp. 282–296

Woodward, J.R., Cirillo, V.P. and Edmunds, L.N., Jr. (1978) Light effects in yeast: Inhibition by visible light of growth and transport in *Saccharomyces cerevisiae* grown at low temperatures. J. Bacteriol. **133**: 692–698

Woolum, J.C. and Strumwasser, F. (1983) Is the period of the circadian oscillator in the eye of *Aplysia* directly homeostatically regulated? J. Comp. Physiol. **151**: 253–259

Wu, J.-T., Tischner, R. and Lorenzen, H. (1986) A circadian rhythm in the number of daughter cells in synchronous *Chlorella fusca* var *vacuolata*. Plant Physiol. **80**: 20–22

Wurtman, R., Axelrod, J. and Kelly, D. (1968) The Pineal. Academic Press, New York and London

Yada, T., Oiki, S., Ueda, S. and Okada, Y. (1986) Synchronous oscillation of the cytoplasmic Ca^{2+} concentration and membrane potential in cultured epithelial cells (Intestine 407). Biochim. Biophys. Acta **887**: 105–112

Yen, A. and Pardee, A.B. (1979) Role of nuclear size in cell growth initiation. Science **204**: 1315–1317

Yeung, S.J. and Eskin, A. (1987) Involvement of a specific protein in the regulation of a circadian rhythm in *Aplysia* eye. Proc. Natl. Acad. Sci. USA **84**: 279–283

Yoneda, M., Ikeda, M. and Washitani, S. (1978) Periodic change in the tension at the surface of activated non-nucleate fragments of sea-urchin eggs. Dev. Growth Diff. **20**: 329–336

Young, M.W., Jackson, F.R., Shin, H.-S. and Bargiello, T.A. (1985) A biological clock in *Drosophila*. Cold Spring Harbor Symp. Quant. Biol. **50**: 865–875

Yu, Q., Jacquier, A.C., Citri, Y., Hamblen, M., Hall, J.C. and Rosbash, M. (1987a) Molecular mapping of point mutations in the period gene that stop or speed up biological clocks in *Drosophila melanogaster*. Proc. Natl. Acad. Sci. USA **84**: 784–788

Yu, Q., Colot, H.V., Kyriacou, C.P., Hall, J.C. and Rosbash, M. (1987b) Behaviour modification by *in vitro* mutagenesis of a variable region within the *period* gene of *Drosophila*. Nature **326**: 765–769

Yuwiler, A. (1983) Light and agonists alter pineal *N*-acetyltransferase induction by vasoactive intestinal polypeptide. Science **220**: 1082–1083

Zatz, M. and Moskal, J.R. (1986) ^{3}H-Melatonin rhythm in cultured chick pineal cells. Soc. Neurosci. Abst. **12**: 844 (No. 233.7)

Zehring, W.A., Wheeler, D.A., Reddy, P., Konopka, R.J., Kyriacou, C.P., Rosbash, M. and Hall, J.C. (1984) P-Element transformation with *period* locus DNA restores rhythmicity to mutant, arrhythmic *Drosophila melanogaster*. Cell **39**: 369–376

Zeuthen, E., ed. (1964) Synchrony in Cell Division and Growth. Wiley-Interscience, New York

Zeuthen, E. (1971) Recent developments in the synchronization of *Tetrahymena* cell cycle. In: Advances in Cell Biology. D.M. Prescott, L. Goldstein and E. McConkey, eds. Appleton-Century-Crofts-Meredith, New York, Vol. 2, pp. 111–152

Zeuthen, E. (1974) A cellular model for repetitive and free-running synchrony in *Tetrahymena* and *Schizosaccharomyces*. In: Cell Cycle Controls. G.M. Padilla, I.L. Cameron and A.M. Zimmerman, eds. Academic Press, New York, pp. 1–30

Zimmerman, N.H. and Menaker, M. (1975) Neural connections of sparrow pineal: Role in circadian control of activity. Science **190**: 477–479

Zimmerman, N.H. and Menaker, M. (1979) The pineal: A pacemaker within the circadian system of the house sparrow. Proc. Natl. Acad. Sci. USA **76**: 999–1003

Zimmerman, W.F. and Goldsmith, T.H. (1971) Photosensitivity of the circadian rhythm and of visual receptors in carotenoid-depleted *Drosophila*. Science **171**: 1167–1169

Zimmerman, W.F., Pittendrigh, C.S. and Pavlidis, T. (1968) Temperature compensation of the circadian oscillation in *Drosophila pseudoobscura* and its entrainment by temperature cycles. J. Insect Physiol. **14**: 669–684

Zorn, G.A. and Smith, B. (1984) Cell cycle clocks and aging. In: Cell Cycle Clocks. L.N. Edmunds Jr., ed. Marcel Dekker, New York, pp. 557–579

Author Index

Subject Index